Molecular Neurobiology of the Olfactory System

Molecular, Membranous, and Cytological Studies

Molecular Neurobiology of the Olfactory System

Molecular, Membranous, and Cytological Studies

Edited by
Frank L. Margolis
Roche Institute of Molecular Biology
Nutley, New Jersey

and
Thomas V. Getchell
Wayne State University School of Medicine
Detroit, Michigan

PLENUM PRESS • NEW YORK AND LONDON

Library of Congress Cataloging in Publication Data

Molecular neurobiology of the olfactory system: molecular, membranous, and cytological studies / edited by Frank L. Margolis and Thomas V. Getchell.
p. cm.
Includes bibliographies and index.
ISBN-13: 978-1-4612-8280-8 e-ISBN-13: 978-1-4613-0989-5
DOI: 10.1007/978-1-4613-0989-5
1. Smell. 2. Molecular neurobiology. 3. Chemoreceptors. I. Margolis, Frank L. II. Getchell, Thomas V.
QP458.M64 1988 88-14821
599'.01826–dc19 CIP

Softcover reprint of the hardcover 1st edition 1988
A Division of Plenum Publishing Corporation
233 Spring Street, New York, N.Y. 10013

Contributors

Richard A. Akeson • Division of Basic Science Research, Children's Hospital Research Foundation, Cincinnati, Ohio 45229

Harriet Baker • Laboratory of Molecular Neurobiology, The Burke Rehabilitation Center, Cornell University Medical College, White Plains, New York 10605

Peter C. Barber • Department of Morbid Anatomy and Histology, Addenbrooke's Hospital, Cambridge CB2 2QQ, England

Klaus Buchner • Institut für Zoologie, Technische Universität München, D-8046 Garching, Federal Republic of Germany

Alan R. Dahl • Inhalation Toxicology Research Institute, Lovelace Biomedical and Environmental Research Institute, Albuquerque, New Mexico 87185

Ted M. Dawson • Department of Neurology, Hospital of the University of Pennsylvania, Philadelphia, Pennsylvania 19143

Valina L. Dawson • Departments of Pharmacology and Psychiatry, University of Utah School of Medicine, Salt Lake City, Utah 84132

John A. DeSimone • Department of Physiology, Virginia Commonwealth University, Richmond, Virginia 23298

Vincent E. Dionne • Division of Pharmacology, Department of Medicine, University of California, San Diego, La Jolla, California 92093

Richard L. Doty • Smell and Taste Center and Department of Otorhinolaryngology and Human Communication, School of Medicine, University of Pennsylvania, Philadephia, Pennsylvania 19104

Albert I. Farbman • Department of Neurobiology and Physiology, Northwestern University, Evanston, Illinois 60208

Marilyn L. Getchell • Department of Anatomy and Cell Biology, Wayne State University School of Medicine, Detroit, Michigan 48201

Thomas V. Getchell • Department of Anatomy and Cell Biology, Wayne State University School of Medicine, Detroit, Michigan 48201

Gerard L. Heck • Department of Physiology, Virginia Commonwealth University, Richmond, Virginia 23298

Steen Jensen • Institute of Anatomy B, University of Aarhus, Aarhus 8000-C, Denmark

Doron Lancet • Department of Membrane Research, The Weizmann Institute of Science, Rehovot 76 100, Israel

Frank L. Margolis • Department of Neuroscience, Roche Institute of Molecular Biology, Roche Research Center, Nutley, New Jersey 07110

Robert A. Maue • Division of Molecular Medicine, New England Medical Center, Boston, Massachusetts 02111

James I. Morgan • Department of Neuroscience, Roche Institute of Molecular Biology, Roche Research Center, Nutley, New Jersey 07110

Randall B. Murphy • Department of Chemistry, New York University, New York, New York 10003; and Department of Psychiatry, Cornell University Medical College-Westchester Division-New York Hospital, White Plains, New York 10605

Krishna C. Persaud • Department of Physiology, Virginia Commonwealth University, Richmond, Virginia 23298

Jonathan Pevsner • Departments of Neuroscience, Pharmacology and Molecular Sciences, and Psychiatry and Behavioral Sciences, Johns Hopkins University School of Medicine, Baltimore, Maryland 21205

Pamela B. Sklar • Departments of Neuroscience, Pharmacology and Molecular Sciences, and Psychiatry and Behavioral Sciences, Johns Hopkins University School of Medicine, Baltimore, Maryland 21205

James B. Snow, Jr. • Smell and Taste Center and Department of Otorhinolaryngology and Human Communication, School of Medicine, University of Pennsylvania, Philadelphia, Pennsylvania 19104

Solomon H. Snyder • Departments of Neuroscience, Pharmacology and Molecular Sciences, and Psychiatry and Behavioral Sciences, Johns Hopkins University School of Medicine, Baltimore, Maryland 21205

James K. Wamsley • Departments of Pharmacology and Psychiatry, University of Utah School of Medicine, Salt Lake City, Utah 84132

Dieter G. Weiss • Institut für Zoologie, Technische Universität München, D-8046 Garching, Federal Republic of Germany

Barbara Zielinski • Department of Anatomy and Cell Biology, Wayne State University School of Medicine, Detroit, Michigan 48201

Preface

The sense of smell and the olfactory system have been a subject of intrinsic interest for millenia. Inquiry into the structure and function of the olfactory system is based on a long tradition that dates back at least to the ancient Greeks. The mechanistic basis for the sensitivity and selectivity of this chemosensory detection system has always posed a challenge and remained largely a mystery. Recently, there has been a renaissance of interest in it and especially in the application of contemporary techniques of biochemistry and cellular and molecular biology. In this volume, current research utilizing these approaches is discussed in depth by a group of scientists who are among the current leaders in the applications of these techniques to the olfactory system. These authors address a wide range of questions that bear directly on the olfactory system but have broader biological implications as well. The various chapters have been grouped into five broad subject areas that emphasize diverse but related questions.

"Transduction and Ligand–Receptor Interactions" considers the biochemical bases of stimulus access, interaction, transduction, elimination, and information processing. What are the current views of the mechanisms that regulate the arrival of odorant molecules at the sensory cells? What role is played by the mucus interface and how is its composition regulated? Having gained access, how do odorant molecules interact with the olfactory neurons and convey their information to the cell? How is this information processed and transmitted to the central nervous system, where it will be used to form the basis for organismal response? Finally, what is known of the enzymatic or physical mechanisms that eliminate prior stimuli?

"Molecular Biophysics and Membrane Function" recognizes that the stimulus interactions with the olfactory neurons take place at, and induce initial changes in, the plasma membrane of the receptor cell. What are the biophysical changes associated with these initial events? How can they be studied at various levels within the tissue? What ion fluxes can be identified in the intact tissue? To what extent are they dependent on identified components of the transduction system? Which cell types are responsible for expressing the conductance changes and how are they regulated? Do membrane fragments manifest the entire phenomenon and can the individual molecular components be disassembled and reconstituted for detailed mechanistic studies?

"Biochemical–Molecular Biological Studies" addresses how the highly plastic properties of this system might be regulated. How do second-order neurons respond to disruption of normal input from the olfactory receptor cells? What happens to their ability to maintain normal cellular and synaptic function? Do these target cells recover when the receptor neuron population is reconstituted? What are the agents that are responsible for these events? How are trophic agents and other components delivered to the distant terminals and targets of these unmyelinated nerves? How is the presynaptic cell body informed about events at the synaptic terminals? Are the properties of these chemoreceptor neurons and the tissue in which they exist dependent upon the selective expression of a particular subset of genes? If so, how can they be identified and characterized?

"Development and Differentiation" considers normal olfactory ontogeny as well as the ability of the receptor neurons to be replaced from an endogenous population of stem cells. Is this property an example of tissue-specific neoteny in an adult animal? How can the various cell types be discriminated? Do the various cell types exhibit biochemical/immunological hallmarks? Are the cells of a given morphological/functional set all identical? Can they develop/survive/function in an ectopic location? What are the cellular and environmental requirements for morphological and functional development of the neurons?

"Biological Relevance of Olfactory Function" illustrates that, in addition to presenting a challenging series of questions in model systems, understanding the olfactory system has relevance for the human condition. What are the effects of age-dependent changes in olfactory function? Why is one of the first events associated with the onset of Alzheimer's-type dementia loss of the ability to identify odors? Will this information help us to create diagnostic, therapeutic, or research models of the disease based on its link to olfactory function?

Our intent in editing this volume has been threefold: First, to focus attention on the functional questions in the olfactory system that are currently in need of answers; second, to illustrate how contemporary techniques in biochemistry and cellular and molecular biology are being used to address these questions; and third, to identify future directions for research in which studies of the olfactory system will also contribute to the solution of broad biological problems. Any success this volume has in achieving these goals accrues to the chapter authors, who have labored creatively and diligently in pursuit of our mutual goal.

Frank L. Margolis
Thomas V. Getchell

Nutley and Detroit

Contents

I. TRANSDUCTION AND LIGAND–RECEPTOR INTERACTIONS

1. Olfactory Receptor Mechanisms: Odorant-Binding Protein and Adenylate Cyclase

Solomon H. Snyder, Pamela B. Sklar, and Jonathan Pevsner

2. Molecular Components of Olfactory Reception and Transduction

Doron Lancet

3. The Effect of Cytochrome P-450-Dependent Metabolism and Other Enzyme Activities on Olfaction

Alan R. Dahl

4. Odorant and Autonomic Regulation of Secretion in the Olfactory Mucosa

Marilyn L. Getchell, Barbara Zielinski, and Thomas V. Getchell

5. *Autoradiographic Localization of Drug and Neurotransmitter Receptors in the Olfactory Bulb*

Valina L. Dawson, Ted M. Dawson, and James K. Wamsley

II. MOLECULAR BIOPHYSICS AND MEMBRANE FUNCTION

6. *Membrane Probes in the Olfactory System: Biophysical Aspects of Initial Events*

Randall B. Murphy

7. *Membrane Properties of Isolated Olfactory Receptor Neurons*

Robert A. Maue and Vincent E. Dionne

8. *Voltage-Clamp Studies of the Isolated Olfactory Mucosa*

John A. DeSimone, Krishna C. Persaud, and Gerard L. Heck

III. BIOCHEMICAL–MOLECULAR BIOLOGICAL STUDIES

9. Neurotransmitter Plasticity in the Juxtaglomerular Cells of the Olfactory Bulb

Harriet Baker

10. Axoplasmic Transport in Olfactory Receptor Neurons

Dieter G. Weiss and Klaus Buchner

13. Primary Olfactory Neuron Subclasses

Richard A. Akeson

14. Cellular Interactions in the Development of the Vertebrate Olfactory System

Albert I. Farbman

15. Olfactory Tissue Interactions Studied by Intraocular Transplantation

Peter C. Barber and Steen Jensen

V. BIOLOGICAL RELEVANCE OF OLFACTORY FUNCTION

16. Age-Related Alterations in Olfactory Structure and Function

Richard L. Doty and James B. Snow, Jr.

I

Transduction and Ligand–Receptor Interactions

1

Olfactory Receptor Mechanisms
Odorant-Binding Protein and Adenylate Cyclase

Solomon H. Snyder, Pamela B. Sklar, and Jonathan Pevsner

1. INTRODUCTION

The olfactory system displays a dramatically sensitive and selective recognition system for odorants. Animals can detect and discriminate thousands of odorants, some at ambient concentrations as low as one part per trillion (1 ppt). While numerous theories have been advanced to explain this sensitivity and selectivity (Amoore, 1982), remarkably few direct experimental data support any specific molecular mechanism. It is not known whether the nose possesses thousands of distinct olfactory receptors or whether the influence of odorants on a small population of receptors can account for our ability to discriminate among large numbers of odorants.

In our laboratory, two major approaches to these problems have been adopted. The first approach is derived from our extensive experience with neurotransmitter and drug receptor binding systems (Snyder, 1984). Using a receptor-binding paradigm, we have measured the binding of radiolabeled odorants to preparations of olfactory mucosa (Pevsner *et al.*, 1985). Our second approach has been to characterize the odorant-sensitive adenylate cyclase in isolated frog olfactory cilia described by Lancet and collaborators (Pace *et al.*, 1985). We have also characterized transducing GTP-binding proteins (G proteins) in the frog olfactory epithelium using subunit specific antisera (Anholt *et al.*, 1987).

2. AN ODORANT-BINDING PROTEIN

Our initial studies focused on the binding of the bell pepper odorant [^{3}H]-2-isobutyl-3-methoxypyrazine ([^{3}H]-IBMP) (Pevsner *et al.*, 1985), an approach employed independently by Pelosi and collaborators (Pelosi and Pisanelli, 1981; Pelosi *et al.*, 1982;

Solomon H. Snyder, Pamela B. Sklar, and Jonathan Pevsner • Departments of Neuroscience, Pharmacology and Molecular Sciences, and Psychiatry and Behavioral Sciences, Johns Hopkins University School of Medicine, Baltimore, Maryland 21205.

Table 1. Regional distribution of [^{3}H]-IBMP Binding[a,b]

Region	Soluble fraction		Membrane bound	
	Bovine	Rat	Bovine	Rat
Olfactory bulb	<0.003	<0.003	<0.003	<0.003
Olfactory epithelium	2.6	5.2	<0.004	<0.004
Respiratory epithelium	3.9	0.6	<0.004	<0.004

[a]From Pevsner *et al.* (1985).
[b]Results are expressed as pmoles bound per mg protein. Levels of specific binding <0.003 pmoles bound per mg protein were measured in adrenal gland, brain stem, cerebral cortex, heart, intestine, kidney, liver, lung, pancreas, and spleen in bovine and rat soluble and membrane-bound fractions. Binding was performed using 3 nM [^{3}H]-IBMP and 0.1–0.5 mg tissue.

Topazzini *et al.*, 1985; Bignetti *et al.*, 1985). IBMP was used because it is one of the most potent odorants, having a human detection threshold of two parts per trillion in ambient air (Maga and Sizer, 1973). We investigated the binding of [^{3}H]-IBMP to crude homogenates of bovine nasal epithelium and detected saturable binding. Subcellular fractionation of the epithelium reveals that the binding is confined to the soluble supernatant fractions. [^{3}H]-IBMP binding to a protein found in the soluble fraction can be measured by a simple and sensitive technique similar to that employed in drug and neurotransmitter receptor binding. Following incubation with [^{3}H]-IBMP, the labeled protein is filtered over glass fiber filters pretreated with 0.3% polyethylenimine to trap soluble proteins. In such preparations, nonspecific binding, estimated with 1 mM unlabeled IBMP, is only about 5% of total binding detectable.

The first evidence that [^{3}H]-IBMP binding is relevant to olfaction comes from studies of the regional distribution of binding, which is highly localized to the nasal epithelium of the cow and rat. Essentially no binding is detected in the olfactory bulb or other parts of the brain, adrenal gland, heart, intestine, kidney, liver, lung, pancreas, and spleen of the cow or rat (Table 1). In the rat, there appears to be more binding in olfactory than respiratory nasal epithelium, while in the cow comparable levels of binding are observed in both olfactory and respiratory epithelium.

A protein which accounts for the [^{3}H]-IBMP binding detected in the soluble fractions can be purified to apparent homogeneity in only three steps (Table 2, Fig. 1). A 130-fold

Table 2. Purification of the IBMP Binding Protein from Bovine Nasal Epithelium[a,b]

Fraction	Total protein (mg)	[^{3}H]-IBMP bound (nmoles)	Recovery of binding activity (%)	Specific [^{3}H]-IBMP binding (pmoles/mg protein)
Crude supernatant	12,848	1.5	100	0.1
Ammonium sulfate	1,590	1.4	95	0.9
DEAE chromatography	296	0.5	39	2.0
Hydroxylapatite	65	0.8	22	12.9

[a]From Pevsner *et al.* (1985).
[b]Values are from a typical purification procedure.

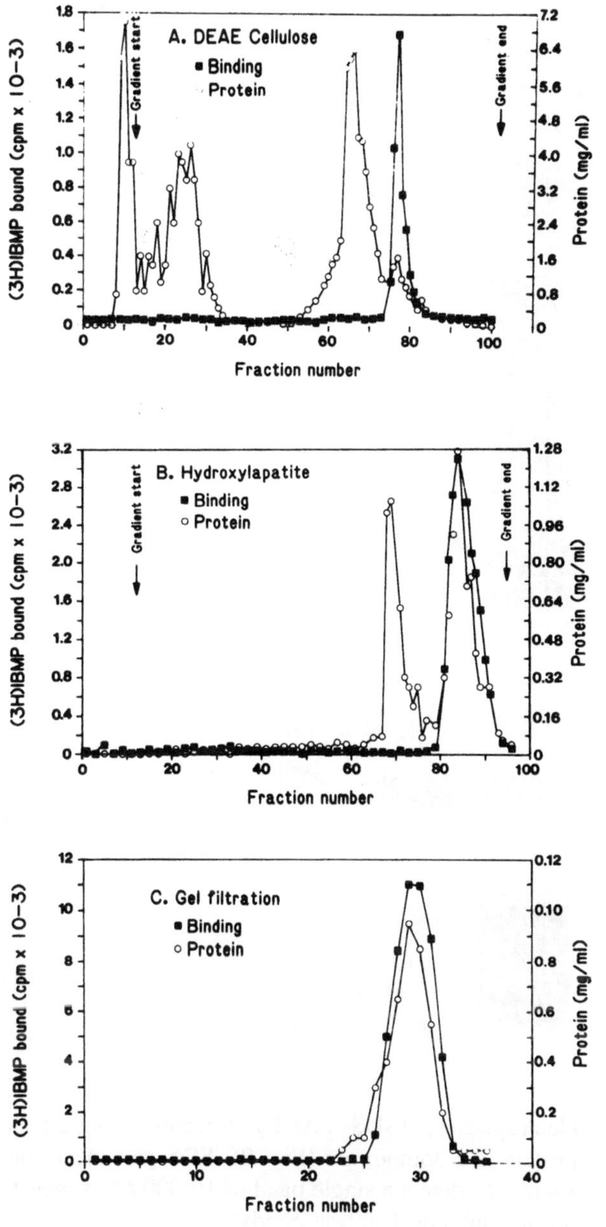

Figure 1. Purification of IBMP-Binding Protein. The protein was purified by sequential column chromatography. (A) DEAE cellulose column chromatography. (B) Hydroxyapatite chromatography. (C) Sephadex G-200 gel filtration. Void volumes (V_0) and (V_i) were at fractions 19 and 51, respectively. Gel filtration led to a single peak of [^{3}H]-IBMP binding activity that coeluted with protein. All fractions were assayed for binding activity using 3 nM [^{3}H]-IBMP in a 0.1-ml volume incubated for 1 hr at 4°C. (From Pevsner *et al.*, 1985.)

purification from bovine nasal epithelium indicates that the odorant-binding protein (OBP) accounts for 0.5% of the total soluble protein. At each step of purification, only a single peak of IBMP binding activity can be detected.

Biochemical characterization of OBP demonstrates a sedimentation coefficient $S_{20,w}$ of 3.2 ± 0.1 S and a Stokes radius determined from gel filtration of 2.7 ± 0.1 nm. Assuming a partial specific volume of 0.74 cc/g, the molecular weight of OBP is 37,000 M_r and the frictional coefficient f/f_0 is 1.1. Sodium dodecyl sulfate–polyacrylamide gel

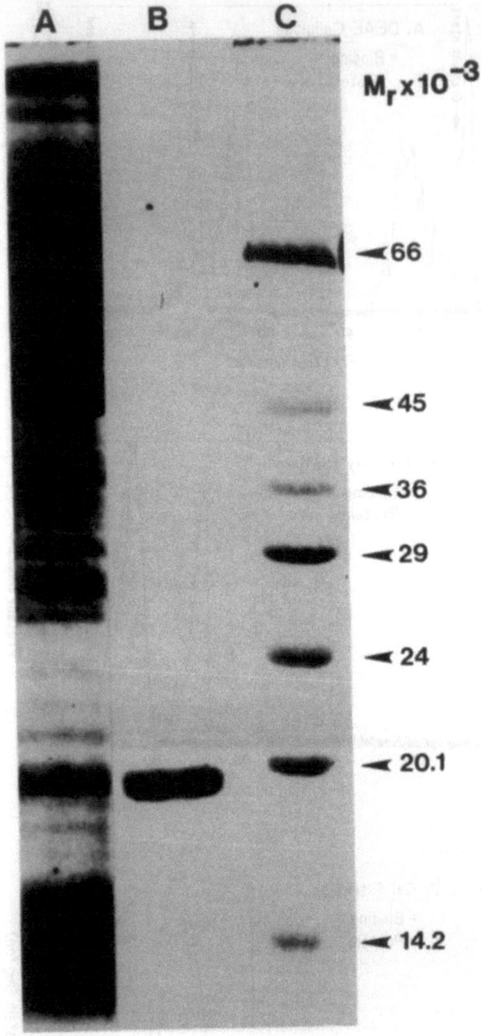

Figure 2. Sodium dodecyl sulfate–polyacrylamide gel electrophoresis (SDS–PAGE) of purified odorant-binding protein and crude soluble bovine nasal epithelium. The samples were electrophoresed into a 12% polyacrylamide gel in the presence of 2-mercaptoethanol and stained with Coomassie blue. (A) Crude nasal epithelium, 75 μg protein. (B) Purified OBP (following gel filtration), 1 μg. (C) Molecular-weight markers. (From Pevsner *et al.*, 1985.)

electrophoresis (SDS–PAGE) indicates a single band of 19,000 M_r, suggesting that the protein is a homodimer (Fig. 2). SDS gels electrophoresed in the absence of a reducing agent also detect a single band of 19,000 M_r, suggesting that, if OBP is a dimer, there are no intersubunit disulfide bonds.

A saturation analysis of [^{3}H]-IBMP binding to purified OBP indicates high- and low-affinity binding sites with apparent K_D values of 10 nM and 3μM, respectively, and B_{max} values of 0.14 and 25 nmoles/mg protein, respectively (Fig. 3). Because of the greater number of low-affinity sites we could not reproducibly monitor odorant binding at the high-affinity site. Accordingly, odorant potencies are measured at the low-affinity site.

Further evidence that OBP is relevant to olfaction comes from the correlation between the affinities of a series of structurally homologous pyrazines for OBP and human

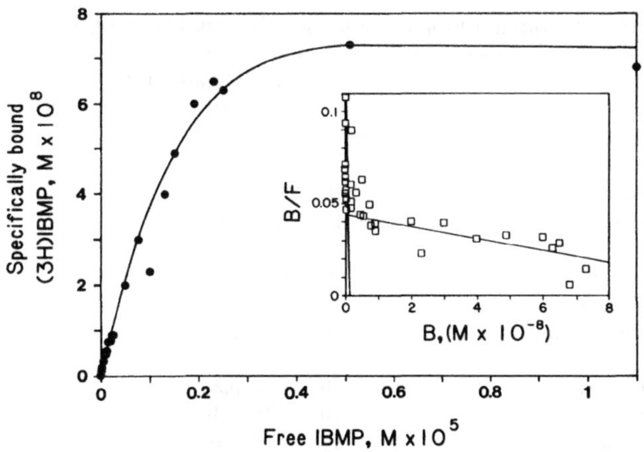

Figure 3. Equilibrium saturation binding of IBMP to purified bovine OBP. OBP was incubated with various concentrations (0.1–10.0 nM) of [^{3}H]-IBMP for 60 min at 4°C. Specific binding was determined by substracting the binding detected in the presence of 1 mM unlabeled IBMP. (Inset) Scatchard plot of the saturation binding data. B, bound ligand; F, free ligand. (From Pevsner *et al.*, 1985.)

odorant-detection thresholds (Table 3). The two most potent odorants in competing for binding, IBMP and 2-isopropyl-3-methoxypyrazine, also have the lowest detection thresholds in behavioral studies. These odorants can be detected at concentrations of ~10 pM. Odorants with higher detection thresholds compete for [^{3}H]-IBMP binding to OBP with weaker affinities. A variety of nonodorous compounds fail to bind to OBP (Table 3).

To ascertain whether OBP is selective for odorant pyrazines, we examined its interactions with five additional tritiated odorants from different structural classes (Table 4). Purified OBP binds [^{3}H]methyldihydrojasmonate ([^{3}H]-MDHJ) and [^{3}H]-3,7-dimethyl-octan-1-ol ([^{3}H]-DMO) with K_D values of 4 μM and 0.5 μM, respectively. Very little binding of [^{3}H]amyl acetate or of [^{3}H]-2-methoxypyrazine to OBP is detectable. In competitive binding studies, each of the unlabeled odorants displays similar potency in competing for the ^{3}H-odorants, suggesting that they are interacting with the same site. Thus, the potency of amyl acetate in inhibiting binding of [^{3}H]-DMO, [^{3}H]-IBMP, and [^{3}H]-MDHJ is similar to the K_D value of [^{3}H]amyl acetate binding to purified OBP (Table 5). Similar competition patterns are apparent for the other odorants. Thus, OBP can bind a wide range of odorants from different structural classes with different odoriferous properties. Conceivably, we have purified a group of structurally similar OBPs that vary in their selectivity for binding odorants. However, amino acid sequence analysis demonstrates the presence of a single amino terminal, suggesting that OBP is a single protein. Furthermore, OBP migrates as a single band by two-dimensional gel electrophoresis.

We immunized rabbits with bovine OBP to produce antisera which could be used for the immunohistochemical localization of OBP (Pevsner *et al.*, 1986). The resultant antiserum displays half-maximal binding to OBP at an antibody dilution of 1 : 8100. To determine the affinity of the antibody for OBP, we examined the binding of ^{125}I-labeled OBP to the antiserum in the presence of unlabeled OBP; 50% of maximal binding is apparent at 0.6 nM OBP. Radioimmunoassay (RIA) shows that the antiserum exhibits

Table 3. Correlation of Human Odorant-Detection Thresholds and Binding Potencies to OBP

Pyrazines	Detection threshold[a,b] (M)	IC_{50}[c] (M)
Highly potent		
2-Methoxy-3-isobutylpyrazine	1.2×10^{-11} (2 ppt)	3×10^{-6}
2-Methoxy-3-isopropylpyrazine	1.3×10^{-11} (2 ppt)	2×10^{-5}
Moderately potent		
2-Methoxy-3-ethypyrazine	2.9×10^{-9} (0.4 ppb)	2×10^{-5}
2-Ethoxy-3-ethylpyrazine	2.6×10^{-9} (0.4 ppb)	7×10^{-5}
2-Methoxy-3-methylpyrazine	3.2×10^{-8} (4 ppb)	1×10^{-4}
Weakly potent		
2-Ethyl-3-methylpyrazine	1.1×10^{-6} (130 ppb)	$>1 \times 10^{-3}$
2-Methoxypyrazine	6.4×10^{-6} (700 ppb)	$>1 \times 10^{-3}$
2,3-Dimethylpyrazine	2.3×10^{-5} (2.5 ppm)	$>1 \times 10^{-3}$
2,5-Dimethylpyrazine	9.3×10^{-6}–3.2×10^{-4} (1–35 ppm)	$>1 \times 10^{-3}$
2,6-Dimethylpyrazine	1.4×10^{-5}–5.0×10^{-4} (1.5–54 ppm)	$>1 \times 10^{-3}$
2-Ethylpyrazine	5.6×10^{-5}–2.0×10^{-4} (6–22 ppm)	$>1 \times 10^{-3}$
2,3,5-Trimethylpyrazine	7.4×10^{-5} (9 ppm)	$>1 \times 10^{-3}$
2,3,5,6-Tetramethylpyrazine	7.4×10^{-5} (10 ppm)	$>1 \times 10^{-3}$
2-Methylpyrazine	6.4×10^{-4}–1.1×10^{-3} (60–105 ppm)	$>1 \times 10^{-3}$
Pyrazine	2.2×10^{-3} (175 ppm)	$>1 \times 10^{-3}$

[a]Human odorant-detection thresholds are from Maga and Sizer (1973), Senf *et al.* (1980), and C. Warren, International Flavors and Fragrances (personal communication). Pyrazines were smelled above aqueous solutions; thresholds are reported as the concentration of the odorant (w/w) in water.
[b]ppt, parts per trillion; ppb, parts per billion; ppm, parts per million in solvent.
[c]IC_{50} values (the concentration of pyrazine that inhibited binding of [^{3}H]-IBMP to OBP 50%) were determined in standard binding assays using purified OBP. Substances inactive (at 1×10^{-6} M–1×10^{-3} M) include *N*-allylnormetazocine (SKF 10,047), caffeine, carnosine, 1-(2-chlorophenyl)-*N*-methyl-*N*-(1-methylpropyl)-3-isoquinolinecarboxamide (PK 111 95), citalopram, clonazepam, cocaine, desmethoxyverapamil (D-888), desmethylimipramine, diazepam, diltiazem, fluspirilene, fucose, GABA, glycine, haloperidol, imipramine, mazindol, methoxyverapamil (D-600), muscimol, naloxone, pimozide, propranolol, serotonin, thioridazine, tiapamil. (From Pevsner *et al.*, 1985.)

considerable sensitivity for OBP, with the lowest detectable level at 30 pM. RIA of rat olfactory mucosal homogenates using the antibody to bovine OBP indicates levels 0.1% of those found in bovine olfactory mucosa, whereas the binding of [^{3}H]-IBMP reveals that the level of OBP is similar in the two species. Thus, there is little antiserum cross-reactivity between cow and rat. The antiserum appears selective for OBP, since immunoblot analysis of crude bovine olfactory and respiratory epithelium displays staining of only a single 19 KDa protein, the same molecular weight as authentic OBP. Moreover, preadsorption of the antiserum with purified bovine OBP eliminates all staining (Fig. 4).

OBP-like immunoreactivity is localized in bovine nasal epithelium to the glands of the lamina propria and along the surface of the epithelium (Fig. 5) (Pevsner *et al.*, 1986). Anatomically, mammalian nasal glands are differentiated into Bowman's glands in the olfactory mucosa and the respiratory glands, which underlie the respiratory mucosa (Getchell *et al.*, 1984). Specific staining for OBP occurs in both types of glands. Preadsorption of the antiserum with purified OBP eliminates glandular staining. By contrast, preadsorption of the antiserum with bovine serum albumin does not alter the staining pattern. The minor band of immunoreactivity along the outer ciliary surface of the olfacto-

Table 4. Binding Constants for Odorants to Purified Bovine OBP and Crude Rat Mucus and Tears[a–c]

	Bovine OBP		Rat mucus		Rat tears	
Odorant	K_D (μM)	B_{max}	K_D (μM)	B_{max}	K_D (μM)	B_{max}
[³H]-amyl acetate	90	22	>1000	n.d.	>1000	n.d.
[³H]-DMO	0.3	30	98	50	100	12.5
[³H]-IBMP	3	27	4	3.6	4.5	0.9
[³H]-MDHJ	8	7	48	6.3	63	0.8
[³H]-2-methoxypyrazine	>1000	n.d.	>1000	n.d.	>1000	n.d.

[a]From Pevsner *et al.* (1986).
[b]n.d., not determined.
[c]Binding assays were performed using bovine OBP, crude rat mucus, or rat tears and 5–15 nM of each tritiated odorant. B_{max} values are nmoles bound per mg protein. No specific binding was detected for the binding of 10 nM [³H]isovaleric acid to bovine OBP or rat mucus. Less than 5 fmoles/mg protein of [³H]-IBMP or [³H]-DMO binding to crude rat saliva was detected.

ry and respiratory epithelium is also observed with normal rat serum and thus is not specific for OBP. Similarly, reaction product deposited in erythrocytes with either OBP antiserum or normal rabbit serum presumably reflects endogenous peroxidase activity.

Immunohistochemical studies do not demonstrate any selective localization of OBP to olfactory cilia, considered the site of odorant receptors, suggesting that the function of OBP is related to its presence in nasal glands. A major role of these glands is the secretion of mucus into the nasal cavity (Widdicombe and Wells, 1982), which suggests that OBP might be secreted in the mucus. Indeed, there is substantial binding of [³H]-DMO, [³H]-IBMP, and [³H]-MDHJ in the nasal mucus and tears, but not the saliva, of rats whose secretions have been stimulated by isoproterenol (Table 4). The binding activity detected in nasal mucus represents authentic OBP. We are able to purify a protein from both rat

Table 5. Inhibition of [³H]Odorant Binding to Purified Bovine OBP[a,b]

	IC_{50} (μM)			
[³H]Odorant	Amyl acetate	DMO	IBMP	MDHJ
[³H]Amyl acetate	—	0.5	18	30
[³H]-DMO	30	—	13	23
[³H]-IBMP	70	0.5	—	13
[³H]-MDHJ	65	2	2	—

[a]From Pevsner *et al.* (1986).
[b]IC_{50} values (concentration of unlabeled odorant that inhibits response by 50%) were calculated from displacement curves using bovine OBP, 30–45 nM [³H]amyl acetate, 3–10 nM [³H]-DMO, 3–10 nM [³H]-IBMP, and 3–10 nM [³H]-MDHJ.

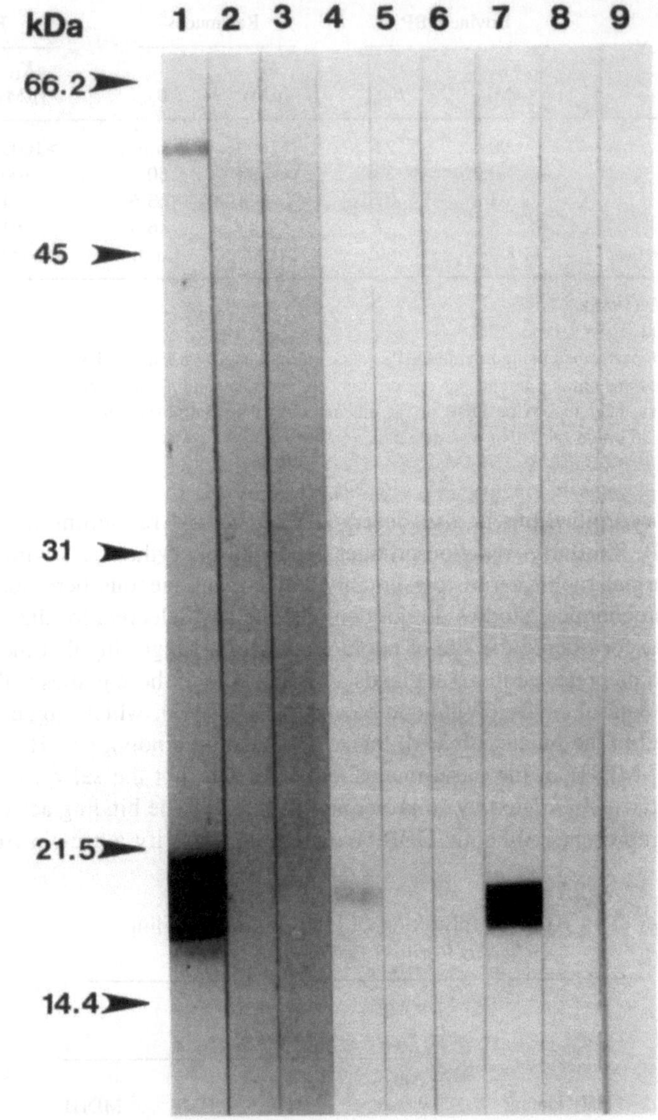

Figure 4. Immunoblot analysis of antiserum to bovine OBP. Samples of purified OBP and bovine nasal homogenates were electrophoresed into 14% sodium dodecyl sulfate–polyacrylamide gels and transferred to nitrocellulose. Purified OBP (20 μg; lanes 1–3), crude nasal olfactory epithelium homogenates (40 μg; lanes 4–6), crude nasal respiratory epithelium homogenates (40 μg; lanes 7–9). Anti-OBP antiserum at a dilution of 1 : 1000 (lanes 1, 4, and 7); anti-OBP antiserum (1 : 1000), preadsorbed for 24 hr with HPLC-purified bovine OBP (50 μg/ml) (lanes 2, 5, and 8); normal rabbit serum (1 : 1000) (lanes 3, 6, and 9). Immunoblots were developed using an avidin–biotin–peroxidase technique. (From Pevsner *et al.*, 1986.)

nasal epithelium and secreted rat mucus by DEAE-chromatography followed by reverse-phase high-pressure liquid chromatography (RP-HPLC). The DEAE and HPLC elution profiles reveal a single peak that comigrates with bovine OBP and displays [^{3}H]-DMO binding activity. SDS–PAGE of rat OBP reveals a single band of 20,000 M_r, similar to bovine OBP. OBP accounts for about 2% of the total protein content of the nasal mucus.

What might be the function of a soluble protein found in nasal mucus that is able to interact with a wide range of odorants? One function may be the concentration and/or transport of odorants from the air to the olfactory epithelium. Potent odorants are routinely detected at nanomolar concentrations in the ambient air. However, electrophysiological (Kashiwayanagi and Kurihara, 1984) and odorant-stimulated adenylate cyclase (Pace *et al.*, 1985) studies reveal a requirement for micromolar concentrations of odorants. There is evidence that odorants are concentrated into nasal mucus to levels two to three orders of magnitude greater than levels in the ambient air (Senf *et al.*, 1980). Simple partitioning into the hydrophobic environment of the nasal mucus might not account for this level of odorant concentration.

To ascertain whether OBP interacts with odorants from the ambient air, we inject rats with isoproterenol to stimulate mucous secretion and then expose the nose, eyes or mouth to [^{3}H]-DMO (Table 6). A substantial concentration of [^{3}H]-DMO is detected in nasal mucus and tears, while markedly lower concentrations of [^{3}H]-DMO accumulate in saliva. We collect the secreted nasal mucus, tears, and saliva from rats which were exposed to [^{3}H]-DMO *in vivo* and find that nasal mucus and tears contain [^{3}H]-DMO trapped on filters, while no odorant binding is present in saliva. Initial experiments suggest that the binding affinity of inhaled [^{3}H]-DMO to nasal mucus is the same as the affinity measured *in vitro*. This implies that [^{3}H]-DMO binds to OBP *in vivo*.

Since OBP binds odorants and is detected in large amounts in nasal mucus, it may play a role in the concentration of odorants from the ambient air into the mucus. A physiological role for OBP in concentrating odorants could explain the parallel between human odorant detection thresholds and affinity for OBP we observe in a series of pyrazines (Table 3). We suggest that the micromolar concentrations of odorants required to stimulate olfactory neuronal receptors might not be attained unless the odorants are concentrated to sufficiently high levels in the nasal mucus. OBP could concentrate odorants within the mucus by a process of facilitated diffusion. This is analogous to the transport of oxygen by hemoglobin and myoglobin, which involves reversible, low-affinity binding of a hydrophobic ligand to a soluble carrier (Mitchell, 1967). It is not known whether OBP concentrates the odorants, transports them through the nose and/or delivers them to odorant receptors. Experiments in progress in our laboratory evaluating the stimulation of ciliary adenylate cyclase by odorants alone or when complexed to OBP may resolve this question.

3. ODORANT SELECTIVITY IN STIMULATION OF ADENYLATE CYCLASE ACTIVITY

Lancet and collaborators reported odorant stimulation of adenylate cyclase in frog olfactory cilia (Pace *et al.*, 1985). We have confirmed and extended these findings and also detected cyclase stimulation in rat olfactory cilia (Sklar *et al.*, 1986). In both species,

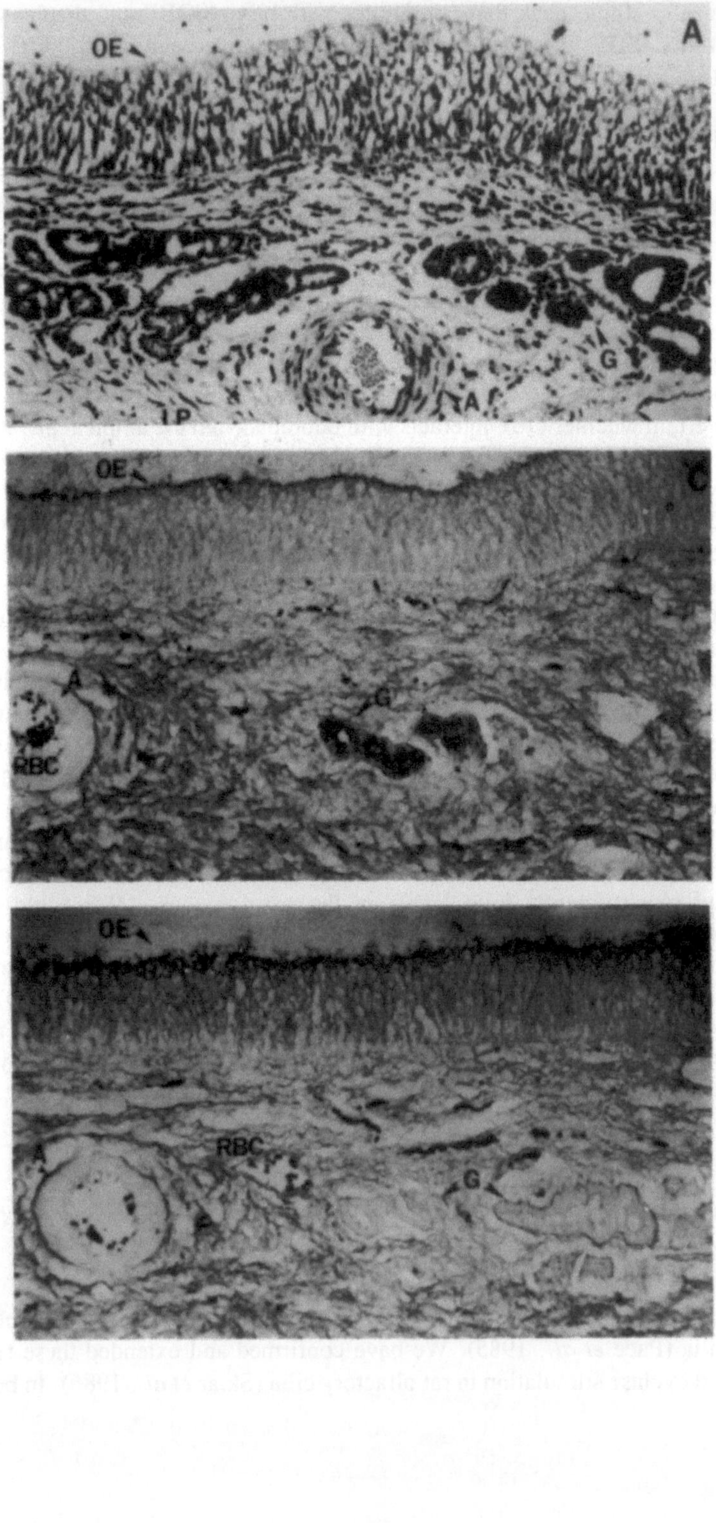
OE
A
G
A
LP
OE
C
A
G
RBC
OE
A
RBC
G

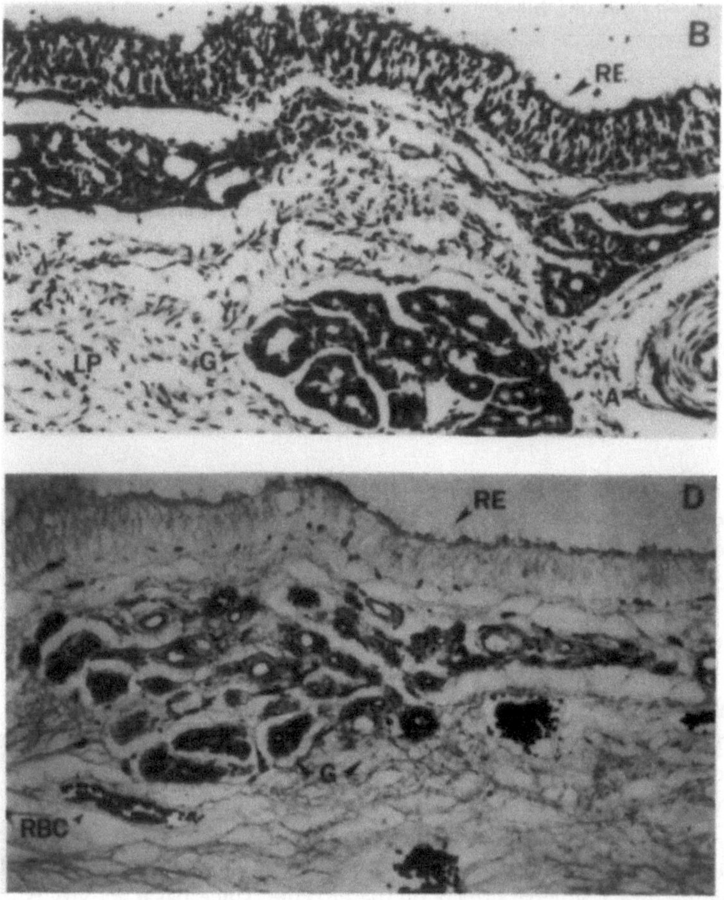

Figure 5. Immunohistochemical localization of bovine OBP. Whole bovine nasal epithelium was glutaraldehyde fixed, cryostat sectioned, and immunohistochemically stained. Bovine olfactory (A, C, E) and respiratory (B, D) epithelium are shown. Normal histology was demonstrated by toluidine blue staining (A, B). The antiserum produced against purified bovine OBP was used at a dilution of 1 : 20,000 and found to stain the glands in the bovine olfactory (C) and respiratory (D) epithelium. (E) Staining of bovine olfactory epithelium by normal rabbit serum (1 : 20,000). OE, olfactory epithelium; RE, respiratory epithelium; RBC, erythrocyte; G, glands; LP, lamina propria; A, arteriole. (From Pevsner *et al.*, 1986.)

basal adenylate cyclase activity is stimulated roughly twofold by GTP and fivefold by GTPγS and forskolin. In addition, odorants stimulate the adenylate cyclase in a tissue-specific and GTP-dependent manner, suggesting that odorant-stimulated signals in the olfactory system may be transduced through regulatory GTP-binding proteins.

Detailed analysis of a large number of odorants reveals that certain odorants enhance adenylate cyclase activity while other odorants do not (Table 7) (Sklar *et al.*, 1986). In general floral, fruity, herbaceous, and minty odorants are active in stimulating the enzyme. Citralva is a particularly potent cyclase stimulator. We therefore, arbitrarily assign

Table 6. Concentration of [^{3}H]-DMO in Rat Secretions after Odorant Exposure *In Vivo*[a,b]

	[^{3}H]-DMO				
	Total		Bound		
Secretion	cpm/μl	concn. (nM)	cpm/μl	concn. (nM)	% Total
Nasal mucus	2280 ± 572 (8)	60 ± 15	539 ± 109 (8)	14 ± 3	24 ± 6
Tears	1942 ± 821 (6)	51 ± 22	304 ± 95 (7)	8 ± 3	16 ± 8
Saliva	16 ± 14 (6)	0.4 ± 0.4	0 ± 4 (6)	0 ± 0.1	0 ± 1

[a]From Pevsner *et al.* (1986).
[b]Individual rats were anesthetized and injected with isoproterenol, and the nose, eyes, or mouth was exposed to odorant (17.5 μM [^{3}H]-DMO; 7.3 × 10^5 cpm/μl). Mucus, tears, or saliva was collected and assayed for [^{3}H]-DMO binding activity.

a standard value of 100% to the effects of 100 μM citralva. This concentration of citralva increases enzymatic activity 55% over the GTP-stimulated basal level. Odorants structurally related to citralva, such as citral dimethyl acetal and citronellal yield about two thirds of the citralva-stimulated effect. Many odorants that are structurally unrelated to citralva, such as methone, D-carvone, L-carvone, 3-hexylpyridine, 2-hexylpyridine, hedione, helional, and coniferan, are also good stimulators of the enzyme. Several odorants from the fruity, floral, minty, and herbaceous classes stimulate activity to an intermediate level. Examples include amyl salicylate, dimethyloctanol, eucalyptol, eugenol, and cinnamic aldehyde. Some odorants among these classes fail to stimulate cyclase activity, such as limonene, lyral, phenylethylalcohol, lilial, and ethyl vanillin (Table 7).

Numerous odorants do not stimulate adenylate cyclase activity even at high concentrations. Inactive odorants tend to be putrid substances, such as isovaleric acid and triethylamine, or chemical solvents, such as pyridine. This apparent lack of stimulation does not involve direct adverse effects on the enzyme, since basal activity is unaffected. Moreover, in the presence of an inactive odorant such as 1 mM triethylamine, 100 μM citralva still produces the same stimulation of adenylate cyclase as when tested alone. Dose–response studies indicate that the odorants citralva, menthone, the stereoisomers D-carvone and L-carvone, and 2-isobutyl-3-methoxypyrazine stimulate adenylate cyclase with similar potencies at 1–100 μM (Fig. 6). By contrast, putrid odorants and odorous chemical solvents, such as isovaleric acid and pyridine, do not stimulate at concentrations up to 10 mM (Fig. 6).

To investigate the differential effects of odorants on adenylate cyclase, we studied homologous series of structurally related odorants (Table 8) (Sklar *et al.*, 1986). Relatively minor variations in the structure of pyrazine, thiazole, and pyridine odorants produce marked alterations in adenylate cyclase stimulation. Enzyme stimulation is maximal with hydrocarbon side chains attached to the ring system. Thus, IBMP and 2-isobutylthiazole, derivatives of methoxypyrazine and thiazole with isobutyl side chains, activate the enzyme, whereas methoxypyrazine and thiazole are inactive. While pyridine alone does not stimulate adenylate cyclase, the addition of hexyl side chains at the 2- or 3-position dramatically increases cyclase activity. Interestingly, addition of the hexyl side chain changes the odor quality from that of a chemical solvent to a floral/fruity quality.

What accounts for the distinction between those odorants that stimulate adenylate cyclase and those that are inactive? Odorants that are the least active are frequently polar charged molecules, whereas the odorants that stimulate adenylate cyclase are frequently nonpolar. In a homologous series of pyrazine, thiazole, and pyridine odorants, compounds with longer hydrocarbon side chains are better able to stimulate adenylate cyclase activity than the unmodified parent compound, suggesting that odorant hydrophobicity may be one factor that determines odorant potency. However, hydrophobicity alone does not explain the differential effects of odorants, since several hydrophobic odorants, such as limonene, lilial, and α-pinene, do not stimulate the enzyme activity.

One possible explanation is that the putrid odorants and odorous chemical solvents act via a different second messenger system than the stimulation of adenylate cyclase. The receptor-mediated breakdown of phosphatidylinositol and the subsequent production of the second messengers diacylglycerol and inositol trisphosphate have recently been shown to be an important mediator of signal transduction (Pfaffinger *et al.*, 1985; Cockcroft and Gomperts, 1985). Diacylglycerol and inositoltrisphosphate affect intracellular metabolism by respectively activating protein kinase C and elevating the intracellular calcium concentration. Like the adenylate cyclase system, phosphoinositide turnover probably requires the mediation of GTP-binding proteins. G_s and G_i proteins are involved, respectively, in the stimulation and inhibition of adenylate cyclase by hormones and neurotransmitters (Gilman, 1984; Schramm and Selinger, 1984). A distinct G_o protein is quantitatively more prominent in the brain than either G_i or G_s (Huff and Neer, 1985). Immunohistochemical studies reveal a similar localization of the G_o protein and protein kinase C suggesting that G_o may be involved in phosphoinositide turnover (Worley, 1986). It is possible that putrid odorants may act through the activation of the phosphoinositide cycle.

4. LOCALIZATION OF GTP-BINDING PROTEINS IN NASAL MUCOSA

To assess the contributions of both the adenylate cyclase and phosphoinositide systems to olfaction, we have employed antibodies that recognize the GTP-binding proteins, G_s, G_i, and G_o in biochemical analyses of isolated olfactory cilia and in immunohistochemical studies of olfactory and respiratory epithelium (Anholt *et al.*, 1986, 1987). The G_s and G_i proteins are respectively associated with stimulation and inhibition of adenylate cyclase (Schramm and Selinger, 1984; Gilman, 1984), while G_o may be linked to the phosphodiesteratic cleavage of phosphatidylinositol bisphosphate to inositol trisphosphate and diacylglycerol. A fourth GTP-binding protein, transducin, has been identified thus far only in the retina and pineal gland (Stryer, 1983; van Veen *et al.*, 1986). Transducin mediates the light-induced phosphodiesteratic cleavage of cyclic guanosine monophosphate (cGMP), which is involved in the initial steps in retinal phototransduction. GTP-binding proteins consist of three subunits, α, β, and γ. The β-subunit is conserved throughout evolution and appears to be similar, if not identical, in G_s, G_i, G_o, and transducin. By contrast, the α-subunits differ substantially between the four G proteins and can be identified by a combination of SDS–PAGE and characteristic labeling by *Vibrio cholerea* and *Bordetella pertussis* toxins.

Using an antiserum that recognizes a common sequence in the α-subunits of G_s, G_i,

Table 7. Stimulation by Odorants of the GTP-Dependent Adenylate Cyclase in Frog Olfactory Cilia[a,b]

Odorant (100 μM)	Stimulation (%)
Odorous chemical solvents	
Toluene	14 ± 8(3)
Ethanol	10 ± 7(3)
Chloroform	5 ± 2(3)
Butanol	4 ± 10(3)
Pyridine	4 ± 22(4)
Xylene	−2 ± 3(2)
Acetic acid	−14 ± 33(3)
Fruity	
Citralva[c]	100
Citral dimethyl acetal[c]	69 ± 10(5)
Citronellal	56 ± 5(3)
β-Ionone	55 ± 5(5)
Citronellyl acetate[c]	50 ± 9(4)
Isoamyl acetate	19 ± 11(8)
Limonene	5 ± 4(5)
Lyral[c]	−4 ± 6(2)
Herbaceous	
D-Carvone	74 ± 4(7)
Eugenol	47 ± 7(5)
Cinnamic aldehyde	34 ± 5(4)
Isoeugenol	31 ± 5(3)
Benzaldehyde	27 ± 4(4)
Ethyl vanillin	−3 ± 6(5)
Putrid	
Furfuryl mercaptan	29 ± 9(3)
Triethylamine	4 ± 7(5)
Phenylethylamine	0 ± 7(3)
Isobutyric acid	−2 ± 7(3)
Pyrollidine	−4 ± 6(2)
Isovaleric acid	−6 ± 8(5)
Minty	
Isomenthone	105 ± 10(3)
L-Carvone	74 ± 31(6)
Menthone	71 ± 3(4)
Eucalyptol	45 ± 8(4)
Floral	
3-Hexyl pyridine	118 ± 10(3)
2-Hexyl pyridine	107 ± 8(3)
Hedione[c] (MDHJ)	63 ± 4(3)
Coniferan[c]	60 ± 20(2)
Geraniol	58 ± 3(3)
Helional	53 ± 6(5)
Decanal	53 ± 8(2)
Amylsalicylate	40 ± 10(2)
Dimethyloctanol	33 ± 9(3)
Acetophenone	30 ± 4(3)
α-Pinene	21 ± 12(3)
Phenylethylalcohol	19 ± 4(3)
Lilial[c]	−1 ± 4(2)

[a]From Sklar *et al.* (1986).

[b]Adenylate cyclase activity was measured as described in Sklar *et al.* (1986). Odorants were tested at 100 μM in the presence of 10 μM GTP. Data are expressed as a percentage of the activity observed in the presence of 100 μM citralva.

[c]Nomenclature for odorants not listed in *Merck Index*, 10th ed. (M. Windholz, S. Budavari, R. F. Blumetti, and E. S. Otterbein, eds.), 1983, Merck & Co., Rahway, New Jersey, is as follows: citralva, 3,7-dimethyl-2,6-octadienenitrile; citral dimethyl acetal, 1,1-dimethoxy-3,7-dimethyl-2,6-octadiene; citronellyl acetate, 3,7-dimethyl-6-octen-1-ol acetate; lyral, 4-(4-hydroxy-4-methylpentyl)-3-cyclohexene-1-carboxaldehyde; hedione, 3-oxo-2-pentylcyclopentaneacetic acid methyl ester; coniferan, 2-*t*-pentylcyclohexenol actate; helional, α-methyl-1,3-benzodioxole-5-propanal; lilial, 4-(1,1)dimethylethyl)-α-methyl-benzenepropanol.

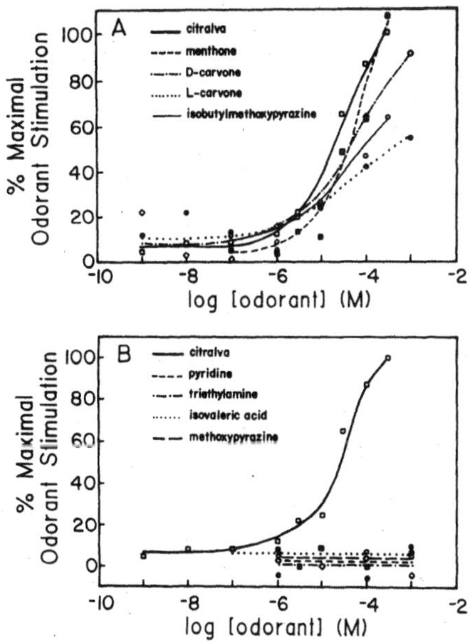

Figure 6. Effect of odorant concentration on adenylate cyclase activity in frog olfactory cilia. (A) Odorants that stimulate adenylate cyclase activity. (B) Odorants that do not stimulate adenylate cyclase activity. Data are expressed as a percentage of the enzymatic activity observed in the presence of 100 μM citralva. (From Sklar *et al.*, 1986.)

G_o, and transducin, we can detect three immunoreactive proteins in immunoblots of frog olfactory nasal tissue (Fig. 7) (Anholt *et al.*, 1987). Isolated olfactory cilia appear to contain primarily G_s and G_o. These proteins are also present in membranes prepared from the olfactory epithelium following cilia removal. The molecular weights for the α-subunits of G_o, G_i, and G_s in these studies are 40,000, 42,000, and 45,000 M_r, respectively. Strikingly, respiratory cilia lack detectable levels of the α-subunit of the three G proteins, G_s, G_i, and G_o. Relative to deciliated membranes, the α-subunit of G_s is present in higher levels in olfactory cilia than the α subunits of either G_o or G_i. The identification of an odorant-sensitive GTP-dependent adenylate cyclase (Pace *et al.*, 1985; Sklar *et al.*, 1986), along with the localization of G_s to the olfactory cilia suggests a role for the GTP-regulated adenylate cyclase in olfaction.

Transducin has a molecular weight similar to that of G_o. To determine whether the 40,000-M_r protein band detected in olfactory cilia contains a transducin-like component as well as G_o, we have employed an antiserum against the alpha subunit of retinal transducin (Fig. 8) (Anholt *et al.*, 1987). Although this antiserum reveals a prominent 40,000-M_r protein in retinal rod outer segments from the frog *Rana catesbeiana,* no immunoreactivity is apparent in either olfactory or respiratory cilia even using 100-fold higher concentrations of antiserum than are needed for the unambiguous detection of retinal transducin. An antiserum reactive with the beta subunit of all G proteins, including transducin, G_i, G_o, and G_s, reveals a single 36,000-M_r band in retinal rod outer segments as well as olfactory cilia. In respiratory cilia, only a very faint band corresponding to the beta subunit is apparent.

Table 8. Stimulation by Odorant Pyrazines, Thiazoles, and Pyridines of the GTP-Dependent Adenylate Cyclase in Frog Olfactory Cilia[a]

Methoxypyrazines		Thiazoles		Alkylpyrazines		Pyridines	
Odorant (100 μM)	Stimulation (%)	Odorant (100 μM)	Stimulation (%)	Odorant (100 μM)	Stimulation (%)	Odorant (100 μM)	Stimulation (%)
Methoxypyrazine	−5 ± 5(8)	Thiazole	−18	2-Ethylpyrazine	−7 ± 8(2)	Pyridine	4 ± 10(4)
2-Methyl-3-methoxypyrazine	9 ± 3(3)	2-Acetylthiazole	3 ± 7(5)	2,3-Dimethylpyrazine	−6 ± 8(3)	2-Hexylpyridine	107 ± 8(3)
2-Ethyl-3-methoxypyrazine	20 ± 8(5)	2,4-Dimethyl-5-acetylthiazole	14 ± 6(2)	2,3,5-Trimethylpyrazine	−3 ± 1(3)	3-Hexylpyridine	118 ± 10(3)
2-Isopropyl-3-methoxypyrazine	36 ± 8(5)	2-Acetylthiazole	25 ± 2(2)	2-Ethyl-3-methylpyrazine	−2 ± 7(3)		
2-Isobutyl-3-methoxypyrazine	53 ± 4(10)	2-Isobutylthiazole	59 ± 5(4)	Pyrazine	0		
				2,3,5,6-Tetramethylpyrazine	0 ± 11(3)		
				2-Methylpyrazine	7 ± 6(3)		
				2,3-Diethyl-5-methylpyrazine	16 ± 9(3)		

[a]Adenylate cyclase activity was as measured as described in Sklar *et al.* (1986). Pyrazine, thiazole, and pyridine odorants were tested at 100 μM in the presence of 10 μM GTP. The data are expressed as a percentage of the activity observed in the presence of 100 μM citralva. (From Sklar *et al.*, 1986.)

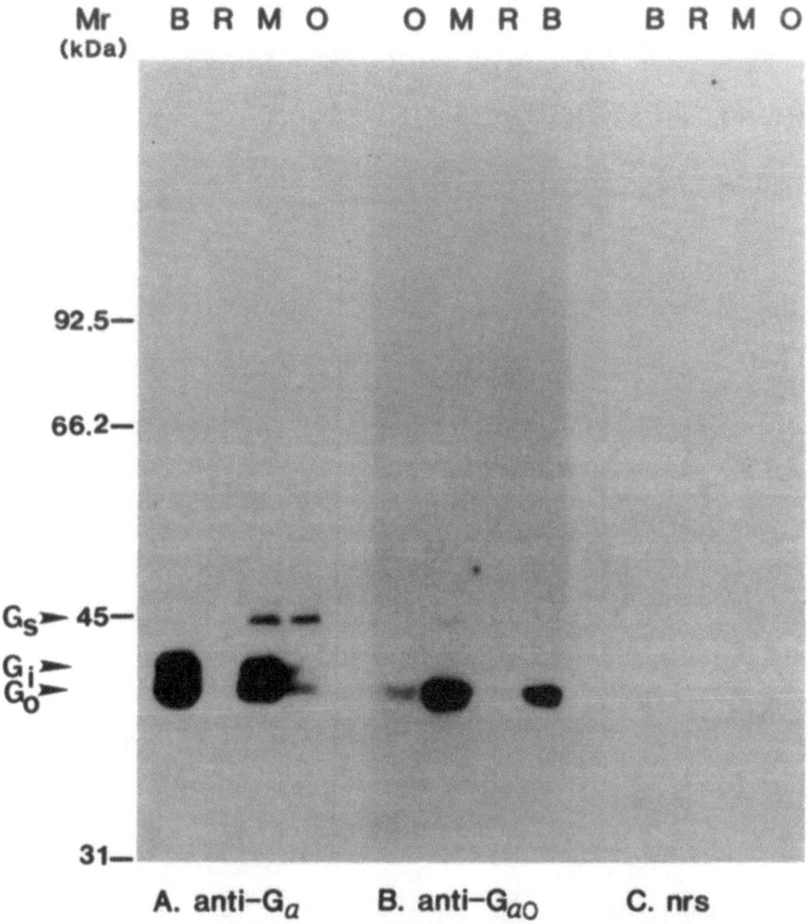

Figure 7. Immunochemical identification of the α-subunits of G-proteins after polyacrylamide gel electrophoresis in SDS and electrophoretic transfer. The lanes were loaded with 100 μg protein of B, brain membranes, R, respiratory cilia, M, deciliated olfactory epithelial membranes, or O, olfactory cilia. (A) Antiserum reactive with a common region of the α-subunits of all transductory G-proteins (anti G_α, antiserum A-569) in a 1 : 2500 dilution. (B) Antiserum specific for the α-subunits of G_O (anti $G_{\alpha O}$, antiserum B-770) in a 1 : 500 dilution. (C) Normal rabbit serum (nrs) in a 1 : 500 dilution. Bound antibody was visualized by autoradiography following incubation with [^{125}I]protein A. The specificities of antibodies A-569 and B-770 have been previously described (Mumby *et al.*, 1986). (From Anholt *et al.*, 1986.)

Immunohistochemical studies utilizing the antiserum recognizing the common beta subunit of G proteins reveals staining of the ciliary surface of the olfactory epithelium and axon bundles on the lamina propria (Fig. 9 B, D, F) (Anholt *et al.*, 1987). By contrast, an antiserum against the common sequence of the α-subunits preferentially stains the cell membranes of the olfactory cells and the acinar cells of Bowman's glands and the deep submucosal glands (Fig. 9 D, C, E). Prolonged incubation periods with these antisera tends to obliterate these differences giving rise to staining by both antisera of the ciliary surface, the olfactory receptor cell membranes, the axon bundles, and the acinar cells of

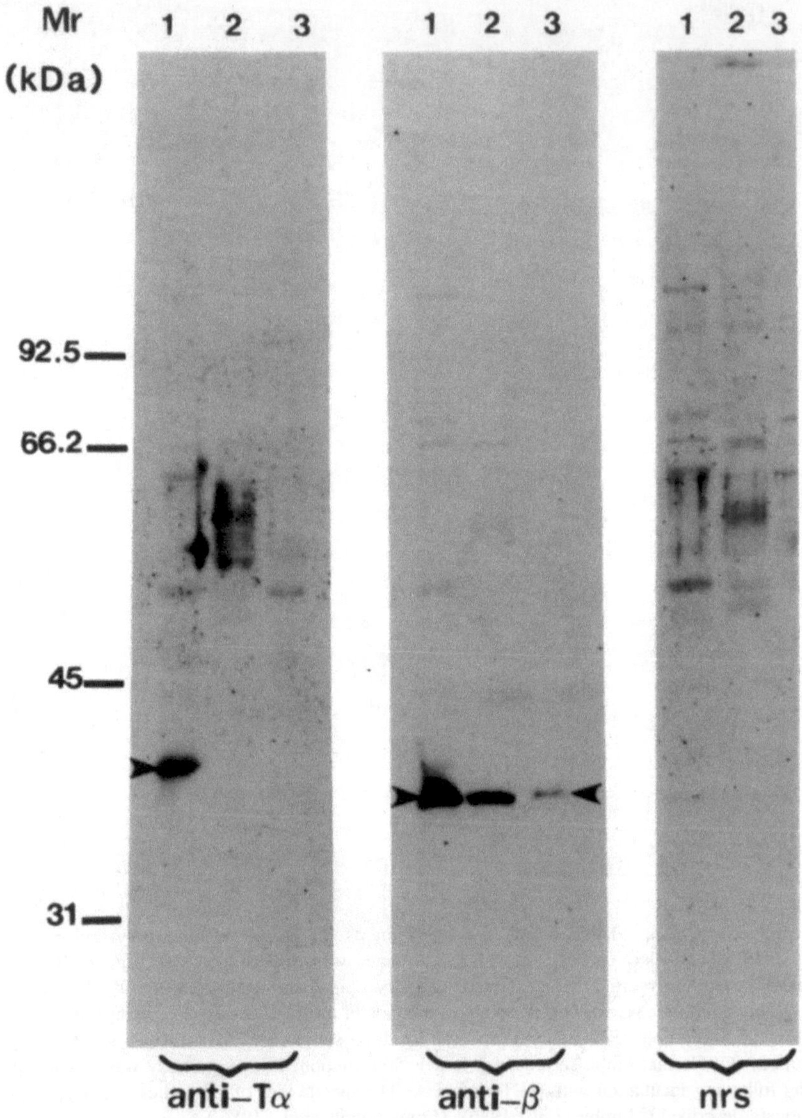

Figure 8. Immunochemical identification of G proteins in retinal rod outer segments, olfactory cilia, and respiratory cilia with an antiserum against the α–subunits of transducin and with an antiserum against the β-subunits of all G proteins. The lanes were loaded with 30 μg protein of retinal rod outer segments (lane 1), olfactory cilia (lane 2), or respiratory cilia (lane 3); immunoblotting was performed with a 200-fold dilution of an antiserum monospecific for the α-subunit of transducin (anti-T_{α}, antiserum U-42), a 10,000-fold dilution of an antiserum reactive with the β-subunit of transductory G proteins (anti-β, antiserum U-49), or a 200-fold dilution of normal rabbit serum (nrs). Bound antibody was visualized with an avidin–biotin–peroxidase technique. The T_{α} band and the β-bands are indicated by arrowheads. The specificities of antibodies U-42 and U-49 have been previously described (Mumby *et al.*, 1986). (From Anholt *et al.*, 1986.)

the glands. Unfortunately, staining is not detected in the paraffin-embedded sections of formalin-fixed decalcified tissue or fixed and unfixed frozen sections using the antibodies to a synthetic peptide of the α-subunits of G_o.

5. CONCLUSIONS

It is hoped that the combination of ligand-binding studies, adenylate cyclase measurements, and immunohistochemical studies of G proteins will provide a clarification of the molecular properties of olfactory transduction. Radiolabeled odorant binding failed to detect specific olfactory receptors on the cilia of olfactory neuronal dendrites. Perhaps the lack of odorant binding to membrane-bound olfactory receptors stems from the low sensitivity postulated for odorant receptors (Lancet, 1986). Conventional neurotransmitter and drug receptor binding is most successful for receptors having high affinity for ligands.

Nonetheless, binding of ^{3}H-odorants to a soluble protein synthesized in nasal glands and secreted into the nasal mucus may reflect a novel regulator of the olfactory process. The properties of OBP suggest that it may represent the initial site for interaction of odorants with the olfactory system and may be able to influence which odorants are detected. Conceivably, human variation in levels or properties of OBP may be responsible for certain forms of anosmia. Since OBP is secreted into the nasal mucus, it can be easily monitored by noninvasive techniques. Presently, we are initiating studies of OBP secretion into the mucus of human subjects.

Our studies of odorant-sensitive adenylate cyclase reveal that odorants can be categorized into two classes: (1) odorants that stimulate adenylate cyclase activity, including many fruity, floral, minty, and herbaceous odorants; and (2) odorants that do not stimulate activity of this enzyme, including putrid odorants and odorous chemical solvents. This suggests that cAMP may only function as the second messenger for selected odorants. The selective localization of the G_o proteins to olfactory and not respiratory cilia suggests a role for the phosphoinositide cycle as a second messenger for odorants that do not stimulate the adenylate cyclase, a possibility we are examining.

Thus far, none of the accumulated information breaks the olfactory code. We still do not know how many different classes of odorant receptors exist. Strategies that we are employing now may be able to clarify this question. We are seeking odorants that are able to antagonize the stimulation of other odorants in the adenylate cyclase assay. Antagonists that block the effects of only selected odorant molecules may help us define specific classes of odorant receptors. With a sufficiently large battery of odorant antagonists, it might be feasible to identify a substantial number of distinct odorant receptor subtypes.

ACKNOWLEDGMENTS. This work was supported by a grant from International Flavors and Fragrances, Inc. We would like to thank Dr. Robert Anholt for valuable advice and critical reading of the manuscript. Dr. Anholt initiated the studies on isolated olfactory cilia from *Rana catesbeiana* in our laboratory and was directly involved in the work on GTP-binding proteins and adenylate cyclase described in this chapter. We also thank Adele Snowman for excellent technical assistance and Nancy Bruce and Rita Hollingsworth for secretarial assistance.

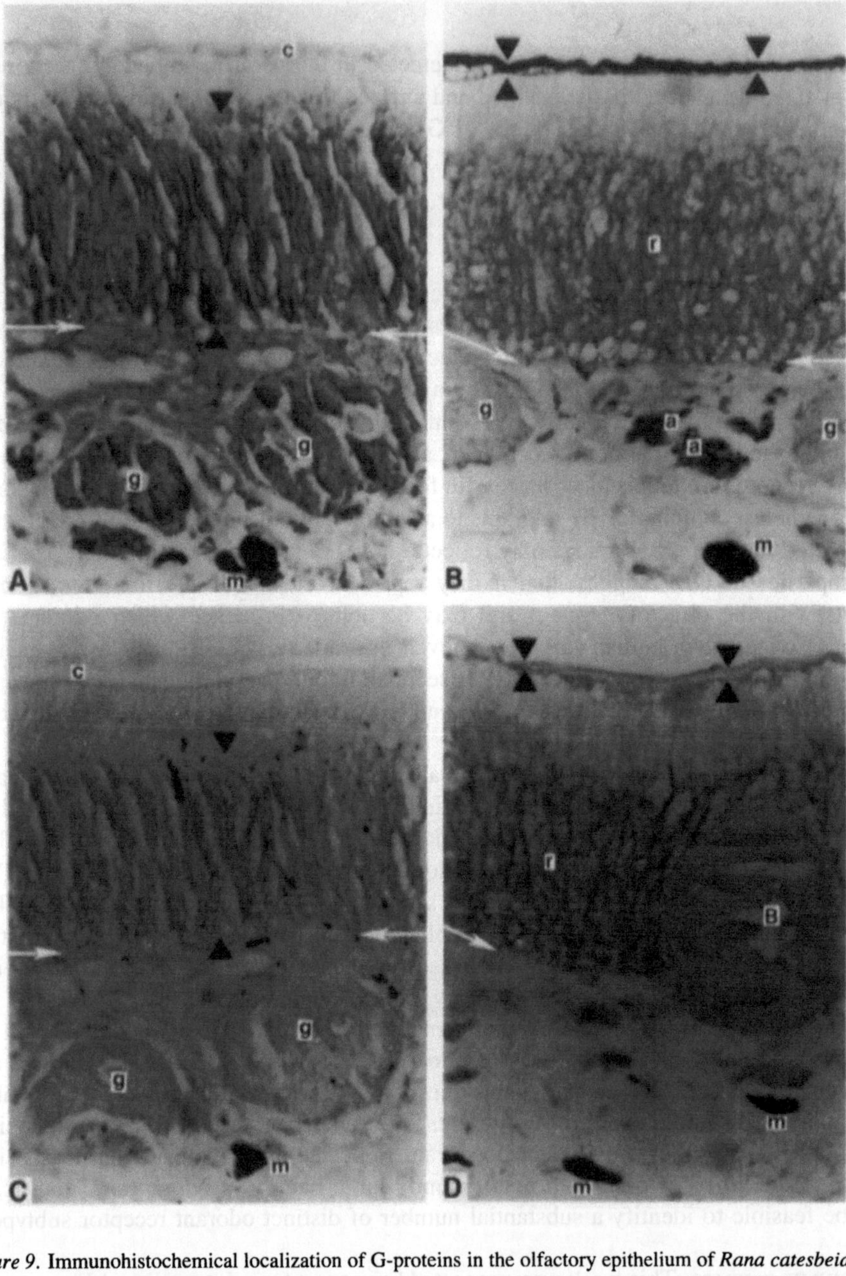

Figure 9. Immunohistochemical localization of G-proteins in the olfactory epithelium of *Rana catesbeiana*. (A) Common anti-G_α antiserum (A-569). (B) Anti-G_β antiserum (U-49). (C) Anti-G_α antiserum (A-569) after 18-hr preincubation at 4°C of the antiserum with 5 mg/ml of the synthetic peptide against which it was generated. (D) Anti-G_β antiserum (U-49) after preincubation at 4°C of the antiserum with 3 mg/ml of the synthetic peptide against which it was generated. (E,F) Preimmune serum. The sections were pretreated with 0.1% SDS before immunohistochemistry. All sera were used at 1000-fold dilution. a, axon bundles in the lamina propria; c, ciliary surface; g, submucosal glands; m, melanocyte; r, olfactory receptor cell layer; B, Bowman's gland. Horizontal arrows designate the location of the basement membrane. (a,c) Arrowheads designate areas of intense staining. The bar represents 100 μm. The specificities of the antibodies A-569 and U-49 have been previously described (Mumby *et al.*, 1986). (From Anholt *et al.*, 1986.)

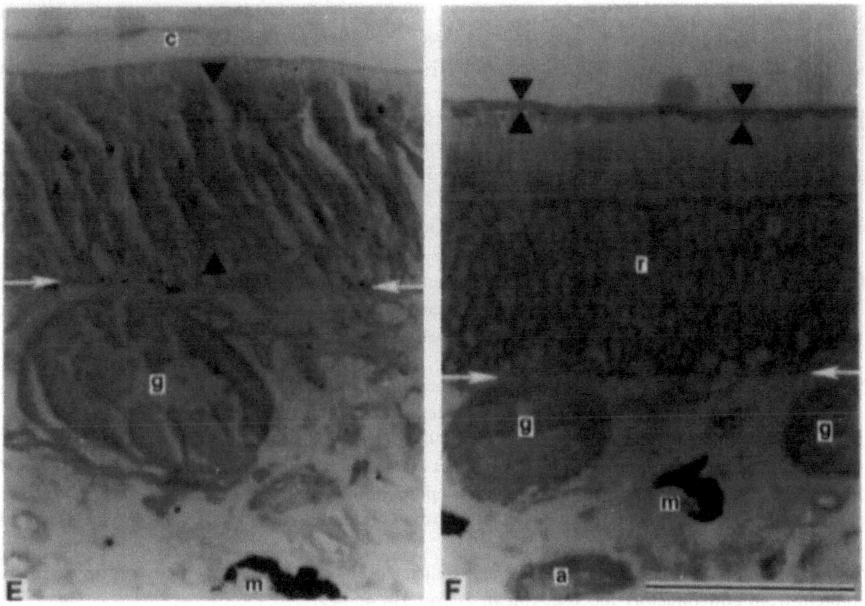

Figure 9. (*continued*)

REFERENCES

Amoore, J. M., 1982, Odor theory and odor classification, in: *Fragrance Chemistry: The Science of the Sense of Smell* (E. T. Theimer, ed.), pp. 27–76, Academic, New York.

Anholt, R. R. H., Aebi, U., and Snyder, S. H., 1986, A partially purified preparation of isolated chemosensory cilia from the olfactory epithelium of the bullfrog, *Rana catesbeiana*, *J. Neurosci.* **6:** 1962–1969.

Anholt, R. R. H., Mumby, S. M., Stoffers, D. A., Girard, P. R., Kuo, J. F., and Snyder, S. H., 1987, Transduction proteins of olfactory receptor cells: Identification of guanine nucleotide binding proteins and protein kinase C, *Biochemistry* **26:** 788–795.

Bignetti, E., Cavaggioni, A., Pelosi, P., Persaud, K. C., Sorbi, R. T., and Tirindelli, R., 1985, Purification and characterization of an odorant-binding protein from cow nasal tissue, *Eur. J. Biochem.* **149:** 227–231.

Cockcroft, S., and Gomperts, B. D., 1985, Role of guanine nucleotide binding protein in the activation of polyphosphoinositide phosphodiesterase, *Nature (Lond.)* **314:** 534–536.

Getchell, T. V., Margolis, F. L., and Getchell, M. L., 1984, Perireceptor and receptor events in vertebrate olfaction, *Prog. Neurobiol.* **23:** 317–345.

Gilman, A. G., 1984, G proteins and dual control of adenylate cyclase, *Cell* **36:** 577–579.

Huff, R. M., Axton, J. M., and Neer, E. J., 1985, Physical and immunological characterization of a guanine nucleotide-binding protein purified from bovine cerebral cortex, *J. Biol. Chem.* **260:** 10864–10871.

Kurihara, K., and Koyama, N., 1972, High activity of adenyl cyclase in olfactory and gustatory organs, *Biochem. Biophys. Res. Commun.* **48:** 30–33.

Lancet, D., 1986, Vertebrate olfactory reception, *Annu. Rev. Neurosci.* **9:** 329–355.

Maga, J. A., and Sizer, C. E., 1973, Pyrazines in foods, *CRC Rev. Food Technol.* **16:** 1–48.

Mitchell, P., 1967, Translocations through natural membranes, *Adv. Enzymol.* **29:** 33–87.

Mumby, S. M., Kahn, R. A., Manning, D. R., and Gilman, A. G., 1986, Antisera of designated specificity for subunits of guanine nucleotide-binding regulatory proteins, *Proc. Natl. Acad. Sci. USA* **83:** 265–269.

Pace, U., Hanski, E., Salomon, Y., and Lancet, D., 1985, Odorant-sensitive adenylate cyclase may mediate olfactory reception, *Nature (Lond.)* **316:** 255–258.

Pelosi, P., and Pisanelli, A. M., 1981, Binding of [^{3}H]-2-isobutyl-3-methoxypyrazine to cow olfactory mucosa, *Chem. Senses* **6:** 77–85.

Pelosi, P., Baldaccini, N. E., and Pisanelli, A. M., 1982, Identification of a specific olfactory receptor for 2-isobutyl-3-methoxypyrazine, *Biochem. J.* **201:** 245–248.

Pevsner, J., Trifiletti, R. R., Strittmatter, S. M., and Snyder, S. H., 1985, Isolation and characterization of an olfactory receptor protein for odorant pyrazines, *Proc. Natl. Acad. Sci. USA* **82:** 3050–3054.

Pevsner, J., Sklar, P. B., and Snyder, S. H., 1986, Odorant binding protein: Localization to nasal glands and secretions, *Proc. Natl. Acad. Sci. USA* **83:** 4942–4946.

Pfaffinger, P. J., Martin, J. M., Hunter, D. D., Nathanson, N. M., and Hille, B., 1985, GTP-binding proteins couple cardiac muscarinic receptors to a K channel, *Nature (Lond.)* **317:** 536–538.

Schramm, M., and Selinger, Z., 1984, Message transmission: Receptor controlled adenylate cyclase system, *Science* **225:** 1350–1356.

Senf, W., Menco, B. P. M., Punter, P. H., and Duyvesteyn, P., 1980, Determination of odor affinities based on the dose–response relationships of the frog's electro-olfactogram, *Experentia* **36:** 213–215.

Sklar, P. B., Anholt, R. R. H., and Snyder, S. H., 1986, The odorant-sensitive adenylate cyclase of olfactory receptor cells: Differential stimulation by distinct classes of odorants, *J. Biol. Chem.* **26:** 15538–15543.

Snyder, S. H., 1984, Drug and neurotransmitter receptors in the brain, *Science* **224:** 22–31.

Stryer, L., 1983, Transducin and the cyclic GMP phosphodiesterase: Amplifer proteins in vision, *Cold Spring Harbor Symp. Quant. Biol.* **47:** 841–852.

Topazzini, A., Pelosi, P., Pasqualetto, P. L. and Baldaccini, N. E., 1985, Specificity of a pyrazine binding protein from cow olfactory mucosa, *Chem. Senses* **10:** 45–49.

van Veen, T., Ostholm, T., Gierschik, P., Spiegel, A., Somers, R., Korf, H. W., and Klein, D. C., 1986, α-Transdusin immunoreactivity in retinae and sensory pineal organs of adult vertebrates, *Proc. Natl. Acad. Sci. USA* **83:** 912–916.

Widdicombe, J. G., and Wells, U. M., 1982, Airway secretions, in: *The Nose—Upper Airway Physiology and the Atmospheric Environment* (D. F. Proctor and I. B. Andersen, eds.), pp. 215–244, Elsevier, New York.

Worley, P. F., Baraban, J. M., Van Dop, C., Neer, E. J., and Snyder, S. H., 1986, G_0, a guanine nucleotide-binding protein: Immunohistochemical localization in rat brain resembles distribution of second messengers, *Proc. Natl. Acad. Sci. USA* **83:** 4561–4565.

2

Molecular Components of Olfactory Reception and Transduction

Doron Lancet

1. INTRODUCTION

During the past few years, considerable progress has been attained in our understanding of molecular events underlying odor recognition (Price, 1981; Dodd and Persaud, 1981; Lancet, 1984, 1986; Getchell *et al.*, 1984; Getchell, 1986; Anholt, 1987; Lancet and Pace, 1987) (see also Chapters 1 and 8, this volume). Although the receptor proteins remain elusive, important information has been obtained on the possible transduction mechanism. Whereas earlier theories concentrated on *ad hoc* mechanisms, mainly aimed at explaining the relations between the physics and chemistry of odorants and their perceived quality or threshold (Davies, 1971), the weight has now shifted toward relating odor reception to universal membrane transduction processes elsewhere in biology (cf. Venter and Harrison, 1984; Kleinzeller and Martin, 1983; Strosberg, 1984). This chapter is guided by the latter approach and focuses on a description of results and hypotheses that form our current view of olfactory mechanisms. Special emphasis is placed on open questions that have yet to be resolved.

2. OLFACTORY CILIA: THE SITE OF OLFACTORY RECEPTION AND TRANSDUCTION

2.1. Olfactory Cilia

Olfactory sensory neurons are equipped with long dendritic extensions in the form of specialized cilia (Reese, 1965; Moulton and Beidler, 1967; Menco, 1980, 1983, 1984; Adamek *et al.*, 1984). In other cells (e.g., those lining respiratory epithelium), similar cilia appear as motile cellular elements, whose function is concentrated primarily in the cytoskeletal core—the ciliary axoneme (Sleigh, 1974; *J. Cell Biol.*, 1981; Baccetti and Gibbons, 1983). In sensory cells, olfactory neurons included, cilia have acquired a

Doron Lancet • Department of Membrane Research, The Weizmann Institute of Science, Rehovot 76 100, Israel

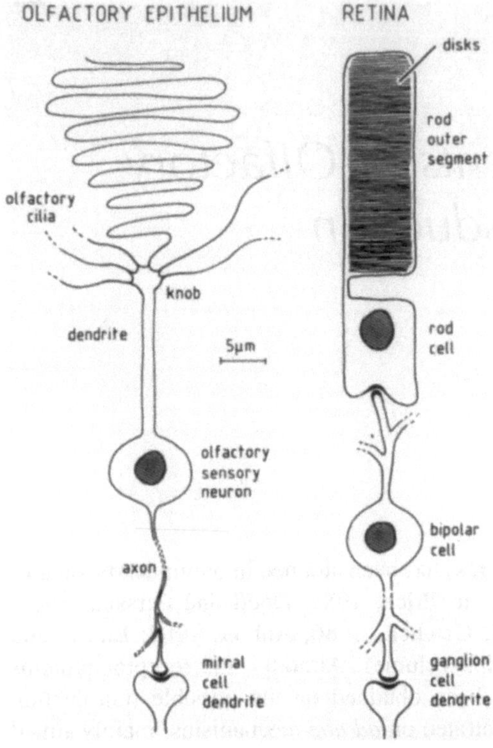

Figure 1. Olfactory and visual sensory cells. Schematic representation of the structures involved in the transmission of sensory signals in the peripheral olfactory and visual systems, drawn approximately to scale. In both, transduction occurs in modified ciliary structures, olfactory cilia, and retinal rod outer segments, the main purpose of which appears to be increased receptive membrane area. In the olfactory system, the passive membrane events in the cilia and dendrite give rise to action potentials along the axon of the same sensory neurons, reaching a secondary projection neuron (mitral or tufted cell) along a monosynaptic pathway. In the case of retinal rods, action potentials are only produced in the projection neurons (ganglion cells); passive membrane events are transmitted across two synapses before reaching these neurons.

secondary function whereby a key role is played by the enclosing lipid bilayer, the axoneme becoming an inert scaffold (Lidow and Menco, 1984). Sensory cilia are often modified in accordance with their function (Vinnikov, 1982), a prominent example being the elaborate membranous disc structures of retinal rod outer segment (O'Brien, 1982) (Fig. 1). It is now commonly accepted that the membranes of olfactory cilia, similar to those of rod outer segments, are the main site of the molecular machinery of sensory transduction (Rhein and Cagan, 1981; Getchell *et al.*, 1984; Lancet, 1986) (Fig. 2).

2.2. Isolated Cilia Preparations

One of the important recent advances in the biochemical studies of olfaction has been the development of isolated olfactory cilia preparations. While such preparations do not usually consist of pure cilia, they are enriched with ciliary membranes with respect to other olfactory epithelial membrane fractions (Rhein and Cagan, 1980, 1981; Chen and Lancet, 1984; Chen *et al.*, 1986*c*; Anholt *et al.*, 1986*a*). An isolated olfactory cilia preparation from the fish, the first to be developed, has been shown to possess binding activity toward amino acid odorants (Rhein and Cagan, 1980, 1981).

A preparation of olfactory cilia from the frog *Rana ridibunda* has paved the way to more recent biochemical studies of olfactory mechanisms (Chen and Lancet, 1984). The frog preparation is advantageous because of the ease of epithelial dissection and because the cilia are very long and can be quantitatively removed. Furthermore, the frog affords a unique background of electrophysiological characterization of its olfactory responses (Moulton and Beidler, 1967; Gesteland, 1976; Holley and MacLeod, 1977; Getchell, 1986). Another olfactory cilia preparation, from the bullfrog *Rana catesbeiana*, provides larger amounts of cilia per frog (Anholt *et al.*, 1986*a*; Sklar *et al.*, 1986). Similar preparations have also been obtained in the rat (Sklar *et al.*, 1986; Chen *et al.*, 1986*a*; Lancet and Pace, 1986; Shirley *et al.*, 1986) and in the cow (Chen *et al.*, 1986*a*).

Olfactory cilia preparations from different species were shown to contain transductory enzymes such as adenylate cyclase (Pace *et al.*, 1985), cyclic nucleotide phosphodiesterase (Shirley *et al.*, 1985) signal-coupling guanosine triphosphate (GTP)-binding proteins (Lancet and Pace, 1984; Pace *et al.*, 1985; Anholt *et al.*, 1986*a,b*), Na^+/K^+ ATPase (Anholt *et al.*, 1986*a*) and cyclic adenosine monophosphate (cAMP)-dependent protein kinase (Lancet *et al.*, 1986; Heldman and Lancet, 1986). Olfactory cilia preparations also have unique surface glycoproteins that constitute candidate odorant receptor molecules (Chen and Lancet, 1984; Chen *et al.*, 1986*a,c*). Some of the olfactory cilia components, such as glycoprotein gp95 (Chen *et al.*, 1986*a*), adenylate cyclase (Pace *et al.*, 1985), as well as its odorant-sensitive form (Chen *et al.*, 1986*a*), and the stimulatory GTP-binding protein, G_s (Pace *et al.*, 1985; Anholt *et al.*, 1986*b*; Lancet and Pace, 1986), have been shown to be enriched specifically in isolated cilia compared with other membrane preparations. Two control membrane preparations important in this respect are respiratory cilia (e.g., from palate in frog) and membrane derived from whole olfactory epithelium after cilia removal.

Sensory cilia are isolated from the underlying tissue by procedures originally developed for nonsensory cilia detachment (Linck, 1973; Adoutte *et al.*, 1980). These methods, commonly known as calcium shock techniques, include incubation of the tissue in the presence of EDTA to decrease Ca^{2+} concentration, followed by abrupt transfer to 10 mM Ca^{2+}. The treatment is sometimes accompanied by relatively high concentrations

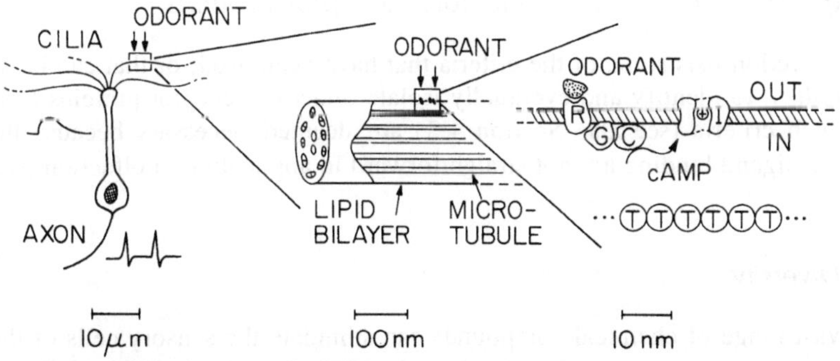

Figure 2. Schematic representation of the olfactory sensory neuron. Odorants interact with olfactory cilia to produce a graded receptor potential in the dendrite of the bipolar neuron. This results in the elicitation of action potentials in the axon. Enlarged, an olfactory cilium is shown to be composed of a membrane (lipid bilayer) sheath, about 200 nm (0.2 μm) in diameter. This structure is supported by a defined microtubular arrangement, typical of all cilia (T-tubulin monomer). Odorants interact with the membrane, which, at an even larger magnification, is depicted schematically, containing the presumed molecular mechanism of olfaction (see Sections 3–5). This includes the receptor protein (R) that binds odorant molecules stereospecifically, the GTP-binding protein (G), and the enzyme adenylate cyclase (C). The latter produces cAMP, which serves as an intracellular second messenger and eventually leads to modulation of ion-channel conductance underlying membrane depolarization.

(up to 10%) of an organic compound (e.g., ethanol or dibucain). Calcium shock with gentle stirring results in the rather quantitative removal of cilia from the underlying tissue. Calcium may disrupt the molecular interactions in the microtubular axonemal base, while the organic compound presumably helps weaken the adhesion forces in the lipid bilayer. The calcium shock method has been successfully applied to all the above-mentioned preparations, as well as in preparing membranes used in single ion-channel recordings in reconstituted membrane patches (Kleene *et al.*, 1985). A different method involving sonication was used by Shirley *et al.* (1986) to obtain a similar olfactory membrane preparation, most probably enriched in cilia.

3. CANDIDATE ODORANT RECEPTOR PROTEINS IN THE SENSORY MEMBRANE

Olfactory receptors* are, by definition, those molecular structures that undertake the initial interaction with the stimulatory ligand. They are presumed to be proteins capable of binding the odorant stereospecifically and of undergoing changes, giving rise to transduction events (Price, 1981; Dodd and Persaud, 1981; Getchell *et al.*, 1984; Lancet, 1986; Getchell, 1986; Anholt, 1987; Lancet and Pace, 1987). This rather universal definition similarly applies to other receptors, e.g., for hormones, neurotransmitters, chemoattractants, growth factors, and antigens (cf. Venter and Harrison, 1984; Cerione *et al.*, 1983; Schlessinger *et al.*, 1983; Strosberg, 1984). Current theory assumes that olfactory receptors are largely analogous to these other receptor types, i.e., membrane proteins that detect extracellular signals and transfer them intracellularly by one of the known transduction mechanisms (Dodd and Persaud, 1981; Lancet, 1986). Finding out the detailed points of similarity and difference is a major present and future goal of olfactory research.

3.1. Expected Properties of Olfactory Receptor Proteins

This section lists some of the criteria that have been used, or that can be applied in future studies, to identify and eventually isolate olfactory receptor proteins (see Lancet, 1986). Such criteria (see also Section 4.3) are deemed necessary because the classic methods of ligand binding are not straightforward in application in olfaction (see Section 3.1.4).

3.1.1. Diversity

A vast range of chemical compounds can stimulate the sensory cells of the nose as well as elicit odor responses (Gesteland *et al.*, 1965; Gesteland, 1976; Holley and MacLeod, 1977; Sicard and Holley, 1984). Practically all volatile organic compounds are odorous, and many nonvolatile ones can give rise to olfactory responses when topically applied (Getchell, 1974). It is reasonable to assume that many receptor types are necessary to accommodate such multitude of chemical substances (cf. Amoore, 1971). Indeed,

*The word *receptor* is used here solely to indicate an odorant-binding molecular structure. The cells that transduce odor signals are called *sensory*, to avoid confusion.

it has been hypothesized that olfactory recognition is performed by a multireceptor family (Goldberg *et al.*, 1979; Boyse *et al.*, 1982; Lancet, 1984, 1986). One or more receptor types may be represented on each sensory neuron (Gesteland, 1976; Lancet, 1986).

The proposed olfactory receptor repertoire may share some properties with products of other multigene families, such as immunoglobulins, T-cell receptors, and histocompatibility antigens (Hood, 1982; Cushley and Williamson, 1982; Kimball and Coligan, 1983; Ploegh *et al.*, 1981; Hendrick *et al.*, 1984; Robertson, 1985). A most economical way for nature to construct a multiprotein family is to duplicate only those parts of the protein that need to vary among members. Thus, immunoglobulins and T-cell receptors have only a few copies of the DNA sequences coding for their common, or constant, regions but a much larger repertoire of variable-region genes (Hood, 1982; Tonegawa, 1983; Robertson, 1985). The entire molecule is constructed by joining a variable and a constant region in a seemingly random fashion. It has been suggested that a similar scenario may hold for the putative family of olfactory receptor proteins (Lancet, 1984, 1986). Diversity, as well as a variable/constant structure, could serve to identify olfactory receptor proteins.

3.1.2. Membrane Specificity and Disposition

Olfactory receptor proteins may be identified and isolated on the basis of several general biochemical criteria derived from the study of other cell-surface receptors (cf. Lancet, 1986). Thus, the chemosensory receptor proteins are expected to be specifically localized in the membrane of olfactory neurons, as compared with other cells in olfactory epithelium or with cells in other tissues. It is not unlikely that they should be concentrated in the sensory extensions—olfactory cilia—relative to other regions of the neuron. To transfer odor signals across the membrane effectively, olfactory receptors should traverse the lipid bilayer, i.e., behave as integral membrane proteins. Olfactory receptors should most probably be glycosylated, i.e., marked with covalently attached oligosaccharide moieties, as with practically all other transmembrane receptors. Finally, it has been suggested that, like nicotinic acetylcholine (ACh) receptors in postsynaptic membranes, or rhodopsin in retinal rod outer segment disc membranes, olfactory receptor molecules should be present in relatively large numbers in the specialized ciliary membrane and could be observed as high densities of freeze-fracture intramembranous particles (Menco, 1980, 1983, 1984).

3.1.3. Interaction with Transduction Components

Olfactory receptors are assumed to be the first elements in a chain of molecular components, e.g., the adenylate cyclase cascade, responsible for chemosensory signal transduction (see Section 4). The receptors could be identified and isolated by virtue of their interaction with these more distal elements, such as the stimulatory binding protein (G_s). They may be assayed by reconstitution experiments in which their ability to activate G_s and adenylate cyclase is examined. Reagents such as antibodies or lectins, that bind specifically to candidate receptor molecules, may be found to modulate the enzymatic activities measured *in vitro* or even the *in vivo* recorded electrophysiological odorant responses. (This approach is detailed in Section 4.3.)

3.1.4. Odorant Binding

While ligand binding is a most natural criterion for identifying olfactory receptors, its application raises several problems not encountered in most other receptor systems (Price, 1981; Lancet, 1986). Odorants are lipophylic and partition into the lipid bilayer, hence give rise to high levels of nonspecific binding. Olfactory receptor proteins may be heterogeneous, hence may display complex, possibly nonsaturable, binding curves. In addition, olfaction is thought to be mediated by loose and fast binding between odorants and receptors, which is detrimental to any ligand-binding measurement (Nachbar and Morton, 1981; Mason and Morton, 1984; Lancet, 1986). Thus, methods that rely on ligand–protein interactions, such as receptor-binding assays, affinity chromatography, and affinity labeling, may not be readily applicable to the identification and isolation of olfactory receptor proteins.

3.2. Olfactory Receptor Candidates

Proteins that fulfill some or all of the criteria delineated in Section 3.1 constitute candidate olfactory receptor proteins, worthy of investigation. Ligand-binding techniques have been used in an attempt to identify the receptors (for review, see Price, 1981; Lancet, 1986) (see also Chapter 1, this volume). This approach led to the identification of several odorant-binding proteins (Goldberg *et al.*, 1979; Fesenko *et al.*, 1979, 1983; Pelosi *et al.*, 1982; Bignetti *et al.*, 1985; Pevsner *et al.*, 1985, 1986). One of these proteins, discovered by virtue of its ability to bind pyrazine derivatives, is a soluble extracellular protein of nasal secretions and glands in olfactory and respiratory epithelia that may serve perireceptor function, e.g., as an odorant carrier (Pelosi *et al.*, 1982; Pevsner *et al.*, 1985, 1986; Bignetti *et al.*, 1985).

A receptor candidate recently identified using some of the above-listed criteria, other than odorant binding, is glycoprotein gp95, a major membrane component of the isolated olfactory cilia preparation in the frog (Chen and Lancet, 1984; Chen *et al.*, 1986*a*). It has a molecular mass of 95kDal and a rather low isoelectric point. Roughly 30% of its apparent molecular mass is due to complex-type oligosaccharide. Histofluorescence localization using lectins and monoclonal antibodies show that gp95 is localized at the ciliary surface layer of frog olfactory epithelium (Chen *et al.*, 1986*b*). Polypeptide gp95 fulfills the five basic biochemical criteria: (1) it is glycosylated; (2) it constitutes a major component of the ciliary membrane; (3) it behaves as a transmembrane protein; (4) it is specific to olfactory cilia when compared with either nonsensory (respiratory) cilia or with membrane derived from deciliated olfactory epithelium (Chen and Lancet, 1984; Chen *et al.*, 1986a), and (5) studies with a lectin, wheat germ agglutinin (WGA), and with a monoclonal antibody (18.1), both of which recognize gp95 specifically (with respect to other ciliary glycoprotein), provide supporting evidence for its role in olfactory reception (Lancet *et al.*, 1986*a,b*; Chen *et al.*, 1985) (see Section 4.3.1). Putative gp95 homologues identified in olfactory cilia preparations from toad, rat, and cow (Chen *et al.*, 1986*a*) require further investigation.

Polypeptide gp95 has a broad electrophoretic mobility on sodium dodecyl sulfate (SDS) polyacrylamide gels. In order to find out whether this is due to its attached carbohydrate, we removed the oligosaccharide by endoglycosidase-F digestion. While this treatment reduced the average molecular mass of gp95 from 95,000 to 60,000 daltons, it did not appreciably alter the electrophoretic heterogeneity (Chen *et al.*, 1986*a*). This result is consistent with, but does not prove, the notion that gp95 represents a group of different polypeptides, another criterion for identifying olfactory receptor protein(s) (see Section 3.1.1).

Additional support for gp95 heterogeneity should be obtained through a detailed protein-chemical investigation. Toward this end, procedures have been developed to obtain gp95 in quantity and purity adequate for gas-phase microsequence analysis (Chen *et al.*, 1986*a*; Lancet *et al.*, 1987). In the near future, gene cloning, aided by protein-sequence information, should permit comprehensive study of gp95 structure, function, and possible diversity.

4. CYCLIC NUCLEOTIDE ENZYMATIC CASCADE IN OLFACTORY TRANSDUCTION

4.1. Role of cAMP in Olfaction

4.1.1. Historical Perspective

During the past two decades, several investigators pointed out the possibility that a cyclic nucleotide cascade mechanism is involved in olfactory transduction (Kurihara and Koyama, 1972; Minor and Sakina, 1973; Menevse *et al.*, 1977; Villet, 1978; Dodd and Persaud, 1981). The modulation of cyclic nucleotide second-messenger concentrations is a widespread receptor-coupled control mechanism in biology. It mediates signal transfer in many hormone and neurotransmitter receptors (Robinson *et al.*, 1971; Daly, 1977; Smigel *et al.*, 1984; Schramm and Selinger, 1984; Cooper and Seamon, 1985) as well as in photoreception (cf. Stryer, 1986; Kaupp and Koch, 1986). This mechanism results in large amplification factors, as one stimulated receptor protein can lead to the activation of up to several hundred enzyme molecules, each of which generates many second-messenger molecules (Chabre, 1985; Stryer, 1986; Kaupp and Koch, 1986). Such amplification is in addition to that afforded by the fact that a single open channel can pass thousands of ions, a common feature of all membrane conductance mechanisms, whether involving second messenger or not. Thus, a cyclic nucleotide mechanism can naturally account for the relatively high sensitivity of olfactory detection (cf. Lancet, 1986).

Several lines of evidence have been produced to support the role of cAMP as an olfactory second messenger, in accordance with the criteria originally delineated by Sutherland, the discoverer of the role of cAMP in biological transduction (Robinson *et al.*, 1971) (Table 1). Kurihara and Koyama (1972) were the first to measure the activity of adenylate cyclase in olfactory epithelium; these workers found that it was comparable to

Table 1. Sutherland's Criteria for cAMP Involvement in a Receptor Mechanism[a]

1.	Presence of the enzyme adenylate cyclase in the tissue or membranes
2.	Modulation or imitation of the ligand-activated response by cyclic nucleotide analogues and phosphodiesterase inhibitors
3.	Activation of adenylate cyclase by the physiologically active ligands

[a]Adapted from Robinson *et al.* (1971).

that of brain and other tissues rich in the transductory enzyme, thus fulfilling the first criterion.

4.1.2. Electrophysiological Studies

Minor and Sakina (1973) carried out an elegant study involving an apparatus, the electro-olfactogram, for recording olfactory summated receptor potentials (Ottoson, 1971) under fluid. Using this technique, they were able to monitor responses to fluid-applied pulses of both odorants and modulators. They reported that 3′,5′-cAMP and, more efficiently, its membrane-penetrable analogue dibutyryl cAMP, gave long-lasting (several seconds) odorant-like responses. 2′,3′-cAMP was ineffective, and 5′-AMP was much less potent than 3′,5′-cAMP. Odorant stimulation during the time of 3′,5′-cAMP application gave rise to responses of lower amplitude, the sum of cAMP and odorant responses being roughly constant. Minor and Sakina note that this indicates a common mechanism for both stimulants. A second part of the study shows that prior incubation of olfactory epithelium with the phosphodiesterase inhibitor theophylline, which blocks the cellular enzymes that degrade cyclic nucleotides, leads to a prolongation of the odorant responses, concomitant with some lowering of its amplitude. Imidazole, a phosphodiesterase stimulator, has an opposite effect.

In a later study, Dodd and colleagues (Menevse *et al.*, 1977) confirmed and complemented Minor and Sakina's results. These workers used the standard electro-olfactogram recording setup, whereby recording olfactory responses are carried out from the naturally air-exposed epithelial surface and can therefore be carried out only after removal of excess aqueous modulator solution. They report a decrease of the electro-olfactogram odorant response amplitude following preincubation with both dibutyryl 3′5′-cAMP and several phosphodiesterase inhibitors. These results are largely consistent with Minor and Sakina's findings. Taken together, the results from both laboratories are in fulfillment of the second of Sutherland's criteria (Table 1), and should be taken as important indications for cAMP involvement in olfactory reception.

4.1.3. Odorant-Sensitive Adenylate Cyclase in Isolated Cilia

The third and most rigorous criterion of Sutherland proved more difficult to fulfill in olfaction. In earlier work, Dodd and Margolis and their groups (Menevse *et al.*, 1977; Margolis, 1975) were unable to demonstrate conclusively odorant activation of adenylate cyclase in crude olfactory epithelial preparations. Only recently has it become possible to demonstrate such odorant stimulation of the transductory enzyme (Pace *et al.*, 1985; Sklar *et al.*, 1986; Shirley *et al.*, 1986). Several parameters appear to have contributed to this more recent success:

1. *The use of preparations of isolated olfactory cilia:* It appears that while the cilia are enriched with odorant-sensitive adenylate cyclase, the rest of the epithelium contains appreciable amounts of the enzyme, which is odorant insensitive (Chen *et al.*, 1986*c*).
2. *The inclusion of GTP in the reaction mixture:* This fulfills the substrate requirements of the signal-coupling GTP-binding protein or G protein (see Section 4.2).

CITRAL

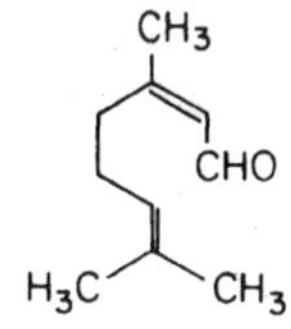

L-CARVONE

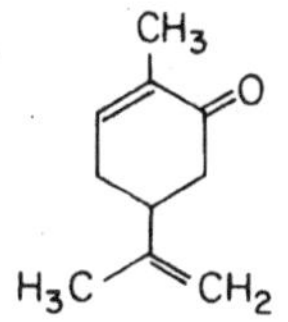

1,8-CINEOLE

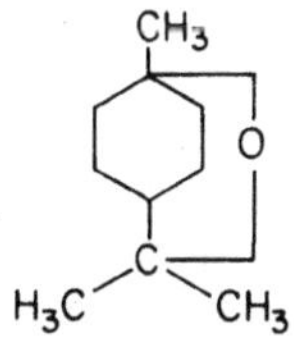

n-AMYL ACETATE

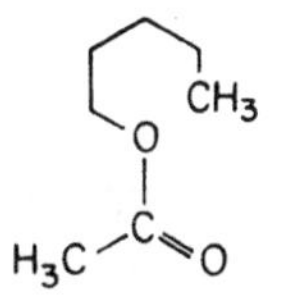

Figure 3. Odorant structure. Structure of the four odorants used in our studies of the biochemical mechanisms of olfaction. These are mostly used in an equimolar mixture, but have also been examined individually. Citral appears to be the most potent of this group, and one of the most potent odorants in general. All four odorants (except citral) are relatively water soluble and can be made into aqueous stocks of up to 10^{-2} M. All four odorants are well-defined olfactory stimulants in humans (smelling as lemon, caraway seeds, camphorlike, and banana). They have been shown to be adequate stimulants in animals and have been used in psychophysical behavioral and electrophysiological studies.

3. *The correct choice of odorants:* Not all electrophysiologically tested odorants are equally effective in activating adenylate cyclase in cell-free experiments.

We have introduced the use of an odorant mixture (Fig. 3) to increase the probability of activation. Such odorants as citral and cineole (and their homologues) have been found to be more effective than others (Pace *et al.*, 1985; Sklar *et al.*, 1986).

Data on odorant activation of adenylate cyclase in olfactory cilia preparations are comprehensive. Data from our laboratory (Pace *et al.*, 1985; Pace and Lancet, 1986) show that isolated olfactory cilia preparations from frog, toad, and rat are extremely rich in adenylate cyclase; their specific activity (per protein) is 5–15 times higher than that of brain membranes and up to 100 times higher than that of respiratory cilia. The olfactory enzyme, like that of many other membrane preparations, is stimulated by guanine nucleotides and fluoride ions (via a G protein) and by forskolin (a direct activator of the cyclase catalytic unit). Most important, adenylate cyclase in all olfactory cilia preparations can be activated by odorants, when presented either individually or in a mixture. Maximal activation (about ×2.5) is obtained at an odorant concentration of 1 mM, but detectable activation occurs at a concentration as low as a few micromolar (μM) for some odorants. Agonists of other receptors, such as isoproterenol (a β-adrenergic agonist) or prostaglandin E_1 (PGE_1) are ineffective (Pace *et al.*, 1985; Shirley *et al.*, 1986).

The relatively high odorant concentrations used in these experiments are not in disagreement with calculated odorant concentrations in the aqueous mucous phase for suprathreshold activation in most reported electrophysiological experiments, where typical air-phase odorant concentrations are 10^{-7}–10^{-4} M (Ottoson, 1971; Getchell, 1974; Senf *et al.*, 1980; Gesteland, 1976; Holley and MacLeod, 1977; Getchell *et al.*, 1984). Considering that the molar partition coefficient (water–air) for most odorants is 10–100, a rather good agreement obtains (cf. Lancet, 1986). The Hill coefficient of the dose–response curve is low (~0.4), consistent with the possibility of heterogeneous receptor population (Pace *et al.*, 1985).

Sklar *et al.* (1986) carried out a study of odorant stimulation in a preparation of olfactory cilia from the frog *Rana catesbeiana*. Their findings are in good agreement with those of Pace *et al.* (1985) concerning the high adenylate cyclase activity and its sensitivity to odorants and guanine nucleotides. They also report that calcium inhibits the olfactory adenylate cyclase with a half-maximal effect at 10 mM. These investigators screened a large number of odorous compounds and report that odorants of certain quality classes (fruity, minty, floral, and herbaceous) are stimulatory, while putrid odorants (mainly aliphatic carboxylic acids and organic amine bases) are ineffective. In homologous series, compounds with longer side chains give better stimulation. The observation that some physiologically active odorants do not stimulate adenylate cyclase was taken as an indication for the existence of an additional olfactory transduction mechanism.

Shirley *et al.* (1985, 1986) reported that adenylate cyclase in a rat olfactory epithelial membrane preparation and in its cilia-enriched fraction show pronounced sensitivity to odorants. Olfactory cAMP phosphodiesterase was measured as well and found to be abundant but odorant insensitive. Guanylate cyclase and cGMP phosphodiesterase, which were also present in the rat olfactory preparations, were odorant insensitive as well (Shirley *et al.*, 1985). These findings underscore the suggested role of adenylate cyclase, rather than other cyclic nucleotide-processing enzymes, in olfaction.

A candidate transduction pathway could be odorant stimulation of phosphatidyl inositol turnover, a mechanism that yields two second messengers (inositol trisphosphate and diacyl glycerol), possibly through the mediation of a yet-undefined GTP-binding protein (Berridge and Irvine, 1984). Indeed, Huque and Bruch (1986) reported preliminary evidence for activation of phosphatidyl inositol hydrolysis by an odorant (L-alanine) or by GTP in fish olfactory cilia. Diacyl glycerol is known to activate protein kinase C (Kishimoto *et al.*, 1980), an enzyme that can also be stimulated by phorbol esters. Anholt *et al.* (1987) report radiolabeled phorbol ester binding in olfactory cilia. However, these workers find similar binding activity in respiratory cilia as well and suggest that it may be unrelated to sensory function. This is in agreement with the finding that phorbol ester, in the presence of phosphatidyl serine and 10^{-5} M Ca^{2+}, shows no detectable activation of protein phosphorylation in frog olfactory cilia (Heldman and Lancet, 1986), suggesting that protein kinase C may not be active in this preparation. The possible function of the other branch of the inositol lipid mechanism, the release of calcium from internal stores by inositol trisphosphate, has yet to be examined.

Another possible mechanism for olfactory transmembrane signaling is direct odorant activation of a transduction component. Some odorants could directly interact with membrane ion channels (see Anholt, 1987), similar to the mechanism of the nicotinic ACh receptor (Changeux *et al.*, 1984). Evidence for this direct gating mechanism has been provided through single ion-channel recordings from artificial lipid bilayers reconstituted with components from rat olfactory membranes (Vodyanoy and Murphy, 1983) or from frog olfactory cilia (Labarca *et al.*, 1986). Other odorants could interact with transduction enzymes in a receptor-independent manner. Thus, acidic or basic odorants could exert their action through pH-induced changes. Sulfhydryl-containing odorant could directly activate a transduction enzyme, such as the G protein or adenylate cyclase, through their functionally important SH groups (cf. Schramm and Selinger, 1984). Some odorants may have fortuitously pharmacological effects; e.g., pyrazine compounds could be sterically related to some phosphodiesterase inhibitors. In some cases, odorants may act both via a receptor-mediated stereospecific mechanism and through general functional group-related activation. In these cases, odorants will self-synergize, leading to high detection sensitivity. Such synergism could account for the potency of certain odorant classes, e.g., thiols or pyrazine derivatives.

4.1.4. Future Studies

While the evidence for the involvement of adenylate cyclase in olfaction is rather strong, future studies should extend the presently available indications. Any of several approaches could be used.

4.1.4.a. Biochemical Modulation of Electrophysiological Responses. These experiments would constitute extensions of the studies used to fulfill Sutherland's second criterion (see Section 4.1.2). Additional electrophysiological techniques may be used, including single-cell (extracellular and intracellular) recordings, summated nerve responses, transepithelial short-circuit current measurements, and single ion-channel recordings in natural and reconstituted membrane patches. All these techniques have either been extensively explored or begun to be applied to olfactory sensory cells (Gesteland *et al.*, 1965; Sicard and Holley, 1984; Getchell, 1977; Trotier and MacLeod, 1983; Masukawa *et al.*, 1985; Maue and Dionne, 1984; Firestein and Werblin, 1985; Maue and Dionne, 1986; Vodyanoy and Murphy, 1983; Kleene *et al.*, 1985; recently reviewed by Getchell, 1986) (see also Chapter 8).

In addition to the reagents used in the earlier studies (cAMP and analogues, phosphodiesterase inhibitors), it may be possible to apply reagents such as forskolin (a direct adenylate cyclase activator) and guanine nucleotides. Studies employing such reagents have recently been carried out. Persuad *et al.* (unpublished) reported the modulation of odorant responses and the induction of odorantlike activation in recordings of transepithelial short-circuit currents from isolated frog olfactory mucosa in an Ussing chamber (reviewed in Chapter 8). Trotier and MacLeod (1986) showed that these modulators can produce membrane currents recorded in the whole-cell patch-clamp configuration from salamander isolated olfactory neurons. Additional reagents that have effects on olfactory responses and could be explored in the future are bacterial toxins, which activate via a covalent modification of G proteins (Gilman, 1984; Spiegel *et al.*, 1985; Schramm and Selinger, 1984; Levitzki, 1986), as well as the catalytic and regulatory subunits of cAMP-dependent protein kinase and protein kinase inhibitor (cf. Browning *et al.*, 1985; Nestler *et al.*, 1984; Levitan, 1985). Naturally, consideration should be given to the question of membrane penetrability. For example, the enzyme protein kinase can only be injected intracellularly or applied to the cytoplasmic side of excised membrane patches because of their inability to cross the lipid bilayer (Levitan, 1985).

4.1.4.b. Biochemical Modulation of Behavioral Responses. Some of the reagents listed in Section 4.1.4.a can be used in an attempt to modulate olfactory perception and behavioral responses. For example, forskolin or cholera toxin, when topically applied, should elicit an intense odorantlike response. In the case of human subjects, these experiments may be rather complex, hard to control, and sometimes dangerous. They could possibly be carried out in animals, such as rat, in which adequate behavioral paradigms have been developed (e.g., see Slotnick and Schoonover, 1984).

4.1.4.c. Correlation to Odorant Sensitivity in Anosmic Strains. Mouse strains that show decreased sensitivity to particular odorants (hyposmic or partly anosmic) have been reported (Wysocki *et al.*, 1977; Price, 1977). Similarly, *Drosophila* mutants defective in their ability to respond to certain odorants have been described (Rodrigues, 1980; Rodrigues and Siddiqi, 1978; Venard and Pichon, 1984). Careful comparison of these deficient animals with wild-type animals, in terms of olfactory threshold to a given odorant and, in parallel, in terms of the ability of the same odorant to activate adenylate cyclase (or other molecular mechanisms) in their cilia preparations, could provide important evidence in support of a proposed transduction mechanism.

4.1.4.d. Odorant Pharmacology. Similar correlation may be obtained for animals with normal sensory acuity, using a pharmacological approach. The question to be asked is whether behaviorally or electrophysiologically potent odorants in a given species are also more efficient activators of transduction enzyme in olfactory cilia derived from the same species. However, in the case of behavioral testing, problems may be encountered because of the complex amplification and processing events that link biochemical transduction to the measured response.

4.2. Olfactory GTP-Binding Protein

While the involvement of adenylate cyclase in olfactory transduction has long been suspected, practically no mention was made of the possibility that a GTP-binding coupler protein is also operative, as in many other transmembrane receptor systems (cf. Gilman, 1984; Spiegel *et al.*, 1985; Schramm and Selinger, 1984; Stryer, 1986; Kaupp and Koch, 1986). The first clue came from [^{32}P]-ADP ribosylation experiments in which the frog isolated olfactory cilia preparation was exposed to cholera toxin and [^{32}P]-NAD (Lancet and Pace, 1984). The covalent modification catalyzed by the bacterial toxin is known specifically to affect the heavy (α)-subunit of G_s, the stimulatory G protein (see Table 2). Our further experiments showed that olfactory cilia contain a G_s-like protein by this criterion (Pace *et al.*, 1985; Pace and Lancet, 1986), corroborated and complemented by Anholt *et al.* (1986*b*). Additional evidence for the involvement of G_s in olfaction came from the finding that odorant activation of ciliary adenylate cyclase in the test tube requires the addition of GTP (Pace *et al.*, 1985; Sklar *et al.*, 1986; Pace and Lancet, 1986;

Shirley *et al.*, 1986). The activation by nonhydrolyzable GTP analogues, such as GTPγS and GppNHp, as well as by fluoride ions and cholera toxin, is also consistent with the modulation of olfactory adenylate cyclase by interaction with a stimulatory GTP-binding protein (Pace *et al.*, 1985; Pace and Lancet, 1986; Sklar *et al.*, 1986; Shirley *et al.*, 1986).

4.2.1. Biochemical Properties

The olfactory G_s α-subunit, identified through cholera toxin-catalyzed ADP ribosylation (Pace *et al.*, 1985; Anholt *et al.*, 1986*b*; Pace and Lancet, 1986) and by reactivity with specific antisera (Mumby *et al.*, 1986; Lancet *et al.*, 1987), is a polypeptide roughly comigrating with G_s α-subunit in other tissues. It is estimated to constitute about 1% of the protein amount in the frog olfactory cilia preparation, a relatively high concentration (Pace and Lancet, 1986). Olfactory G_s is much more concentrated on a per-protein basis in the olfactory cilia preparation compared with membranes from deciliated epithelium and is found in rather small amounts in the respiratory cilia preparation (Pace *et al.*, 1985; Anholt *et al.*, 1986*b*; Pace and Lancet, 1986). Olfactory cilia also contain an inhibitory G protein (G_i) and a possible homologue of the brain G protein (G_0), identified by undergoing ADP-ribosylation by pertussis toxin and by reactivity with specific antisera (Pace *et al.*, 1985; Anholt *et al.*, 1986*b*; Lancet *et al.*, 1987). However, these proteins do not appear to be specifically enriched in olfactory cilia compared with control membrane preparations (Pace *et al.*, 1985; Anholt *et al.*, 1986*b*); their functional significance remains undetermined, as no adenylate cyclase inhibition by odorants or guanine nucleotides has been observed (Pace and Lancet, 1986) (see Section 4.2.2).

An interesting question is whether olfactory G_s is identical to G_s in other tissues. The limited data available suggest a high degree of similarity. However, it is possible that the α-subunit of the olfactory protein has a slightly different apparent molecular weight, and the olfactory G_s may have higher functional affinity toward guanine nucleotides than its counterparts in brain or liver (Pace *et al.*, 1985; Pace and Lancet, 1986). This question could be addressed by molecular cloning of the olfactory signal transducing protein (Jones and Reed, 1987), in parallel with similar work currently carried out for G_s in other tissues (Harris *et al.*, 1985).

Human patients with the genetic disease pseudohypoparathyroidism (PHP) type Ia are deficient in G_s and therefore respond less than normally to parathyroid and other hormones. An intriguing question is whether PHP patients may have impaired olfactory capabilities. This will depend on whether olfactory G_s is identical to hormonal G_s or is under a common genetic control. Since PHP patients are expected to have normal olfactory receptors but abnormal amounts of a transductory protein, it is anticipated that they would be capable of detecting most or all odorants, except with abnormally high thresholds. Indeed, Henkin (1968) reported that for three odorants, olfactory thresholds were up to 10^6 higher in PHP patients as compared with normal subjects. A more recent and extensive study employing 10 odorants was performed by Weinstock *et al.* (1986), who reported impairment of olfactory identification capabilities in G_s-deficient PHP patients but not in patients with another PHP type having normal G_s levels. Thus, it remains to be conclusively established whether the olfactory deficits of PHP patients are due to change in sensitivity, discrimination ability, or both.

4.2.2. Significance

The importance of the discovery of olfactory GTP-binding protein lies in the following factors:

1. It underscores the homology between olfaction and other receptor mechanisms (Fig. 4). While many of the G-protein-containing receptor mechanisms use cAMP as a second messenger (Gilman, 1984; Spiegel *et al.*, 1985; Schramm and Selinger, 1984), other G-protein-related mechanisms involve the activation of different enzymes. Notably, photoreception in both vertebrates and invertebrates includes light activation of G proteins, in turn activating enzymes other than adenylate cyclase: cGMP phosphodiesterase in vertebrates (Stryer, 1986; Kaupp and Koch, 1986; Chabre, 1985) and, most notably, phosphatidyl inositol phospholipase-C in invertebrates (Berridge and Irvine, 1984; Vandenberg and Montal, 1984). Thus, the common link between photoreception and chemoreception resides in the homology between the coupling G polypeptide (Fig. 4).
2. Olfactory G protein is the first molecular element in the odor-detection mechanism that is known as a distinct polypeptide. In general, appreciably more is known, at the molecular level, about G proteins (Table 2) than about the enzymes they activate. The discovery of a G protein in olfaction provides a molecular component of the chemosensory mechanism that can be readily studied and analyzed, aided by information available on its homologues. This includes prospective protein sequence analysis and the use of DNA probes from other G-protein genes for the isolation and molecular cloning of olfactory G_s gene(s) (Jones and Reed, 1987).
3. Judging by the mode of action of other receptors (e.g., β-adrenergic receptor or visual rhodopsin), the G protein interacts directly with the receptor protein. A G protein therefore affords a direct probe for the study of the yet undiscovered olfactory receptor protein(s) (cf. Section 4.3).

The olfactory G protein may be important for signal amplifications, as it is in other receptor systems. This is because one molecule of activated receptor can catalyze the exchange of GDP to GTP in many G-protein molecules (Stryer, 1986). A full understanding of the molecular properties of olfactory G protein may help in studying, or even modulating, a molecular mechanism that underlies olfactory sensitivity. The G-protein pathway could also provide dual control of the enzyme adenylate cyclase (cf. Gilman, 1984; Spiegel *et al.*, 1985), through the existence in olfactory cilia of both the stimulatory and inhibitory G proteins. The latter, when activated by a receptor, can inhibit adenylate cyclase. An intriguing possibility is that inhibitory ligands endogenously produced by epithelial secretory cells, (supporting cells and cells of Bowman's glands) can vary the gain of the olfactory response (Fig. 5).

Since G proteins and adenylate cyclase contain functionally important SH groups (Schramm and Selinger, 1984), it is possible that thiol-specific reagents, which have been reported to inhibit olfactory function (Getchell and Gesteland, 1972; Shirley *et al.*, 1983*a*), act through reaction with transduction proteins, rather than with the receptor. Indeed, we have found that *N*-ethylmaleimide (NEM) blocks the forskolin and GTPγS

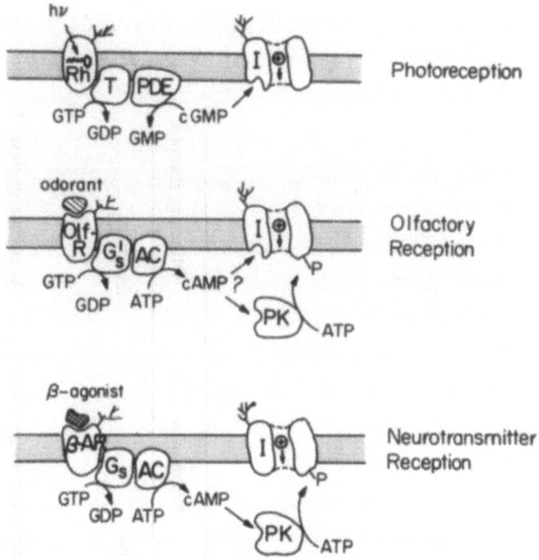

Figure 4. Molecular transduction mechanisms. Comparison of the proposed molecular components involved in visual photoreception, olfactory reception, and reception of a neurotransmitter. The olfactory mechanism may be intermediary between those of vision and neurotransmission. This is because the olfactory neuron is both a sensory cell and a bona fide nerve cell that responds to chemical stimulation at its dendrite by membrane depolarization and action potentials. The known facts are that all three mechanisms use a coupling GTP-binding protein: transducin (T) in vision, a stimulatory G protein (G_s) for β-adrenergic receptors, and a possible G_s variant (G_s') in olfaction. All three G proteins convert GTP into GDP and phosphate. All three transduction mechanisms also appear to have a cyclic nucleotide second messenger: cGMP for photoreceptors, cAMP for the other two mechanisms. An important difference is that in vision the cyclic nucleotide is broken down by transduction enzyme phosphodiesterase (PDE), while in olfaction and β-adrenergic reception adenylate cyclase (AC) synthesizes cAMP. The visual receptor protein is rhodopsin, which contains retinal that absorbs photons of visible light (hv). The β-adrenergic receptor (β-AR) has been extensively studied, purified and its amino acid has been determined through gene cloning. It appears to share overall structural features with rhodopsin. Both receptors are transmembrane glycoproteins, and are capable of G-protein activation. Thus, it is not unreasonable to predict that olfactory receptor proteins will resemble the other two and that all three receptor protein types will be found to be members of an even larger G-protein-activating transmembrane receptor family. The distal action of cyclic nucleotide in photoreceptors has recently been elucidated: cGMP directly interacts with, and leads to, the opening of ion channels (I) in the rod outer segment membrane. The light-induced decrease of cGMP therefore leads to ion-channel closure. In several neuronal systems, it has been shown that ion channel conductance can be modulated by cAMP-induced protein phosphorylation mediated by protein kinases (PK). Olfactory sensory neurons appear to use the first mechanism.

stimulation of adenylate cyclase more efficiently than it inhibits odorant activation, suggesting that a transduction protein may be the primary modification targets (Lancet *et al.*, 1985).

4.3. Transduction Components as Odorant Receptor Probes

The elucidation of a transduction mechanism in olfactory neurons is important not only as a goal in itself, but also because it provides a means of identifying and isolating the receptor proteins. In other receptor systems, it has been possible to identify the receptor by virtue of its ability to specifically bind a ligand. However, such an approach is not easily applicable in olfaction (see Section 3.1.4). Olfactory receptor proteins may present an unusual case in which reconstitution with transduction proteins may serve not only to confirm receptor function (Smigel *et al.*, 1984; Cerione *et al.*, 1983), but also the method of choice for receptor identification. It should be pointed out that olfactory neurons are not the only cells whose functional receptors could not be isolated by conven-

Table 2. GTP-Binding Proteins[a]

Receptor	R_s	R_i	Rhodopsin	Olfactory receptor
Agonist	β-Adrenergic Glucagon ACTH PGE_1	Opiates α_2-Adrenergic D_2-Dopaminergic Musc-cholinergic	Light	Odorants
Regulated enzyme	Adenylate cyclase ↑	Adenylate cyclase ↓	cGMP phosphodiesterase ↑	Adenylate cyclase ↑
G protein	G_s	G_i	Transducin	Olfactory G protein
Molecular weight of G_α	43,000–45,000	41,000	39,000	42,000
Modifying toxin	*V. cholera*	*B. pertussis*	*B. pertussis* (*V. cholera*)	*V. cholera*
Amplification	?	?	4×10^5	?

[a]Summary of properties of various G proteins.

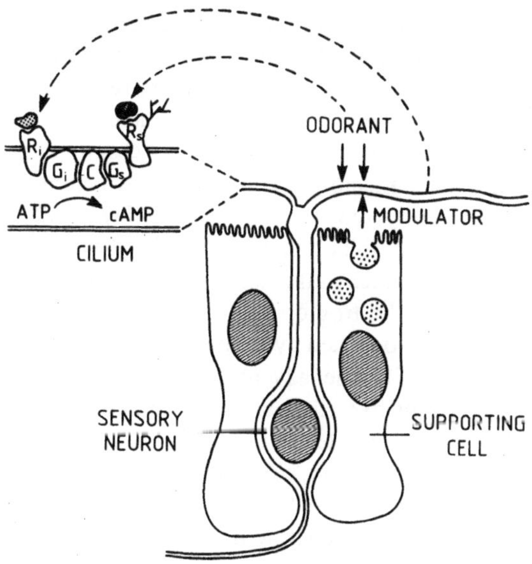

Figure 5. Dual modulation of olfactory adenylate cyclase. Hypothetical mechanism of dual modulation of adenylate cyclase in olfactory neurons. Most odorants appear to depolarize (activate) the sensory neurons and to stimulate adenylate cyclase (C), presumably through the action of a stimulatory receptor (R_s) that activate the stimulatory G protein (G_s). It is proposed that modulators, which may be endogenously secreted by supporting cells and by subepithelial glands (not shown), can gauge the olfactory response by interacting with receptors (R_i) that activate the inhibitory G protein (G_i).

tional ligand-binding methods: the antigen-binding receptor of T lymphocytes is a well-known example in which indirect methods were necessary for receptor identification (Hendrick *et al.*, 1984). Several strategies are available for identifying olfactory receptor through their interaction with transduction components:

4.3.1. *Modulation of the Transductory Responses by Reagents That Bind Specifically to Surface Determinants of Membrane Receptor Candidates*

An example of this approach is the work done using specific antibodies to the nicotinic ACh receptor or to epidermal growth factor (EGF) receptor (Schlessinger *et al.*, 1983; Greaves, 1984; Changeux *et al.*, 1984). The results expected for such modulation may vary according to the receptor type and its transduction mechanism. Practically all cell-surface receptors are glycoproteins that can interact with both antibodies and lectins. While antibodies can bind either to the carbohydrate moiety or to the polypeptide portion, lectins will only recognize the oligosaccharide. Both antibodies and lectins are usually bi- or polyvalent, and their binding can lead to receptor aggregation. For receptors that are activated by an aggregative process (e.g., the immunoglobulin-E–Fc receptor complex or EGF receptor), antibodies and lectins are often found to be agonistic; they can mimic the functional ligands through their ability to crosslink the receptors (Schlessinger *et al.*, 1983; Greaves, 1984). For this type of receptor, inhibition will only be observed in specific (and consequently rare) cases of antibodies that block the ligand binding site or modify receptor conformation. Lectins will inhibit only if they happen to interfere with ligand binding because of the proximity of a carbohydrate attachment point and the ligand-binding site.

Another class of receptors are those whose activation is conformational, e.g., nicotinic ACh receptor (Changeux *et al.*, 1984) or the β-adrenergic receptor (Cerione *et al.*, 1983; Smigel *et al.*, 1984; Levitzki, 1986). Here, agonistic effects will be more rarely observed, since they require that antibody binding will mimic the conformational effect of the agonistic ligand. Lectins will practically never have an agonist effect. On the other hand, since many of these receptors (e.g., β-adrenergic receptor, rhodopsin) activate G proteins through a catalytic "touch-and-go" mechanism that depends on lateral diffusion in the plane of the lipid bilayers, crosslinking by antibodies and lectins may lead to inhibition. Olfactory receptor proteins belong, in all probability, to the latter class.

Recently reported studies (Shirley *et al.*, 1983*b*; Chen *et al.*, 1985) suggest that lectins can inhibit olfactory reactivity *in vivo*. Lectins are applied topically to the olfactory epithelial surface immediately prior to recording electro-olfactogram responses to various odorants in the frog or rat. The lectin WGA produces a highly efficient inhibition in the frog that is abolished by the soluble sugar *N*-acetylglucosamine (Chen *et al.*, 1985; Lancet *et al.*, 1987). The advent of a cell-free assay for olfactory reception in the form of odorant-sensitive adenylate cyclase in isolated olfactory cilia (Chen *et al.*, 1985) prompted us to examine lectin inhibition using such assay. It was found that WGA produced a similar inhibition *in vitro* and that this effect was also reversed in the presence of *N*-acetylglucosamine (Chen *et al.*, 1985; Lancet *et al.*, 1986, 1987). This inhibitory effect suggests that a glycoprotein that reacts with WGA may be involved in olfactory reception. The transmembrane ciliary glycoprotein gp95 is interesting in this respect, since it is the major WGA-reactive protein of the sensory organelles.

4.3.2. Interaction with Transduction Components

The presumed partner of olfactory receptor proteins in the transduction chain is G_s, the stimulatory G protein. Can the interaction of candidate receptors be monitored with this transduction component? Under steady-state operating conditions, the interaction of a receptor with the G protein is a complex chain of dynamic steps that include the association and dissociation of the receptor–G-protein pair, coupled to changes of conformation in both proteins, the dissociation of G-protein subunit, the binding of GTP, its hydrolysis to GDP, and dissociation of the latter (Gilman, 1984; Schramm and Selinger, 1984; Spiegel *et al.*, 1985). Yet, under appropriate conditions, detergent-solubilized receptor proteins have been reported to associated with G proteins (Limbrid *et al.*, 1980). This interaction is enhanced by agonist binding to the receptor and is reversed by GppNHp, a nonhydrolyzable GTP analogue. A similar approach may be useful for olfactory receptor studies.

Receptor–G-protein interaction can also be monitored indirectly, by studying the modulation by the activated receptor of G-protein function: GTPase activity and guanine nucleotide binding. We have recently studied odorant activation of GTPγS binding in isolated olfactory cilia. Binding was measured by incubation of isolated olfactory cilia with GTPγ^{35}S, followed by rapid filtration on nitrocellulose filters. Small but reproducible odorant effects were seen, suggesting that guanine nucleotide binding can serve as an additional *in vitro* assay for the olfactory response. The advantage of the nucleotide binding assay is that it depends only on binary receptor–G-protein interactions rather than on the integrity of a three-component system that also includes the adenylate cyclase catalytic unit. A possible disadvantage is the lack of the second amplification step afforded by monitoring synthesis of cAMP.

Another way to observe the relationship of a candidate receptor with the transduction apparatus is through study of its indirect physical association with the cyclase catalytic subunit. The G-protein–cyclase pair has been suggested to be tightly bound (Levitzki, 1986), so it is conceivable that a receptor G-protein association will be manifested as a receptor–cyclase link. We have found that a monoclonal antibody that immunoprecipitates gp95 also bring down practically all the adenylate cyclase activity in a detergent solution of frog olfactory cilia (Lancet *et al.*, 1987). Thus, it is possible that a gp95–G_s–cyclase ternary complex is present in this solution.

4.3.3. Reconstitution of Odorant-Activated Adenylate Cyclase

In view of the evidence for the function of cAMP as an olfactory second messenger, it may be legitimate to regard the ability to confer odorant sensitivity on adenylate cyclase as a functional definition of the receptor. Solubilized fractions of olfactory cilia can be assayed for their ability to induce odorant sensitivity on adenylate cyclase, from either a homologous source (olfactory cilia) or a heterologous source (e.g., erythrocytes or liver). The methods necessary for such functional reconstitution have been developed for other receptors, notably the β-adrenergic receptor (Smigel *et al.*, 1984; Cerione *et al.*, 1983), and involve solubilization of the components in the appropriate detergents and detergent removal in the presence of phospholipids. In principle, this procedure can be used to detect olfactory receptor proteins and may form a basis for a receptor-purification protocol.

5. POSSIBLE MECHANISMS OF OLFACTORY ION-CHANNEL MODULATION BY CYCLIC NUCLEOTIDES

Odorant stimulation of olfactory sensory neurons leads to depolarization and firing of action potentials. An intriguing open question is how changes in the intracellular concentrations of the proposed second messenger, cAMP, could result in membrane conductance changes. Two possible mechanisms are (1) cAMP-dependent phosphorylation, and (2) direct cyclic nucleotide gating of ion channels.

5.1. Protein Phosphorylation in Olfactory Epithelium

The intracellular effects of cAMP are practically always expressed in changes of protein phosphorylation, mediated by cAMP-dependent protein kinases. Thus, phosphorylation or dephosphorylation, either of a channel protein or of proteins associated with the channel, may underlie receptor-induced cAMP-mediated conductance changes in olfactory neurons, as they do in several other neuronal systems (Siegelbaum and Tsien, 1983; Browning *et al.*, 1985; Nestler *et al.*, 1984). A prerequisite for the validity of such mechanism is that the relevant enzyme and its specific substrates will be present in olfactory neurons. We have examined this question by studying protein phosphorylation in isolated frog olfactory cilia (Heldman and Lancet, 1986; Lancet *et al.*, 1987). When incubated with [^{32}P]-ATP, olfactory cilia can catalyze the phosphorylation of exogenous substrate (e.g., histone); this kinase activity is enhanced by cAMP, either added (at concentrations of ≤ 1 μM) or endogenously generated by adenylate cyclase activation with GTPγS. Incubation of isolated olfactory cilia with [^{32}P]-ATP *without* exogenous substrate results in the incorporation of [^{32}P]phosphate into ciliary polypeptides in a cAMP-dependent manner. Notable among these polypeptides are the phosphoproteins pp24 and pp26, which are specific to olfactory cilia as compared with respiratory cilia. It appears that the cilia preparation contains both the enzymatic activity and the specific substrates that could underlie the proposed phosphorylation-mediated sensory transduction steps.

It should be noted that both cAMP-dependent protein kinase and its specific substrates are also present in membranes from deciliated olfactory epithelium, hence may be localized also in the sensory dendritic shaft. Since adenylate cyclase and olfactory G_s are highly enriched in the ciliary extensions, it is possible that the cilia serve as cAMP generators and that the second messenger diffuses into the dendrite, to exert at least part of its effects there. The time course of the olfactory response (a few hundred milliseconds (Getchell *et al.*, 1984; Ottoson, 1971) is not inconsistent with this notion.

Phosphoproteins pp24 and pp26 belong to a large group of polypeptides discovered by virtue of their cAMP-dependent phosphorylation (cf. Nestler *et al.*, 1984). Considerable research will be needed to establish any role they may have in chemosensory mechanism. First, it will be necessary to obtain corroborative evidence through studying olfactory protein phosphorylation *in vivo*. This could be done by incubating olfactory epithelial tissue explants in short-term organ culture with [^{32}P]phosphate. Such organ culture has been previously shown to be viable for 72 hr and to be capable of incorporating [^{35}S]methionine into newly biosynthesized proteins (Chen *et al.*, 1986*c*). Subsequently, it will be necessary to demonstrate odorant-related changes in protein phosphorylation. Later, it may become possible to purify the phosphoproteins and prepare antibodies against them (cf. Browning *et al.*, 1985; Nestler *et al.*, 1984). Such antibodies would be useful in further probing the significance of protein phosphorylation in olfactory transduction. Finally, it may become possible to demonstrate the involvement of cAMP-dependent kinase in ion-channel modulation through direct application of its activated catalytic subunit to membrane patches upon recording of single ion-channel currents.

Protein phosphorylation may be involved in processes other than the direct line of transduction, e.g., desensitization or adaptation of the olfactory response. The molecular basis of such desensitization phenomena has been investigated in other G-protein-related receptors, such as β-adrenergic receptor (Sibley and Lefkowitz, 1985) or visual photoreceptor (Chabre, 1985; Stryer, 1986). Protein phosphorylation has been suggested to play a central role in these mechanisms. Future studies should indicate whether the putative olfactory receptor proteins undergo ligand-related phosphorylation, analogous to that found in the other receptor mechanisms.

5.2. Direct Ion-Channel Gating

It has recently been demonstrated that in visual transduction, cGMP interacts directly with ion channels in the rod outer segment membrane (Fesenko *et al.*, 1985; Stryer, 1986; Kaupp and Koch, 1986). This is a novel mechanism for the mode of action of cyclic nucleotides, and its occurrence in retina may not be exclusive. Direct ion-channel modulation may be an alternative mechanism for the action of cyclic nucleotide on neuronal ion channels. Olfactory neurons may provide a useful model system for studying such a putative general mechanism. Direct modulation could be best demonstrated through recordings from isolated membrane patches, devoid of nucleotides and cytoplasmic enzymes, such as cAMP-dependent protein kinase. Specific modulation of ion conductance by application of cAMP can serve to support the direct modulation hypothesis. Such experiments may be carried out using ciliary membranes (cf. Kleene *et al.*, 1985), assuming that the sensory ionic mechanism resides, at least partially, in the cilia or on excised membrane patches derived from intact chemosensory neurons (cf. Maue and Dionne, 1984, 1986; Firestein and Werblin, 1985).

Evidence for direct cAMP gating has been produced through recordings in artificial lipid bilayers reconstituted with rat olfactory epithelial membranes (Vodyanoy and Vodyanoy, 1986). More recently, in an elegant study of excised membrane patches from the cilia of isolated toad olfactory neurons, Nakamura and Gold (1987) provided clear evidence for a cyclic nucleotide-gated cation conductance. Micromolar concentrations of both cAMP and cGMP increase this conductance, which appears to resemble the cGMP-gated conductance of retinal rod outer segments. The comparable potency of cGMP and cAMP in activating conductance changes in the olfactory membranes has been observed in other studies as well (Vodyanoy and Vodyanoy, 1986; Trotier and MacLeod, 1986) (see Chapter 8), but in all cases 2′,3′-cAMP was found to be much less potent.

6. CONCLUSIONS

Future studies may lead to a complete elucidation of olfactory transduction mechanisms. It is not impossible that in many ways taste transduction, at least for sweet/bitter ligands, will be found to involve similar transmembrane events (Kurihara and Koyama, 1972; Price, 1973; Cagan, 1976; Schiffman *et al.*, 1985; Wieczorek and Schweikl, 1985; Bryant *et al.*, 1986; Striem *et al.*, 1986). In both systems, the understanding of molecular transduction could lead to identification and isolation of the receptor proteins themselves. Molecular cloning techniques, such as those used to isolate the T-lymphocyte receptor genes (Hendrick *et al.*, 1984) or color photoreceptor genes (Nathans *et al.*, 1986), will probably be useful in this respect. G-protein adenylate-cyclase coupling in reconstituted membrane vesicles (Lancet *et al.*, 1987) or odor-induced ion conductances in messenger RNA (mRNA)-injected *Xenopus* oocytes (Dascal *et al.*, 1986) can serve as functional assays for the expressed gene products.

The isolation of olfactory receptor genes could help resolve questions related to olfactory processing in the general molecular–neurobiological context (Lancet, 1986; Lancet and Pace, 1987). Thus, understanding the generation of receptor diversity and of the control of receptor expression in individual olfactory cells could further our understanding of synaptic contact specification and network formation in chemosensory organs, as well as in other neuronal arrays.

REFERENCES

Adamek, G. D., Gesteland, R. C., Mair, R. G., and Oakley, B., 1984, Transduction physiology of olfactory receptor cilia, *Brain Res.* **310**:87–97.

Adoutte, A., Ramanathan, R., Lewis, R. M., Dute, R. R., Ling, K. Y., Kung, C., and Nelson, D. L., 1980, Biochemical studies of the excitable membrane of *Paramecium tetraaurelia*, *J. Cell Biol.* **84:**717–738.

Amoore, J. E., 1971, Olfactory gentics and anosmia, in: *Handbook of Sensory Physiology*, Vol. 4, Part 1 (L. M. Beidler, ed.), pp. 245–256, Springer-Verlag, New York.

Anholt, R. R. H., 1987, Primary events in olfactory reception, *Trends Biochem. Sci.* **12:**58–62.

Anholt, R. R. H., Aebi, U., and Snyder, S., 1987, A partially purified preparation of isolated chemosensory cilia from the olfactory epithelium of the bullfrog, *Rana catesbeiana*, *J. Neurosci.* **6:**1962–1969.

Anholt, R. R. H., Mumby, S. M., Stoffers, D. A., Girard, P. R., Kno, J. F., and Snyder, S. H., 1987, Transductory proteins of olfactory receptor cells: Identification of guanine nucleotide binding proteins and protein kinase C, *Biochemistry* **26:**788–795.

Bacatti, B., and Gibbons, I. (eds.), 1983, International conference on development and function in cilia and flagella, *J. Submicros. Cytol.* **15:**1–374.

Berridge, M. J., and Irvine, R. F., 1984, Inositol trisphosphate, a novel second messenger in cellular signal signal transduction, *Nature (Lond.)* **312:**315–321.

Bignetti, E., Cavaggioni, A., Pelosi, P., Persaud, K. C., Sorki, R. T., and Trindelli, R., 1985, Purification and characterization of an odorant-binding protein from cow nasal tissue, *Eur. J. Biochem.* **149:**227–231.

Boyse, E. A., Beauchamp, G. K., Yamazaki, K., Bard, J., and Thomas, L., 1982, Chemosensory communication: A new aspect of the major histocompatibility complex and other genes in the mouse, *Oncodev. Biol. Med.* **4:**101–116.

Browning, M. D., Huganir, R., and Greengard, P., 1985, Protein phosphorylation and neuronal function, *J. Neurochem.* **45:**11–23.

Bryant, B. P., Brand, J. G., Kalinoski, D. L., Bruch, R. C., and Cagan, R. H., 1986, Use of monoclonal antibodies to characterize amino acid taste receptors in catfish: Effect on binding and neural responses, *Chem. Senses* **11:**586.

Cagan, R. H., 1976, Biochemical studies of taste sensation: Labeling of cyclic AMP of bovine taste papillae in response to sweet and bitter stimuli, *J. Neurosci. Res.* **2:**363–371.

Cerione, R. A., Strulovici, B., Benovic, J. L., Lefkovitz, R. J., and Caron, M. G., 1983, Pure β-adrenergic receptor: The single polypeptide confers catecholamine responsiveness to adenylate cyclase, *Nature (Lond.)* **306:**562–566.

Chabre, M., 1985, Trigger and amplification mechanisms in visual phototransduction, *Annu. Rev. Biophys. Biophys. Chem.* **14:**331–360.

Changeux, J.-P., Devillers-Thiery, A., and Chemouilli, P., 1984, Acetylcholine receptor: An allosteric protein, *Science* **225:**1335–1345.

Chen, Z., and Lancet, D., 1984, Membrane proteins unique to vertebrate olfactory cilia: Candidates for sensory receptor molecules, *Proc. Natl. Acad. Sci. USA* **81:**1859–1863.

Chen, Z., Greenberg, M., Pace, U., and Lancet, D., 1985, A cell free assay for olfactory reactivity reveals properties of odorant receptor molecules, *Soc. Neurosci. Abst.* **11:**970–970.

Chen, Z., Pace, U., Ronen, D., and Lancet, D., 1986*a*, Polypeptide gp95: A unique glycoprotein of olfactory cilia with transmembrane receptor properties, *J. Biol. Chem.* **261:**1299–1305.

Chen, Z., Ophir, D., and Lancet, D., 1986*b*, Monoclonal antibodies to glycoproteins of frog olfactory cilia, *Brain Res.* **368:**329–338.

Chen, Z., Pace, U., Heldman, J., Shapira, A., and Lancet, D., 1986*c*, Isolated frog olfactory cilia: A preparation of dendritic membranes from chemosensory neurones, *J. Neurosci.* **6:**2146–2154.

Cooper, D. M. F., and Seamon, K. B. (eds.), 1985, *Advances in Cyclic Nucleotide and Protein Phosphorylation Research*, Vol. 19, Raven, New York.

Cushley, W., and Williamson, A. R., 1982, Expression of immunoglobulin genes, *Essays Biochem.* **18:**1–39.

Daly, J., 1977, *Cyclic Nucleotides in the Nervous System*, Plenum, New York.

Davies, J. T., 1971, Olfactory theories, in: *Handbook of Sensory Physiology*, Vol. 4, Part 1 (L. M. Beidler, ed.), pp. 322–350, Springer-Verlag, New York.

Dascal, N., Heldman, J., Gershon, E., and Lancet, D., 1986, Possible expression of odorant receptor proteins in *Xenopus* oocytes injected with rat olfactory epithelial mRNA, *Soc. Neurosci. Abst.* **12:**1354.

Dodd, G., and Persaud, K., 1981, Biochemical mechanisms in vertebrate primary olfactory neurons, in: *Biochemistry of Taste and Olfaction* (R. H. Cagan and M. R. Kare, eds.), pp. 333–358, Academic, New York.

Fesenko, E. E., Novoselov, V. I., and Krapivinskaya, L. D., 1979, Molecular mechanisms of olfactory reception. IV. Some biochemical characteristics of the camphor receptor from rat olfactory epithelium, *Biochim. Biophys. Acta* **587**:424–433.

Fesenko, E. E., Novoselov, V. I., Krapivinskaya, L. D., Mjasoedov, N. F., and Zolotarev, J. A., 1983, Molecular mechanisms of odor sensing. VI. Some biochemical characteristics of a possible receptor for amino acids from the olfactory epithelium of the skate *Dasyatis pastinaca* and Carp *Cyprinus carpio*, *Biochim. Biophys. Acta* **759**:250–256.

Fesenko, E. E., Kolesnikov, S. S., and Lyubarsky, A. L., 1985, Induction by cyclic GMP of cationic conductance in plasma membranes of retinal rod outer segment, *Nature (Lond.)* **313**:310–312.

Firestein, S., and Werblin, F., 1985, Electrical properties of olfactory cells isolated from the epithelium of the tiger salamander, *Soc. Neurosci. Abst.* **11**:970–970.

Gesteland, R. C., 1976, Physiology of olfactory reception, in *Frog Neurobiology* (R. Llinas and W. Precht, eds.), pp. 234–249, Springer-Verlag, New York.

Gesteland, R. C., Lettvin, J. Y., and Pitts, W. H., 1965, Chemical transmission in the nose of the frog, *J. Physiol. (Lond)* **181**:525–559.

Getchell, T. V., 1974, Unitary responses in frog olfactory epithelium to sterically related molecules at low concentrations, *J. Gen. Physiol.* **64**:241–261.

Getchell, T. V., 1977, Analysis of intracellular recordings from Salamander olfactory epithelium, *Brain Res.* **123**:275–286.

Getchell, T. V., 1986, Functional properties of vertebrate olfactory receptor neurons, *Physiol. Rev.* **66**:772–817.

Getchell, M. L., and Gesteland, R. C., 1972, The chemistry of olfactory reception: Stimulus-specific protection from sulphydryl reagent inhibition, *Proc. Natl. Acad. Sci. USA.* **69**:1494–1498.

Getchell, T. V., Margolis, F. L., and Getchell, M. L., 1984, Perireceptor and receptor events in vertebrate olfaction, *Prog. Neurobiol.* **23**:317–345.

Gilman, A. G., 1984, G proteins and dual control of adenylate cyclase, *Cell* **36**:577–579.

Goldberg, S. J., Turpin, J., and Price, S., 1979, Anisole binding protein from olfactory epithelium: Evidence for a role in transduction. *Chem. Senses Flavour* **4**:207–213.

Greaves, M. F. (ed.), 1984, Monoclonal antibodies to receptors: Probes for receptor structure and function, in: *Receptors and Recognition:* Ser. B, Vol. 17, Chapman and Hall, London.

Harris, B. A., Robishaw, J. D., Mumby, S. M., and Gilman, A. G., 1985, Molecular cloning of cDNA for the α subunit of the G-protein that stimulates adenylate cyclase, *Science* **229**:1274–1277.

Heldman, J., and Lancet, D., 1986, Cyclic AMP dependent protein phosphorylation in chemosensory neurons: Identification of cyclic nucleotide regulated phosphoproteins in olfactory cilia, *J. Neurochem.* **47**:1527–1533.

Hendrick, S. M., Cohen, D. I., Nielsen, E. A., and Davis, M. M., 1984, Isolation of cDNA clones encoding T cell-specific membrane-associated proteins, *Nature (Lond.)* **308**:149–160.

Henkin, R. I., 1968, Impairment of olfaction and of the taste of sour and bitter in pseudohypoparathyroidism, *J. Clin. Endocr.* **28**:624–628.

Holley, A., and MacLeod, P., 1977, Transduction et codage des informations olfactives chez les vertebres, *J. Physiol. (Paris)* **73**:725–828.

Hood, L., 1982, Antibody genes: Arrangements and rearrangements, in: *Molecular Genetic Neuroscience* (F. O. Schmitt, S. J. Bird, and F. E. Bloom, eds.), pp. 75–85, Raven, New York.

Huque, T., and Bruch, R. C., 1986, Odorant- and guanine nucleotide-stimulated phosphinositide turnover in olfactory cilia, *Biochem. Biophys. Res. Commun.* **137**:36–43.

J. Cell Biol., 1981, **91(3)**.

Jones, D., and Reed, R., 1987, Molecular cloning of five GTP-binding protein-DNA species from rat olfactory neuroepithelium, *J. Biol. Chem.* **262**:14241–14249.

Kaupp, U. B., and Koch, K.-W., 1986, Mechanism of photoreception in vertebrate vision, *Trends Biochem. Sci.* **11**:43–47.

Kimball, E. S., and Coligan, J. E., 1983, Structure of class I major histocompatibility antigens, *Contemp. Topics Mol. Immunol.* **9**:1–63.

Kishimoto, A., Takai, Y., Mori, T., Kikkawa, U., and Nishizuka, Y., 1980, Activation of calcium and phospholipid dependent protein kinase by diacylglycerol, its possible relation to phosphatidylinositol turnover, *J. Biol. Chem.* **255**:2273–2276.

Kleene, S. J., MacDonald, R. C., Lidow, M. S., and Gesteland, R. C., 1985, Membrane channels of olfactory cilia, *Chem. Senses* **10**:393–393.

Kleinzeller, A., and Martin, B. R. (eds.), 1983, *Current Topics in Membranes and Transport,* Vol. 18: *Membrane Receptors,* Academic, New York.

Kurihara, K., and Koyama, N., 1972, High activity of adenyl cyclase in olfactory and gustatory organs, *Biochem. Biophys. Res. Commun.* **48**:30–34.

Labarca, P., Simon, S. A., and Anholt, R. R. H., 1986, Activation by odorants of a voltage dependent, cation-selective channel in ciliary membranes from olfactory epithelium in the bullfrog, *Soc. Neurosci. Abst.* **12**:1178–1178.

Lancet, D., 1984, Molecular view of olfactory reception, *Trends Neurosci.* **7**:35–36.

Lancet, D., 1986, Vertebrate olfactory reception, *Annu. Rev. Neurosci.* **9**:329–355.

Lancet, D., and Pace, U., 1984, Proteins of olfactory cilia that may be involved in cyclic nucleotide-mediated sensory transduction, *Soc. Neurosci. Abst.* **10**:655–655.

Lancet, D., and Pace, U., 1987, The molecular basis of odor recognition, *Trends Biochem. Sci* **12**:63–66.

Lancet, D., Heldman, J., and Pace, U., 1985, Adenylate cyclase and G-protein mediate sensory transduction in olfactory neurons, *Soc. Neurosci. Abst.* **11**:815–815.

Lancet, D., Chen, Z., Heldman, J., and Pace, U., 1986, Cyclic nucleotide cascade in olfactory transduction, *Biophys. J.* **49**:183a.

Lancet, D., Chen, Z., Ciobotariu, A., Heldman, J., Ophir, D., Pines, M., and Pace, U., 1987, Towards a comprehensive molecular analysis of olfaction, *Ann. NY Acad. Sci.* **510**:27–32.

Levitan, I., 1985, Phosphorylation of ion channels, *J. Membr. Biol.* **87**:177–190.

Levitzki, A., 1986, β-Adrenergic receptors and their mode of coupling to adenylate cyclase, *Physiol. Rev.* **66**:819–854.

Lidow, M. S., and Menco, B. P. M., 1984, Observations on axonemes and membranes of olfactory and respiratory cilia in frogs and rats using tannic acid-supplemented fixation and photographic rotation, *J. Ultrastruct. Res.* **86**:18–30.

Limbrid, L. E., Gill, D. M., and Lefkowitz, R. J., 1980, Agonist-promoted coupling of the β-adrenergic receptor with the guanine nucleotide regulatory protein of the adenylate cyclase system, *Proc. Natl. Acad. Sci. USA* **77**:775–779.

Linck, R. W., 1973, Comparative isolation of cilia and flagella from the lamellibranch mollusc *Acquipecten irradians, J. Cell Sci.* **12**:345–367.

Margolis, F. L., 1975, Biochemical studies of the primary olfactory pathway, in: *Soc. for Neurosci. Symp.* Vol. 3, (J. A. Ferrendelli, ed.), pp. 167–188, Society for Neuroscience, Bethesda (MD).

Mason, J. R., and Morton, T. H., 1984, Fast and loose covalent binding of ketones as a molecular mechanism in vertebrate olfactory receptors, *Tetrahedron* **40**:483–492.

Masukawa, L. M., Hedlund, J. S., and Shepherd, G. M., 1985, Electrophysiological properties of identified cells in the *in vitro* olfactory epithelium of the tiger salamander, *J. Neurosci.* **5**:128–135.

Maue, R. A., and Dionne, V. E., 1984, Ion channel activity in isolated murine olfactory receptor neurons, *Soc. Neurosci. Abst.* **10**:655–655.

Maue, R. A., and Dionne, V. E., 1986, Membrane conductance mechanisms in neonatal and embrionic mouse olfactory neurons, *Biophys. J.* **49**:556–556a.

Menco, B. P. M., 1980, Qualitative and quantitative freeze-fracture studies on olfactory and nasal respiratory epithelial surfaces of frog, ox, rat and dog. II. Cell apices, cilia and microvilli, *Cell Tissue Res.* **211**:5–29.

Menco, B. P. M., 1983, The Structure of olfactory and nasal respiratory epithelial surfaces, in: *Nasal Tumors in Animals and Man,* Vol. 1 (G. Reznik and S. F. Stinton, eds.), pp. 45–102, CRC Press, Boca Raton, Florida.

Menco, B. P. M., 1984, Ciliated and microvillar structures of rat olfactory and nasal respiratory epithelia, *Cell Tissue Res.* **235**:225–241.

Menevse, A., Dodd, G., and Poynder, T. M., 1977, Evidence for the specific involvement of cAMP in the olfactory transduction mechanism, *Biochem. Biophys. Res. Commun.* **77**:671–677.

Minor, A. V., and Sakina, N. L., 1973, Role of cyclic adenosine-3′,5′-monophosphate in olfactory reception, *Neurofysiologiya* **5**:415–422.

Moulton, D., and Beidler, L. M., 1967, Structure and function in the peripheral olfactory system, *Physiol. Rev.* **47**:1–52.

Mumby, S. M., Kahn, R. A., Manning, D. R., and Gilman, A. G., 1986, Antisera of designed specificity for subunits of guanine nucleotide-binding regulatory proteins, *Proc. Natl. Acad. Sci. USA* **83**:265–269.

Nachbar, R. B., and Morton, T. H., 1981, A gas chromatographic (GLPC) model for the sense of smell: Variation of olfactory sensitivity with conditions of stimulation, *J. Theor. Biol.* **89:**387–407.

Nakamura, T., and Gold, G. H., 1987, A cyclic-nucleotide gated conductance in olfactory receptor cilia, *Nature (Lond.)* **325:**442–444.

Nathans, J., Thomas, D., and Hogness, D. S., 1986, *Science* **232:**193–202.

Nestler, E. J., Walaas, S. I., and Greengard, P., 1984, Neuronal phosphoproteins: Physiological and clinical implications, *Science* **225:**1357–1364.

O'Brien, D. F., 1982, The chemistry of vision, *Science* **218:**961–966.

Ottoson, D., 1971, The electro-olfactogram, in: *Handbook of Sensory Physiology*, Vol. 4, Part 1 (L. M. Beidler, ed.), pp. 95–131, Springer-Verlag, New York.

Pace, U., and Lancet, D., 1986, Olfactory GTP-binding protein: Signal transducing polypeptide of vertebrate chemosensory neurons, *Proc. Natl. Acad. Sci. USA* **83:**4947–4951.

Pace, U., Hanski, E., Salomon, Y., and Lancet, D., 1985, Odorant-sensitive adenylate cyclase may mediate olfactory reception, *Nature (Lond.)* **316:**255–258.

Pelosi, P., Baldaccini, N. E., and Pisanelli, A. M., 1982, Identification of a specific olfactory receptor for 2-isobutyl-3-methoxypyrazine, *Biochem. J.* **201:**245–248.

Persaud, K. C., Getchell, T. V., Heck, G. L., and DeSimone, J. A., 1987, Electrophysiological evidence for a G-protein mediated adenylate cyclase in olfactory transduction, unpublished.

Pevsner, J., Trifiletti, R. R., Strittmatter, S. M., and Snyder, S. H., 1985, isolation and characterization of an olfactory receptor protein for odorant pyrazines, *Proc. Natl. Acad. Sci. USA* **82:**3050–3054.

Pevsner, J., Sklar, P. B., and Snyder, S. H., 1986, Odorant binding protein: Localization to nasal glands and secretions, *Proc. Natl. Acad. Sci. USA* **83:**4942–4946.

Ploegh, H. L., Orr, H. T., and Strominger, J. L., 1981, Major histocompatibility antigens: The human (HLA-A, -B, -C) and murine (H-2K, H-2D) class I molecules, *Cell* **24:**287–299.

Price, S., 1973, Phosphodiesterase in tongue epithelium: Activation by bitter taste stimuli, *Nature (Lond.)* **241:**54–55.

Price, S., 1977, Specific anosmia to geraniol in mice, *Neurosci. Lett.* **4:**49–50.

Price, S., 1981, Receptor proteins in vertebrate olfaction, in: *Biochemistry of Taste and Olfaction* (R. H. Cagan and M. R. Kare, eds.), pp. 69–84, Academic, New York.

Reese, T. S., 1965, Olfactory cilia in the frog, *J. Cell Biol.* **25:**209–230.

Rhein, L. D., and Cagan, R. H., 1980, Biochemical studies of olfaction: Isolation characterization, and odorant binding activity of cilia from rainbow trout olfactory rosettes, *Proc. Natl. Acad. Sci. USA* **77:**4412–4416.

Rhein, L. D., and Cagan, R. H., 1981, Role of cilia in olfactory recognition, in: *Biochemistry of taste and Olfaction* (R. H. Cagan and M. R. Kare, eds.), pp. 47–68, Academic, New York.

Robertson, M., 1985, T-cell receptor: The present state of recognition, *Nature (Lond.)* **317:**768–771.

Robinson, G. A., Butcher, R. W., and Sutherland, E. W., 1971, *Cyclic AMP*, Academic, New York.

Rodrigues, V., 1980, Olfactory behavior of Drosophila melanogaster, in: *Development and Neurobiology of Drosophila* (O. Siddiqi, P. Babu, L. M. Hall, and J. C. Hall, eds.), pp. 361–371, Plenum, New York.

Rodrigues, V., and Siddiqi, O., 1978, Genetic analysis of chemosensory pathway, *Proc. Indian Acad. Sci.* **87B:**147–160.

Schiffman, S. S., Gill, J. M., and Diaz, C., 1985, Methyl xanthines enhance taste: Evidence for modulation of taste by adenosine receptors, *Pharmacol. Biochem. Behav.* **22:**195–203.

Schlessinger, J., Schreiber, A. B., Lieberman, T. A., Lax, I., Avivi, A., and Yarden, Y., 1983, Polypeptide-hormone-induced receptor clustering and internalization, in: *Cell Membranes: Methods and Reviews*, Vol. 1 (E. Elson, W. Frazier, and L. Glaser, eds.), pp. 117–149, Plenum, New York.

Schramm, M., and Selinger, Z., 1984, Message transmission: Receptor controlled adenylate cyclase system, *Science* **225:**1350–1356.

Senf, W., Menco, B. P. M., Punter, P. H., and Duyventeyn, P., 1980, Determination of odour affinities on the dose–response relationships of the frog's electro-olfactogram, *Experientia* **36:**213–215.

Shirley, S., Polak, E., and Dodd, G., 1983*a*, Chemical-modification studies on rat olfactorv mucosa using a thiol-specific reagent and enzymatic iodination, *Eur. J. Biochem.* **132:**485–494.

Shirley, S., Polak, E., and Dodd, G., 1983*b*, Selective inhibition of rat olfactory receptors by Concanavalin A, *Biochem. Soc. Trans.* **11:**780–781.

Shirley, S., Robinson, J., Dickenson, K., and Rajinder, A., 1985, Rat olfactory adenylate cyclase, in: *NATO Advanced Research Workshop Abstracts* (G. Dodd, ed.), p. 16, University of Warwick, Coventry, England.

Shirley, S., Robinson, J., Dickinson, K., Aujla, R., and Dodd, G. H., 1986, Olfactory adenylate cyclase of the rat, *Biochem. J.* **240**:605–607.

Sibley, D. R., and Lefkowitz, R. J., 1985, Molecular mechanisms of receptor desensitization using the β-adrenergic receptor-coupled adenylate cyclase system as a model, *Nature (Lond.)* **317**:124–129.

Sicard, G., and Holley, A., 1984, Receptor cell responses to odorants: Similarities and differences among odorants, *Brain Res.* **292**:283–296.

Siegelbaum, S. A., and Tsien, R. W., 1983, Modulation of gated ion channels as a mode of transmitter action, *Trends Neurosci.* **6**:307–313.

Sigal, N. H., and Klinman, N. R., 1978, The B-cell clonotype repertoire, *Adv. Immunol.* **26**:255–337.

Sklar, P. B., Anholt, R. R. H., and Snyder, S. H., 1986, The odorant-sensitive adenylate cyclase of olfactory receptor cells: Differential stimulation by distinct classes of odorants, *J. Biol. Chem.* **261**:15538–15543.

Sleigh, M. A. (ed.), 1974, *Cilia and Flagella,* Academic, New York.

Slotnick, B. M., and Schoonover, F. W., 1984, Olfactory thresholds in unilaterally bulbectomized rats, *Chem. Senses* **9**:325–340.

Smigel, M. D., Ross, E. M., and Gilman, A. G., 1984, Role of the β-adrenergic receptor in the regulation of adenylate cyclase, in: *Cell Membranes: Methods and Reviews,* Vol. 2 (E. Elson, W. Frazier, and L. Glaser, eds.), pp. 247–296, Plenum, New York.

Spiegel, A. M., Gierschick, P., Levine, M. A., and Downs, R. W., 1985, Clinical implications of guanine nucleotide-binding proteins as receptor–effector couplers, *N. Engl. J. Med.* **312**:26–33.

Striem, B. J., Pace, U., Zehavi, U., Naim, M., and Lancet, D., 1986, Is adenylate cyclase involved in sweet taste transduction?, *Chem. Senses* **11**:669.

Strosberg, A. D., 1984, Receptors and recognition: From ligand binding to gene structure, *Trends Biochem. Sci.* **9**:166–169.

Stryer, L., 1986, Cyclic GMP cascade of vision, *Annu. Rev. Neurosci.* **9**:87–119.

Tonegawa, S., 1983, Somatic generation of antibody diversity, *Nature (Lond.)* **302**:575–581.

Trotier, D., and MacLeod, P., 1983, Intracellular recordings from salamander olfactory receptor cells, *Brain Res.* **268**:225–237.

Trotier, D., and McLeod, P., 1986, cAMP and cGMP open channels and depolarize olfactory receptor cells, *Chem. Senses* **11**:674.

Vandenberg, C. A., and Montal, M., 1984, Light-regulated biochemical events in invertebrate photoreceptors: Light regulated phosphorylation of rhodopsin and phosphoinositides in squid photoreceptor membranes, *Biochemistry* **23**:2347–2352.

Venard, R., and Pichon, Y., 1984, Electrophysiological analysis of the peripheral response to odours in wild type and smell-deficient olf C mutant of *Drosophila melanogaster, J. Insect Physiol.* **30**:1–5.

Venter, C. J., and Harrison, L. C. (eds.), 1984, *Receptor Biochemistry and Methodology,* Vols. 1–3, Liss, New York.

Villet, R. H., 1978, Mechanism of insect sex pheromone sensory transduction: Role of adenyl cyclase, *Comp. Biochem. Physiol.* **61**:389–394.

Vinnikov, Y. A., 1982, *Molecular Biology Biochemistry and Biophysics,* Vol. 34: *Evolution of Receptor Cells,* Springer-Verlag, New York.

Vodyanoy, V., and Murphy, R. B., 1983, Single channel fluctuations in bimolecular lipid membranes induced by rat olfactory epithelial homogenates, *Science* **220**:717–719.

Vodyanoy, V., and Vodyanoy, I., 1986, Electrical properties of chemical-gated cation channels from olfactory epithelium homogenates, *Biophys. J.* **49**,521a.

Weinstock, R. S., Wright, H. N., Spiegel, A. M., Levine, M. A., and Moses, A. M., 1986, Olfactory dysfunction in humans with deficient guanine nucleotide-binding protein, *Nature (Lond.)* **322**:635–636.

Wieczorek, H., and Schweikl, H., 1985, Concentrations of cyclic nucleotides and activities of cyclases and phosphodiesterases in an insect chemosensory organ, *Insect Biochem.* **15**:723–728.

Wysocki, C. J., Whitney, G., and Tucker, D., 1977, Specific anosmia in the laboratory mouse, *Behav. Genet.* **7**:171–188.

3

The Effect of Cytochrome P-450-Dependent Metabolism and Other Enzyme Activities on Olfaction

Alan R. Dahl

1. RELATIONSHIPS AMONG THE FIELDS OF INHALATION TOXICOLOGY, FOREIGN COMPOUND METABOLISM, AND OLFACTORY PHYSIOLOGY

The scientific fields of inhalation toxicology, foreign compound metabolism, and olfactory physiology share broad interests in the interaction of foreign materials with biological systems. Inhalation toxicologists and olfactory physiologists share a narrower common interest in the effects of foreign compounds on nasal tissues. Inhalation toxciologists are interested in toxic effects, and olfactory physiologists concern themselves with stimulation to produce an olfactory response. The third field to be included here, that concerned with foreign compound metabolism, has until recently been of little interest to inhalation toxicologists (and that interest largely confined to the metabolism of inhaled compounds in the lung and liver) and has been of even less concern to olfactory physiologists. But, the increasing frequency with which nasal cancers and other lesions have been reported in laboratory animals exposed to toxicants by inhalation has led to increased research into the response of nasal tissue to inhaled materials. One recent finding is that the nasal tissues have high concentrations of enzymes that metabolize xenobiotics (Dahl, 1985*a–c*).

To the toxicologist, metabolism resulting in the chemical transformation of an inhaled compound to one or more different compounds may signal either decreased or increased toxicity. For example, enzymatic oxidation of the common air pollutant, benzo(a)pyrene, in the noses of Syrian hamsters resulted in seven identified products (metabolites) in addition to the 3-hydroxybenzo(a)pyrene shown in chemical equation (1) (CE1) of Fig. 1. Included among these identified metabolites was the immediate precursor to the potent carcinogen, 7,8-dihydro-7,8 dihyroxy-9,10-epoxybenzo(a)pyrene. Production of this metabolite within the nasal cavity is probably the cause of nasal cancer in hamsters that inhale benzo(a)pyrene (Dahl *et al.*, 1985). By contrast, enzymes that catalyze the hydrolysis of potentially carcinogenic β-lactones (CE2, Fig. 1) are also present in the

Alan R. Dahl • Inhalation Toxicology Research Institute, Lovelace Biomedical and Environmental Research Institute, Albuquerque, New Mexico 87185.

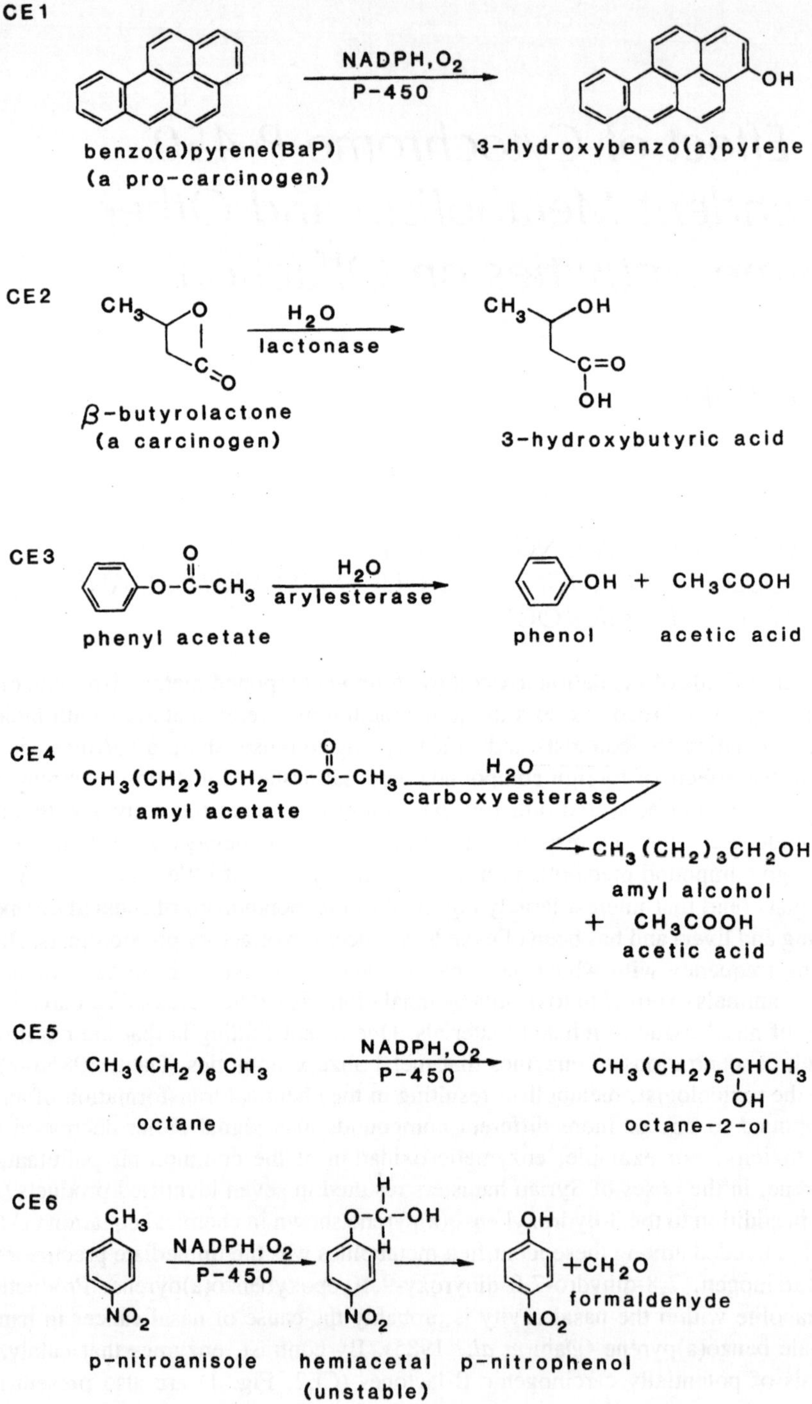

Figure 1. Chemical transportations catalyzed by olfactory mucosal enzymes.

CE7

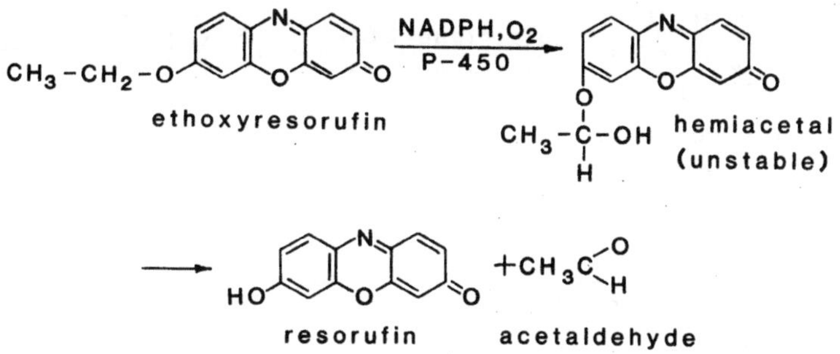

CE8

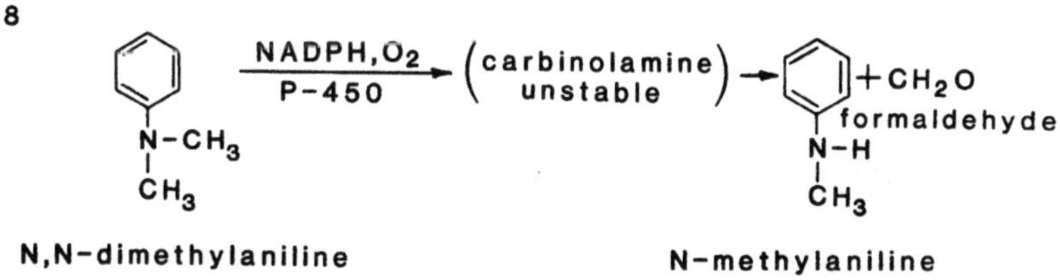

CE9

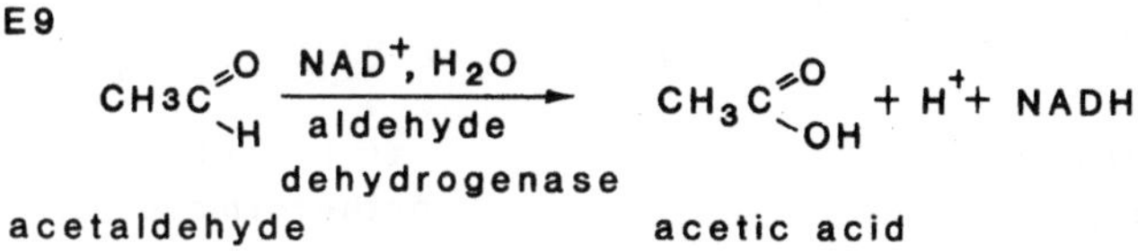

CE10

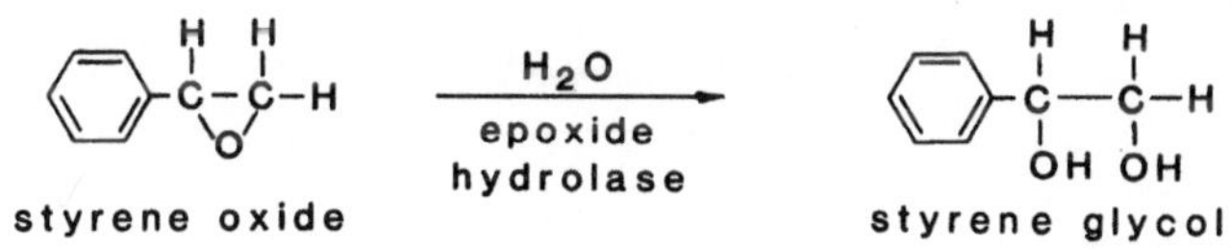

CE11

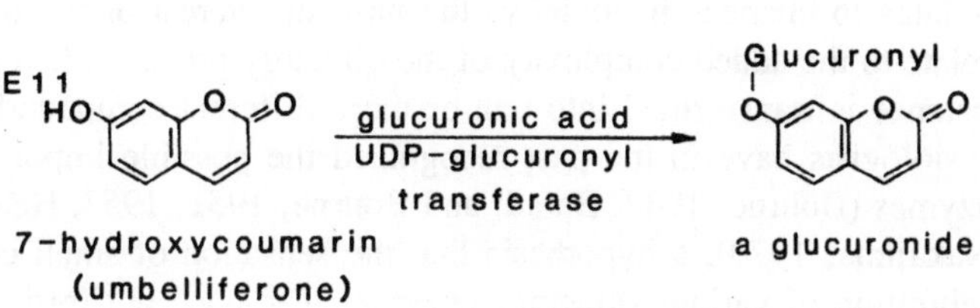

Figure 1. (*continued*)

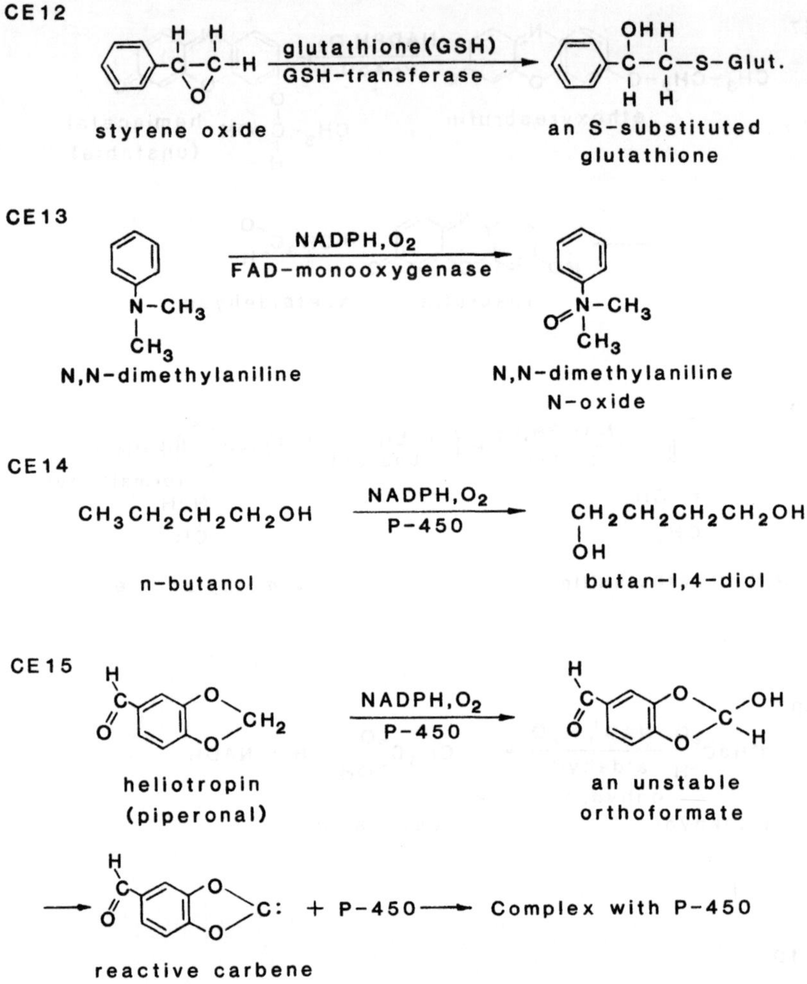

Figure 1. (continued)

nasal cavities of rats, rabbits, hamsters, and probably other species (Dahl *et al.*, 1987). In this case, metabolism results in less toxic products.

Whereas the inhalation toxicologist's interest in nasal xenobiotic metabolism of inhaled compounds relates to changes in toxicity, the nascent interest of the olfactory physiologist should relate to the added complexity of the olfactory process when a single inhaled chemical substance is transformed into one or more different compounds. To a degree, olfactory physiologists have in the past recognized the possible importance of olfactory mucosal enzymes (Bourne, 1948; Baradi and Bourne, 1951, 1953; Heberhold, 1968; Shantha and Nakajima, 1970); a hypothesis that the sensation of smell could be related to selective inhibition of various olfactory enzymes was once put forth (Kistiakowsky, 1950). However, the enzymes investigated by these scientists, with the exception of simple esterases, were enzymes involved with normal physiological processes and not enzymes that catalyze chemical changes in inhaled odorants and other xenobiotics. The possible importance of xenobiotic metabolizing enzymes in the nasal cavity was not recognized.

More recently, Getchell and Getchell (1977) commented that esterases, if present in mucus, might make important contributions to the quality of scent of inhaled esters. Getchell *et al.* (1984) suggested that certain common odorants might be catabolized by olfactory P-450 enzymes. This possibility offered some support to the suggestion by Dahl *et al.* (1982) that one role of the newly observed nasal cytochrome P-450-dependent monooxygenases might be to destroy lipophilic odorants and thus facilitate removal from the sensory area. Commenting on the same enzyme system, Price (1983) ventured that the effect of degradation of odorants such as *p*-nitroanisole (CE6, Fig. 1) would be to produce metabolites that would contribute to the characteristic odor of a compound. Indeed, preliminary evidence that odorants were transformed in the nose has been put forth (Hornung and Mozell, 1977). In addition, interesting studies describing the metabolism of the boar steroid 5α-androsterone by sow nasal epithelial enzymes have been published (Gennings *et al.*, 1974; Gower *et al.*, 1981).

It is reasonable to postulate that olfactory xenobiotic metabolizing enzymes would have an effect on the characteristic odors of compounds. However, none of the many theories of olfaction, e.g., those listed by Benignus and Prah (1982), appears able to take into account metabolite contributions to characteristic odors. Thus, these theories need to be reconsidered.

In order to provide a clear framework to begin consideration of odorant metabolites in olfactory theory, this chapter first reviews what is known regarding nasal xenobiotic metabolizing enzymes, followed by a somewhat speculative foray into the possible effects of metabolism on olfaction from the author's viewpoint. Finally, before summarizing, some suggestions are provided as to how research efforts might be directed to provide data on the role of nasal metabolism in olfaction.

2. XENOBIOTIC METABOLIZING ENZYMES IDENTIFIED IN THE NOSE AND THEIR FUNCTION

2.1. Nasal Enzymes

Xenobiotic metabolizing enzymes located in the nasal cavity include cytochromes P-450 (Table 1), flavin-containing monooxygenase (FCM) and aldehyde dehydrogenases (ADH) (Table 2), epoxide hydrolases and so-called phase 2 enzymes (Table 3), and carboxylesterases (Tables 4 and 5). In general, these enzymes are present in the nasal mucosa at concentrations comparable to levels found in liver. Nasal tissue metabolic rates for xenobiotics often exceed those of liver (e.g., Tables 6 and 7), generally far exceeding those of other extrahepatic tissues (e.g., Table 8).

In contrast to the substrate specifities exhibited by most cellular enzymes, enzymes that are chiefly responsible for the metabolism of xenobiotics often accept a wide variety of molecules as substrates. In fact, such nonspecificity is a hallmark for xenobiotic metabolizing enzymes.

Table 1. Cytochrome P-450 in Nasal Mucosa

Animal	Concn. cytochrome P-450 (nmoles P-450/mg microsomal protein)			
	Respiratory mucosa	Olfactory mucosa	Nasal mucosa (average)	Total nasal P-450 (nmoles)
Mouse[a]	—[d]	—	0.07	0.03
Syrian hamster[a]	—	—	0.46	0.61
Rat	0.04[b]	0.25[b]	0.11[a]	0.18[a]
Guinea pig[a]	—	—	0.09	0.40
Rabbit	0.25[b]	0.45[b]	0.35[a]	9.5[a]
Cynomolgus monkey[b]	0.00	0.40	0.12	0.42
Dog[a,c]	0.04	0.25	0.11	10.0

[a]Hadley and Dahl (1983).
[b]A. R. Dahl (unpublished results).
[c]Dahl *et al.* (1982).
[d]—, no data.

The activities reported usually represent maximum rates of metabolism obtained *in vitro* when the concentration of substrate is not rate limiting. Such high substrate concentrations probably occur rarely *in vivo* in the massive liver but can readily occur in nasal tissues because they are directly exposed to substrate-containing air and have relatively small masses (Hadley and Dahl, 1983).

The olfactory and respiratory epithelia of the nasal cavity often have very different metabolic capabilities. The yellow-brown olfactory epithelium covering the ethmoturbinates usually has substantially higher metabolic capabilities than does the pinkish-white

Table 2. FAD-Containing Monooxygenase, Formaldehyde Dehydrogenase, and Aldehyde Dehydrogenase Activities in Rat Nasal and Liver Tissue

Enzyme	Substrate	Tissue		
		Respiratory	Olfactory	Liver
FAD-Monooxygenase (nmoles product/mg microsomal protein/min)[a]	*N,N*-Dimethylaniline	2.51 ± 0.26	6.04 ± 1.19	4.92 ± 0.28
	Dimethylamine	—[b]	0.17	0.23
Formaldehyde dehydrogenase (nmoles product/mg protein/min)[c]	Formaldehyde	0.90	1.77	—[d]
Aldehyde dehydrogenase (nmoles product/mg protein/min)[c]	Formaldehyde	4.07	4.39	—[d]
	Acetaldehyde	129	30	—[d]

[a]McNulty and Heck (1983). Products were *N*-oxide for *N,N*-dimethylaniline; formaldehyde for dimethylamine.
[b]McNulty *et al.* (1983).
[c]Casonova-Schmitz and Heck, 1984.
[d]Data not included in cited references.

Table 3. Activities of Some Rat Nasal Tissue Phase 2 Enzymes and Epoxide Hydrolase[a]

Enzyme assay	Substrate	Specific activity[b]
Glutathione transferase	Styrene oxide	24.8 ± 1.0
UDP-Glucuronyl transferase	7-Hydroxycoumarin	20.4 ± 1.2
Epoxide hydrolase	Styrene oxide	6.4 ± 1.3

[a]From Bond (1983*a*).
[b]Units are nmole product/mg S-9 protein/min.

Table 4. Nasal Mucosal Carboxylesterase Activities[a,b]

Animal species	EGMEAC per nose[c] (μmole/min)	Substrate[d]					
		EGMEAC	EGEEAC	PGMEAC	MA	EA	BA
Mice	2.2	560 ± 23	475	52	157	369	92
Rats	7.4	456 ± 31	—	—	—	—	—
Rabbits	23.8	88 ± 12	—	—	—	—	—
Dogs	158.8	635 ± 79	—	—	—	—	—

[a]Adapted from Stott and McKenna (1985).
[b]EGMEAC, ethyleneglycol monomethyl ether acetate; EGEEAC, ethyl ether analogue; PGMEAC, propyleneglycol analogue; MA, EA, and BA, methyl, ethyl, and butyl acrylate; —, no data available.
[c]Estimate based on available S-5 protein in entire nasal mucosa (unpublished data).
[d]Units are nmoles product/mg S-5 protein/min (±SE, where noted).

Table 5. Nasal Olfactory Mucosal Carboxylesterase Activities[a,b]

Substrate	Animal species		
	Rat	Rabbit	Hamster
Amyl acetate	120 ± 5	190 ± 2	1300 ± 33
Phenyl acetate	250 ± 30	740 ± 20	1700 ± 110
β-Butyrolactone	64 ± 4	380 ± 10	260 ± 20

[a]From Dahl *et al.* (1987).
[b]In nmoles carboxylic acid formed/mg S-9 protein/min ±SE.

Table 6. Liver and Nasal Metabolism of Polycyclic Aromatics, Nitrosamines, Phenacetin, and Butanol[a]

		Tissue	
Compound	Species	Liver	Nasal
1-Nitropyrene[b]	Rat	307 ± 19	681 ± 31
Benzo(a)pyrene[b,c]	Rat	49[c]	23 ± 2[b]
Benzo(a)pyrene[d]	Hamster	—	7 ± 2
N-Nitrosodiethylamine[e]	Hamster	396 ± 45	895 ± 131
N-Nitrosodiethylamine[f]	Mouse	150 ± 25	350 ± 35
N-Nitrosopyrrolidine[f]	Rat	356 ± 128	99 ± 26
N-Nitropyrrolidine[f]	Mouse	900 ± 167	149 ± 44
N-Nitrosodibutylamine[g]	Rat	107 ± 27	343 ± 49
Phenacetin[h]	Rat	1.59 ± 0.13	10.76 ± 1.03
Phenacetin[h]	Rabbit	0.51 ± 0.08	3.37 ± 0.85
Butanol[i]	Rabbit	—	1500

[a]In pmoles metabolites/mg S-9 protein/min for NP and nasal-BaP; pmoles metabolite/mg microsomal protein/min for liver-BaP and butanol; dpm $^{14}CO_2$/mg wet tissue/hr for the nitroso compounds; pmoles CO_2/mg wet tissue/hr for phenacetin.
[b]Bond (1983*a, b*).
[c]Vaino and Hietanan (1980).
[d]Dahl *et al.* (1984).
[e]Löfberg and Tjälve (1985).
[f]Brittebo *et al.* (1981).
[g]Brittebo and Tjälve (1982).
[h]Brittebo and Ahlman (1984).
[i]Ding *et al.* (1985).

respiratory epithelium covering the maxilloturbinates. This is illustrated by the data in Table 1 and other published data (Dahl *et al.* 1982; Dahl, 1985*a–c*). A notable exception is reported for acetaldehyde dehydrogenase (Table 2).

Cell types containing nasal xenobiotic metabolizing enzymes have not been clearly identified. One form of rat P-450 and the reductase associated with P-450 activities was identified, using immunohistochemical techniques, in Bowman's glands as well as in olfactory and respiratory epithelia (Voigt *et al.*, 1985). Similar techniques showed that carbonic anhydrase, the zinc-containing enzyme that accelerates the rate of formation of CO_2 from carbonic acid, is present in high concentrations in rat olfactory cells but absent in supporting cells (Brown *et al.*, 1984). Besides its normal function with carbonic acid as a substrate, carbonic anhydrase exhibits considerable simple esterase activity (Pocker *et al.*, 1977) and may be responsible for a major part of the olfactory esterase activity.

Cytochrome P-450 is actually a family of closely related enzymes that have different substrate specificities. The individual enzymes, termed isozymes, are present in variable ratios in different tissues. The isozymes of P-450 present in nasal tissue have been partly characterized. Substrate specificities indicate that the isozymes or ratios of isozymes present in the nasal cavity are different from those, for example, in lung or liver. Antibodies to rabbit liver types LM2 and LM4 P-450 were not effective inhibitors of rabbit nasal P-450-dependent activities (Dahl *et al.*, 1983), but antibody against rat ethanol-inducible type 3a was an effective inhibitor of rat butanol oxidase (Ding *et al.*, 1985). This finding indicates that a large fraction, possibly a preponderance, of the rat nasal cytochrome P-450, is the type 3a isozyme. This isozyme is thought to metabolize nitrosamines (Yang *et al.*, 1985), and it has been reported that carcinogenic nitrosamines are rapidly metabolized by nasal tissue (Brittebo *et al.*, 1981).

2.2. Types of Metabolic Transformation

In view of the foregoing discussion of the types and locations of nasal enzymes, it might appear that foreign compound metabolism is so complex and detailed that little understanding useful to olfactory physiologists can be gained without inordinate (and perhaps begrudging) effort. In fact, simple rules can be adopted to facilitate conceptualization of the possible fate of an odorant, although prediction of actual products in the right ratios is indeed complex and, so far, not within reach.

Table 7. Rates of Formaldehyde Formation with Rat Microsomal Enzymes[a,b]

	Substrate	Nasal microsomes	Liver microsomes
Solvents	*N,N*-Dimethylaniline	3360 ± 1360	4520 ± 1360
	Hexamethylphosphoramide	3680 ± 400	1360 ± 270
	Ethyleneglycol dimethylether	290 ± 60	0
	Methyal	270 ± 90	Trace
Air pollutants	Diesel soot extract	530 ± 80	0
Essences	Dimethyl anthranilate	1660 ± 150	3890 ± 120
	p-Methoxyacetophenone	1330 ± 420	500 ± 200
Nasal decongestants	Methamphetamine	410 ± 110	630 ± 130
	Propylhexedrine	300 ± 40	400 ± 50
	Pyrilamine	300 ± 40	1780 ± 230
	Carbinoxamine	210 ± 50	2760 ± 340
Other Drugs	Cocaine	3100 ± 500	2600 ± 500
	Nicotine	500 ± 140	1150 ± 860
Dye intermediates	*p*-Cresidine[c]	780 ± 35	500 ± 20
	o-Cresidine[c]	707 ± 70	420 ± 20
	m-Cresidine[c]	100 ± 70	70 ± 20
P-450 test substrates	Benzphetamine[d]	1200 ± 60	7970 ± 750
	p-Nitroanisole[e]	570 ± 70	220 ± 20
	Aminopyrine	550	1050

[a]Dahl and Hadley (1983*a*), except where noted otherwise.
[b]In pmoles formaldehyde/mg microsomal protein/min ±SD.
[c]A. R. Dahl (unpublished data); Dahl and Hadley (1983*b*).
[d]McNulty *et al.* (1983).
[e]Hadley and Dahl (1982).

Table 8. Metabolites of Benzo(a)pyrene Produced by Tissue Homogenates from Hamsters[a]

Metabolites	Maxilloturbinates	Ethmoturbinates	Nasal membrane	Esophagus	Forestomach	Trachea	Larynx	Lung
Water soluble[b]	1900 ± 310	3900 ± 380	3900 ± 630	1300 ± 150	1800 ± 380	380 ± 27	670 ± 80	370 ± 130
Tetrols	1000 ± 20	1400 ± 170	1700 ± 210	150 ± 3	160 ± 45	2300 ± 2100	170 ± 20	200 ± 110
9,10-Diol	450 ± 50	43 ± 43	630 ± 440	0	0	220 ± 220	31 ± 31	10 ± 10
4,5- and 7,8-Diols	860 ± 430	630 ± 570	650 ± 470	49 ± 32	250 ± 180	300 ± 130	220 ± 140	210 ± 90
Quinones and oxides	790 ± 140	300 ± 240	770 ± 660	240 ± 240	120 ± 120	130 ± 130	100 ± 40	1100 ± 150
3- and 9-Phenols	760 ± 470	820 ± 800	690 ± 580	95 ± 18	200 ± 140	530 ± 120	380 ± 350	700 ± 60

[a]Dahl *et al.* (1985). Data obtained using S-9 homogenate normalized for equivalent tissue mass (pmoles/g tissue per hr ±SE).
[b]Not extracted into ethyl acetate.

Enzymes important in the metabolism of odorants, or any xenobiotic, can be classified into three broad categories:

1. *Enzymes that catalyze oxidation or reduction:* These include cytochromes P-450 and FAD-containing monooxygenases and dehydrogenases. Representative reactions catalyzed by these enzymes are illustrated by CE1, 5–9, and 13 of Fig. 1, and CE2, 3, and 5 of Fig. 2.
2. *Enzymes that catalyze hydrolysis:* These include epoxide hydrolase, esterases (lactonases are a type of esterase), aryl sulfatases, and β-glucuronidases. Reactions catalyzed by some of these enzymes are illustrated in CE2–4 and 10 of Fig. 1 and CE7 of Fig. 2.
3. *Enzymes that catalyze conjugation of two molecules:* These are illustrated by CE11 and 12 of Fig. 1 and CE1, 4, and 6 of Fig. 2.

Although more than one enzyme may be involved in metabolizing a given type of molecule, for simple molecules that have few functional groups, the possibilities are limited. Thus, an alkane such as *n*-octane (CE5, Fig. 1) is metabolized initially by oxidation to an alcohol. There are, however, a number of isomeric alcohols that can be produced, only one of which is illustrated in CE5 of Fig. 1. After an alcohol is produced, further metabolic transformation can take place. For example, the octane-2-ol illustrated in CE5 of Fig. 1 can be further oxidized to a number of isomeric diols or to a ketone (CE8, Fig. 2). Thus, the ultimate number of products can be very large. Examples of possible metabolites that may be formed from odorants in the olfactory mucosa are illustrated in Table 9. Jakoby *et al.* (1982) reviewed xenobiotic metabolism, much of which is applicable to the metabolism of odorants.

The examples presented in Table 9 show that metabolites are themselves often substrates for further enzymatic transformations. This observation raises an important question: How likely is secondary and tertiary metabolism of an odorant or other xenobiotic to contribute to the compounds present on the olfactory mucosa? Detailed answers to this question are not well in hand, but some information and tentative conclusions are given here and in Sections 2.4 and 2.5.

2.3. Capacity of Nasal Enzymes

Table 10 calculates the maximum air concentrations of three substances that could be completely metabolized in the nasal cavities of six animal species. That is, the nasal metabolic capacity would not be exceeded by inhalation of these compounds at the calculated concentrations. The range is 0.1–4.4 ppm for the P-450 substrates *p*-nitroanisole and aniline and 700–2800 ppm for ethyleneglycol monomethyl ether acetate (EGMEAC), an esterase substrate. Higher air concentrations than those shown may be required to achieve maximum metabolic rates, since not all the inhaled vapor will necessarily be available to the enzymes. The calculated values relate to measured values and represent the best available estimates at this time. An obvious implication from these calculations is that animals inhaling vapors at about the concentrations listed in Table 10 will have substantial concentrations of metabolites in their nasal cavities. The metabolite concentration may even equal or exceed the concentration of the inhaled vapor molecules.

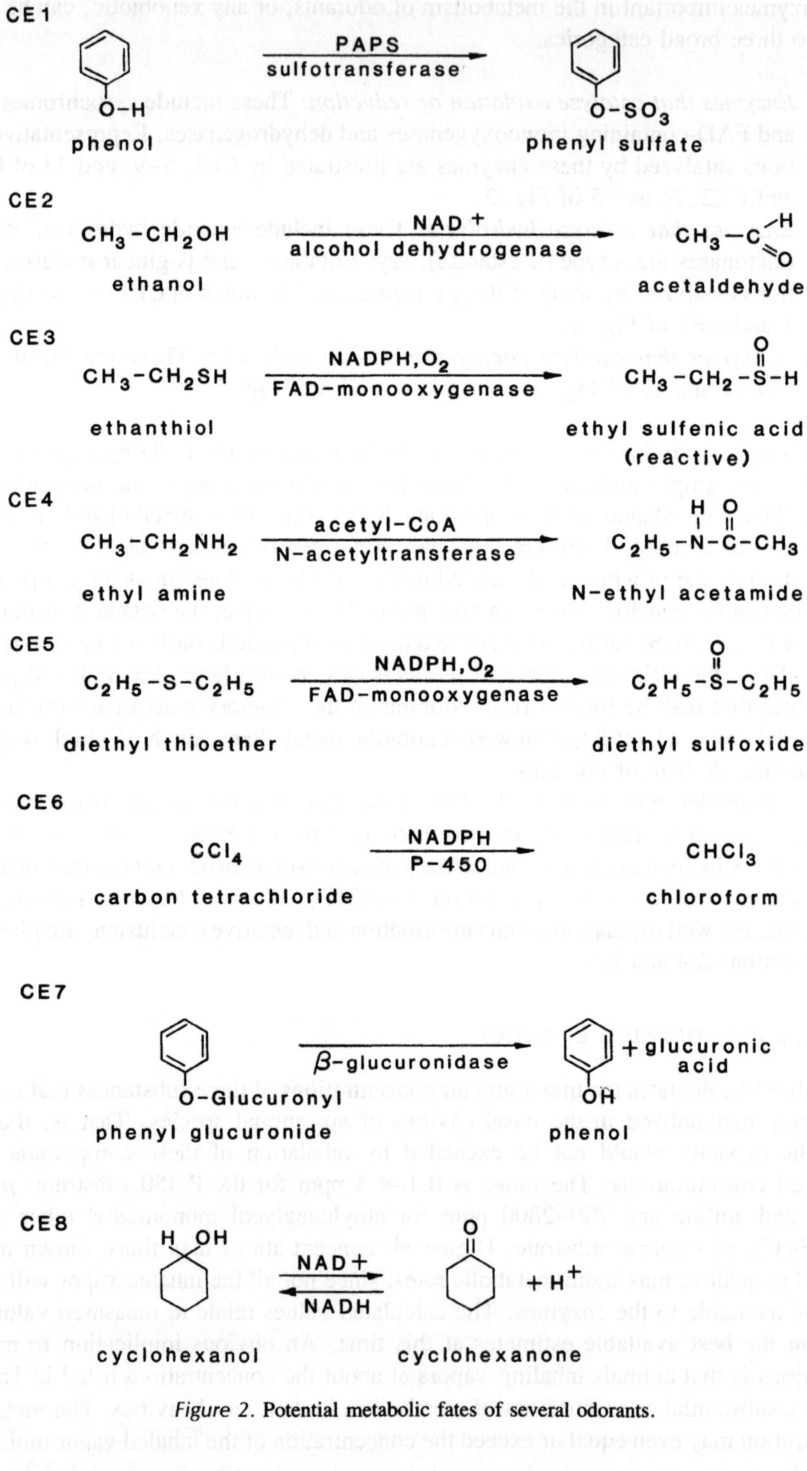

Figure 2. **Potential metabolic fates of several odorants.**

Table 9. Illustrations of Possible Metabolism of Odorants to Other Odorants

Illustration No.	Initial odorant	Metabolic Fate (Enzyme)	Metabolites
1	Alkanes	Oxidation (P-450)	1°, 2°, or 3° alcohols
2	Alkenes	Oxidation (P-450)	Epoxides
3	Alkynes (terminal)	Oxidation (P-450)	Carboxylic acids?
4	Epoxides	Hydrolysis (epoxide hydrolase)	Diols
5	Alcohols	Oxidation (dehydrogenase)	Aldehydes, ketones
6	Aldehydes	Oxidation (dehydrogenase)	Carboxylic acids
7	Aryl esters	Hydrolysis (aryl esterases)	Phenols and carboxylic acids
8	Alkyl esters	Hydrolysis (carboxylesterase)	Alcohols and carboxylic acids
9	Thioethers	Oxidation (FAD mono oxygenase)	Sulfoxides
10	Mercaptans	Oxidation (FAD monooxygenase)	Mixed disulfides
11	Aliphatic amines (1°, 2°)	*N*-Hydroxylation (FAD monooxygenase)	Hydroxyamines
12	Aliphatic amines (3°)	*N*-Oxidation (FAD monooxygenase)	*N*-Oxides
13	Aromatic amines (1°)	Conjugation (*N*-acetyltransferase)	Acetamides
14	Aliphatic amines (2°, 3°)	α-Carbon oxidation (P-450)	1° or 2° amines and aldehydes
15	Aromatic amines (1°)	*N*-Oxidation (P-450, FAD-monooxygenase)	Nitroso or nitro aromatics
16	Aliphatic ethers (P-450)	α-Carbon oxidation	Alcohols and aldehydes
17	Halogenated alkanes ($C_nX_mH_{2n+2-m}$)	Reduction (P-450)	$C_nX_{m-1}H_{2n+3-m}$
18	Aromatic hydrocarbons	Oxidation (P-450)	Epoxides
19	Ketones	Reduction (?)	2° alcohols
20	Phenols	Conjugation (various)	Sulfates, glucuronidates
21	Aliphatic phosphines	*P*-Oxidation (FAD monooxygenase?)	Phosphinites? (ultimately phosphates?)
22	Carboxylic acids	Conjugation (various)	Glucuronates, other soluble nonvolatile conjugates
23	Cyanide ion	Oxidation, sulfation (Rhodanese)	Thiocyanate

Raising the concentration of vapor above the maximum will result in an increased ratio of vapor to metabolite.

In practical terms, the calculations indicate that for a material such as *p*-nitroanisole (see CE6, Fig. 1) inhaled at ~1 ppm, the nasal cavity could contain about equal amounts of *p*-nitroanisole and its metabolites, *p*-nitrophenol and formaldehyde. Whether these metabolites will actually appear in the mucus that bathes the olfactory cilia, and thus affect the olfactory sense, is a question for which a definitive answer is not yet available. However, some relevant data are presented in Section 2.5.

Table 10. Approximate Air Concentrations That Could Be Completely Metabolized in the Nose

		Substrate					
		p-Nitroanisole		Aniline		EGMEAC[d]	
Animal species	$\dot{V}_E$[a]	Rate[b]	ppm_{max}[c]	Rate[b]	ppm_{max}[c]	Rate[b] ($\times 10^3$)	ppm_{max}[c] ($\times 10^3$)
Dog	2	23.1	0.3	22.8	0.3	158.8	2.0
Rabbit	0.8	47.1	1.5	16.6	0.5	23.8	0.7
Guinea pig	0.15	4.6	0.8	3.0	0.5	—	—
Rat	0.2	0.9	0.1	1.0	0.1	7.4	0.9
Syrian hamster	0.05	8.7	4.4	5.6	2.8	—	—
Mouse	0.02	0.3	0.4	0.8	1.0	2.2	2.8

[a]Approximate minute volume (liters/min) from Likens and Mauderly (1979) or Muggenburg and Mauderly (1974).

[b]Rate of nasal metabolism (nmoles/min/animal). Calculated from Hadley and Dahl (1983) or Stott and McKenna (1985).

[c]Maximum air concentration that could be completely metabolized in the nose, calculated from rates and minute volumes as follows:

$$ppm_{max} = \frac{\text{rate } (\mu\text{moles/min/animal}) \times 25 \text{ (liters air/mole air)}}{\dot{V}_E \text{ (liters air/min/animal)}}$$

[d]EGMEAC is ethyleneglycol monomethyl ether acetate.

2.4. *Interactions of Two or More Compounds with Nasal Enzymes*

Enzymes of many types are inhibited by certain compounds. The inhibition may be of a simple competitive type that occurs when two molecules of different compounds compete for the same enzyme site. In this case, the relative concentrations of the two compounds and the affinity of the enzyme for one or the other will determine the extent and nature of the inhibition. Such competitive inhibition will occur with olfactory tissue enzymes whenever mixtures of compounds are inhaled.

Of greater interest to biochemical toxicologists and, perhaps, olfactory physiologists alike, is noncompetitive inhibition. In this case, the inhibiting molecule forms an essentially irreversible complex with the inhibited enzyme and, in effect, reduces the number of enzyme molecules available to catalyze chemical transformation. A well-known example from non-nasal tissues is found in the inhibition of acetylcholinesterase (AChE) at nerve junctions by phosphorus esters. This inhibiting interaction is the basis for many insecticides and the nerve gases used in warfare.

Inhibition of esterases by phosphorus esters provides one of the only published examples of nasal enzyme inhibition *in vivo* (Stott and McKenna, 1984). Rats treated with triorthocresyl phosphate (TOCP) were compared with control rats with regard to nasal absorbtion of inhaled ethyl acrylate. At ethyl acrylate air concentrations of >200 ppm, the nasal cavities of the TOCP-treated rats absorbed 0.250 μmoles/min less ethyl acetate than did those of control rats. Only about 4% of the total metabolic capability of the nasal estrases was inhibited, assuming EGMEAC and ethyl acetate metabolism rates in rat

Table 11. Inhibition of Rabbit Nasal Cytochrome P-450-Dependent HMPA N-Demethylase Activity by Methylenedioxyphenyl Inhibitors

Inhibitor	I_{50}[a] (μm)
Heliotropin	2.4
Isosafrole	4.4
Dihydrosafrole	6.2
Safrole	7.0
Piperonyl butoxide	10.6

[a] I_{50} values are the concentrations that inhibited the membrane-bound (microsomal) enzymes by 50% *in vitro*.

noses are about equal (Tables 4, 5, and 10). Higher doses of TOCP than those used could not be used to achieve more inhibition because of adverse side effects on the rats.

Results of *in vitro* tests with the commonly encountered class of compounds called methylenedioxyphenyl (MDP) compounds (e.g., heliotropin, CE15, Fig. 1) indicate that these compounds may inhibit nasal cytochrome P-450-dependent metabolic capabilities *in vivo* (Dahl, 1982; Dahl and Brezinski, 1985). MDP compounds are widely distributed in nature. Occurring as natural products in many plants, they turn up in a great variety of foods and spices, including nutmeg (safrole, myristicin), sassafras oil, camphor oil, and ocotea pretiosa oil (safrole), black pepper (safrole, piperonal, piperine, myristicin), mace, parsley, dill, laurel, parsnip, and carrots (myristicin), and sesame oil (sesamolin, sesamin) (Hodgson and Philpot, 1974). The MDP compound piperonal, also called heliotropin, is commonly used in artificial cherry and vanilla flavoring (Bedoukian, 1967). Piperonal is also one of the most potent *in vitro* nasal P-450 inhibitors (Table 11). In addition to their occurrence in food, MDP compounds are commonly used along with other compounds in insecticide aerosol formulations, e.g., piperonyl butoxide (Table 11). MDP compounds function to inhibit insect P-450 and thereby slow the rate of which the toxic components of the formulations are metabolically detoxified. If inhaled, they may inhibit mammalian nasal P-450 as well.

The inhibition of nasal xenobiotic metabolizing enzymes may have important effects on olfaction if metabolites of odorants contribute substantially to odor quality. However, except for the single case with esterase inhibition by TOCP, there appear to be no reported investigations in the area of *in vivo* inhibition of nasal enzymes.

2.5. Fate of Inhaled Materials: Are Metabolites of Odorants Present in Mucus?

There appear to be no reports of xenobiotic metabolizing enzymes in nasal mucus. However, the enzymes themselves need not necessarily be in the mucus to affect olfaction, as suggested by Getchell and Getchell (1977). Instead, it is only necessary that their metabolic products be present. Because olfactory P-450 enzymes, and probably other enzymes, are located within cells (Voight *et al.*, 1985), for metabolites to appear in mucus would require absorption and metabolism of the odorant in the epithelium, and release of the metabolites back into the mucus rather than their clearance by the circulatory system.

Apparently only the report by Dahl *et al.* (1985) describes the detailed metabolic fate of a compound placed on nasal mucosa. In this case, the substrate was not an odorant but was rather the carcinogen benzo(a)pyrene (BaP) (CE1, Fig. 1); however, the results may well be the same for certain odorants. In the case of BaP, about 50% of the material placed on the nasal mucosa was metabolized, with essentially all the metabolites appearing in the mucus. Virtually none of the metabolites or parent compound was absorbed into the blood.

While the results for BaP may parallel those for some odorants, it is by no means clear that all odorants or other xenobiotics will follow the same pattern. Dahl and Bechtold (1985) and Anik *et al.* (1983) suggested that clearance of substances from the nose into the blood is highly variable, depending on the particular compound considered. Thus, ergotamine tartrate (Hussain *et al.*, 1984), insulin (Pontiroli *et al.*, 1982), and propranolol (Hussain *et al.*, 1980) tend to clear from the nose into the blood. Progesterone cleared from rat noses into blood quickly, and without undergoing metabolism (Hussain *et al.*, 1981), but nafarelin acetate seemed to clear poorly, perhaps with metabolism (Anik *et al.*, 1983). By contrast, very little luteinizing hormone-releasing hormone (LH-RH) (Fink *et al.*, 1974), bisulfate (Dahl *et al.*, 1983), the iron atom on ferrocene (Dahl and Briner, 1980), ruthenium introduced as ruthenium tetroxide (Snipes, 1981), and BaP (Dahl *et al.*, 1985) cleared from the nose into the blood.

In the absence of a systematic study of the disposition of nasally instilled materials, the fate of inhaled xenobiotics, including odorants, cannot be accurately predicted. The best that can be said at this time is that some odorants probably share a fate analogous to that of BaP, and their metabolites will appear in the mucus. Others probably clear mainly by the blood.

3. POSSIBLE EFFECTS OF NASAL METABOLISM ON OLFACTION

Metabolism of odorants in the nasal cavity may include the following effects on olfaction:

1. *Conversion of a nonodorant into one or more odorants:* An example might be found, for instance, in a reaction such as that illustrated in CE7 of Fig. 2. The enzyme β-glucuronidase cleaves the bulky highly water-soluble glucuronyl group from glucuronidated alcohols. If the alcohol is an odorant, this would result in production of an odorant from a nonodorant, since highly water-soluble glucuronides are unlikely to be odorants. However, this effect is hypothetical, since the presence of β-glucuronidase in nasal tissue has not been demonstrated.
2. *Conversion of odorants to nonodorants:* This effect may occur by reactions according to CE11, 12, and 13 of Fig. 1, CE1 of Fig. 2, or possibly illustration number 22 of Table 9. All involve conjugation of odorants with large water-

soluble molecules. The resultant bulky soluble molecules will probably be nonodorants.

3. *Transformation of odorants to other odorants:* There are many examples in Figs. 1 and 2 and Table 9.
4. *Transformation of a lipophilic compound into a more water-soluble one:* Similarly, this effect is illustrated by many examples. In fact, most xenobiotic metabolism results in metabolites that are more water soluble than the parent compound.
5. *Inhibition of the metabolizing enzymes:* The last anticipated effect of nasal metabolism, which may alter effects 1–4, is illustrated in by CE15 of Fig. 1 with heliotropin as substrate. Although *in vivo* inhibition of nasal P-450 by heliotropin and similar compounds has not been demonstrated, *in vitro* studies indicate that such inhibition might occur.

No data appear to be available to support any of the possible effects listed. However, these hypotheses are amenable to experimental testing. Effects 1–4 would be difficult to demonstrate were it not for the possibility of effect 5. Thus, if the enzymes responsible for metabolism could be effectively inhibited *in vivo,* clear-cut demonstrations of effects 1–4 might be possible.

4. RESEARCH NEEDED TO RELATE NASAL METABOLISM OF ODORANTS TO OLFACTION

Neither toxicologists nor olfactory physiologists should be content with the amount of data available describing the metabolic capability of the olfactory mucosa. The data available provide tantalizing insights into the possible fates of inhaled materials, including odorants, but definitive research with specific compounds is scarce and there is certainly not enough to provide the framework for an overall description encompassing various classes of compounds. Added to the shortage of data for specific compounds is the fact that those that have been obtained were with a variety of animal species. Based on these considerations, the first priority for research would seem to be to obtain more hard data regarding the detailed fate of compounds deposited on the olfactory epithelum.

Although much has been learned about the types of enzymes present in the nasal cavity and despite the interspecies comparisons that have been made, Fig. 2 shows some important enzyme systems that have not been looked for in the nasal mucosa. Moreover, of those identified in the nasal cavity, only cytochromes P-450 have been examined in detail including many substrates and animal species. Thus, a second area of research in need of expansion is that of nasal tissue enzymology.

Finally, some initial research by olfactory physiologists would seem to be in order. Comparisons of olfactory neuron stimulation with mixtures of nonodorants and odorants (e.g., that might compete for the same enzyme systems) might yield illuminating data. Thus, the third area of research to be initiated is an attempt to link metabolism of xenobiotics with effects on olfaction in direct experiments.

5. SUMMARY AND CONCLUSIONS

The olfactory mucosa is highly active with regard to the capacity to metabolize inhaled xenobiotics, including odorants. Demonstrated nasal enzymes include cytochromes P-450, FAD-containing monooxygenases, aldehyde dehydrogenases, esterases, epoxide hydrolases, and conjugating enzymes. A number of enzymatic transformations that metabolize odorants have been shown to be catalyzed by olfactory tissue. *In vivo* experiments indicate that the products of such metabolism can appear in the mucus and may therefore be detected as odorants. This was demonstrated in the Syrian hamster using the nonodorant, benzo(a)pyrene, as substrate. However, the possible role of such metabolites in contributing to odor quality is not addressed by current theories of olfaction.

It seems possible, perhaps even likely, that an olfactory theory that satisfactorily relates chemical structure to odor quality and intensity will need to encompass olfactory metabolic capability toward odorants. The data to date are insufficient to permit adequate predictions of the fate of inhaled xenobiotics, including odorants. However, techniques are available to test reasonable hypotheses regarding the effects of metabolism of odorants on olfaction, and such data could be obtained.

ACKNOWLEDGMENTS. This research was supported by the U. S. Department of Energy under Contract DE-AC04-76EV01013 in facilities fully accredited by the American Association for the Accredidation of Laboratory Animal Care.

REFERENCES

Anik, S. T., McRae, G., Nerenberg, C., Worden, A., Foreman, J., Hwang, J. Y., Kushinky, S., Jones, R. E., and Vickery, B., 1983, Nasal absorption of nafarelin acetate, the decapeptide [D-Nal(2)^6]-LHRH, in Rhesus monkeys. I., *J. Pharmacol. Sci.* **73:** 684–685.

Baradi, A. F., and Bourne, G. H., 1951, Localization of gustatory and olfactory enzymes in the rabbit, and the problems of taste and smell, *Nature (Lond.)* **168:** 977–979.

Baradi, A. F., and Bourne, G. H., 1953, Gustatory and olfactory epithelia, *Intl. Rev. Cytol.* **11:** 289–330.

Bedoukian, P. Z., 1967, *Perfumery and Flavoring Synthetics,* 2nd ed., Elsevier, New York.

Benignus, V. A., and Prah, J. D., 1982, Olfaction: Anatomy, physiology and behavior, *Environ. Health Perspect.* **44:** 15–21.

Bond, J. A., 1983*a,* Some biotransformation enzymes responsible for polycyclic aromatic hydrocarbon metabolism in rat nasal turbinates; effects on enzyme activities of *in vitro* modifiers and intraperitoneal and inhalation exposure of rats to inducing agents, *Cancer Res.* **43:** 4804–4811.

Bond, J. A., 1983*b,* Bioactivation and biotransformation of l-nitropyrene in liver, lung and nasal tissue of rats, *Mutat. Res.* **124:** 315–324.

Bourne, G. H., 1948, Alkaline phosphatase in taste buds and nasal mucosa, *Nature (Lond.)* **161:** 445–446.

Brittebo, E. B., and Ahlman, M., 1984, Metabolism of a nasal carcinogen, phenacetin, in the mucosa of the upper respiratory tract, *Chem. Biol. Interact.* **50:** 233–245.

Brittebo, E. B., and Tjälve, H., 1982, Tissue specificity of N-nitrosodibutylamine metabolism in Sprague-Dawley rats, *Chem. Biol. Interact.* **38:** 231–247.

Brittebo, E., Löfberg, B., and Tjälve, H., 1981, Extrahepatic sites of metabolism of N-nitrosopyrrolidine in mice and rats, *Xenobiotica* **11:** 619–625.

Brown, D., Garcia-Segura, L.-M., and Orci, L., 1984, Carbonic anhydrase is present in olfactory receptor cells, *Histochemistry* **80:** 307–309.

Casanova-Schmitz, M., David, R. M., and Heck, H. D., 1984, Oxidation of formaldehyde and acetaldehyde by

NAD+-dependent dehydrogenases in rat nasal mucosal homogenates, *Biochem. Pharmacol.* **33:** 1137–1142.

Dahl, A. R., 1982, The inhibition of rat nasal cytochrome P-450-dependent monooxygenase by the essence heliotropin (piperonal), *Drug Metab. Dispos.* **10:** 553–554.

Dahl, A. R., 1985*a*, Selective activation of carcinogens and other toxicants by nasal mucosae, in: *Proceedings of the Ninth IUPHAR Congress of Pharmacology,* Vol. 1 (W. Paton, J. Mitchell, and P. Turner, eds.), pp. 203–208, Macmillan, London.

Dahl, A. R., 1985*b*, Activation of carcinogens and other xenobiotics by nasal cytochromes P-450, in: *Microsomes and Drug Oxidations* (A. R. Boobis, J. Caldwell, F. deMatteis, and C. R. Elcombe, eds.), pp. 299–309, Taylor and Francis, Philadelphia.

Dahl, A. R., 1985*c*, Possible consequences of cytochrome P-450-dependent monooxygenases in nasal tissues, in: *Toxicology of the Nose* (C. S. Barrow, ed.), pp. 263–271, Chemical Industry Institute for Toxicology, Research Triangle Park, NC.

Dahl, A. R., and Bechtold, W. E., 1985, Deposition and clearance of a water-reactive vapor, methylphosphonic difluoride (difluoro), inhaled by rats, *Toxicol. Appl. Pharmacol.* **81:** 58–66.

Dahl, A. R., and Brezinski, D. A., 1985, The inhibition of rabbit nasal and hepatic cytochrome P-450-dependent hexamethylphosphoramidc (HMPA) N-demethylase by methylenedioxyphenyl compounds, *Biochem. Pharmacol.* **34:** 632–636.

Dahl, A. R., and Briner, J. T., 1980, Biological fate of a representative lipophilic metal compound (ferrocene) deposited by inhalation in the respiratory tract of rats, *Toxicol. Appl. Pharmacol.* **56:** 232–239.

Dahl, A. R., and Hadley, W. M., 1983, Formaldehyde production promoted by rat nasal cytochrome P-450-dependent monooxygenases with nasal decongestants, essences, solvents, air polluttans, nicotine and cocaine as substrates, *Toxicol. Pharmacol.* **67:** 200–205.

Dahl, A. R., and Hadley, W. M., (1983b), The relationship between nasal cancer and formaldehyde production by the action of nasal cytochrome P-450 dependent monooxygenase, *Toxicol. Letts. Supplement 1.* **18:** 137.

Dahl, A. R., Hadley, W. M., Hahn, F. F., Benson, J. M., and McClellan, R. O., 1982, Cytochrome P-450-dependent monooxygenases in olfactory epithelium of dogs; possible role in tumorigenicity, *Science* **216:**57–59.

Dahl, A. R., Hall, L., and Hadley, W. M., 1983, Characterization and partial purification of rabbit nasal cytochrome P-450, *Toxicologist* **3:**91.

Dahl, A. R., Coslett, D. S., Bond, J. A., and Hesseltine, G. R., 1985, Exposure of the hamster alimentary tract to benzo(a)pyrene metabolites produced in the nose, *J. Natl. Cancer Inst.* **75:** 135–139.

Dahl, A. R., Miller, S. C., and Petridou-Fischer, J., 1987, Carboxylesterases in the respiratory tracts of rabbits, rats and Syrian hamsters, *Toxicol. Lett.* **36:** 129–136.

Ding, X., Koop, D. R., and Coon, M. J., 1985, Extrahepatic identification of ethanol-inducible rabbit cytochrome P-450 isozyme 3a, *Fed. Proc.* **44:** 1449.

Fink, G., Gennser, G., Liedholm, P., Thorell, J., and Mulder, J., 1974, Comparison of plasma levels of luteinizing hormone releasing hormone in men after intravenous or intranasal administration, *J. Endocrinol.* **63:** 351–360.

Gennings, J. N., Gower, D. B., and Bannister, L. H., 1974, Studies on the metabolisms of the odiferous ketones, 5α-androst-16-n-3-one and 4,16-androstadien-3-one by the nasal epithelium of the mature and immature sow, *Biochim. Biophys. Acta* **369:** 294–303.

Getchell, T. V., and Getchell, M. L., 1977, Early events in vertebrate olfaction, *Chem. Senses Flavor* **2:**313–326.

Getchell, T. V., Margolis, F. L., and Getchell, M. L., 1984, Perireceptor and receptor events in vertebrate olfaction, *Prog. Neurobiol.* **23:** 317–345.

Gower, D. B., Hancock, M. R., and Bannister, L. H., 1981, Biochemical studies on the boar pheromones, 5α-androst-16-en-3-one and 5α-androst-16-en-3α-01, and their metabolism by olfactory tissue, in: *Biochemistry of Taste and Olfaction* (R. H. Cagan and M. R. Kare, eds.), pp. 8–28, Academic, New York.

Hadley, W. M., and Dahl, A. R., 1982, A cytochrome P-450-dependent monooxygenase activity in rat nasal epithelial membranes, *Toxicol. Lett.* **10:** 417–422.

Hadley, W. M., and Dahl, A. R., 1983, Cytochrome P-450-dependent monooxygenase activity in nasal membranes of six species, *Drug Metab. Dispos.* **11:** 275–276.

Heberhold, C., 1968, Vergleichende histochemische Untersuchungen am peripheren Riechorgan von Säugetieren und Fischen, *Arch. Klin. Ohren, Nasen Kehlkopfheilkd.* **190:** 166–182.

Hodgson, E., and Philpot, R. M., 1974, Interaction of methylenedioxyphenyl (1,3-benzodioxole) compounds with enzymes and their effects on mammals, *Drug Metab. Rev.* **3:** 231–301.

Hornung, D. E., and Mozell, M. M., 1977, Preliminary data suggesting alteration of odorant molecules by interaction with receptors, in: *Olfaction and Taste* (J. Le Magnen and P. MacLeod, eds.), p. 63, Information Retrieval, London.

Hussain, A., Foster, T., Hirai, S., Kashihara, T., Batenhoist, R., and Jones, M., 1980, Nasal absorption of propranolol in humans, *J. Pharmacol. Sci.* **69:** 1240–1242.

Hussain, A. A., Hirai, S., and Banarshi, R. C., 1981, Nasal absorption of natural contraceptive steroids in rats—Progesterone absorption, *J. Pharmacol. Sci.* **70:** 461–467.

Jakoby, W. B., Bend, J. R., and Caldwell, J. (eds.), 1982, *Metabolic Basis of Detoxification: Metabolism of Functional Groups,* Academic, New York.

Kistiakowsky, G. B., 1950, On the theory of odors, *Science* **112:** 154–155.

Likens, S. A., and Mauderly, J. L., 1979, Repiratory measurements in small laboratory mammals: A literature review, Lovelace Inhalation Toxicology Research Institute, Report No. LF-68, National Technical Information Service, U.S. Department of Commerce, Springfield, Virginia.

Löfberg, B., and Tjälve, H., 1984, The disposition and metabolism of N-nitrosodiethylamine in adult, infant and faetal tissue of the Syrian golden hamster, *Acta Pharmacol. Toxicol.* **54:** 104–114.

Hussan, A., Kimurd, R., Huang, C-H., and Mustafa, R., 1984, Nasal absorption of ergotamine tartrate in rats, *Int. Journal of Pharmaceutics* **21:** 289–294.

McNulty, M. J., and Heck, H. D., 1983, Disposition and pharmacokinetics of inhaled dimethylamine in the Fischer-344 rat, *Drug Metab. Dispos.* **11:** 417–420.

McNulty, M. J., Casanova-Schmitz, M., and Heck, H. D., 1983, Metabolism of dimethylamine in the nasal mucosa of the Fischer-344 rat, *Drug Metab. Dispos.* **11:** 421–425.

Muggenburg, B. A., and Mauderly, J. L., 1974, Cardiopulmonary function of awake, sedated, and anesthetized Beagle dogs, *J. Appl. Physiol.* **37:** 152–157.

Pocker, Y., Bjorkquist, L., and Bjorkquist, D. W., 1977, Zinc and cobalt bovine carbonic anhydrases. Comparative studies and esterase activity, *Biochemistry* **16:** 3967–3973.

Pontiroli, A. E., Aberetto, M., Secchi, A., Dossi, G., Bosi, I., and Pozza, G., 1982, Insulin given intranasally induces hypoglycemia in normal and diabetic subjects, *Br. Med. J.* [*Clin. Res.*] **284:** 303–306.

Price, S., 1983, Mechanisms of stimulation of olfactory neurons: An essay, *Chem. Senses* **8:** 341–354.

Shantha, T. R., and Nakajima, Y., 1970, Histological and histochemical studies on the Rhesus monkey (*Macaca mulatta*) olfactory mucosa, *Z. Zellforsch.* **103:** 291–319.

Snipes, M. B., 1981, Metabolism and dosimetry of ^{106}Ru inhaled as $^{106}RuO_4$ by Beagle dogs, *Health Phys.* **41:** 303–317.

Stott, W. T., and McKenna, M. J., 1984, The comparative absorbtion and excretion of chemical vapors by the upper, lower and intact respiratory tract of rats, *Fundam. Appl. Toxicol.* **4:** 594–602.

Stott, W. T., and McKenna, M. J., 1985, Hydrolysis of several glycol ether acetates and acrylate esters by nasal mucosal carboxylesterase *in vitro, Fundam. Appl. Toxicol.* **5:** 399–404.

Vaino, H., and Hietanen, E., 1980, Role of extrahepatic metabolism in drug disposition and toxicity, in: *Concepts in Drug Metabolism,* Part A (P. Jenner and B. Testa, eds.), pp. 251–284, Dekker, New York.

Voigt, J. M., Guengerich, F. P., and Baron, J., 1985, Localization of xenobiotic-metabolizing enzymes in nasal tissues of untreated and 3-methylcholanthrene (MC) and Aroclor 1254 (A) pretreated rats, *Toxicologist* **5:** 162.

Yang, C. S., Koop, D. R., Wang, T., and Coon, M. J., 1985, Immunochemical studies on the metabolism of nitrosamines by ethanol-inducible cytochrome P-450, *Biochem. Biophys. Res. Commun.* **128:** 1007–1013.

4

Odorant and Autonomic Regulation of Secretion in the Olfactory Mucosa

Marilyn L. Getchell, Barbara Zielinski, and
Thomas V. Getchell

1. INTRODUCTION

The mechanisms associated with sensory transduction and subsequent membrane events in sensory reception are influenced by activity in ancillary cells within the sense organs. It is well documented that sheath cells in the pacinian corpuscle, Müller cells in the retina, and sustentacular cells in the organ of Corti influence the corresponding sensory receptor cells in somatosensory, visual, and auditory systems. The influence of perireceptor events in chemoreception (i.e., the olfactory, vomeronasal, and gustatory systems) is an emerging area of intense research activity.

The purpose of this chapter is to review the experimental evidence that (1) identifies the extrinsic innervation of the olfactory epithelium from a neuroanatomical and immunocytochemical perspective, (2) characterizes the secretory activity in the olfactory mucosa from a neuropharmacological perspective, and (3) proposes cellular mechanisms by which secretory and olfactory receptor cell activities are integrated.

2. ORGANIZATION AND CHARACTERIZATION OF THE CELLS IN THE OLFACTORY MUCOSA

The olfactory mucosa of vertebrates (Fig. 1A) consists of two layers, a superficial sensory neurepithelium (OE) and a deeper lamina propria (LP). The former is essentially an avascular pseudostratified columnar epithelium containing three morphologically identified cell types: sustentacular cells (SC), olfactory receptor neurons (ORC), and basal cells (BC). The nuclei of these cell types form three rather distinct layers. The lamina propria contains several elements including: multicellular glands (BG) (Fig. 1F), blood vessels (BV Fig. 1E), bundles of olfactory receptor axons (NB, Fig. 1G), and an occasional bundle of myelinated fibers (MN, Fig. 1H) in addition to loose connective tissue.

Marilyn L. Getchell, Barbara Zielinski, and Thomas V. Getchell • Department of Anatomy and Cell Biology, Wayne State University School of Medicine, Detroit, Michigan 48201.

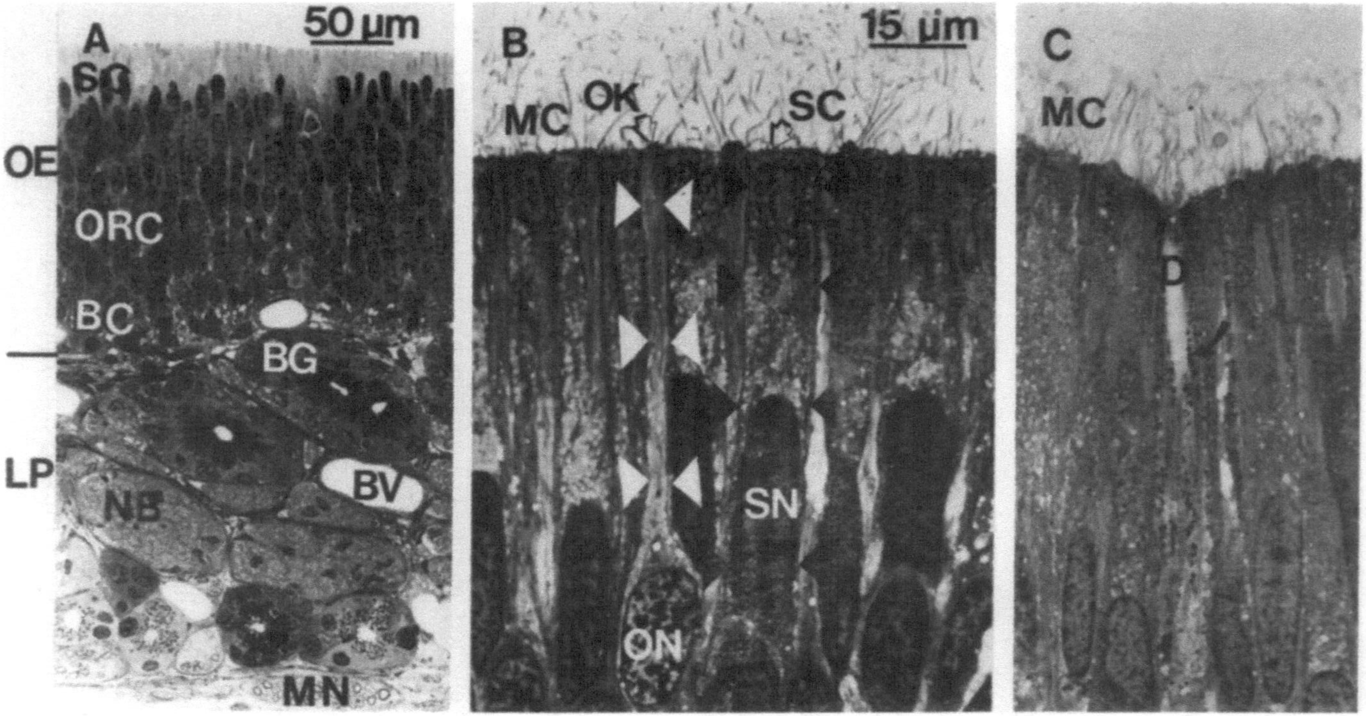

Figure 1. Morphology of the olfactory mucosa in the adult salamander. Calibration bars in (A) and (B); bar in (B) is for B–H. (A) Organization of olfactory mucosa. Sensory neurepithelium (OE) contains olfactory receptor neurons (ORC), sustentacular cells (SC), and basal cells (BC), whose nuclear strata are indicated at left. Underlying lamina propria (LP) contains Bowman's glands (BG), blood vessels (BV), nonmyelinated (NB) and myelinated (MN) nerve bundles, and loose connective tissue. (B) Surface of the olfactory epithelium. Olfactory receptor dendrites (white arrowheads) terminate in ciliated knobs (OK) that project above the level of microvillar sustentacular cells (SC, black arrowheads) into mucociliary complex (MC). ON, nucleus of olfactory receptor neuron; SN, nucleus of sustentacular cell. (C) Olfactory gland duct. Duct (D), lined by flattened, granule-containing cells (arrow), opens at surface of the epithelium between sustentacular cells. MC, mucociliary complex. (D) Olfactory epithelium–lamina propria interface. Basement membrane (arrows) separates foot processes of sustentacular cells and basal cells from blood vessels (BV) and glands (BG). (E) Vasculature of olfactory mucosa. Blood cells lie within blood vessel in lamina propria. (F) Bowman's gland. Pyramidal acinar cells surround central lumen (L); secretory granules (SG) occupy apical poles.

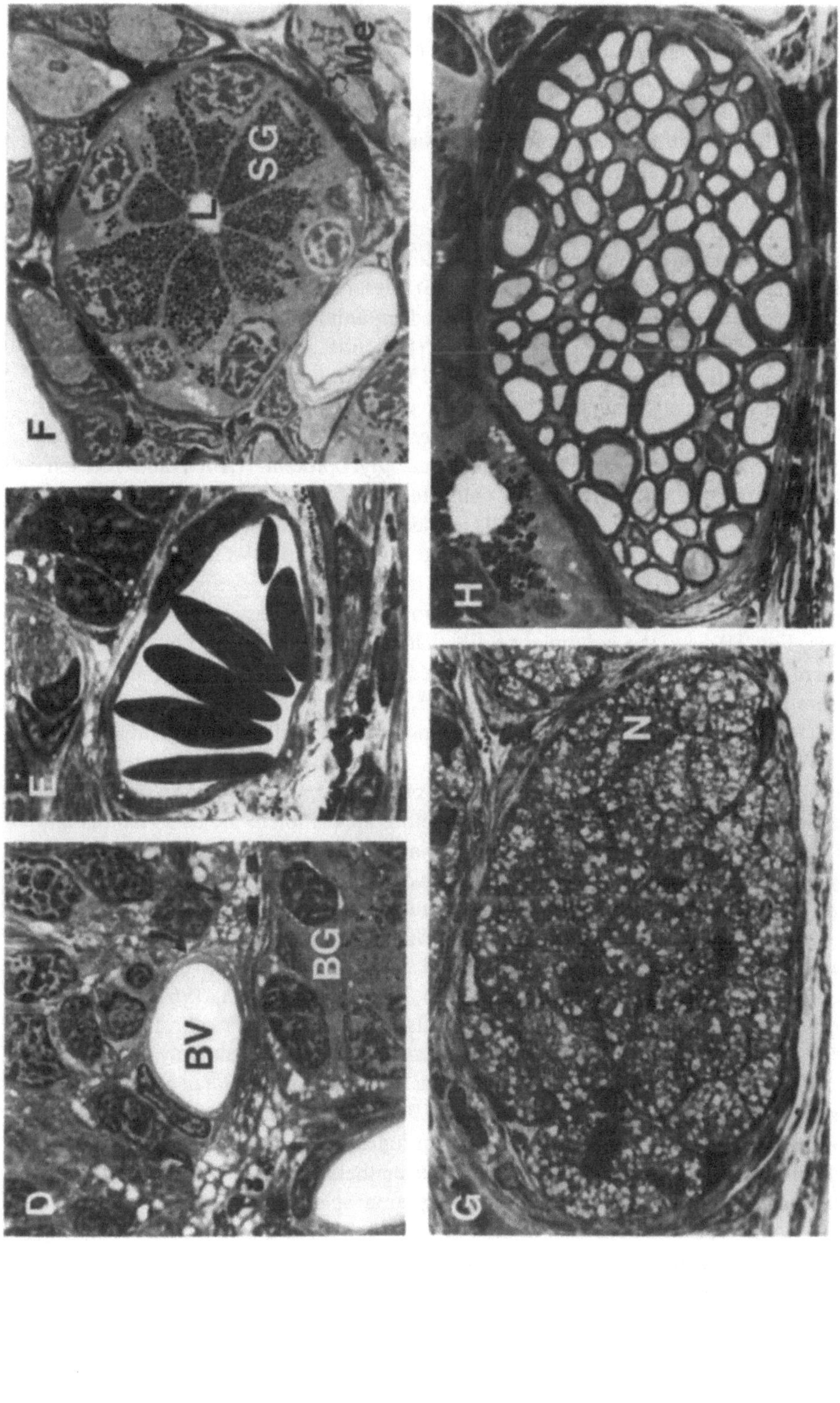

Note close proximity of blood vessels, processes of melanocytes (Me), and nonmyelinated nerve bundles. (G) Olfactory nerve bundle. Fascicles containing axons of olfactory receptor cells are enclosed by Schwann cells. N, Schwann cell nucleus. (H) Myelinated nerve bundle, found primarily at the base of mucosa.

The olfactory receptor cell (white arrowheads, Fig. 1B) is a bipolar neuron whose primary function is to detect odorants and to transmit sensory information directly to the olfactory bulb in the brain. The dendritic knob at the surface of the epithelium bears cilia of variable length and number; the axon passes without collaterals to the olfactory bulb. There is substantial evidence indicating that olfactory receptor neurons undergo continual cell renewal throughout an animal's life and are replaced through mitosis and maturation of cells in the basal region of the epithelium (Nagahara, 1940; Moulton *et al.*, 1970; Graziadei and Metcalf, 1971; Harding *et al.*, 1978; Simmons and T. V. Getchell, 1981; Cancalon, 1982; Graziadei and Monti Graziadei, 1983; Hinds *et al.*, 1984; Farbman, 1986). The molecular events (Lancet, 1986), receptor and perireceptor events associated with sensory transduction (T. V. Getchell *et al.*, 1984) and the neurophysiological properties of the receptor neuron (T. V. Getchell, 1986) have recently been reviewed.

The olfactory receptor neuron is characterized immunocytochemically by the presence of a unique olfactory marker protein (OMP) (Margolis, 1972, 1980) and by vimentin-like immunoreactive material (Schwob *et al.*, 1986). Monoclonal antibodies selective for olfactory receptor cells and their subclasses (Fujita *et al.*, 1985; Hempstead and Morgan, 1985*a,b;* Imamura *et al.*, 1985; Mori *et al.*, 1985) have been demonstrated, including one, Mab-2B8, specific for cell-surface glycoproteins (Allen and Akeson, 1985*a,b*). Also, a population of receptor cells stains positively for carbonic anhydrase (Brown *et al.*, 1984). Olfactory receptor cells in rodents stain selectively with fluorescein-conjugated concanavalin A (Con A) (Hempstead and Morgan, 1983); neurons of the olfactory system in *Xenopus* stain selectively with soybean lectin (Key and Giorgi, 1986).

The major secretory element in the epithelium *per se* is the sustentacular cell (black arrowheads, Fig. 1B). It has a columnar profile that extends from the epithelial surface to the basement membrane. Two morphological types have recently been described: type I, which resembles ependymal tanycytes, and type II, which resembles velate protoplasmic astrocytes (Rafols and T. V. Getchell, 1983). Sustentacular cell nuclei lie most superficially of the three nuclear strata in the epithelium. The distal apical membrane terminates in numerous short microvilli that project into the overlying mucus. The proximal region terminates in a basilar expansion that is found in close apposition to the basal lamina and to blood vessels and glands in the lamina propria (Rafols and T. V. Getchell, 1983). Secretory vesicles or granules and other organelles suggestive of mucous secretion have been identified in the supranuclear region of the sustentacular cells of many vertebrates, including fish (Zeiske *et al.*, 1976); amphibia (Bloom, 1954; Reese, 1965; Graziadei, 1971; Farbman and Gesteland, 1974; Okano and Takagi, 1974; M. L. Getchell *et al.*, 1984); reptiles (Graziadei, 1971; Kratzing, 1975); birds (Wesolowski, 1967; Graziadei, 1971; Müller *et al.*, 1979); and certain mammals such as mice (Frisch, 1967), cows (Gladysheva and Martynova, 1982), koalas (Kratzing, 1984), and monkeys (Saini and Breipohl, 1976). Sustentacular cells do not appear to possess the cellular machinery for secretion in either rabbits or bats (Yamamoto, 1976) or humans (Moran *et al.*, 1982; Jafek, 1983). Sustentacular cells may also perform functions such as isolation of receptor neurons from one another, glial-like activity, transepithelial transport of molecules and electrolytes, and guidance of the postmitotic migration of newly differentiated receptor neurons.

The secretory material in sustentacular cells generally appears vesicular and stains positively for the presence of acidic, sulfated and neutral mucopolysaccharides in a variety of vertebrates (see M. L. Getchell *et al.*, 1984, for review, and Gladysheva *et al.*,

1986). No further analysis of the secretory products of these cells has been reported. Neutral and compound lipids (Herberhold, 1968; Shantha and Nakajima, 1970) as well as phospholipids (Herberhold, 1968) have been identified in the supranuclear region of sustentacular cells. Monoclonal antibodies have been generated against sustentacular cells (Hempstead and Morgan, 1983*b*, 1985*a*). One in particular, designated SUS-1, reacts specifically with mature sustentacular cells, does not cross react with antibodies to OMP and is insensitive to olfactory nerve section. The presence of S-100 protein, a glial marker, and the absence of OMP has been reported in sustentacular cells (Hirsch and Margolis, 1979). The presence of enzymes of carbohydrate metabolism, such as glucose 6-phosphate dehydrogenase, succinic dehydrogenase, cytochrome oxidase, aldolase, and lactic dehydrogenase; a marker for the Golgi complex, thiamine pyrophosphatase; nonspecific esterase and lipase has been demonstrated by enzyme histochemical techniques (Baradi and Bourne, 1953; Mira, 1963; Zinnin, 1964; Herberhold, 1968; Shantha and Nakajima, 1970; Cuschieri, 1974).

The tubuloacinar Bowman's glands (BG, Fig. 1A, F) in the lamina propria are thought to produce most of the mucus that covers the olfactory epithelium. In the salamander, the animal in which the studies on neural regulation (see Section 5) have been performed, Bowman's glands appear to be typical serous glands. The acini are composed of pyramidal polarized cells containing membrane-bound secretory granules located supranuclearly. The ducts (D, Fig. 1C) of the glands, lined with flat, secretory-granule containing epithelial cells, pass through the basement membrane of the epithelium to open at the surface of the mucosa between sustentacular cells.

The monoclonal antibody SUS-1 that reacts with mature sustentacular cells also stains structures in the lamina propria tentatively identified as Bowman's glands (Hempstead and Morgan, 1983*b*). Carnosinase has been immunocytochemically localized preferentially in the acini and ducts of Bowman's glands, with the sustentacular cells displaying only a weak diffuse immunoreactivity (Farbman and Margolis, 1982; Margolis *et al.*, 1983). Nerve growth factor-like immunoreactivity has been localized in duct and occasional acinar cells of amphibian Bowman's glands (M. L. Getchell *et al.*, 1987*a*). Both S-100 protein and OMP are absent from gland cells. In addition, the acini and ducts stain preferentially with fluorescein-conjugated peanut lectin (Hempstead and Morgan, 1983*a*). In mice, substance P-like immunoreactivity has been found in nerve fibers associated with Bowman's glands; terminals tentatively identified as peptidergic have been shown to terminate in association with acinar cells (Papka and Matulionis, 1983). In amphibians, VIP-like and substance P-like immunoreactivity have been demonstrated in fibers around Bowman's gland acini (M. L. Getchell *et al.*, 1987*a*).

The epithelial surface is bathed with a mobile layer of mucus derived from olfactory and nasal glands as well as sustentacular and respiratory goblet cells. The mucus layer together with the cilia of the receptor neurons and microvilli of the sustentacular cells has been called the mucociliary complex (MC, Fig. 1B, C). This complex stains intensely with fluorescent poke weed mitogen (Hempstead and Morgan, 1983*a*).

3. EXTRINSIC INNERVATION OF THE OLFACTORY MUCOSA

Four neural systems associated with the nasal and olfactory mucosae develop embryonically from the olfactory placode (Bojsen-Møller, 1975). They are the olfactory, septal, vomeronasal and terminal nerves. In certain species, such as the rat, the fibers of the vomeronasal and terminal nerves are reported to be enveloped within the same dural sheath as the axons of olfactory receptor cells in the olfactory nerve (Bojsen-Møller, 1975). The respiratory part of the nasal mucosa (DeLong and T. V. Getchell, 1987) and the olfactory mucosa (see Fig. 8) are supplied with fibers of the trigeminal nerve. In addition, blood vessels and glands in the lamina propria of the mucosae receive autonomic innervation.

The olfactory sensory mucosa is the discrete region of the nasal epithelium that subserves the primary sense of smell. It is characterized by the presence of olfactory receptor neurons. The distal process of the bipolar neuron composed of the initial axon segment, cell body, and dendrite with apical cilia is localized in the olfactory epithelium *per se*. The proximal nonmyelinated axonal process, in association with many others, forms olfactory nerve fascicles and bundles in the lamina propria (see Fig. 1G), the olfactory nerve proper and the olfactory nerve layer in the olfactory bulb. The olfactory nerve is generally considered to consist of a homogeneous population of fibers that have similar neuroanatomical, ultrastructural, and neurophysiological characteristics.

Several observations are reported in the literature indicating that fiber types other than those of olfactory receptor neurons may be found in the olfactory mucosa and nerve. These studies have utilized primarily Golgi and silver impregnation techniques and electron microscopy. Because each of these techniques has well-documented limitations (e.g., incomplete staining of neurons and limited sample size, respectively) and because young olfactory receptor neurons continually differentiate from a population of basal cells, the general validity of the observations has been held in question. Van Gehuchten (1890) and Graziadei and Gagne (1973) have reviewed the early literature on the controversy associated with the interpretation of Golgi-impregnated free nerve endings in the olfactory mucosa. For example, Grassi and Castronovo (1889) were the first to report the occurrence of Golgi-impregnated branched nerve fibers with dilations in the olfactory mucosa that appeared to project from a larger diameter fiber in the lamina propria and to terminate at various levels in the olfactory epithelium *per se* (Fig. 2A, left). They also reported the occurrence of Golgi-impregnated nonbranched fibers with varicosities, bipolar olfactory receptor neurons, and sustentacular cells in the olfactory mucosa (Fig. 2A, right). Van Gehuchten (1890) also reported the occurrence of Golgi-impregnated nonbranched fibers without and with varicosities (Fig. 2B, left) in the olfactory mucosa that appeared to terminate in the deeper region of the epithelium in addition to bipolar olfactory receptor neurons (Fig. 2B, middle) and sustentacular cells (Fig. 2B, right). Van Gehuchten interpreted the nonbranched fibers to represent the axonal processes of incompletely impregnated receptor neurons. Branched fibers that appeared to penetrate the olfactory mucosa and appeared to be similar to those described originally by Grassi and Castronovo (1889) were described subsequently by von Brunn (1892) and Ramón y Cajal (1911). Ramón y Cajal confirmed von Brunn's observations and presented data that these fibers terminated at the surface of the olfactory epithelium (Fig. 2C, arrow). In contrast to von Brunn's (1892) and Kolmer's (1927) interpretation that these free nerve endings represented branches of the trigeminal nerve, Ramón y Cajal (1911) was uncertain of their origin and stated: "Les fibres terminent à la surface libre, mais on ignore leur origine." He called these von Brunn fibers.

Using electron microscopic techniques, Graziadei and Gagne (1973) reported that

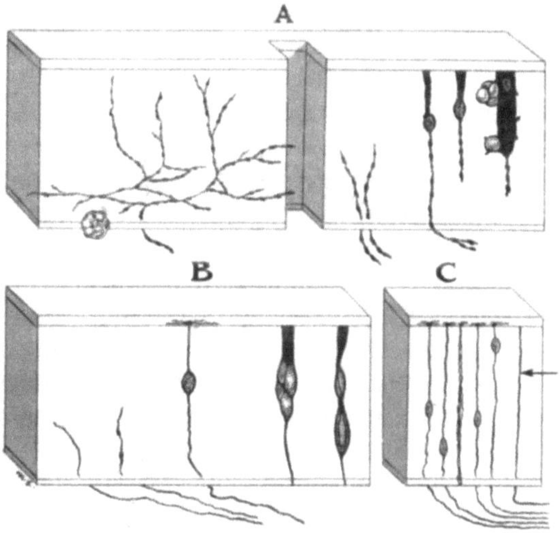

Figure 2. Extrinsic and intrinsic innervation of the olfactory mucosa. (A) The Golgi-impregnated cells in the olfactory mucosa were selected and redrawn from plate XXI of Grassi and Castronovo (1889). The multibranched fiber shown on the left arises from a single fiber in the lamina propia and penetrates the olfactory epithelium, terminating at various depths in the epithelium *per se*. The neuron is probably a primary sensory afferent fiber of the trigeminal nerve. The unbranched fibers with varicosities shown on the right probably represent the axons of olfactory receptor cells in which the more distal somas and dendrites were not Golgi impregnated. The neuron with a beaded axon is an olfactory receptor cell. The other cells are probably sustentacular cells. (B) The Golgi-impregnated cells in the olfactory mucosa were selected and redrawn from Figs. 1–14 in van Gehuchten (1890). The single fibers without and with varicosities shown on the left are presumably axons of olfactory receptor neurons; a bipolar receptor neuron is shown in the center; sustentacular cells are shown on the right. (C) The Golgi-impregnated cells in the olfactory mucosa were redrawn from Ramón y Cajal (1911). Several olfactory receptor neurons and one sustentacular cell are shown. The arrow identifies a single nonbranched fiber that projects to the epithelial surface; the fiber is presumably a primary sensory neuron of the trigeminal nerve.

both nonmyelinated and myelinated nerve fibers are present in the lamina propria and that these fibers terminate as free nerve endings in the olfactory epithelium ***per se***. Based on the association of the fiber with the Schwann cell, the fiber diameter and ultrastructural characteristics, nonmyelinated fibers that contained dark-core vesicles were postulated to be catecholaminergic and therefore to represent postganglionic sympathetic neurons. This observation suggests that sympathetic fibers innervate structures in the olfactory epithelium *per se*. The occurrence of noradrenaline and dopamine in the olfactory mucosa was demonstrated by Kawano and Margolis (1985); unilateral superior cervical ganglionectomy resulted in reduced levels of these catecholamines in the ipsilateral mucosa, suggesting that the ganglion is the source of these fibers. Although the immunocytochemical localization and neurophysiological characterization of these fibers have not been investigated extensively, the most likely candidates to receive innervation would be the apical acinar and neck cells of Bowman's glands as the acini penetrate the basal region of the epithelium to form ducts and blood vessels near the neck of the glands. This innervation would presumably serve to regulate secretion and other metabolic functions, such as electrolyte and water balance in gland cells, and blood flow in the associated vessels. The myelinated fibers lose their myelin as they penetrate the base of the olfactory epithelium and branch subsequently to terminate as free nerve endings at various depths within the epithelium. They presumably resemble the ones described by Grassi and Castronovo (1889), von Brunn (1892), and Ramón y Cajal (1911). This observation suggests that these fibers are a component of the cerebrospinal system (Graziadei and Gagne, 1973), most probably associated with the trigeminal nerve. Although the immunocytochemical identification (see Section 3), and neurophysiological characterization of these free nerve endings and fibers has not been extensively investigated, they would presumably represent a general sensory component of the trigeminal nerve. They may also participate in secretomotor and other reflex activity (Rasmussen, 1932; Alarie, 1973, Tucker, 1971). In contrast to these findings, Bojsen-Møller (1975) reported that the respiratory part of the nasal mucosa, but not the olfactory mucosa, is supplied with afferent and efferent fibers of the trigeminal nerve. In addition, based on results obtained with the Falck-Hillarp fluorescence technique, catecholaminergic nerves were not associated with glands in the respiratory or olfactory regions but with blood vessels. The sensory qualities associated with stimulation of trigeminal afferents are pain, temperature, touch, and proprioception (Tucker, 1971).

Using several neuroanatomical techniques, Bojsen-Møller (1975) demonstrated that the bundles of the olfactory nerve as they pass through the cribriform plate of the ethmoid bone in the rat contain fibers of the vomeronasal and terminal nerves in addition to the axons of olfactory receptor neurons. He also noted that catecholamine-containing fibers and cholinesterase-positive fibers were found in association with the blood vessels that traveled along with olfactory nerve bundles as they passed through the cribriform plate. These important observations suggested that the primary olfactory pathway, which was regarded generally as being purely olfactory, contained fibers associated with other cranial nerves. This study also established the base on which to investigate the composition of fibers in the primary olfactory system using contemporary immunocytochemical techniques.

The immunoreactive (ir) properties of neurons in the primary olfactory system have been investigated using standard immunocytochemical procedures for olfactory marker protein (OMP), calcitonin gene-related peptide (CGRP), luteinizing hormone-releasing hormone (LHRH), acetylcholinesterase (AChE), vasoactive intestinal peptide (VIP), and substance P in a variety of vertebrate species. The distribution of these immunoreactive fibers is shown in Table 1. Mature olfactory receptor neurons are characterized by the presence of a low-molecular-weight protein called olfactory marker protein (Margolis, 1972, 1980). The immunoreactivity of the protein is distributed throughout the dendrite, soma, and axon of the bipolar neuron. In addition, the primary sensory neurons of the vomeronasal nerve are also OMP positive (Raisman, 1985). CGRP-ir fibers have also been reported to occur in the olfactory nerve proper and among fibers in the superficial nerve layer that project into the glomeruli of the olfactory bulb (Rosenfeld *et al.*, 1983). Although their peripheral distribution has not been reported, the authors suggest that because of their localization and central glomerular projection, they represent a subset of primary olfactory fibers.

Recently, LHRH-ir fibers, some with varicosities, have been found in the olfactory mucosa and nerve (Wirsig and T. V. Getchell, 1986). It is unlikely that these fibers represent a subset of olfactory receptor neurons because (1) the LHRH-ir soma of the fiber is found in the olfactory nerve and not in the epithelium *per se*; (2) the soma is approximately 35 × 12 μm in size, much larger than the receptor soma; and (3) OMP-positive somas with these dimensions were not observed in the olfactory nerve in parallel studies. In addition, AChE histochemistry demonstrated the presence of AChE-positive fibers and somas in the olfactory nerve proper but not the epithelium. Because of the occurrence of neurons with similar LHRH-ir- and AChE-staining properties in the terminal nerve and its central projection in lower vertebrates, the results suggest that these neurons represent

Table 1. Localization of Immunoreactivity in Neural Elements of the Primary Olfactory System[a]

	Olfactory mucosa				
	Epithelium	Lamina propria	Olfactory nerve	Species	Reference
OMP	Soma, dendrites	Fibers	Fibers	Mouse, rat	Margolis (1972) Farbman (1986) Raisman (1985)
LHRH	Fibers	Fibers	Soma, fibers	Bullfrog, salamander	Wirsig and T. V. Getchell (1986)
AChE	—	—	Soma, fibers	Bullfrog, salamander	Wirsig and Getchell (1986)
CGRP	—	—	Fibers	Rat	Rosenfeld *et al.* (1983)
	Fibers	Fibers	Fibers	Frog, salamander	M. L. Getchell *et al.* (1987*a*)
SP	—	Fibers	—	Mouse	Papka and Matulionis (1983)
	Fibers	Fibers	—	Frog	Bouvet *et al.* (1987)
	Fibers	Fibers	Fibers	Frog, salamander	M. L. Getchell *et al.* (1987*a*)
VIP	Fibers	Fibers	—	Bullfrog, grass frog, salamander	M. L. Getchell *et al.* (1987*a*)

[a]OMP, olfactory marker protein; LHRH, luteinizing hormone releasing hormone; AChE, acetylcholinesterase; CGRP, calcitonin gene-related peptide; SP, substance P; VIP, vasoactive intestinal peptide.

fibers of the terminal nerve that comingle with olfactory receptor neurons in the peripheral olfactory system.

Substance P-ir fibers were observed in the lamina propria of the olfactory mucosa of mice in close association with Bowman's glands and blood vessels; substance P-ir fibers were not found in the olfactory epithelium or the olfactory nerve *per se* (Papka and Matulionis, 1983). In salamanders and frogs, varicose substance P-ir fibers were observed both in association with Bowman's glands and passing through the olfactory epithelium to the mucosal surface (M. L. Getchell *et al.*, 1987*a*). By analogy with the similar immunoreactive properties of sensory neurons in the trigeminal nerve and dorsal root ganglia, these workers suggest that the substance P-ir fibers are likely to be primary sensory afferents of the trigeminal nerve.

Finally, VIP-ir fibers were identified both in association with Bowman's glands and blood vessels in the lamina propria and extending through the olfactory epithelium to the mucosal surface. Their origin has not been determined. However, the coexistence of VIP and ACh in autonomic fibers innervating salivary glands (Lundberg *et al.*, 1981) and the demonstration of cholinergic responsivity of Bowman's glands (M. L. Getchell and T. V. Getchell, 1984*b*) suggest the possibility that VIP exists in parasympathetic fibers innervating olfactory gland and possibly duct cells.

In summary, there is a growing body of evidence derived from morphological and immunohistochemical studies indicating that the primary olfactory system is not a homo-

Table 2. Characterization of Elements in and Associated with the Olfactory Nerve

Neural elements
Olfactory receptor neurons
OMP+
Vomeronasal fibers
OMP+
Terminal nerve neurons
LHRH+
AChE+
Olfactory fibers
OMP+
CGRP+
SP+
Neuronal elements associated with blood vessels and glands
Autonomic fibers
Catecholamine+
Cholinesterase+
Substance P+
VIP+
CGRP+

geneous neural system (Table 2). It is clear that olfactory receptor neurons have a number of distinctive immunoreactive properties that may reflect different subsets of receptor neurons or different stages in their life cycle. It is also clear that fascicles of fibers in the olfactory nerve may also contain fibers of the vomeronasal and terminal nerves in certain species. Neuroanatomical and immunohistochemical data suggest that free nerve endings of trigeminal afferents are found in the olfactory epithelium and lamina propria. Adrenergic and cholinergic fibers traveling with olfactory nerve bundles as they pass through the cribriform plate may be associated with glands and blood vessels.

On the basis of cytological observations and the results obtained electrophysiologically by Bouvet *et al.* (1984, 1986), it is possible to propose a model that integrates the physiological activities of olfactory receptor neurons, trigeminal afferents, and autonomic fibers in the regulation of secretion from Bowman's glands (Fig. 3). Odorants, particularly those associated with noxious chemical stimulation (Cain, 1976; Doty, 1975; Doty *et al.*, 1978), are postulated to interact with free nerve endings of trigeminal sensory fibers located in the nasal and olfactory mucosae (closed arrowheads) in addition to activating olfactory receptor neurons (open arrowheads). These interactions would result in (1) stimulation of the olfactory receptor cells, (2) release of substance P from the sensory terminals of the trigeminal afferents, and (3) the initiation of action potentials in trigeminal sensory fibers. Substance P released from the trigeminal terminals could interact with possible substance P receptors (1) on olfactory neurons to modulate their electrical activity, and (2) on sustentacular cells and acinar cells of Bowman's glands to stimulate secretion. In addition to these two intrinsic mechanisms, a mechanism extrinsic to the olfactory mucosa (i.e., one involving a secretomotor autonomic reflex) would be activated by the transmission of action potentials in the trigeminal sensory fibers to the brain; this reflex would initiate further secretion from the gland cells. As a result, changes in the

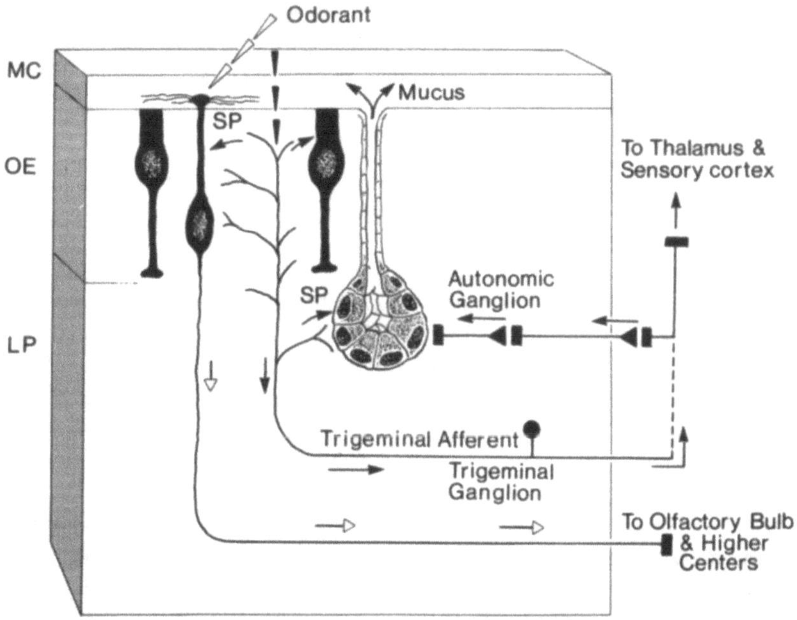

Figure 3. Schematic representation of a secretomotor reflex. Odorants interact with olfactory receptor neurons (open arrowheads) and with free nerve endings of trigeminal afferents (closed arrowheads). Neural activity resulting from odorant interaction with olfactory receptor neurons travels in the olfactory nerve to the olfactory bulb and higher centers. Neural activity resulting from odorant interaction with trigeminal afferents travels in the trigeminal nerve to the central nervous system, where neural activity is initiated in and transmitted by autonomic nerves innervating Bowman's glands, and to the thalamus and sensory cortex. Neural activity in trigeminal afferents may also cause the intraepithelial release of substance P (SP) that interacts with possible receptors on olfactory receptor neurons, modulating their neural activity, and on sustentacular cells and Bowman's glands, stimulating mucous secretion. MC, mucociliary complex; OE, olfactory epithelium; LP, lamina propria.

thickness and composition of the mucus layer and in the rate of transport in the mucociliary complex would be enhanced to mitigate the effects of noxious stimulation.

4. CELLULAR ASPECTS OF SECRETION IN SUSTENTACULAR CELLS

4.1. Mucous Secretion

The release of mucus from sustentacular cells resembles secretory processes in goblet cells. Two modes of release of secretory product have been described in goblet cells (Forstner, 1978): a low-level, continuous release of mucus from the cell surface (Neutra and Leblond, 1966; Specian and Neutra, 1980) that presumably replenishes the protective mucous covering and a rapid apocrine discharge of mucus (Specian and Neutra, 1980) that provides a means by which the epithelium can react quickly and on a large scale to noxious stimuli. Both types of secretion have been observed from sustentacular cells. Bloom (1954) and Okano and Takagi (1974) described morphological signs of baseline release of mucus from the sustentacular cells of amphibia. The synchronous discharge of large quantities of mucus from the surface of the sustentacular cells has been observed in response to stimulation by odorants (Graziadei, 1970, 1971; Mair *et al.*, 1982; M. L. Getchell *et al.*, 1987*b*), solvents (Okano and Takagi, 1974; Ekblom *et al.*, 1984), and irritants (Wesolowski, 1967; Graziadei, 1971, 1973).

4.2. Electrolyte and Water Transport and Secretion

An important function of airway epithelia is the secretion of water from the vascular system into the luminal mucus for protection against dehydration and for maintenance of the efficacy of mucociliary clearance. The basic mechanisms of transepithelial electrolyte and water secretion have been elucidated by studies on tracheal (Olver *et al.*, 1975; Welsh *et al.*, 1980) and intestinal (Frizzell *et al.*, 1979) epithelia. Water secretion appears to depend on Cl^- secretion from the apical epithelial cell surface. Cl^- is transported into epithelial cells by a $Na^+-2Cl^--K^+$ cotransporter located in the basolateral membrane. Upon stimulation by secretagogues, the apical membrane conductance for Cl^- increases, and Cl^- exits passively into the lumen, followed by water. Passive Na^+ entry through the apical membrane also occurs. The Na^+-K^+-ATPase located in the basolateral epithelial cell membrane is involved in the maintenance of the electrochemical gradients to which Cl^- movement is linked.

Electrogenic ion transport has been demonstrated across amphibian olfactory mucosa; the nonstimulated short-circuit current was furosemide sensitive, suggesting the involvement of Na^+-Cl^- cotransport (Heck *et al.*, 1984; Persaud *et al.*, 1987). There is also morphological evidence that sustentacular cells are involved in electrolyte and water secretion. Certain stimuli cause the appearance of clear, membrane-bound vacuoles within the cytoplasm of these cells (arrows, Fig. 4A, C, E); the vacuoles are generally concentrated subjacent to the apical surface membranes but have also been observed in the nuclear region and in the basilar expansions on the basement membrane. Vacuoles such as these have been identified in secretory cells stimulated with pharmacological agents and have been shown to represent the uptake of water as a result of ion fluxes within secretory cells (see Section 5.1.2, for a more detailed discussion).

4.3. Agents That Induce Secretion from Sustentacular Cells

Modulators of neural (prostaglandin $F_{2\alpha}$ and substance P) and vasomotor (vasoactive intestinal peptide) activity, a second messenger system [cyclic adenosine monophosphate (cAMP) and theophylline], and a calcium channel blocker (cobalt chloride) have been observed to activate mucous and/or electrolyte and water secretion from sustentacular cells. For example, topical application of 50 μg/ml prostaglandin $F_{2\alpha}$ ($PGF_{2\alpha}$) caused the formation of blebs from the apical surface of the sustentacular cell (Fig. 4A); the blebs contained mucous vesicles and large vacuoles, which were also observed deeper in the epithelium. By contrast, the same concentration of PGE_1 (Fig. 4B) and PGE_2 had relatively little effect on sustentacular cells; these observations make it unlikely that the morphological signs of secretory activity observed in these studies were due to effects of the carrier solution (0.9% saline) or the fixation procedures, which were identical for the three PGs. (For details of tissue preparation for the experiments described in this chapter,

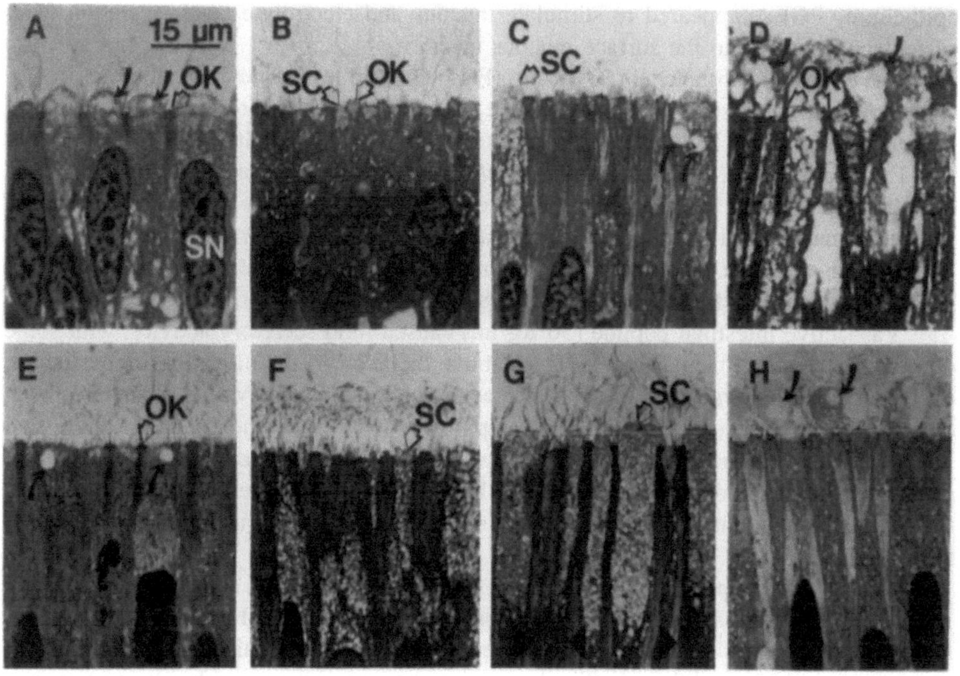

Figure 4. Effects of drugs and odorants on sustentacular cell secretion. All substances were applied topically for 15 min, except where indicated. Calibration bar appears in A. (A) Effects of 50 µg/ml $PGF_{2\alpha}$. Small blebs containing mucous vesicles and vacuoles (arrows) protrude from sustentacular cell surface. Ciliated surfaces of olfactory receptor cell knobs (OK) lie below the level of sustentacular cell surfaces. SN, nucleus of sustentacular cell. (B) Effects of 50 µg/ml PGE_1. Olfactory knobs (OK) lie above the level of sustentacular cell surfaces (SC), indicating relatively little sustentacular cell secretory activity. (C) Effects of 50 µg/ml of substance P. Domes at surfaces of sustentacular cells (SC) and appearance of vacuoles (arrows) within sustentacular cell cytoplasm are indicative of secretory activity. (D) Effects of 5 mM theophylline. Extensive vacuolation (arrows) and blebbing indicate vigorous secretion from sustentacular cells. Olfactory receptor knobs (OK) appear compressed. (E) Effects of 1 mM guaiacol for 1 min. Vacuoles (arrows) appear under surface membranes of sustentacular cells. OK, olfactory knob. (F) Effects of 1 mM guaiacol for 7.5 min. Small domes appear at sustentacular cell surface (SC). (G) Effects of 1 mM guaiacol for 15 min. Domes at surface of sustentacular cells (SC) are larger, and secretory vesicles have aggregated apically, leaving areas within supranuclear region devoid of secretory material (arrowheads). (H) Effects of 1 mM guaiacol for 30 min. Large blebs containing vacuoles (arrows) and mucous secretory vesicles protrude from surface of sustentacular cells.

see M. L. Getchell and T. V. Getchell, 1984*a*, and M. L. Getchell *et al.*, 1987*b*.) Endogenous PGs of the E and F series (Adaikian *et al.*, 1979) as well as the enzymes responsible for PGE synthesis and degradation (Bedwani *et al.*, 1984) have been demonstrated in nasal mucosa. $PGF_{2\alpha}$ caused increased secretory rates in human bronchial mucosa (Lopez-Vidriero *et al.*, 1977; Peatfield *et al.*, 1983); PGE_1 had no effect (Peatfield *et al.*, 1983). However, PGE_1 stimulated mucous secretion from goblet cells in cat and rat trachea and bronchi (Iravani *et al.*, 1976). $PGF_{2\alpha}$ and PGE_1 stimulated electrolyte and water transport in tracheal mucosa (Al-Bazzaz *et al.*, 1981). In the tracheal epithelium, PGE_1 caused ion secretion mediated by cAMP; the mechanism of $PGF_{2\alpha}$-induced ion secretion could not be identified (Smith *et al.*, 1982). In the olfactory epithelium, $PGF_{2\alpha}$ appeared to stimulate mucous and electrolyte/water secretion from sustentacular cells into the surface mucous layer.

The effects of substance P (50 μg/ml) (Fig. 4C) were less marked than those of $PGF_{2\alpha}$ (Fig. 4A) but were still clearly suggestive of mucous and electrolyte/water secretion. Substance P-ir fibers have been demonstrated in the olfactory epithelium of amphibians (M. L. Getchell *et al.*, 1987*a*). Bouvet *et al.* (1984, 1986) demonstrated that application of substance P to the surface of frog olfactory mucosa resulted in slow potentials; these potentials may result in part from secretory activity in sustentacular cells. Substance P has been shown to affect electrolyte and water secretion in the small intestine by a direct action on mucosal cells (Kachur *et al.*, 1982; Keast *et al.*, 1985).

Vasoactive intestinal peptide (50 μg/ml) caused mild secretory activity, with some doming of the sustentacular cell surfaces above the level of the olfactory knobs and the occurrence of small vacuoles in the sustentacular cell cytoplasm. VIP-ir fibers have been observed in the olfactory epithelium of amphibians (M. L. Getchell *et al.*, 1987*a*), and VIP-ir neurons originating from the pterygopalatine ganglion have been demonstrated in nasal epithelium (Uddman *et al.*, 1980). VIP increased Cl^- secretion into the lumen of the canine trachea, but its role in regulation of fluid secretion was judged to be that of fine tuning (Nathanson *et al.*, 1983).

cAMP and theophylline, a phosphodiesterase inhibitor, each at a concentration of 5 mM, elicited vigorous secretory activity from sustentacular cells (Fig. 4D). Both mucous and electrolyte/water secretion were affected, and vacuolation within the epithelium was pronounced. cAMP and theophylline have been shown to increase chloride ion transport (Widdicombe and Welsh, 1980; Al-Bazzaz *et al.*, 1981) and fluid secretion (Welsh *et al.*, 1980) across tracheal epithelium. cAMP-mediated chloride ion secretion has been identified as the mechanism for transepithelial salt and water transport across the surface epithelium of the trachea for several secretagogues (Smith *et al.*, 1982). Electrolyte/water transport appeared to be localized to Cl^--secreting columnar cells within the epithelium. High levels of adenyl cyclase have been found in the membrane fractions from olfactory mucosa (Kurihara and Koyama, 1972; Pace *et al.*, 1985; Anholt *et al.*, 1986). Thus, it is possible that in sustentacular cells, cAMP mediates ion-transport processes leading to water secretion into the mucous layer.

Cobalt chloride (0.9 mM) also elicited active secretion of mucous vesicles and water vacuoles from sustentacular cells. Such activity had been previously observed by Mair *et al.* (1982) and was ascribed to the calcium channel-blocking properties of the cobalt ion. In hen trachea, both increasing and decreasing calcium ion concentration appeared to activate ion pumps in the epithelium (Richardson *et al.*, 1985). The calcium ionophore A23187 caused Cl^- secretion across the tracheal epithelium, and Ca^{2+}-dependent Cl^- secretion has been described for a variety of epithelia (Smith *et al.*, 1982). Thus cobalt-elicited secretion from sustentacular cells is suggestive of a role for calcium.

Odorants applied topically in solution also elicited secretory responses from sustentacular cells. Morphological signs of secretion were related to both the concentration of the odorant and the duration of exposure (M. L. Getchell *et al.*, 1987*b*). For example, a 1-min exposure to 10^{-3}M guaiacol caused the appearance of vacuoles indicative of electrolyte and water secretion just under the surface membranes of the sustentacular cells (arrows, Fig. 4E). Small blebs containing vacuoles and mucoid secretory material formed at the surface of the cells after a 7.5-min exposure (Fig. 4F); a 15-min exposure caused concentration of the secretory material in the supranuclear part of the cells (Fig. 4G). Exposure to the odorant for 30 min resulted in the formation of large blebs containing mucoid material and vacuoles and the depletion of much of the secretory material from the supranuclear portion of the cells (Fig. 4H). A similar progression of secretory effects was observed as the concentration of the odorant 2-isobutyl-3-methoxypyrazine (IBMP, galbazine) was increased from 10^{-5} to 10^{-3}M. IBMP has been shown to stimulate a furosemide-sensitive active ion transport across the olfactory mucosa (DeSimone and T. V. Getchell, 1985).

Like goblet cells, sustentacular cells do not appear to receive direct efferent innervation. In general, mature goblet cells did not respond to pharmacological stimulation that caused secretion from nearby multicellular glands (Stahl and Ellis, 1973; Gallagher *et al.*, 1975; Specian and Neutra, 1980). Similarly, β-adrenergic agonists, which caused secretion from Bowman's glands, had no effect on mucous or water secretion from sustentacular cells (M. L. Getchell and T. V. Getchell, 1984*a*).

4.4. Effects of Olfactory Nerve Section

Olfactory nerve transection also affects sustentacular cell secretory activity. Following transection, there appears to be more secretory material in most sustentacular cells as evaluated electron microscopically (Okano and Takagi, 1974) and histochemically (M. L. Getchell *et al.*, 1984*a*). The effect of olfactory nerve transection may be due to transcellular effects of molecules released by the degenerating olfactory receptor neurons (for discussion, see M. L. Getchell *et al.*, 1984*a*); these molecules may stimulate synthesis of mucopolysaccharides in sustentacular cells.

5. NEUROPHARMACOLOGICAL REGULATION OF GLANDULAR SECRETION

5.1. Agonist-Induced Secretion

5.1.1. Secretory Granule Depletion

Secretory granules in the glands of the respiratory mucosa contain the glycoproteins (Spicer *et al.*, 1983) responsible for the viscous properties of mucus (Litt and Khan, 1976; Bloomfield, 1983). Stimulation of respiratory glands by both β-adrenergic and muscarinic cholinergic agonists causes production of a secretion with a relatively high mucin and protein content (Nadel, 1983; Quinton, 1979) compared with the secretion that results from α-adrenergic secretion.

The β-adrenergic agonist isoproterenol caused a concentration-dependent reduction in the secretory granule content of the acinar cells of Bowman's glands (Fig. 5A). At the lowest concentration, 3 mg/kg, the agonist reduced the granule content by 23%, whereas at a higher concentration, 30 mg/kg, a marked 65% reduction occurred as shown in Figs. 5A and 6B.

Agonist-induced granule depletion is blocked by pretreatment of the olfactory mucosa with the antagonist propranolol. The results of these experiments (reported by M.

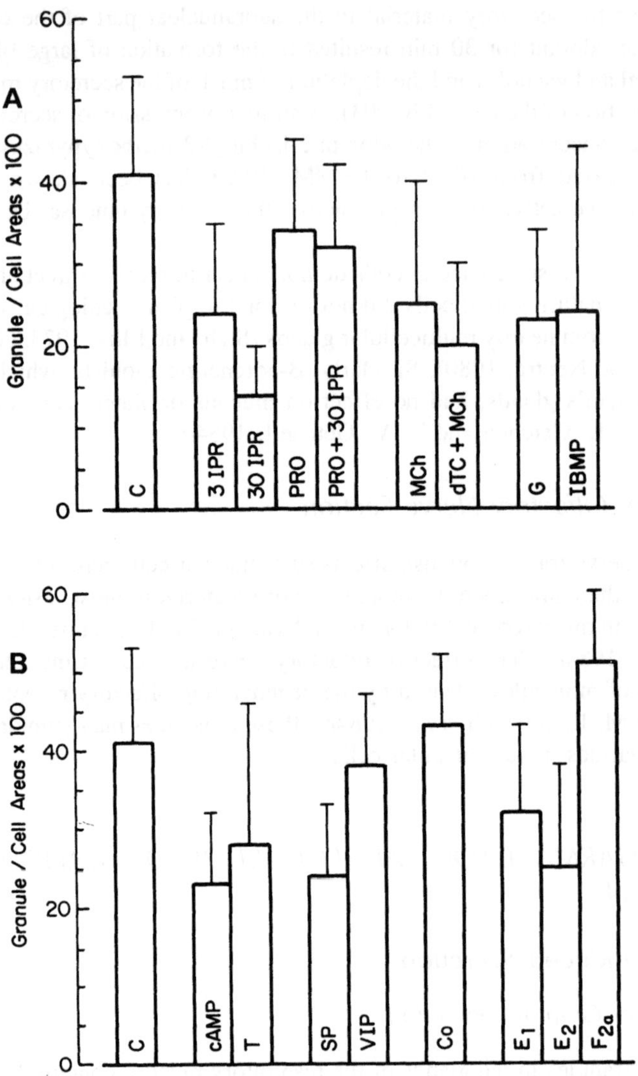

Figure 5. Morphometric analysis of the effects of drugs, odorants, second messengers, and modulators on secretion from Bowman's glands. Measurements and statistics were performed with the Bioquant program (for details, see M. L. Getchell and T. V. Getchell, 1984). Secretory granule content is expressed as the percentage of the cell area occupied by secretory granules. (A) Effects of drugs and odorants on secretory granule content. C, control (0.9% saline); 3 IPR, 3 mg/kg isoproterenol; 30 IPR, 30 mg/kg isoproterenol; PRO, 42 mg/kg propranolol; PRO, + 30 IPR, 42 mg/kg propranolol followed by 30 mg/kg isoproterenol; MCh, 5 mg/ml methacholine; dTC + MCh, 0.6 mg *d*-tubocurarine followed by 5 mg/ml methacholine; G, 10^{-3}M guaiacol; IBMP, 10^{-3} M 2-isobutyl-3-methoxypyrazine. Measured number of cells per treatment ranged from 24 to 38. Statistical significance in A and B is indicated by A, $p < 0.001$; B, $p < 0.01$; c, $p < 0.05$. (B) Effects of second messengers and modulators on secretory granule content. C, control (0.9% saline); cAMP, 5×10^{-3}M cAMP; T, 5×10^{-3}M theophylline; SP, 50 μg/ml substance P; VIP, 50 μg/ml vasoactive intestinal peptide; Co, 9×10^{-4}M cobalt chloride; E_1, 50 μg/ml prostaglandin E_1; E_2, 50 μg/ml prostaglandin E_2; $F_{2\alpha}$, 50 μg/ml prostaglandin $F_{2\alpha}$. Measured number of cells per treatment ranged from 15 to 41.

L. Getchell and T. V. Getchell, 1984*a*), indicate that secretion from Bowman's glands is regulated partially by sympathetic postganglionic neurons. The muscarinic cholinergic agonist methacholine, applied topically at 5 mg/kg, was associated with a 37% depletion in the area occupied by secretory granules in the acinar cells (M. L. Getchell and T. V. Getchell, 1984*b;* T. V. Getchell and M. L. Getchell, 1984), as shown in Figs. 5a and 6c. Methacholine-induced granule depletion was reduced by the muscarinic antagonist scopolamine but not by the nicotinic antagonist curare suggesting that glandular secretion is also regulated partially by parasympathetic postganglionic neurons. Topical application of the α-adrenergic agonist phenylephrine at the concentration used therapeutically, 0.25% (12.25 mM), was also associated with secretory granule depletion (Fig. 6A) (T. V. Getchell and M. L. Getchell, 1984). Interpretation of these observations is somewhat ambiguous because application of the agonist may also be associated with changes in the distribution of pigment granules in the melanocytes that often partially envelop the acinus (Figs. 1F and 6A, Me). Further experiments are required to document quantitatively the α-adrenergic regulation of glandular secretion.

5.1.2. Electrolyte and Water Secretion

Mucus contains a large amount of water, approximately 95% in bronchial mucus (Medici and Radielovic, 1979). In salivary glands, water is thought to be transported from the blood stream to salivary gland cells by the establishment of an osmotic gradient based on sustained Ca^{2+}-dependent K^+ efflux from the gland cells; this K^+ efflux is stimulated by muscarinic cholinergic and α-adrenergic agonists (Putney, 1976). Accordingly, muscarinic cholinergic and α-adrenergic stimulation are associated with high secretion flow rates compared with the effects of β-adrenergic stimulation (Quinton, 1979; Nadel, 1983). Stimulation of secretory cells with α-adrenergic agonists (Batzri *et al.*, 1971; Mills and Quinton, 1981; Quinton, 1981) and cholinergic agonists (Schramm and Selinger, 1974; Garrett *et al.*, 1978; Quinton, 1981) results in the appearance of vacuoles that have been correlated with the transport of ions (Bogart and Picarelli, 1978) and the obligatory movement of water (Quinton, 1981). Cl^- secretion has also been implicated in ion transport across frog skin glands; the channels for ion movement in gland cells, which are amiloride insensitive, appear to be pharmacologically different from those in epithelial cells, which are amiloride sensitive (Mills, 1985).

Muscarinic cholinergic and, to a lesser extent, α-adrenergic, stimulation of Bowman's glands resulted in the appearance of vacuoles in the acinar cells. In addition, the α-adrenergic antagonist phentolamine not only blocked granule depletion caused by phenylephrine but also enhanced the formation of vacuoles within acinar cells treated with both phenylephrine and phentolamine. Stimulus-induced vacuoles occurred in the basal region of the gland cell, among the secretory granules, and/or adjacent to the lumen. Vacuoles often contained pale secretory granules, and secretory granules in the vicinity of vacuoles often appeared swollen. These observations are consistent with the conclusion of Mills and Quinton (1981) that secretory granules accumulate water to form vacuoles.

5.2. Second Messengers and Modulators of Agonist-Induced Secretion

β-Adrenergic agonists such as isoproterenol stimulate secretion from gland cells through the cAMP second messenger system (Baum *et al.*, 1981). When 5 mM dibutyryl

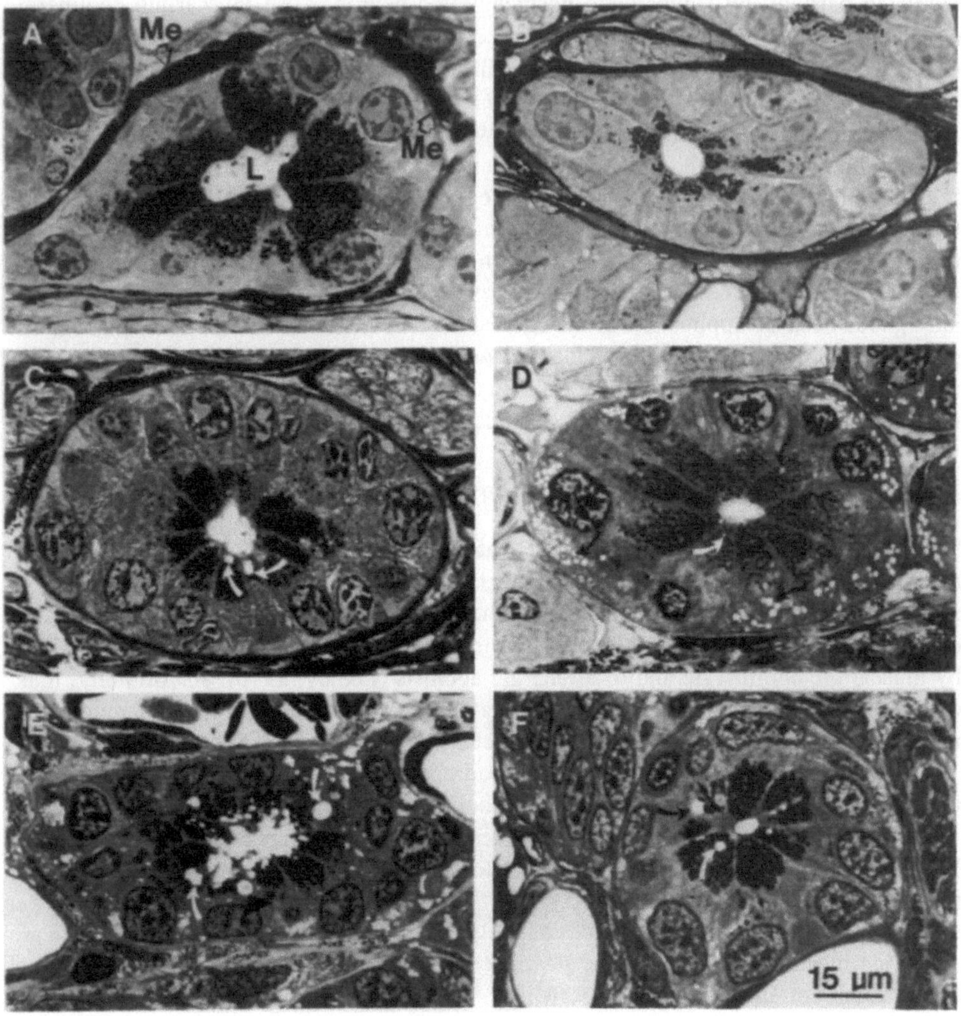

Figure 6. Effects of drugs and modulators on secretion from Bowman's glands. Drugs were applied topically for 15 min, except isoproterenol, which was injected 2 hr prior to perfusion of animal. Calibration bar appears in (F). (A) Effects of 0.25% phenylephrine. Secretory granule content of acinar cells is reduced; lumen (L) is dilated. Note prominence of melanocytes (Me). (B) Effects of 30 mg/kg isoproterenol. Substantial reduction in secretory granule content is obvious. (C) Effects of 5 mg/ml methacholine. In addition to the reduction in secretory granule content, the lumen appears dilated, and vacuoles (arrows) appear in the apical acinar cell area. (D) Effects of 50 μg/ml substance P. Secretory granule content is reduced, and small vacuoles (arrows) appear basally and among the secretory granules. (E) Effects of 5×10^{-3}M dibutyryl cAMP. Large vacuoles (arrows) appear throughout the cells. Many secretory granules appear swollen; others appear to be dissolving in vacuoles (arrowhead). Secretory granule content is reduced. (F) Effects of 5×10^{-3} M theophylline. Reduction in secretory granule content is observed along with the appearance of vacuoles (arrows) and swollen secretory granules.

cAMP or 5 mM theophylline was applied to the surface of the olfactory mucosa for 15 min, statistically significant secretory granule depletion occurred (Figs. 5B and 6E, F). In addition, the perimeter of the lumens of the acini measured for the statistical analysis was significantly increased ($p<0.01$) by cAMP, indicating fusion of the secretory granules with the luminal membranes of the acinar cells. Both cAMP and theophylline also caused the formation of vacuoles (arrows, Fig. 6E, F) within the acinar cells. cAMP did not elicit vacuole formation in parotid gland cells (Batzri *et al.*, 1971); however, VIP, which increased cAMP concentration in trachea (Kitamura *et al.*, 1980), also caused vacuole formation in olfactory gland cells (see below). Thus, there may be a cAMP-dependent pathway for fluid secretion in airways that does not exist in salivary glands.

Substance P-ir fibers have been observed in close association with Bowman's gland cells (Papka and Matulionis, 1983; M. L. Getchell *et al.*, 1987*a*) (see Section 3). Topical application of 50 μg/ml substance P for 15 min caused significant granule depletion from acinar cells (Figs. 5B and 6D); in addition, small vacuoles appeared primarily in the basal region of the acinar cells. A specific substance P receptor has been demonstrated on parotid cell membranes (Liang and Cascieri, 1981).

Although topical application of VIP did not cause secretory granule depletion (Fig. 5B), numerous small vacuoles appeared basally and among the secretory granules, indicating activation of ion and water transport. In ferret tracheal glands, VIP stimulated mucin secretion but not ion transport (Peatfield *et al.*, 1983). VIP-ir fibers have been observed in close association with Bowman's glands in amphibian olfactory mucosa (M. L. Getchell *et al.*, 1987*a*). VIP has also been demonstrated in cholinergic nerves innervating nasal (Lundberg *et al.*, 1981) and salivary (Lundberg *et al.*, 1980) glands; although VIP seems to mediate primarily vasomotor and not secretory responses, the effect of simultaneous cholinergic and VIP stimulation was synergistic, suggesting a possible involvement of VIP in secretory processes (Lundberg *et al.*, 1980).

The effects of α-adrenergic agonists appear to be mediated via Ca^{2+} as a second messenger (Jones and Michell, 1978). In addition, α- and β-adrenergic and cholinergic agonists activate calcium channels in parotid gland cells (Putney, 1979). Although topical application of 0.9 mM cobalt chloride, a calcium channel blocker, to the olfactory mucosa had no effect on the secretory granule content of the gland cells (Fig. 5B), the effect of cobalt chloride on agonist-induced secretion has not yet been tested.

When applied to the surface of the olfactory mucosa, the three prostaglandins tested (E_1, E_2, and $F_{2\alpha}$), resulted in the appearance of vacuoles indicative of electrolyte and water secretion. Although $PGF_{2\alpha}$ did not cause secretory granule depletion (Figs. 5b and 7a), PGE_1 and PGE_2 did (Figs. 5B, 7B, C); PGE_2 caused significantly more secretory granule release than PGE_1. Prostaglandin E_2 stimulates gastric mucus secretion via a pathway involving cAMP as a second messenger (Bersimbaev *et al.*, 1985). Prostaglandin E_1 has been shown to be involved in the release of transmitters from postganglionic sympathetic nerves in lacrimal glands (Pholpramool and Tangkrisanavinont, 1983).

5.3. Odorant-Induced Secretion

The odorants guaiacol and IBMP caused concentration- and time-dependent reductions in the secretory granule content of acinar cells when applied topically in solution to the surface of the olfactory mucosa (M. L. Getchell *et al.*, 1987*b*). At concentrations of

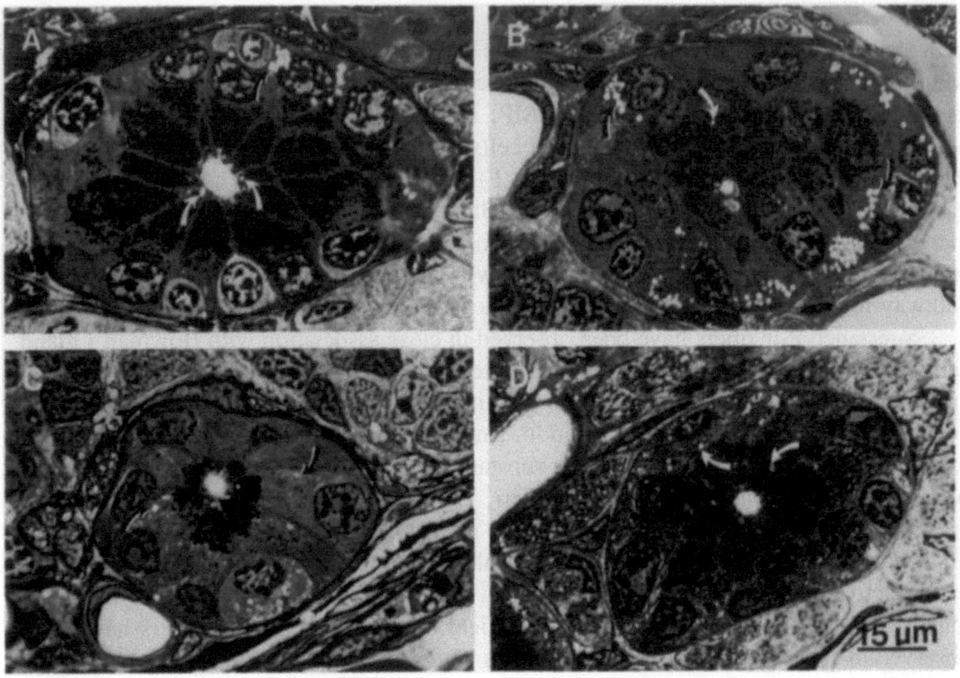

Figure 7. Effects of modulators and odorants on secretion from Bowman's glands. All substances were applied topically for 15 min. Calibration bar appears in (D). (A) Effects of 50 µg/ml $PGF_{2\alpha}$. Although secretory granule content is not reduced significantly, small vacuoles (arrows) appear apically and basally within the acinar cells. (B) Effects of 50 µg/ml PGE_1. A reduction in secretory granule content is accompanied by the appearance of vacuoles (arrows) basally and among the secretory granules. (C) Effects of 50 µg/ml PGE_2. There is a substantial reduction in secretory granule content as well as the appearance of vacuoles (arrows) within the acinar cells. (D) Effects of 10^{-3}M guaiacol, an odorant. Vacuoles (arrows) and a reduction in secretory granule content are observed.

IBMP ranging from 10^{-5} to 10^{-3} M, the secretory granule content of the acinar cells was significantly reduced by 18–37%; vacuoles were observed at all three concentrations. Application of 1 mM guaiacol for periods of from 1–15 min resulted in significant reduction of secretory granule content by 45–48%; a 30-min application resulted in a secretory granule content equivalent to that of control glands. Vacuoles in the acinar cells were observed for all time periods. Prior application of scopolamine, but not propranolol, lessened the amount of secretory granule depletion induced by IBMP. Thus odorants induce mucous secretion as well as electrolyte and water secretion from Bowman's glands. Odorants may stimulate gland secretion by interaction with presynaptic nerve terminals on the glands to release neurotransmitters, by interacting directly with pharmacological receptors on the acinar cells or by stimulation of free nerve endings in the epithelium to activate a secretomotor reflex; the effect of scopolamine is suggestive of neural involvement.

5.4. Neural Pathways of Agonist-Induced Secretion

The primary extrinsic regulatory influence on secretion from Bowman's glands is presumably mediated by the autonomic nervous system. This innervation represents the

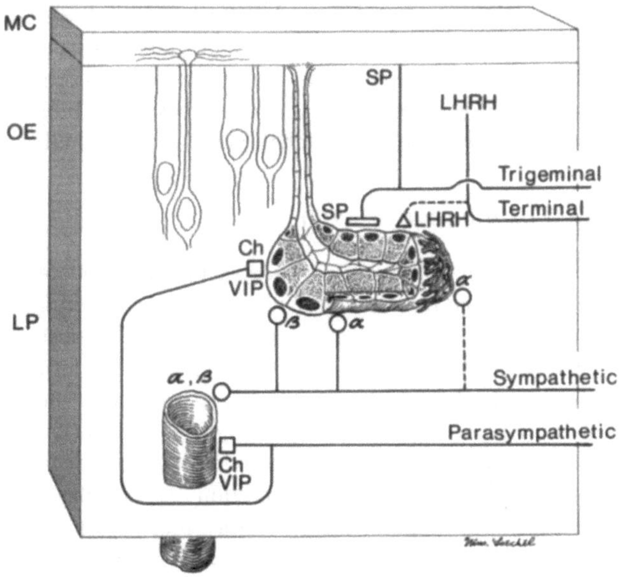

Figure 8. Proposed model for the regulation of secretion from Bowman's glands. Sympathetic (α- and β-adrenergic) neurons and parasympathetic (cholinergic) neurons may terminate directly on gland cells. In addition, there may be neural influences on blood vessels and melanocytes (○) in the mucosa as indicated. Both VIP from cholinergic nerves and substance P from trigeminal nerves may also affect gland cells; VIP may affect blood vessels as well. LHRH-ir fibers from the terminal nerve may also be associated with gland cells. Second messenger systems involving cAMP and calcium operate within the gland cells (see Fig. 5B). α, α-adrenergic; β, β-adrenergic; Ch, cholinergic; LP, lamina propria; MC, mucociliary complex; OE, olfactory epithelium; SP, substance P; VIP, vasoactive intestinal peptide.

efferent component of a secretomotor reflex (see Fig. 3). Secretory granule depletion in the acinar cells of olfactory glands is caused by adrenergic and cholinergic agonists, which suggests a role for the extrinsic autonomic regulation of glandular secretion. β-Adrenergic and cholinergic secretogogues appear to act directly on the acinar cells but may also affect the vasculature in the lamina propria. α-Adrenergic agonists may act directly on acinar cells or on vasculature; in addition, there were morphological suggestions that granule depletion could be associated with changes in melanocytes surrounding the acini. Melanocytes in the inner ear may be involved in the regulation of microvasculature or basement membrane permeability and/or of epithelial cell function (LaFerriere *et al.*, 1974; Gussen, 1978). The proposed mechanisms responsible for the neural regulation of secretion from Bowman's glands are summarized in Fig. 8.

6. IMPLICATIONS FOR SENSORY TRANSDUCTION

6.1. Prereceptor Events

The activation of ion transport and fluid and mucous secretion in sustentacular cells and olfactory glands presumably changes the composition and physical properties of the mucus in which odorant binding, the first step in sensory transduction, occurs. Thus, access of the odorant to receptor sites on olfactory receptor neurons and accumulation of odorant molecules at the receptor sites to the level required to reach threshold concentrations for neural activity will be influenced by the depth and viscosity of the mucus. In addition, binding of odorant molecules to receptor sites on olfactory neurons and possibly to macromolecules in the mucous layer, and the driving forces for ion movement provided by ionic gradients across receptor cell membranes will be influenced by the composition of the mucous layer.

In an attempt to begin to define the effect of secretion on olfactory sensory activity, stimulation of secretion from Bowman's glands by the β-adrenergic agonist isoproterenol was shown to decrease the amplitude of the odorant-elicited electro-olfactogram ($V_{eog(-)}$), a summated receptor potential; the decrease, was blocked by propranolol, which blocked secretion as well (M. L. Getchell and T. V. Getchell, 1984*a*).

6.2. Postinteractive Events

Secretion presumably influences the rate of adaptation of olfactory receptors through an effect on the rate at which odorants are removed from the vicinity of receptor sites on olfactory neurons. Odorants may be taken up by sustentacular cells, possibly in a secretion-linked process, and may be inactivated by intracellular enzyme systems or transported to the basement membrane for transfer to the circulatory system. Alternatively, they may be bound to substances in the secretion (Kaissling, 1986; Pevsner *et al.*, 1986) or swept away by mucociliary activity enhanced because of the lowering of the viscosity of the mucus due to water secretion. By similar mechanisms, secretion presumably influences the sensitivity of olfactory receptors to subsequent stimulation by aiding or impeding clearance of odorants from the vicinity of the receptor sites on olfactory neurons. (For a more detailed discussion of these concepts, see T. V. Getchell *et al.*, 1984.)

7. CONCLUSIONS AND RESEARCH NEEDS

Activation of secretory processes by a variety of pharmacological agonists, modulators, and mediators of cellular processes has been demonstrated in sustentacular cells and gland cells in the olfactory mucosa. Comparison of the action of these agents in other secretory systems has been discussed to support and amplify the conclusions drawn from primarily morphological data. Finally, possible roles for these secretory processes in olfactory sensory function have been suggested.

Research needs in this area are numerous. Although several studies have reported the relationships between the ontogeny of olfactory receptor cells and secretory elements (Breipohl, 1972; for review, see M. L. Getchell *et al.*, 1986), further investigations with immunohistochemical, biochemical, and electrophysiological techniques will clarify the relative timing of the onsets of receptor function and secretory activity. A more complete histochemical, biochemical, and immunohistochemical analysis of the secretory products of sustentacular cells and olfactory glands, as well as how the composition of the secretion changes under different stimulatory conditions, will clarify some of the roles of mucus in receptor functions and the relative contribution of sustentacular versus gland cells to the mucus surrounding the receptor neurons. The neuropharmacological studies discussed represent an initial approach to identifying factors regulating secretion. A more thorough analysis of the effects of these secretagogues will amplify understanding of regulatory mechanisms and interactions between various regulatory pathways.

Immunohistochemical investigations of the innervation of the epithelium and of the glands and vasculature of the lamina propria are needed to clarify the pharmacological nature and distribution of autonomic fibers in the olfactory mucosa. Additional studies involving olfactory and trigeminal axotomy combined with immunohistochemical techniques presumably will identify sources of innervation and address the question of regulation of receptor site number and distribution following nerve section in the olfactory mucosa. Further investigation of the mechanisms associated with active ion transport (i.e., identification of the ions and direction of fluxes carrying the current, location of ion pumps, ion conductance changes at various transporting membranes, and pathways of water movement through the olfactory mucosa) will lead to an understanding of the regulation of the ion and water content of the mucus, which in turn influences the rate of diffusion of odorants through the mucus, the rate of mucociliary transport, and the ionic gradients across the olfactory receptor cell membrane. In addition, techniques involving the measurement of short-circuit current will provide a means of measuring electric activity due to secretory processes and thus provide a means of quantifying secretory activity in relationship to olfactory stimulation and receptor cell activity.

Pursuit of these avenues of research will undoubtedly yield further insight into the interaction between secretion and olfactory sensory processes.

ACKNOWLEDGMENTS. The research discussed in this chapter was supported by grant BNS-85-07949 from the National Science Foundation (M.L.G.) and by grant NS-16340 from the National Institutes of Health (T.V.G.). The authors, especially T. V. Getchell, wish to thank Dr. P. P. C. Graziadei for stimulating and nurturing their interests in the topic of extrinsic innervation of the olfactory mucosa. Thanks are also due to William Lochell, for the schematic translations of the neuroanatomical data shown in Figs. 2, 3, and 8, and to Tracy Miele, for her excellent secretarial assistance.

REFERENCES

Adaikan, P. G., Karim, S. M. M., and Kunaratnam, N., 1979, Prostaglandins and nasal patency, in: *Practical Applications of Prostaglandins and Their Synthesis Inhibitors* (S. M. M. Karim, ed.), pp. 25–38, University Park Press, Baltimore.

Alarie, Y., 1973, Sensory irritation of the upper airways by airborne chemicals, *Toxicol. Appl. Pharmacol.* **24:** 279–297.

Al-Bazzaz, F., Yadava, V. P., and Westenfelder, C., 1981, Modification of Na and Cl transport in canine tracheal mucosa by prostaglandins, *Am. J. Physiol.* **9:** F101–F105.

Allen, W. K., and Akeson, R., 1985*a*, Identification of a cell surface glycoprotein family of olfactory receptor neurons with a monoclonal antibody, *J. Neurosci.* **5:** 284–296.

Allen, W. K., and Akeson, R., 1985*b*, Identification of an olfactory receptor neuron subclass: Cellular and molecular analysis during development, *Dev. Biol.* **109:** 393–401.

Anholt, R. R. H., Aebi, U., and Snyder, S. H., 1986, A partially purified preparation of isolated chemosensory cilia from the olfactory epithelium of the bullfrog, *Rana catesbeiana, J. Neurosci.* **6**(7): 1962–1969.

Baradi, A. F., and Bourne, G. H., 1953, Gustatory and olfactory epithelia, *Int. Rev. Cytol.* **11:** 289–330.

Barber, P. C., and Lindsay, R. M., 1982, Schwann cells of the olfactory nerves contain glial fibrillary acidic protein and resemble astrocytes, *Neuroscience* **7:** 3077–3090.

Batzri, S., Amsterdam, A., Selinger, Z., Ohad, I., and Scramm, M., 1971, Epinephrine-induced vacuole formation in parotid gland cells and its independence of the secretory process, *Proc. Natl. Acad. Sci. USA* **68:** 121–123.

Baum, B. J., Freiberg, J. M., Ito, H., Roth, G. S., and Filburn, C. R., 1981, β-Adrenergic regulation of protein phosphorylation and its relationship to exocrine secretion in dispersed rat parotid gland acinar cells, *J. Biol. Chem.* **256:** 9731–9736.

Bedwani, J. R., Eccles, R. and Jones, A. S., 1984, A study of the synthesis and inactivation of prostaglandin E by pig nasal mucosa, *Acta Otolaryngol. (Stockh.)* **98:** 308–314.

Bersimbaev, R. I., Tairov, M. M., and Salganik, R. I., 1985, Biochemical mechanisms of regulation of mucus secretion by prostaglandin E in rat gastric mucosa, *Eur. J. Pharmacol.* **115:** 259–266.

Bloom, G., 1954, Studies on the olfactory epithelium of the frog and the toad with the aid of light and electron microscopy, *Z. Zellforsch.* **41:** 89–100.

Bloomfield, V. A., 1983, Hydrodynamic properties of mucous glycoproteins, *Biopolymers* **22:** 2141–2154.

Bogart, B. I., and Picarelli, J., 1978, Agonist-induced secretions and potassium release from rat submandibular gland slices, *Am. J. Physiol.* **235:** C256–C268.

Bojsen-Møller, F., 1975, Demonstration of terminalis, olfactory, trigeminal and perivascular nerves in the rat nasal septum, *J. Comp. Neurol.* **159:** 245–256.

Bouvet, J.-F., Delaleu, J.-C., and Holley, A., 1984, Electrical responses by acetylcholine and substance P in the frog's olfactory mucosa, *C. R. Acad. Sc. (Paris)* **298**(6): 169–172.

Bouvet, J.-F., Delaleu, J.-C., and Holley, A., 1986, Olfactory receptor cell functioning affected by trigeminal nerve activity and substance P, *Chem. Senses* **11:** 583 (abst.).

Bouvet, J.-F., Delaleu, J. C., and Holley, A., 1987, Olfactory receptor cell function is affected by trigeminal nerve activity, *Neurosci. Lett.* **77:** 181–186.

Breipohl, W., 1972, Licht- und elektronenmikroskopische Befunde zur Struktur der Bowmanschen Drüsen im Riechepithel der weissen Maus, *Z. Zellforsch.* **131:** 329–346.

Brown, D., Garcia-Segura, L.-M., and Orci, L., 1984, Carbonic anhydrase is present in olfactory receptor cells, *Histochemistry* **80:** 307–309.

Cain, W. S., 1976, Olfaction and the common chemical sense: Some psychophysical contrasts, *Sens. Processes* **1:** 57–67.

Cuschieri, A., 1974, Enzyme histochemistry of the olfactory mucosa and vomeronasal organ in the mouse, *J. Anat.* **118:** 477–498.

DeLong, R. E., and Getchell, T. V., 1987, Nasal respiratory function—Vasomotor and secretory regulation: A review, *Chem. Senses* **12:** 3–36.

DeSimone, J. A., and Getchell, T. V., 1985, Odorant-stimulated current transients across the frog olfactory mucosa *in vitro,* in: *Proceedings of the Nineteenth International Symposium on Taste and Smell* (S. Kimura, A. Miyoshi, and I. Shimada, eds.), pp. 46–48, Asahi University Press, Gifu, Japan.

Doty, R. L., 1975, Intranasal trigeminal detection of chemical vapors by humans, *Physiol. Behav.* **14:** 855–859.

Doty, R. L., Brugger, W. E., Jurs, P. C., Orndorff, M. A., Snyder, P. J., and Lowry, L. D., 1978, Intranasal trigeminal stimulation from odorous volatiles: Psychometric responses from anosmic and normal humans, *Physiol. Behav.* **20:** 175–185.

Ekblom, A., Flock, A., Hansson, P., and Ottoson, D., 1984, Ultrastructural and electrophysiological changes in the olfactory epithelium following exposure to organic solvents, *Acta Otolaryngol. (Stockh.)* **98:** 351–361.

Farbman, A. I., 1986, Prenatal development of mammalian olfactory receptor cells, *Chem. Senses* **11:** 3–18.

Farbman, A. I., and Gesteland, R. C., 1974, Fine structure of olfactory epithelium in the mud puppy, *Necturus maculosus, Am. J. Anat.* **139:** 227–244.

Farbman, A. I., and Margolis, F. L., 1982, Immunohistochemical localization of carnosinase in olfactory and other tissues in the mouse, *Anat. Rec.* **202:** 53A.

Forstner, J. F., 1978, Intestinal mucins in health and disease, *Digestion* **17:** 234–263.

Frisch, D., 1967, Ultrastructure of mouse olfactory mucosa, *Am. J. Anat.* **121:** 87–120.

Frizzell, R. A., Field, M., and Schultz, S. G., 1979, Sodium-coupled chloride transport by epithelial tissues, *Am. J. Physiol.* **236:** F1–F8.

Fujita, S. C., Mori, K., Imamura, K., and Obata, K., 1985, Subclasses of olfactory cells and their segregated central projections demonstrated by a monoclonal antibody, *Brain Res.* **326:** 192–196.

Gallagher, J. T., Kent, P. W., Passatore, M. Phipps, R. J., and Richardson, P. S., 1975, The composition of tracheal mucus and the nervous control of its secretion in the cat, *Proc. R. Soc. Lond. Ser. B* **192:** 49–76.

Garrett, J. R., Thulin, A., and Kidd, A., 1978, Variation in parasympathetic secretory and structural responses resulting from differences in the pre-stimulation state of parotid acini in rats, *Cell Tissue Res.* **188:** 235–250.

Getchell, M. L., and Getchell, T. V., 1984*a,* β-Adrenergic regulation of the secretory granule content of acinar cells in olfactory glands of the salamander, *J. Comp. Physiol. A* **155:** 435–443.

Getchell, M. L., and Getchell, T. V., 1984*b*, Cholinergic control of glandular secretion in the olfactory mucosa of the salamander, in: *Proceedings of the Sixth Annual Meeting of the Association of Chemoreception Sciences Sarasota, Florida* (abst. 51).

Getchell, M. L., Rafols, J. A., and Getchell, T. V., 1984, Histological and histochemical studies of the secretory components of the salamander olfactory mucosa: Effects of isoproterenol and olfactory nerve section, *Anat. Rec.* **208:** 553–565.

Getchell, M. L., Zielinski, B., and Getchell, T. V., 1986, Ontogeny of the secretory elements in vertebrate olfactory mucosa, in: *Ontogeny of Olfaction* (W. Breipohl, ed.), pp. 71–82, Springer-Verlag, Berlin.

Getchell, M. L., Finger, T. E., and Getchell, T. V., 1987*a*, Localization of NGF-like, VIP-like, substance P-like and CGRP-like immunoreactivity in the olfactory mucosae of the salamander, bullfrog and grass frog, in: *Proceedings of the Ninth Annual Meeting of the Association of Chemoreception Sciences, Sarasota, Florida* (abst. 158).

Getchell, M. L., Zielinski, B., DeSimone, J. A., and Getchell, T. V., 1987*b*, Odorant stimulation of secretory and neural processes in the salamander olfactory mucosa, *J. Comp. Physiol. A* **160:** 155–168.

Getchell, T. V., 1986, Functional properties of vertebrate olfactory receptor neurons, *Physiol. Rev.* **66:** 772–818.

Getchell, T. V., and Getchell, M. L., 1977, Histochemical localization and identification of secretory products in salamander olfactory epithelium, in: *Olfaction and Taste* (J. Le Magnen and P. MacLeod, eds.), pp. 105–112, IRL Press, London.

Getchell, T. V., and Getchell, M. L., 1984, Adrenergic and cholinergic modulation of secretory granule content in the olfactory glands of the salamander, in: *Proceedings of the Sixth Annual Meeting of the European Chemoreception Organization, Lyon, France,* p. 55.

Getchell, T. V., Margolis, F. L., and Getchell, M. L., 1984, Perireceptor and receptor events in vertebrate olfaction, *Prog. Neurobiol.* **23:** 317–345.

Gladysheva, O., and Martynova, G., 1982, The morpho-functional organization of the bovine olfactory epithelium, *Gegenbaurs Morph. Jahrb.* **128:** 78–83.

Gladysheva, O., Kukushkina, D., and Martynova, G., 1986, Glycoprotein composition of olfactory mucus in vertebrates, *Acta Histochem.* **78:** 141–146.

Grassi, B., and Castronovo, A., 1889, Beitrag zur Kenntniss des Geruchsorgans des Hundes, *Arch. Mikrosk. Anat.* **34:** 385–390.

Graziadei, P. P. C., 1970, The mucous membranes of the nose, *Ann. Otol. Rhinol. Laryngol.* **79:** 443–452.

Graziadei, P. P. C., 1971, The olfactory mucosa of vertebrates, in: *Handbook of Sensory Physiology*. Vol. IV: *Chemical Senses. Part* I (L. M. Beidler, ed.), pp. 27–58, Springer-Verlag, New York.

Graziadei, P. P. C., 1973, The ultrastructure of vertebrate olfactory mucosa, in: *The Ultrastructure of Sensory Organs*, (I. Friedmann, ed.), pp. 269–305, Elsevier, New York.

Graziadei, P. P. C., and Gagne, H. T., 1973, Extrinsic innervation of olfactory epithelium, *Z. Zellforsch.* **138:** 315–326.

Graziadei, P. P. C., and Metcalf, J. F., 1971, Autoradiographic and ultrastructural observations on the frog's olfactory mucosa, *Z. Zellforsch.* **116:** 305–318.

Graziadei, P. P. C., and Monti-Graziadei, G. A., 1983, Regeneration in the olfactory system of vertebrates, *Am. J. Otolaryngol.* **4:** 228–233.

Gussen, R., 1978, Melanocyte system of the endolymphatic duct and sac, *Ann. Otol.* **87:** 175–179.

Harding, J. W., Getchell, T. V., and Margolis, F. L., 1978, Denervation of the primary olfactory pathway in mice. V. Longterm effect of intranasal $ZnSO_4$ irrigation on behavior, biochemistry and morphology, *Brain Res.* **140:** 271–287.

Heck, L., DeSimone, J. A., and Getchell, T. V., 1984, Evidence for electrogenic active ion transport across the frog olfactory mucosa in vitro, *Chem. Senses* **9:** 272–283.

Hempstead, J. L., and Morgan, J. I., 1983*a*, Fluorescent lectins as cell-specific markers for the rat olfactory epithelium, *Chem. Senses* **8:** 107–120.

Hempstead, J. L., and Morgan, J. I., 1983*b*, Monoclonal antibodies to the rat olfactory sustentacular cell, *Brain Res.* **288:** 289–295.

Hempstead, J. L., and Morgan, J. I., 1985*a*, A panel of monoclonal antibodies to the rat olfactory epithelium, *J. Neurosci.* **5:** 438–449.

Hempstead, J. L., and Morgan, J. I., 1985*b*, Monoclonal antibodies reveal novel aspects of the biochemistry and organization of olfactory neurons following unilateral olfactory bulbectomy, *J. Neurosci.* **5:** 2382–2387.

Herberhold, D., 1968, Comparative histochemical analysis of the peripheral olfactory organ of mammals and fish, *Arch. Klin. exper. Ohren. Nasen. Kehlkopfheilkd.* **190:** 166–182.

Hinds, J. W., Hinds, P. L., and McNelly, N. A., 1984, An autoradiographic study of the mouse olfactory epithelium: Evidence for long-lived receptors, *Anat. Rec.* **210:** 375–383.

Hirsch, J. D., and Margolis, F. L., 1979, Cell suspensions from rat olfactory neuroepithelium: Biochemical and histochemical characterization, *Brain Res.* **161:** 277–291.

Imamura, K., Mori, K., Fujita, S. C., and Obata, K., 1985, Immunochemical identification of subgroups of vomeronasal nerve fibers and their segregated terminations in the accessory olfactory bulb, *Brain Res.* **328:** 362–366.

Iravani, J., Melville, G. N., and Richter, H.-G., 1976, Mucus production influenced by drugs: An electron microscopic study, *Pneumonologie (Suppl.)* **1976:** 267–273.

Jafek, B. W., 1983, Ultrastructure of human nasal mucosa, *Laryngoscope* **93:** 1576–1599.

Jones, L. M., and Michell, R. H., 1978, Stimulus–response coupling at α-adrenergic receptors, *Biochem. Soc. Lond. Trans.* **6:** 673–688.

Kachur, J. F., Miller, R. J., Field, M., and River, J., 1982, Neurohumoral control of ileal electrolyte transport. II. Neurotensin and substance P, *J. Pharmacol. Exp. Ther.* **220:** 456–463.

Kaissling, K.-E., 1986, Chemo-electrical transduction in insect olfactory receptors, *Annu. Rev. Neurosci.* **9:** 121–145.

Kawano, T., and Margolis, F. L., 1985, Catecholamines in olfactory mucosa decline following superior cervical ganglionectomy, *Chem. Senses* **10:** 353–356.

Keast, J. R., Furness, J. B., and Costa, M., 1985, Different substance P receptors are found on mucosal epithelial cells and submucous neurons of the guinea pig small intestine, *Naunyn Schmiedebergs Arch. Pharmacol.* **329:** 382–387.

Key, B., and Giorgi, P. P., 1986, Soybean lectin binding to the olfactory system of Xenopus, in: *Molecular Aspects of Neurobiology* (R. L. Montalcini, P. Calissano, E. R. Kandel, and A. Maggi, eds.), pp. 172–175, Springer-Verlag, Berlin.

Kitamura, S., Ishihara, Y., and Said, S. I., 1980, Effect of VIP, phenoxybenzamine and prednisolone on cyclic nucleotide content of isolated guinea pig lung and trachea, *Eur. J. Pharamacol.* **67:** 219–223.

Kolmer, W., 1927, D. Geruchsorgan, in: *Handbuch der Microskopischen Anatomie des Menschen Bd. 3* Pt. 1, (H. Mollendorff, ed.), pp. 192–249, Springer-Verlag, Berlin.

Kratzing, J. E., 1975, The fine structure of the olfactory and vomeronasal organs of a lizard (*Tiliqua scincoides scincoides*), *Cell Tissue Res.* **156:** 239–252.

Kratzing, J. E., 1984, The anatomy and histology of the nasal cavity of the koala (*Phascolarctos cinereus*), *J. Anat.* **138:** 55–65.

Kurihara, K., and Koyama, N., 1972, High activity of adenyl cyclase in olfactory and gustatory organs, *Biochem. Biophys. Res. Commun.* **48:** 30–34.

LaFerriere, K. A., Arenberg, I. K., Hawkins, J. E., Jr., and Johnsson, L. G., 1974, Melanocytes of the vestibular labyrinth and their relationship to the microvasculature, *Ann. Otol.* **83:** 685–694.

Lancet, D., 1986, Vertebrate olfactory reception, *Annu. Rev. Neurosci.* **9:** 329–355.

Liang, T., and Cascieri, M. A., 1981, Substance P receptor on parotid cell membranes, *J. Neurosci.* **1:** 1133–1141.

Litt, M., and Khan, M. A., 1976, Mucus rheology: Relation to structure and function, *Biorheology* **13:** 37–48.

Lopez-Vidriero, M. T., Das, I., Smith, A. P., Picot, R., and Reid, L., 1977, Bronchial secretion from normal human airways after inhalation of prostaglandin $F_{2\alpha}$, acetylcholine, histamine, and citric acid, *Thorax* **32:** 734–739.

Lundberg, J. M., Anggard, A., Fahrenkrug, J., Hökfelt, T., and Mutt, V., 1980, Vasoactive intestinal polypeptide in cholinergic neurons of exocrine glands: Functional significance of coexisting transmitters for vasodilation and secretion, *Proc. Natl. Acad. Sci. USA* **77:** 1651–1655.

Lundberg, J. M., Anggard, A., Emson, P., Fahrenkrug, J., and Hökfelt, T., 1981, Vasoactive intestinal polypeptide and cholinergic mechanisms in cat nasal mucosa: Studies on choline acetyltransferase and release of vasoactive intestinal polypeptide, *Proc. Natl. Acad. Sci. USA* **78:** 5255–5259.

Mair, R. G., Gesteland, R. C., and Blank, D. L., 1982, Changes in morphology and physiology of olfactory receptor cilia during development, *J. Neurosci.* **7:** 3091–3103.

Margolis, F. L., 1972, A brain protein unique to the olfactory bulb, *Proc. Natl. Acad. Sci USA* **69:** 1221–1224.

Margolis, F. L., 1980, A marker protein for the olfactory chemoreceptor neuron, in: *Proteins of the Nervous System* (Bradshaw, R. A., and Schneider, D., eds.), pp. 59–84, Raven Press, New York.

Margolis, F. L., Grillo, M., Grannot-Reisfeld, N., and Farbman, A. I., 1983, Purification, characterization and immunocytochemical localization of mouse kidney carnosinase, *Biochim. Biophys. Acta* **744:** 237–248.

Medici, T. C., and Radielovic, P., 1979, Effects of drugs on mucus glycoproteins and water in bronchial secretion, *J. Int. Med. Res.* **7:** 434–442.

Mills, J. W., 1985, Ion transport across the exocrine glands of the frog skin, *Eur. J. Physiol.* **405**(Suppl. 1): S44–S49.

Mills, J. W., and Quinton, P. M., 1981, Formation of stimulus-induced vacuoles in serous cells of tracheal submucosal glands, *Am. J. Physiol.* **241:** C18–C24.

Mira, E., 1963, Oxidative and hydrolytic enzymes in Bowman's glands, *Acta Otolaryngol.* **56:** 706–714.

Moran, D. T., Rowley, J. C. III, Jafek, B. W., and Lovell, M. A., 1982, The fine structure of the olfactory mucosa in man, *J. Neurocytol.* **11:** 721–746.

Mori, K., Fujita, S. C., Imamura, K., and Obata, K., 1985, Immunohistochemical study of subclasses of olfactory nerve fibers and their projections to the olfactory bulb in the rabbit, *J. Comp. Neurol.* **242:** 214–229.

Moulton, D. G., Celebi, G., and Fink, R. P., Olfaction in mammals—two aspects: Proliferation of cells in the olfactory epithelium and sensitivity to odors, in: *Taste and Smell in Vertebrates* (G. E. W. Wolstenholme and J. Knight, eds.), pp. 227–250, J. and A. Churchill, London.

Müller, H., Drenckhahn, D., and Haase, L., 1979, Comparative and quantitative ultrastructural studies of the olfactory organs of four strains of domestic pigeons, *Z. Mikrosk. Anat. Forsch.* **93:** 888–900.

Nadel, J. A., 1983, Neural control of airway submucosal gland secretion, *Eur. J. Respir. Dis.* **64**(Suppl 128): 322–326.

Nagahara, Y., 1940, Experimentelle Studien über die histologischen Veränderungen des Geruchsorgans nach der Olfactoriusdurchschneidung. Beiträge zur Kenntnis des feineren Baus der Geruchsorgan, *Japn. J. Med. Sci. V. Pathol.* **5:** 165–199.

Nathanson, I., Widdicombe, J. H., and Barnes, P. J., 1983, Effect of vasoactive intestinal peptide on ion transport across dog tracheal epithelium, *J. Appl. Physiol. Respir. Environ. Exercise Physiol.* **55:** 1844–1848.

Neutra, M., and Leblond, 1966, Synthesis of the carbohydrate of mucus in the Golgi complex as shown by electron microscope radioautography of goblet cells from rats injected with glucose-H^3, *J. Cell Biol.* **30:** 119–136.

Okano, M., and Takagi, S. F., 1974, Secretion and electrogenesis of the supporting cell in the olfactory epithelium, *J. Physiol. (Lond.)* **242:** 353–370.

Olver, R. E., Davis, B., Marin, M. G., and Nadel, J. A., 1975, Active transport of Na^+ and Cl^- across canine tracheal epithelium, *Am. Rev. Resp. Dis.* **112:** 811–815.

Pace, U., Hanski, E., Salomon, Y., and Lancet, D., 1985, Odorant-sensitive adenylate cyclase may mediate olfactory reception, *Nature (Lond.)* **316:** 255–258.

Papka, R. E., and Matulionis, D. H., 1983, Association of substance-P-immunoreactive nerves with the murine olfactory mucosa, *Cell Tissue Res.* **230:** 517–525.

Peatfield, A. C., Barnes, P. J., Bratcher, C., Nadel, J. A., and Davis, B., 1983, Vasoactive intestinal peptide stimulates tracheal submucosal gland secretion in ferret, *Am. Rev. Respir. Dis.* **128:** 89–93.

Persaud, K. C., DeSimone, J. A., Getchell, M. L., Heck, G. L., and Getchell, T. V., 1987, Ion transport across the frog olfactory mucosa: the basal and odorant-stimulated states, *Biochim. Biophys. Acta* **902:** 65–79.

Pevsner, J., Sklar, P. B., and Snyder, S. H., 1986, Odorant-binding protein: Localization to nasal glands and secretions, *Proc. Natl. Acad. Sci. USA* **83:** 4942–4946.

Pholpramool, C., and Tangkrisanavinont, V., 1983, Evidence for the requirement of sympathetic activity in the PGE_1-induced lacrimal secretion in rabbits, *Arch. Int. Pharmacol. Ther.* **265:** 128–137.

Putney, J. W., Jr., 1976, Biphasic modulation of potassium release in rat parotid gland by carbachol and phenylephrine, *J. Pharmacol. Exp. Ther.* **198:** 375–384.

Putney, J. W., Jr., 1979, Receptor regulation of calcium channels in exocrine gland cells, *Proc. West. Pharmacol.* **22:** 295–299.

Quinton, P. M., 1979, Composition and control of secretions from tracheal bronchial submucosal glands, *Nature (Lond.)* **279:** 551–552.

Quinton, P. M., 1981, Possible mechanisms of stimulus-induced vacuolation in serous cells of tracheal secretory glands, *Am. J. Physiol.* **241:** C25–C32.

Rafols, J. A., and Getchell, T. V., 1983, Morphological relations between the receptor neurons, sustentacular cells and Schwann cells in the olfactory mucosa of the salamander, *Anat. Rec.* **206:** 87–101.

Raisman, G., 1985, Specialized neuroglial arrangement may explain the capacity of vomeronasal axons to reinnervate central neurons, *Neuroscience* **14:** 237–254.

Ramón y Cajal, S., 1911, *Histologie du Système Nerveux del l'Homme et des Vértebrés,* Vol. II. (trans. L. Azoulay), pp. 647–655, C.S.I.C., Madrid.

Rasmussen, A. T., 1932, *The Principal Nervous Pathways,* Macmillan, New York.

Reese, T. S., 1965, Olfactory cilia in the frog, *J. Cell Biol.* **25:** 209–230.

Richardson, P. S., Mian, N., and Balfre, K., 1985, The role of calcium ions in airway secretion, *Br. J. Clin. Pharmacol.* **20:** 275S–279S.

Rosenfeld, M. G., Mermod, J., Amara, S. G., Swanson, L. W., Sawchenko, P. E., Rivier, J., Vale, W. W., and Evans, R. M., 1983, Production of a novel neuropeptide encoded by the calcitonin gene via tissue-specific RNA processing, *Nature (Lond.)* **304:** 129–135.

Saini, K. D., and Breipohl, W., 1976, Surface morphology in the olfactory epithelium of normal male and female rhesus monkeys, *Am. J. Anat.* **147:** 433–446.

Schramm, M., and Selinger, Z., 1974, The function of α- and β-adrenergic receptors and a cholinergic receptor in the secretory cell of rat parotid gland, *Adv. Cytopharmacol.* **2:** 29–32.

Schwob, J. E., Farber, N. B., and Gottlieg, D. I., 1986, Neurons of the olfactory epithelium in adult rats contain vimentin, *J. Neurosci.* **6:** 208–217.

Shantha, T. R., and Nakajima, Y., 1970, Histological and histochemical studies on the rhesus monkey (*Macaca mulatta*) olfactory mucosa, *Z. Zellforsch.* **103:** 291–319.

Simmons, P. A., and Getchell, T. V., 1981, Neurogenesis in olfactory epithelium: Loss and recovery of transepithelial voltage transients following olfactory nerve section, *J. Neurophysiol.* **45:** 516–528.

Smith, P. L., Welsh, M. J., Stoff, J. S., and Frizzell, R. A., 1982, Chloride secretion by canine tracheal epithelium. I. Role of intracellular cAMP levels, *J. Membrane Biol.* **70:** 217–226.

Specian, R. D., and Neutra, M. R., 1980, Mechanism of rapid mucus secretion in goblet cells stimulated by acetylcholine, *J. Cell Biol.* **85:** 626–640.

Spicer, S. S., Schulte, B. A., and Thomopoulos, G. N., 1983, Histochemical properties of the respiratory tract epithelium in different species, *Am. Rev. Respir. Dis.* **128:** S20–S26.

Stahl, G. H., and Ellis, D. B., 1973, Biosynthesis of respiratory-tract mucins. A comparison of canine epithelial goblet-cell and submucosal-gland secretions, *Biochem. J.* **136:** 845–850.

Todd, B., and Bowman, W., 1857, *The Physiological Anatomy and Physiology of Man,* Blanchard & Lea, Philadelphia.

Tucker, D., 1971, Nonolfactory Responses from the Nasal Cavity: Jacobson's Organ and the Trigeminal System, in: *Handbook of Sensory Physiology Vol. IV Chemical Senses Part I* (Beidler, L. M., ed.), pp. 151–181, Springer-Verlag, Berlin.

Uddman, R., Malm, L., and Sundler, F., 1980, The origin of vasoactive intestinal polypeptide (VIP) nerves in the feline nasal mucosa, *Acta Otolaryngol.* **89:** 152–156.

van Gehuchten, A., 1890, Contributions a l'étude de la muqueuse olfactive chez les mammifères, *Cellule* **6:** 395–408.

von Brunn, A., 1892, Beiträge zur mikroskopischen Anatomie der menschlichen Nasenhöhle, *Arch. Mikrosk. Anat.* **39:** 632–651.

Welsh, M. J., Widdicombe, J. H., and Nadel, J. A., 1980, Fluid transport across the canine tracheal epithelium, *J. Appl. Physiol. Respirat. Environ. Exer. Physiol.* **49:** 905–909.

Wesolowski H., 1967, The behaviour of mitochondria and the secretion of the olfactory epithelial cells and the olfactory glands after pyridine stimulation in domestic birds, *Folia Biol.* **15:** 303–324.

Widdicombe, J. H., and Welsh, M. J., 1980, Ion transport by dog tracheal epithelium, *Fed. Proc.* **39:** 3062–3066.

Wirsig, C. R., and Getchell, T. V., 1986, Amphibian terminal nerve: Distribution revealed by LHRH and AChE markers, *Brain Res.* **385:** 10–21.

Yamamoto, M., 1976, An electron microscopic study of the olfactory mucosa in the bat and rabbit, *Arch. Histol. Jpn.* **38:** 359–412.

Zeiske, E., Melinkat, R., Breucker, H., and Kux, J., 1976, Ultrastructural studies on the epithelia of the olfactory organ of cyprinodonts (Teleostei, Cyprinodontoidea), *Cell Tissue Res.* **172:** 245–267.

Zinnin, K., 1964, Histochemical studies of the olfactory mucosa of the rabbit, *J. Wakayama Med. Soc.* **15:** 213–221.

5

Autoradiographic Localization of Drug and Neurotransmitter Receptors in the Olfactory Bulb

Valina L. Dawson, Ted M. Dawson, and James K. Wamsley

1. INTRODUCTION

Olfaction plays a dominant role in most vertebrate species in mediating behaviors such as mating, feeding, maternal care, and social organization. The olfactory bulb is the first synaptic relay in the olfactory pathway; thus, it provides the first region for integration of stimuli generated at the olfactory mucosa receptor cell before progression of the stimulus to the cortical centers (Shepherd, 1977). The olfactory bulb has been shown to be a cornucopia of neuroactive substances that mediate action at this relay between sensory input and projections from the olfactory bulb to central limbic structures, which are thought to modulate behavior (Macrides and Davis, 1983). In addition to containing a wide variety of neuroactive substances, the olfactory bulb is well organized into layers. The distinct lamination is comparable to the level of organization found in the cerebellum and retina. The neurons and synaptic contacts that make up these laminae have been identified and well characterized. The cell types and their relationship to the various laminae are reviewed in Fig. 1.

The principal neurons in the olfactory bulb are Golgi type I cells, also known as mitral cells (Gershon *et al.*, 1981). Mitral cell bodies are located in the mitral cell layer; these cells extend a primary dendrite to the glomerular layer, which ends as a tuft in a glomerulus. The mitral cell tuft forms a primary synaptic contact on the olfactory nerve from receptor cells (Shepherd, 1977). Furthermore, each mitral cell projects horizontal and tangential secondary dendrites into the external plexiform layer (EPL) to form dendrodendritic synapses on granule cells (Gershon *et al.*, 1981). Axons of the mitral cells exit the olfactory bulb by the lateral olfactory tract (LOT) (Shepherd, 1977).

Other neurons in the olfactory bulb are tufted cells, interneurons, and Golgi type II

Valina L. Dawson and James K. Wamsly • Departments of Pharmacology and Psychiatry, University of Utah School of Medicine, Salt Lake City, Utah 84132. *Ted M. Dawson* • Department of Neurology, Hospital of the University of Pennsylvania, Philadelphia, Pennsylvania 19143.

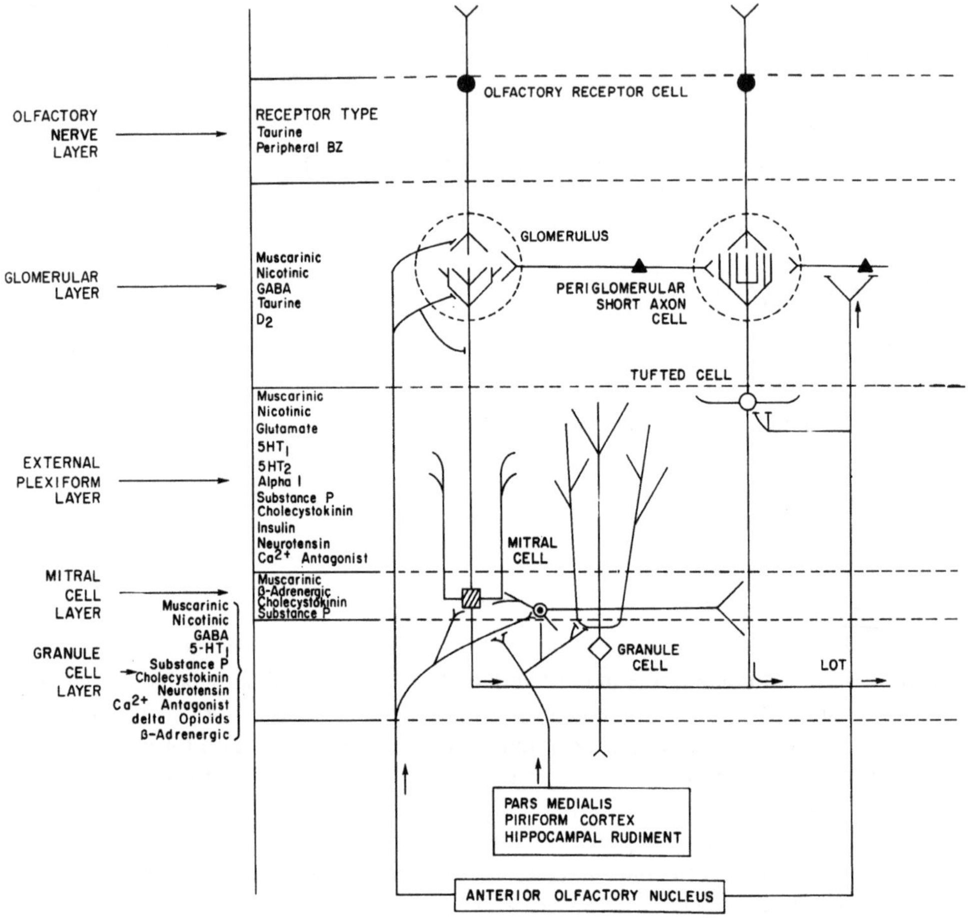

Figure 1. Schematic representation of the structural and cellular interrelationships of the olfactory bulb and the receptor types associated with the various nerve layers of the olfactory bulb. Receptor types are noted on the left-hand side of the figure, as well as the nerve layer in which they have been localized. See text for details concerning the receptors associated with the olfactory bulb and corresponding references. BZ, benzodiazepine; D_2, dopamine type 2; GABA, γ-aminobutryic acid; LOT, lateral olfactory tract; 5-HT_1, serotonin type 1; 5-HT_2, serotonin type 2. (Adapted from Halasz and Shepherd, 1983, and Davis and Macrides, 1981.)

cells—the periglomerular cells and granule cells (Gershon *et al.*, 1981). Tufted cell morphology is similar to that of mitral cells; as such, they are often considered a smaller version of the mitral cells. Most of the tufted cell bodies are located in the middle to outer layers of the EPL and have a primary dendritic projection to a glomerulus and have collateral projections to the EPL. The axons of tufted cells leave the olfactory bulb by the LOT (Shepherd, 1977). Periglomerular cells are short axon cells located in the glomerular layer. The dendrites of periglomerular cells synapse on mitral cell dendrites at one glomerulus with an axon projecting laterally and terminating on mitral cell dendrites in an adjacent glomerulus. These dendrodendritic synapses between the mitral cells and periglomerular cells are reciprocally inhibitory (Gershon *et al.*, 1981). The most numerous cells in the olfactory bulb are the granule cells. These small-bodied cells have no axons.

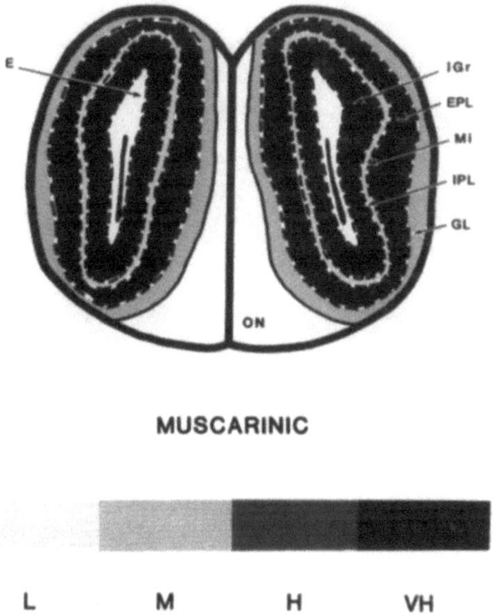

Figure 2. Distribution and density of muscarinic receptors in the laminae of the rat olfactory bulb. Schematic diagram summarizing data obtained by quantitative autoradiography in which [^{3}H]propylbenzilycholine mustard was used to label the muscarinic receptor (Rotter *et al.*, 1979*a*). Densities of muscarinic receptors were arbitrarily assigned values in which very high was 75–100% of the highest reported specific binding in the brain, high ran 50–75%, moderate was 25–50%, and low was less than 25% of the highest reported specific binding. Schematic patterns are shown in the key at the bottom of the figure. Note the very high density of muscarinic receptors in the external plexiform layer and the internal granular layer. E, ependyma and subependymal layer; EPL, external plexiform layer; GL, glomerular layer; H, high; IGr, internal granular layer; L, low; Mi, mitral cell layer; M, moderate; ON, olfactory nerve layer; VH, very high. (Modified from Paxinos and Watson, 1982, abbreviations according to Paxinos and Watson, 1982.)

The highest density of granule cells are located in the granule cell layer, although some granule cell neurons can be located in the mitral cell layer (Shepherd, 1977). The dendrites of the granule cells project deep in the bulb and extend peripherally to the EPL, where they synapse with the secondary dendrites of the mitral/tufted cells. When the mitral or tufted cell is excited, the secondary dendrites excite the granule cells. These granule cells then reciprocally and in an orthograde fashion inhibit the initial mitral/tufted cell as well as other mitral/tufted cells in which the granule cell has synaptic contact (Gershon *et al.*, 1981; Halasz and Shepherd, 1983).

The distinct lamination and segregation of the intrinsic components of the olfactory bulb make this structure ideal not only for neurochemical and electrophysiological studies but for receptor autoradiography as well. The defined internal layers of the olfactory bulb permit visualization of neurotransmitter and drug-binding sites in the individual laminae by autoradiography (Figures 2–8). As research into the receptor populations of the olfactory bulb expands, these data can be interpreted as they pertain to the knowledge now existing on the neuronal morphology and synaptic organization of the olfactory bulb, its connections to the central nervous system (CNS), and its role in regulating behavior and hormonal patterns. This chapter emphasizes recent findings in macrosomatic animals using light microscopic receptor autoradiography.

2. LIGHT MICROSCOPIC RECEPTOR AUTORADIOGRAPHY

Autoradiography is an anatomical technique that permits visualization of the sequestration of radioactive substances into various tissues by juxtaposition of the labeled tissue to a photographic emulsion. Young and Kuhar (1979) were able to apply standard radioligand binding methods (see Bennett and Yamamura, 1985, for review) to slide mounted tissue sections; by apposing these labeled sections to emulsion-coated cov-

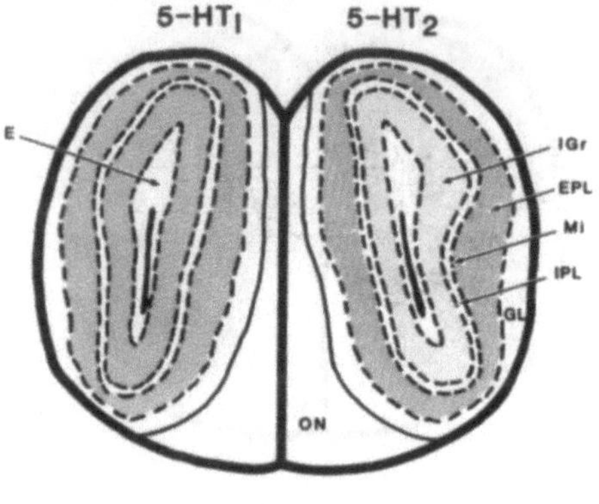

Figure 3. Distribution and density of serotonin receptors in the laminae of the rat olfactory bulb. Schematic diagram representing data obtained by quantitative autoradiography in which 5-HT_1 receptors were labeled with [^{3}H]-5-HT (Pazos and Palacios, 1985) and 5-HT_2 receptors were labeled with [^{3}H]mesulergine (Pazos *et al.*, 1985). For a description of abbreviations used and a key to schematic patterns, see Fig. 2.

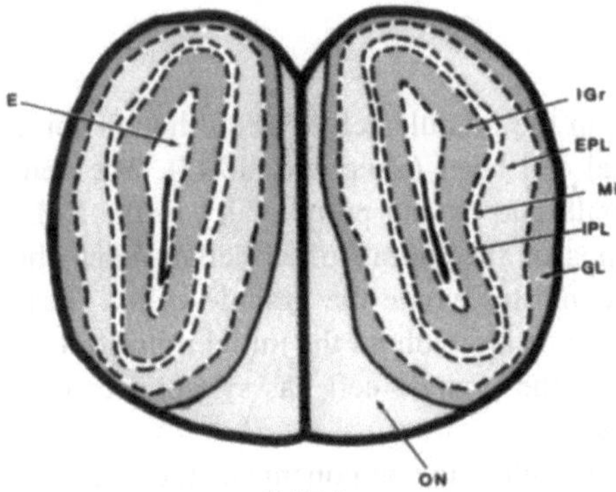

Figure 4. Distribution and density of GABA receptors in the laminae of the rat olfactory bulb. Schematic diagram summarizing data obtained by quantitative autoradiography in which [^{3}H]-GABA was used to label all subtypes of GABA receptor (Jaffe *et al.*, 1983). For a description of abbreviations used and a key to schematic patterns, see Fig. 2.

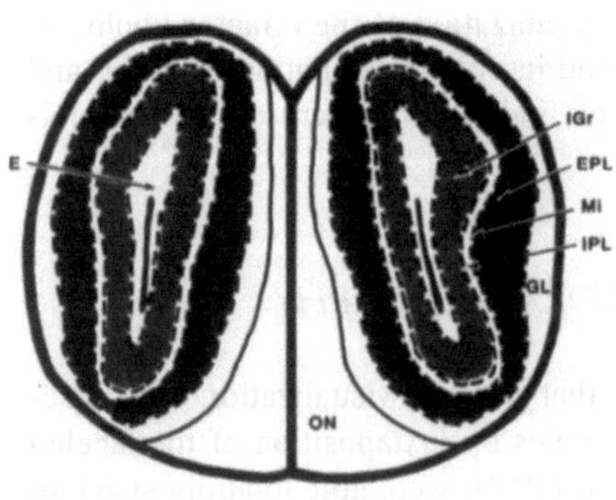

Figure 5. Distribution and density of neurotensin receptors in the laminae of the rat olfactory bulb. Schematic diagram summarizing data obtained by quantitative autoradiography in which [^{3}H]neurotensin was employed to label neurotensin receptors (Young and Kuhar, 1981). For a description of abbreviations used and a key to schematic patterns, see Fig. 2.

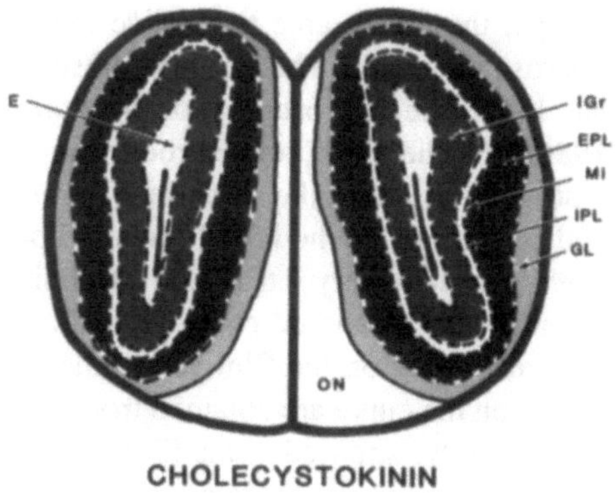

Figure 6. Distribution and density of cholecystokinin receptors in the laminae of the rat olfactory bulb. Schematic diagram summarizing data obtained from quantitative autoradiography in which [^{125}I]cholecystokinin was used to label the receptor. For a description of abbreviations used and a key to schematic patterns, see Fig. 2.

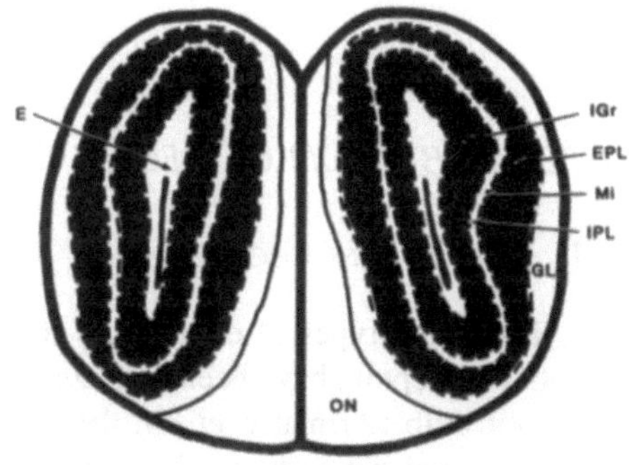

Figure 7. Distribution and density of substance P receptors in the laminae of the rat olfactory bulb. Schematic diagram summarizing data obtained by quantitative autoradiography in which [^{3}H]substance P (Quirion *et al.*, 1984) and [^{125}I]substance P (Rothman *et al.*, 1984) were employed to label the receptor. For a description of abbreviations used and a key to schematic patterns, see Fig. 2.

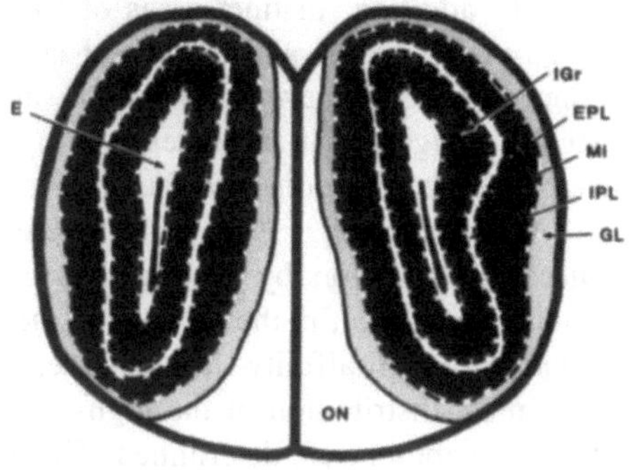

Figure 8. Distribution and density of calcium antagonist binding sites in the laminae of the rat olfactory bulb. This schematic diagram is a summary of data obtained by quantitative autoradiography in which [^{3}H]-PN 200-110 and [^{3}H]-PY 108-068 were employed to label the calcium antagonist binding sites (Cortes *et al.*, 1984). For a description of abbreviations used and a key to schematic patterns, see Fig. 2.

erslips, they were able to provide the first *in vitro* localization of receptors at the light microscopic level. This method subsequently proved a widely applicable technique that has led to the microscopic localization of many different receptor types (Kuhar *et al.*, 1986). Furthermore, the introduction of tritium-sensitive film, LKB Ultrofilm (LKB Instruments, Rockville, Maryland) (Palacios *et al.*, 1981*b*) provided a relatively simple and convenient method for the autoradiographic localization of receptors. Unnerstall *et al.* (1982) subsequently introduced a technique for quantitation of autoradiographic grains by reference to a standard curve plotted as a linear regression line of the natural log of the optical density on the film versus the natural log of the disintegrations per minute. The points on the curve are obtained from tritium brain paste standards with known amounts of radioactivity, apposed to the film, from which the actual amount of ligand bound per mg tissue or protein can be determined in discrete microscopic regions of the brain. Computerized methods are now available for autoradiography and provide a rapid means for quantitation and analysis of autoradiograms (Kuhar, 1985).

A detailed discussion of receptor autoradiography is beyond the scope of this chapter. The reader is referred to Wamsley and Palacios (1983), Kuhar (1985), and Kuhar *et al.* (1986) for discussion of methods and the theoretical limitations of this technique.

3. DISTRIBUTION OF RECEPTOR TYPES IN THE OLFACTORY BULB

3.1. Cholinergic Receptors

Two main types of cholinergic receptors are recognized: nicotinic and muscarinic. Nicotinic receptors have been localized, using [^{125}I]alphabungarotoxin to label the mouse olfactory bulb (Arimatsu *et al.*, 1981), in low to moderate densities over the external plexiform and glomerular layers of the main and accessory bulb. Muscarinic receptors have been localized using the irreversible ligand [^{3}H]propylbenzilylcholine mustard (Rotter *et al.*, 1979*a*). The highest binding occurred in the external plexiform layer, which also had the highest binding in the brain. Furthermore, high levels of binding were observed in the glomerular layer, the mitral cell layer, and the granule cell layer. Ontogeny studies (Rotter *et al.*, 1979*b*) revealed high levels of [^{3}H]propylbenzilylcholine binding in the glomerular layer, external plexiform layer, and granule layer at 1 day after birth. In addition, distinct areas of low-density binding were localized over groups of periglomerular cells and the mitral cell body layer. At day 17, binding was markedly reduced in the glomerular layer and was absent in the periglomerular layer. Rotter *et al.* (1979*b*) suggest that these results imply that receptor synthesis occurs in advance of established connections and therefore may play a role in synaptogenesis.

Wamsley *et al.* (1980, 1983) were able to determine the anatomical distribution of muscarinic receptors by utilizing [^{3}H]-*N*-methylscopolamine. By including a muscarinic agonist, carbachol, in the incubation media for selective displacement of the radiolabeled ligand from high-affinity sites while leaving the binding to the low-affinity sites intact, the anatomical distribution of the high- and low-agonist affinity conformations of the muscarinic receptors were determined. The highest concentration of low-affinity sites found in the brain occurred in the external plexiform layer of the olfactory bulb. In general, other layers of the olfactory bulb also contained high concentrations of low-affinity sites. The lowest level of binding in the olfactory bulb took place in the glomerular layer, with these sites concentrated in the glomeruli. The olfactory bulb had a low density of high-affinity sites in all layers compared with the rest of the brain.

Muscarinic receptors can also be divided into M_1 and M_2 sites on the basis of the affinity of the receptor for the antagonist, pirenzepine (Watson *et al.*, 1983; Hammer *et al.*, 1980). M_1 receptors are labeled using the ligand, [^{3}H]pirenzepine. By including unlabeled pirenzepine in the incubation media of [^{3}H]-*N*-methylscopolamine or [^{3}H]propylbenzilylcholine, the M_2 receptor subtype may be differentiated. The distribution of the M_1 and M_2 receptor subtypes in the olfactory bulb is yet to be determined.

3.2. Biogenic Amines

3.2.1. Dopamine

Two subtypes of dopamine receptors have been shown to exist based on dopamine agonist stimulation of adenylate cyclase (Kebabian and Calne, 1979). Stimulation of dopamine type 1 (D_1) receptors results in the activation of adenylate cyclase, whereas stimulation of dopamine type 2 (D_2) receptors results in decreased adenylate cyclase activity. When serial sections of rat olfactory bulb tissue were incubated with [^{3}H]dopamine, light labeling was observed over the olfactory nerve and external plexiform layers (Priestley *et al.*, 1979). The heaviest labeling was seen in the periglomerular layer and was confined to the glomeruli themselves. The short-axon periglomerular cell dendrites pass deep into the glomerulus and form synaptic arrangements with both mitral and tufted cells. It is these dendrites from the periglomerular cells that the ligand is, at least in part, labeling (Halasz and Shepherd, 1983), suggesting that dopamine is one of the neurotransmitters employed at these synapses.

Selective ligands for the D_1 and D_2 receptor sites have been developed, and autoradiographic techniques have shown that there is no significant labeling of D_1 receptors, labeled with [^{3}H]-SCH 23390, in the olfactory bulb (Dawson *et al.*, 1986). However, very high labeling of D_2 receptors, labeled with [^{3}H]sulpiride (Gehlert and Wamsley, 1985), was observed over the glomerular layer of the olfactory bulb. This finding suggests that the receptors on periglomerular cells labeled with [^{3}H]dopamine may be only D_2 receptors.

3.2.2. Serotonin

Serotonin receptors can be localized in the brain with the use of a variety of ligands. Recent studies, focusing on serotonin subtypes 5-hydroxytryptamine 1 and 2 (5-HT_1 and 5-HT_2), showed a significant localization of serotonin receptors in the olfactory bulb (Pazos and Palacios, 1985). 5-HT_1 sites can be characterized by the binding of [^{3}H]-5-HT, and these sites can be further divided into high-affinity 5-HT_{1A} receptors (labeled with 8-OH-[^{3}H]-DPAT), low-affinity 5-HT_{1B} receptors that are not labeled with [^{3}H]-LSD, and 5-HT_{1C} receptors that label with [^{3}H]mesulergine (Pazos and Palacios, 1985). The external plexiform layer showed intermediate labeling of 5-HT_1 receptors, and the internal granule layer had a moderate density of labeling. The 5-HT receptors in the external plexiform layer were mainly of the 5-HT_{1A} type, while the granule cell layer had mainly 5-HT_{1B} receptors.

Labeling for the low-affinity 5-HT_2 receptor sites showed only low densities in the external plexiform layer and very low to background labeling densities in the other layers of the olfactory bulb (Pazos *et al.*, 1985). Early studies employed [^{3}H]spiperone as a ligand to label dopamine and serotonin receptors (Palacios *et al.*, 1981*a*). Since this compound only binds to the 5-HT_2 subtype of the serotonin receptor, it is easy to understand why no serotonin receptor populations were reported initially in the olfactory bulb. Intraventricular injections of [^{3}H]serotonin into the rat olfactory bulb showed labeling of granule, external plexiform, and glomerular layer nerve terminals (Halasz *et al.*, 1979). Some cells in these layers presumably contain serotonin as their neurotransmitter. A few of these labeled nerve terminals were shown to synapse on periglomerular cells. It has been demonstrated by in vivo techniques that [^{3}H]-5-HT is taken up by cells in the olfactory bulb and transported in a retrograde axonal fashion to the cell body in the midbrain raphé nuclei. At the electron microscopic level, [^{3}H]serotonin was found to be associated with mitochondria, lysosomes, and endoplasmic reticulum. The association with these subcellular compartments suggests that the serotonin is being either degraded or processed for re-release (Araneda *et al.*, 1980).

3.2.3. Norepinephrine

Norepinephrine acts at α- and β-adrenergic receptors. Using the ligand [^{3}H]prazosin, which primarily binds to α_1-receptors, labeling was found to occur in the olfactory bulb; however, the specific lamina where the binding took place was not specified (Dashwood, 1983). Binding in the bulb was also reported using the α_1-antagonist WB-4101, with the highest density of labeling concentrated primarily in the external plexiform layer (Palacios and Wamsley, 1983). The binding of the β-adrenergic receptor ligand, [^{3}H] dihydroalprenolol, in the olfactory bulb of the dog was very low and localized to the mitral cell-granule cell layer (Nadi *et al.*, 1980).

3.2.4. Histamine

Histamine type 1 (H_1) receptors are characterized by their interaction with classical antihistamines and are localized by the ligand [^{3}H]mepyramine (Wamsley and Palacios, 1984*a*). Histamine type 2 (H_2) receptors, although characterized, do not currently have an exclusive ligand and so these receptor populations have not been well defined. With [^{3}H]mepyramine, low to very low labeling was observed in the olfactory bulb (Wamsley and Palacios, 1984*a*). Since biochemical studies show low concentrations of histamine in the olfactory bulb (Schwartz *et al.*, 1970), it may be that histamine is not a major neuroactive substance in the bulb.

3.3. Amino Acids

3.3.1. Glutamate

[^{3}H]Glutamate has been used autoradiographically to label all glutamate receptor subtypes. The molecular layers and nuclei of the olfactory bulb showed high-binding densities; however, binding of [^{3}H]glutamate in the glomerular layer was not above background (Greenamyre *et al.*, 1984). The latter observation stands in contrast to the high concentration of glutamate known to be present in the glomerular layer of the dog olfactory bulb (Nadi *et al.*, 1980), suggesting that in this area either the endogenous amino acid represents a component of a metabolic pool (rather than serves a transmitter function), relates to species differences, or is an example of the mismatch problem wherein distribution of receptors and the endogenous ligand do not correspond anatomically.

Kainic acid is an analogue to glutamate that is postulated to bind to a subset of the glutamate binding sites. The tritiated form of this ligand has been localized to the mitral and granule cell layers of the dog olfactory bulb, where these two cell types reciprocally form dendrodendritic synapses (Nadi *et al.*, 1980). In the rat, Unnerstall and Wamsley (1983) reported low levels of binding in the glomerular layer and a high density of [^{3}H]kainic acid binding in the external plexiform and granule cell layers.

3.3.2. Aspartate

Aspartate has an excitatory effect on the granule cells of the olfactory bulb (Nicoll, 1971) and shows retrograde axonal transport from the bulb to the cell bodies in the anterior olfactory nucleus, pyriform cortex, and nucleus of the lateral olfactory tract (Watanabe and Kawana, 1983). Labeling with the ligand [^{3}H]-*N*-methyl-D-aspartate demonstrates a high density of binding sites in the external plexiform layer (Monaghan and Cotman, 1985).

3.3.3. Glycine

Glycine is proposed to be an important inhibitory transmitter in the CNS. Although small concentrations of glycine are reported in the olfactory bulb of the dog (Nadi *et al.*, 1980), localization of the glycine receptor with [^{3}H]strychnine was either very low (Zarbin *et al.*, 1981) or nonexistent (Wamsley and Palacios, 1984*b*).

3.3.4. Taurine

Taurine is a free amino acid present in most tissues, yet it is not incorporated into proteins. It is present in the developing CNS, but its physiological role is unknown. [^{3}H]Taurine has been localized to the olfactory fiber and glomerular layers of the bulb (Quinn *et al.*, 1981) and has been reported to be evenly distributed throughout the olfactory bulb structure (Nadi *et al.*, 1980; Austin *et al.*, 1978). The olfactory receptor cells undergo periodic turnover (Halasz and Shepherd, 1983), causing the olfactory nerve terminals in the glomerular layer of the olfactory bulb to experience periodic turnover. In this way, the glomerulus may resemble the developing CNS. Thus, the study of taurine in the olfactory bulb may provide evidence for the role of this amino acid in development of the nervous system.

3.3.5. γ-Aminobutyric Acid

γ-Aminobutyric acid (GABA) is an important inhibitory neurotransmitter in the CNS. GABA receptor sites have been localized and differentiated into high- and low-affinity $GABA_A$ sites using the ligands [^{3}H]-GABA (Halasz *et al.*, 1979; Jaffe *et al.*,

1983), [^{3}H]muscimol (Palacios *et al.*, 1981*c*), and [^{3}H]bicuculline methochloride in the presence of thiocyanate (Olsen *et al.*, 1984); and $GABA_B$ receptor sites with [^{3}H]baclofen (Gehlert *et al.*, 1985*c*). [^{3}H]-GABA labels both the high- and low-affinity sites, resulting in a total representation of GABA receptors in a given structure. *In vitro* incubation with [^{3}H]-GABA and β-alanine (to block nonspecific neuropil GABA sites) showed the heaviest labeling in the glomerular and granule cell layers and moderate labeling in the external plexiform layer (Jaffe *et al.*, 1983). Electron microscopic localization, following *in vivo* labeling showed binding of dendritic terminals around unlabeled mitral cells (Jaffe *et al.*, 1983). A few short-axon cell, granule cell, and periglomerular cell somata showed [^{3}H]-GABA labeling (Halasz *et al.*, 1979). High amounts of labeling were observed at granule cell terminals and dendrites with the mitral cell dendrites having the lowest labeling. Binding was observed in the short axon nerve terminals that synapse with granule cell bodies (Halasz *et al.*, 1979).

High- and low-affinity GABA sites have been characterized by autoradiography. Although the high-affinity sites overlap with the low-affinity and benzodiazepine sites in some areas, it appears that it is the low-affinity GABA receptor that is associated with benzodiazepine receptor populations (Wamsley *et al.*, 1986). The high-affinity sites, labeled with [^{3}H]muscimol, are concentrated in the external plexiform layer and, to a lesser extent, in the glomeruli. The other layers of the bulb showed low to background amounts of labeling (Palacios *et al.*, 1981*c;* Nadi *et al.*, 1980). While high-affinity GABA receptor sites are present, the olfactory bulb is devoid of low-affinity GABA receptor sites (Bowery *et al.*, 1984).

3.4. Neuropeptides

3.4.1. Neurotensin

Neurotensin is a tridecapeptide neurotransmitter in the CNS. In the developing brain, one of the first areas to show the presence of neurotensin, as detected by immunohistochemical analysis, is the olfactory bulb (Hara *et al.*, 1982). Using a tritiated form of the peptide, receptor populations for neurotensin have been shown to be localized in very high concentrations in the external plexiform layer of both the olfactory and accessory bulb. This high-density binding was reported to be localized in the internal granule cell layer (Young and Kuhar, 1981). In human tissue, the only olfactory area examined to date is the paraolfactory gyrus cortex, which labeled with very high density when [^{125}I] neurotensin was employed as the ligand (Sarrieau *et al.*, 1985).

3.4.2. Insulin

The role of insulin and insulin receptors in the brain is unknown, but autoradiographic studies and ontogeny studies suggest that insulin may play a role in development. Insulin may also be a neurotransmitter (Young *et al.*, 1980) or neuromodulator, as localization of insulin receptors is similar to that observed for other putative neurotransmitters. The adult rat brain labels relatively evenly with [^{125}I]insulin except for the high-density binding seen in the external plexiform layer of the olfactory bulb. Some binding was noted below the mitral cells, but the rest of the bulb had a background level (Young *et al.*, 1980).

3.4.3. Cholecystokinin

Receptors for the peptide cholecystokinin (CCK), generally recognized as an intestinal hormone involved in regulating gallbladder and pancreatic functions, have been localized in guinea pig and rat brain (Zarbin *et al.*, 1983) using the ligand [^{125}I]-CCK. In the guinea pig olfactory bulb, very dense receptor populations were seen over the external plexiform layer. A high density of these receptors was found in the mitral cell layer and lower, but very dense labeling was still observed in the internal granule cell layer. Moderate density binding was seen in the glomerular layer and the external granular layer. The rat brain showed similar labeling with the exception of a high density of binding in the glomerular layer (Zarbin *et al.*, 1983). Moderate binding of the ligand was also observed in areas containing nerve fibers. These binding sites may be associated with receptors in transport or represent axonal receptors (Zarbin *et al.*, 1983).

3.4.4. Angiotensin II

There is increasing evidence for a discrete renin-angiotensin system in the brain. Using [^{125}I]angiotensin II, receptors for this peptide have been localized in the rat brain. Moderate densities of labeling were confined to the external plexiform and glomerular layers of the olfactory bulb. Other areas of the brain involved with processing sensory information also bound the ligand, suggesting involvement of angiotensin II in the processing of somatic sensory information (Gehlert *et al.*, 1986).

3.4.5. Somatostatin

The tetradecapeptide, somatostatin, is widely distributed in the CNS, where receptors for this peptide have been localized with the radiolabeled ligands, [^{125}I]-tyr^{11}SS (Tran *et al.*, 1984), [^{125}I]-SS14 (Leroux *et al.*, 1985), [^{125}I]-SS28 (Beal *et al.*, 1986; Leroux *et al.*, 1985; Tran *et al.*, 1984), [^{125}I]-204-090 (Ruebi and Maurer, 1985), and [^{125}I]-CGP 23996 (Gulya *et al.*, 1985). In the rat brain, high levels of binding are reported in the olfactory bulb with the ligands [^{125}I]-SS28 and [^{125}I]-SS14 (Leroux *et al.*, 1985). High-density labeling was also reported in the anterior olfactory nucleus and tubercle with the ligand [^{125}I]-204-090 (Reubi and Maurer, 1985). Other sensory and limbic systems were labeled, indicating a role for somatostatin as a possible neurotransmitter or modulator in the somatic sensory system (Beal *et al.*, 1986; Gulya *et al.*, 1985; Leroux *et al.*, 1985; Ruebi and Maurer, 1985; Tran *et al.*, 1984).

3.4.6. Substance P

Historically, studies of substance P have focused on the pain-modulating effects of this neuropeptide, even though substance P has been found to be associated with many unrelated areas of the brain, presumably involved with other sensory functions. Autoradiographic localizations using the ligand, [^{3}H]substance P (Quirion *et al.*, 1984) and [^{125}I] BH-SP (Rothman *et al.*, 1984), showed very high-density binding to the olfactory bulb. The binding was localized in the external plexiform layer, the mitral cell layer, and the adjacent half of the internal granule cell layer.

3.4.7. Thyrotropin-Releasing Hormone

Using the ligand [^{3}H]-MeTRH, thyrotropin-releasing hormone (TRH) receptor populations were localized in rat brain. Very dense labeling was seen in the accessory bulb (Manaker *et al.*, 1985; Mantyh and Hunt, 1985). In the olfactory bulb proper, the inner plexiform and mitral cell layers had moderate- to low-density binding, and the external plexiform and granule cell layers exhibited a low density of receptors (Mantyh and Hunt, 1985).

3.4.8. Opiate Receptors

In the dog olfactory bulb, μ-opiate receptors were localized to the fiber and mitral–granule cell layers using [^{3}H]dihydromorphine (Nadi *et al.*, 1980). Interestingly, the olfactory tubercle shows a preferential localization of Δ-receptors with the radiolabeled enkephalin, [^{125}I]-DADL, in the presence of 1 nM FK 33-824, a μ-receptor selective enkephalin (Goodman *et al.*, 1980). Comparison of the localization of Δ- and μ-receptors shows these two subtypes to be concentrated in distinct areas, reinforcing other data suggesting that the two opioid receptor subtypes subserve different functions (Goodman *et al.*, 1980). Ontogenic autoradiographic studies show high concentrations of μ-receptors existing in the olfactory bulb in the 2-day-old rat. A similar density is found throughout the first 3 weeks, followed by a decline to the moderate to low level of receptors found in the adult rat (Unnerstall *et al.*, 1983). During development, the opiate receptor density is first highest in the external plexiform layer and later appears more concentrated in the glomerular layer, which resembles the pattern seen in the adult animal. Neurogenesis studies in the mouse indicate that mitral cells appear at embryonic days 13–14 and that tufted cells arise at E18 (Hinds, 1968). Therefore, this change in binding sites may be associated with the movement of newly synthesized receptors from the cell body to the developing synaptic zone, or the decrease of receptors in the EPL could be due to dilution as new neurons infiltrate this lamina (Unnerstall *et al.*, 1983).

3.5. Drug-Binding Sites

3.5.1. Benzodiazepine

Benzodiazepine (BZ) receptors are divided into the central BZ_1 and BZ_2 receptor types and the peripheral benzodiazepine receptors. In the olfactory bulb of the dog, the central type of benzodiazepine receptor was localized to the glomerular and mitral cell layers with the ligand [^{3}H]diazepam (Nadi *et al.*, 1980). Ontogeny studies, using the ligand [^{3}H]methylclonazepam, show the appearance of central type BZ receptors early in gestation (day 19), but the receptors do not reach near adult levels until the postnatal period (day 8) (Anholt *et al.*, 1984). The peripheral type of BZ receptor populations has been characterized by two ligands. PK 11195 shows a high level of binding in the olfactory bulb nerve tract (Benavides *et al.*, 1983), and low levels of binding were observed on the glomerular layer of the bulb when using [^{3}H]-RO 5-4864 (Gehlert *et al.*, 1985*b*). Using the latter ligand, substantial densities of binding have been shown in the nasal epithelium, the nerve fiber layer and the adjacent glomerular region. The grain patterns in the olfactory receptor cells in the nasal epithelium suggest that the receptors are distributed over the short process extending into the mucous layer of the nasal cavity and the long axon projecting into the olfactory bulb, making synapses in the glomeruli. The glomeruli are the termination point for the primary olfactory nerves, where they synapse with mitral cells and other interneurons (Anholt *et al.*, 1984). Destruction of the nasal epithelium with $ZnSO_4$ reduces the binding of [^{3}H]-RO 5-4864 to almost background levels in the bulb, suggesting that the action of the peripheral BZ receptor is presynaptic. The same lesion does not, however, alter the binding of the central type BZ receptor ligand [^{3}H]methylclonazepam, indicating that the central BZ receptors are not directly involved with the afferent signals from the olfactory receptor cells (Anholt *et al.*, 1984). Ontogeny studies show that near-adult levels of peripheral type BZ receptors could be localized in the olfactory bulb early in gestation (day 19) (Anholt *et al.*, 1984). Location of the binding of a ligand specific for peripheral-type benzodiazepine receptors suggests a possible role for peripheral type BZ receptors in odor recognition, modulation of olfactory nerves or modulation of neurotransmitter release.

3.5.2. Buspirone

The anxiolytic drug, buspirone, has been shown to interact with dopaminergic and 5-HT_2 sites. It is not surprising that [^{3}H]buspirone localizes with high-density binding in the glomerular layer of the olfactory bulb. This finding does not imply, however, that buspirone acts on dopamine or serotonin receptors but that it is associated with them by binding studies (Kaulen *et al.*, 1985).

3.5.3. Calcium Antagonist-Binding Sites

Calcium antagonist-binding sites (CABS) have pharmacological properties similar to those of drug receptor sites associated with calcium channels. Binding of the ligands [^{3}H]-PN 200-110 and [^{3}H]-PY 108-068 localized with high density in the external plexiform layer, high density in the granule layer, and moderate density in the glomerular layer. The external plexiform layer displayed the highest specific binding in the rat brain, suggesting association of CABS with the dendrites of the mitral and granule cells (Cortes *et al.*, 1984). [^{3}H]Nitrendpine exhibits the same general pattern but shows circular patterns of binding in the glomerular layer, suggesting that the binding is associated with the glomerular tufts, where the olfactory nerves synapse with the dendrites of the mitral cells. It is suggested that CABS are localized to areas of concentrated synaptic interaction and represent sites of receptors for endogenous neurotransmitters whose receptor action is linked to calcium channels (Murphy *et al.*, 1982).

3.6. Miscellaneous

3.6.1. Adenylate Cyclase

A potential tool to investigate the density and distribution of adenylate cyclase is [^{3}H]forskolin (Gehlert *et al.*, 1985*a*). Adenylate cyclase is an important intermediate in the translation of the receptor action on the cell. The olfactory bulb expressed low levels of [^{3}H]forskolin binding, indicating that low levels of activated adenylate cyclase were present, in the external plexiform layer and the glomerular layer. Since many receptor types are localized to these areas, it is possible that adenylate cyclase may play an intermediary role in the translation of sensory information (Gehlert *et al.*, 1985*a*).

4. CONCLUSIONS

The olfactory bulb is a well-laminated structure the cell morphology of which has been well characterized. For these reasons, it is well suited for autoradiography. The technique of *in vitro* receptor autoradiography permits localization of receptors and drug-binding sites to discrete anatomical structures. While in many cases the binding sites correspond with their neurotransmitter distribution as determined by immunohistochemical studies, there are cases in which the neurotransmitter and binding sites do not correspond. This has been termed the mismatch problem (Kuhar *et al.*, 1986) and has been attributed to subtypes of receptors, the occasional cross-reactivity seen with [^{3}H]serotonin, newly synthesized receptors, and receptors in transport to and from the cell body. In addition, immunohistochemical methods that label the neurotransmitter substance identify the chemical not only in the nerve terminal but also in other forms such as vesicular, cytoplasmic, or storage that would contribute to the mismatch problem (Wamsley, 1984).

Localization of receptors or drug-binding sites to a certain area does not indicate the presence of a functional receptor or drug-binding site, although function may be inferred from the known actions of the neurotransmitter at that site or in other areas of the nervous system (Wamsley, 1984). Determination of the biological significance of specific receptor population can be determined by the use of autoradiography in combination with other techniques such as electron microscopy, *in vivo* uptake assay, lesion studies, chemical or electrical stimuli, and electrophysiological recording. Electron microscopy can show the location on the cell membrane of the receptor of interest and thus indicate the presence of a possible physiologically functional site (Priestly *et al.*, 1979). If the receptors are not visualized at the synapses, they might play a metabolic role or are receptors in transport.

Lesion studies use chemicals specific for a neurotransmitter system that cause destruction of specific neurons. Neurochemical, electrical, and surgical lesions can be used to prevent nerve cells from communicating with each other. The changes induced by this destruction of communication give another clue to the function of the specific nerve types, their relationship to each other, the association of a particular receptor type with known neuronal cell populations or how a substance may affect the release of other neurotransmitters. If behavioral studies are performed after the lesions and before the animals are sacrificed, even more can be deduced about the possible function of a specific group of neurons and neurotransmitters. While knowledge exists on the behavioral changes caused by various lesions in the olfactory bulb, few of these studies have involved a combination of behavioral and biological analyses. Animal models of human pathological states, induced with chemicals or electrical stimulation, can help determine the possible role receptors play in disease. *In vivo* studies can give a picture of the role played by specific receptors and transmitters in sensory nervous stimulation and integration of information from the external environment (Halsaz *et al.*, 1981, 1983; Priestley *et al.*, 1979). [^{14}C]-2-Deoxyglucose has been used in this fashion to measure the regional brain energy consumption after stimulation. In several studies involving the olfactory bulb, the animal was injected with the radioactive ligand and exposed to odors; the tissue was then analyzed by autoradiography to determine the distribution of silver grains. This process permitted identification of areas of energy use thought to correspond to the degree of excitation of the specific neurons in the olfactory system (Benson *et al.*, 1985; Greer *et al.*, 1981; Hammer and Herkenham, 1984; Jourdan *et al.*, 1980).

A few problems are associated with the use of autoradiographic techniques. The major drawback is that many of the ligands are radiolabeled with tritium, a low-energy level β-emitter, which results in the possible quenching of emission from white matter areas due to the presence of a high density of tissue (Rainbow *et al.*, 1984). Labeling the compounds with high-energy emitters such as [^{125}I] can substantially reduce this problem (Kuhar *et al.*, 1986).

A vast array of receptor populations and neurotransmitter substances have been localized in the olfactory bulb, making it perhaps the most diversified region of the brain. Because of its location in the rat brain and its strong adherence to the bony tissue, it is somewhat difficult to remove the olfactory bulb without damaging it. However, the rich assortment of potential receptor interaction, the distinct lamination, the defined cell morphology, and the control that olfaction has on the central limbic system, and thus on many aspects of animal behavior, make the olfactory bulb an interesting and important brain structure for further study.

ACKNOWLEDGMENTS. The authors wish to thank Linda Miller for her excellent secretarial assistance.

REFERENCES

Anholt, R. H., Murphy, K. M., Mack, G. E., and Snyder, S. H., 1984, Peripheral type benzodiazepine receptors in the central nervous system: Localization to olfactory nerves, *J. Neurosci.* **4:** 593–603.

Araneda, S., Gamrani, H., Font, C., Calas, A., Pujol, F., and Bobillier, P., 1980, Retrograde axonal transport following injection of [^{3}H]serotonin into the olfactory bulb. II. Radiographic study, *Brain Res.* **196:** 417–427.

Arimatsu, Y., Seto, A., and Amano, T., 1981, An atlas of alpha-bungarotoxin binding sites and structures containing acetylcholinesterase in mouse central nervous system, *J. Comp. Neurol.* **198:** 603–631.

Austin, L., Recasens, M., Mathur, R. L., and Mandel, P., 1978, The distribution of taurine, cysteinylsulphinilic acid decarboxylase and crysteinylsulphinilic acid alpha-ketoglutarate transaminase in the rat olfactory bulb and olfactory nucleus, *Neurosci. Lett.* **9:** 59–63.

Beal, M. F., Tran, V. T., Mazurek, M. F., Chattha, G., and Martin, J. B., 1986, Somatostatin binding sites in human and monkey brain: Localization and characterization, *J. Neurochem.* **46:** 359–364.

Benavides, J., Quateronet, D., Imbault, F., Malgouris, C., Uzan, A., Renault, C., Dubroeucq, M. C., Gueremy, C., and LeFur, G., 1983, Labelling of "Peripheral-type" benzodiazepine binding sites in the rat brain by using [^{3}H]PK 11195, an isoquinoline carboxamide derivative: Kinetic studies and autoradiographic localization, *J. Neurochem.* **41:** 1744–1750.

Bennett, J. P., and Yamamura, H. I., 1985, Neurotransmitter, hormone, or drug receptor binding methods, in: *Neurotransmitter Receptor Binding* (H. I. Yamamura, S. J. Enna, and M. J. Kuhar, eds.), pp. 61–89, Raven, New York.

Benson, T. E., Burd, G. D., Greer, C. A., Landis, D. M. D., and Shepherd, G. M., 1985, High-resolution 2-deoxyglucose autoradiography in quick-frozen slabs of neonatal rat olfactory bulb, *Brain Res.* **339:** 67–78.

Bowery, N. G., Price, G. W., Hudson, A. L., Hill, D. R., Wilkin, G. P., and Turnbull, M. J., 1984, GABA receptor multiplicity: Visualization of different receptor types in the mammalian CNS, *Neuropharmacology* **23:** 219–231.

Cortes, R., Supavilai, P., Karobath, M., and Palacios, J. M., 1984, Calcium antagonist binding sites in the rat brain: Quantitative autoradiographic mapping using the 1,4-dihydropyridines [^{3}H]PN 200-110 and [^{3}H]PY 108-068, *J. Neurol. Trans.* **60:** 169–197.

Dashwood, M. R., 1983, Central and peripheral prazosin binding: An in vitro autoradiographic study in the rat, *Eur. J. Pharmacol.* **86:** 51–58.

Davis, B. J., and Macrides, F., 1981, The organization of centrifugal projections from the anterior olfactory nucleus, ventral hippocampal rudiment, and piriform cortex to the main olfactory bulb in the hamster: An autoradiographic study, *J. Comp. Neurol.* **203:** 475–493.

Dawson, T. M., Gehlert, D. R., McCabe, R. T., Barnett, A. and Wamsley, J. K., 1986, D-1 dopamine receptors in the rat brain: A quantitative autoradiographic analysis, *J. Neurosci.* **8:** 2352–2365.

Gehlert, D. R., and Wamsley, J. K., 1985, Dopamine receptors in the rat brain: Quantitative autoradiographic localization using [^{3}H]-sulpiride, *Neurochem. Int.* **7:** 717–723.

Gehlert, D. R., Dawson, T. M., Yamamura, H. I., and Wamsley, J. K., 1985*a*, Quantitative autoradiography of [^{3}H]-forskolin binding sites in the rat brain, *Brain Res.* **361:** 351–360.

Gehlert, D. R., Yamamura, H. I., and Wamsley, J. K., 1985*b*, Autoradiographic localization of "peripheral-type" benzodiazepine binding sites in the rat brain, heart and kidney, *Naunyn Schmiedeburgs Arch. Pharmacol.* **328:** 454–460.

Gehlert, D. R., Yamamura, H. I., and Wamsley, J. K., 1985*c*, Gamma-aminobutyric acid$_B$ receptors in the rat brain: Quantitative autoradiographic localization using [^{3}H](−)-baclofen, *Neurosci. Lett.* **56:** 183–188.

Gehlert, D. R., Speth, R. C., and Wamsley, J. K., 1986, Distribution of [^{125}I]-angiotensin II binding sites in the rat brain: A quantitative autoradiographic study, *Neuroscience* **18:** 837–856.

Gershon, M. D., Schwartz, J. H., and Kandel, E. R., 1981, Morphology of chemical synapses and patterns of interaction, in: *Principles of Neural Science* (E. R. Kandel and J. H. Schwartz, eds.), pp. 91–105, Elsevier/North-Holland, New York.

Goodman, R. R., Snyder, S. H., Kuhar, M. J., and Young, W. S., 1980, Differentiation of delta and mu opiate receptor localizations by light microscopic autoradiography, *Proc. Natl. Acad. Sci. USA* **77:** 6239–6243.

Greenamyre, J. T., Young, A. B., and Penney, J. B., 1984, Quantitative autoradiographic distribution of L-[^{3}H]glutamate binding sites in rat central nervous system, *J. Neurosci.* **4:** 2133–2144.

Greer, C. A., Stewart, W. B., Kauer, J. S., and Shepherd, G. M., 1981, Topographical and laminar localization of 2-deoxyglucose uptake in rat olfactory bulb induced by electrical stimulation of olfactory nerves, *Brain Res.* **217:** 279–293.

Gulya, K., Wamsley, J. K., Gehlert, D., Pelton, J. T., Duckles, S. P., Hruby, V. J., and Yamamura, H. I., 1985, Light microscopic autoradiographic localization of somatostatin receptors in the rat brain, *J. Pharmacol. Exp. Ther.* **235:** 254–258.

Halasz, N., and Shepherd, G. M., 1983, Neurochemistry of the vertebrate olfactory bulb, *Neuroscience* **10:** 579–619.

Halasz, N., Ljungdahl, A., and Hokfelt, T., 1979, Transmitter histochemistry of the rat olfactory bulb. III. Autoradiographic localization of [^{3}H]-GABA, *Brain Res.* **167:** 221–240.

Halasz, N., Parry, D. M., Blackett, N. M., Ljungdahl, A., and Hokfelt, T., 1981, [^{3}H]gamma-aminobutyrate autoradiography of the rat olfactory bulb: Hypothetical grain analysis of the distribution of silver grain, *Neuroscience* **6:** 473–479.

Halasz, N., Nowycky, M. C., and Shepherd, G. M., 1983, Autoradiographic analysis of [^{3}H] dopamine and [^{3}H] dopa uptake in the turtle olfactory bulb, *Neuroscience* **8:** 705–715.

Hammer, R., Berrie, C. P., Birdsall, N. J. M., Burgen, A. S. V., and Hulme, E. C., 1980, Pirenzepine distinguishes between different subclasses of muscarinic receptors, *Nature (Lond.)* **283:** 90–92.

Hammer, R. P., and Herkenham, M., 1984, Tritiated 2-deoxy-D-glucose: A high resolution marker for autoradiographic localization of brain metabolism, *J. Comp. Neurol.* **222:** 128–139.

Hara, Y., Shiosaka, S., Senba, E., Sakanaka, M., Inagaki, S., Takagi, H., Kawai, Y., Takatsuki, K., Matsuzaki, T., and Tohyama, M., 1982, Ontogeny of neurotensin containing neuron system of the rat: Immunohistochemical analysis. I. Forebrain and diencephalon, *J. Comp. Neurol.* **208:** 177–195.

Hinds, J. W., 1968, Autoradiographic study of the histogenesis in the mouse olfactory bulb. II. Cell proliferation and migration, *J. Comp. Neurol.* **134:** 305–322.

Jaffe, E. H., Cuello, A. C. and Priestly, J. V., 1983, Localization of ^{3}H-GABA in the rat olfactory bulb: An in vivo and in vitro autoradiographic study, *Expt. Brain Res.* **50:** 100–106.

Jourdan, F., Duveau, A., Astic, L., and Holley, A., 1980, Spatial distribution of [^{14}C]2-deoxyglucose uptake in the olfactory bulbs of rats stimulated with two different odours, *Brain Res.* **188:** 139–154.

Kaulen, P., Bruning, G., Schneider, U., and Baumgarten, H-G., 1985, Autoradiographic localization of [^{3}H]buspirone binding sites in rat brain, *Neurosci. Lett.* **53:** 191–195.

Kebabian, J. W., and Calne, D. B., 1979, Multiple receptors for dopamine, *Nature (Lond.)* **277:** 93–96.

Kuhar, M. J., 1985, Receptor localization with the microscope, in: *Neurotransmitter Receptor Binding* (H. I. Yamamura, S. J. Enna, and M. J. Kuhar, eds.), pp. 153–176, Raven, New York.

Kuhar, M. J., De Souza, E. B., and Unnerstall, J. R., 1986, Neurotransmitter receptor mapping by autoradiography and other methods, *Annu. Rev. Neurosci.* **9:** 27–59.

Leroux, P., Quirion, R., and Pelletier, G., 1985, Localization and characterization of brain somatostatin receptors as studies with somatostatin-14 and somatostatin-28 receptor radioautography, *Brain Res.* **347:** 74–84.

Macrides, F., and Davis, B. J., 1983, The olfactory bulb, in: *Chemical Neuroanatomy* (P. C. Emson, ed.), pp. 391–426, Raven, New York.

Manaker, S., Winokur, A., Rostene, W. H., and Rainbow, T. C., 1985, Autoradiographic localization of thyrotropin-releasing hormone receptors in the rat central nervous system, *J. Neurosci.* **5:** 167–174.

Mantyh, P. W., and Hunt, S. P., 1985, Thyrotropin-releasing hormone (TRH) receptors. Localization by light microscopic autoradiography in rat brain using [^{3}H][3-Me-His^2]TRH as the radioligand, *J. Neurosci.* **5:** 551–561.

Monaghan, D. T., and Cotman, C. W., 1985, Distribution of *N*-methyl-D-aspartate sensitive L-[^{3}H]glutamate binding sites in rat brain, *J. Neurosci.* **5:** 2909–2919.

Murphy, K. M., Gould, R. J., and Snyder, S. H., 1982, Autoradiographic visualization of [^{3}H]-nitrendipine binding sites in rat brain: Localization to synaptic zones, *Eur. J. Pharmacol.* **81:** 517–519.

Nadi, N. S., Hirsch, J. D., and Margolis, F. L., 1980, Laminar distribution of putative neurotransmitter amino acid and ligand binding sites in the dog olfactory bulb, *J. Neurochem.* **34:** 138–146.

Nicoll, R. A., 1971, Pharmacological evidence for GABA as the transmitter in granule cell inhibition in the olfactory bulb, *Brain Res.* **35:** 137–149.

Olsen, R. W., Snowhill, E. W., and Wamsley, J. K., 1984, Autoradiographic localization of low affinity GABA receptors with [^{3}H]-bicuculline methochloride, *Eur. J. Pharmacol.* **99:** 247–248.

Palacios, J. M., and Wamsley, J. K., 1983, Microscopic localization of adrenoceptors, in: *Adrenoceptors and Catecholamine Action.* Part B (G. Kunos, ed.), pp. 295–313, Wiley, New York.

Palacios, J. M., Niehoff, D. L., and Kuhar, M. J., 1981*a*, [^{3}H]-spiperone binding sites in brain: Autoradiographic localization of multiple receptors, *Brain Res.* **213:** 277–289.

Palacios, J. M., Niehoff, D. L., and Kuhar, M. J., 1981*b*, Receptor autoradiography with tritium-sensitive film: Potential for computerized densitometry, *Neurosci. Lett.* **25:** 101–105.

Palacios, J. M., Wamsley, J. K., and Kuhar, M. J., 1981*c*, High affinity GABA receptors—Autoradiographic localization, *Brain Res.* **222:** 285–307.

Paxinos, G., and Watson, C., 1982, *The Rat Brain: In Stereotaxic Coordinates*, pp. 13–14, Academic, New York.

Pazos, A., and Palacios, J. M., 1985, Quantitative autoradiographic mapping of serotonin recptors in the rat brain. I. Serotonin-1 receptors, *Brain Res.* **346:** 205–230.

Pazos, A., Cortes, R., and Palacios, J. M., 1985, Quantitative autoradiographic mapping of serotonin receptors in the rat brain. II. Serotonin-2 receptors, *Brain Res.* **346:** 231–249.

Priestley, J. V., Kelly, J. S., and Cuello, A. C., 1979, Uptake of [^{3}H]dopamine in periglomerular cells of the rat olfactory bulb: An autoradiographic study, *Brain Res.* **165:** 149–155.

Quinn, M. R., Sturman, J. A., Wysocki, C. J., and Wen, G. Y., 1981, Accumulation of [^{35}S]taurine in peripheral layers of the olfactory bulb, *Brain Res.* **230:** 378–383.

Quirion, R., Shults, C. W., Moody, T. W., Pert, C. B., Chase, T. N., and O'Donohue, T. L., 1984, Autoradiographic distribution of substance P receptors in rat central nervous system, *Nature (Lond.)* **303:** 714–716.

Rainbow, T. C., Biegon, A., and Berck, D. J., 1984, Quantitative receptor autoradiography with tritium-labeled ligands: Comparison of biochemical and densitometric measurements, *J. Neurosci. Methods* **11:** 231–241.

Reubi, J. C., and Maurer, R., 1985, Autoradiographic mapping of somatostatin receptors in the rat central nervous system and pituitary, *Neuroscience* **15:** 1183–1193.

Rothman, R. B., Herkenhamp, M., Pert, C., Liang, T., and Casieri, M. A., 1984, Visualization of rat brain receptors for the neuropeptide, substance P, *Brain Res.* **309:** 47–54.

Rotter, A., Birdsall, N. J. M., Burgen, A. S. V., Field, P. M., Hulme, E. C., and Raisman, G., 1979*a*, Muscarinic receptors in the central nervous system of the rat. I. Technique for autoradiographic localization of the binding of [^{3}H]propylbenzilylcholine mustard and its distribution in the forebrain, *Brain Res. Rev.* **1:** 141–165.

Rotter, A., Field, P. M., and Raisman, G., 1979*b*, Muscarinic receptors in the central nervous system of the rat. III. Postnatal development of binding of [^{3}H]propylbenzilylcholine mustard, *Brain Res. Rev.* **1:** 185–205.

Sarrieau, A., Javoy-Agid, F., Kitagbi, P., Dussaillant, M., Vial, M., Vincent, J. P., Agid, Y., and Rostene, W. H., 1985, Characterization and autoradiographic distribution of neurotensin binding sites in the human brain, *Brain Res.* **348:** 375–380.

Shepherd, G. M., 1977, The olfactory bulb: A simple system in the mammalian brain, in: *The Handbook of Physiology Section.* I. *The Nervous System.* Vol. I: *Cellular Biology of Neurons* (J. M. Brookhart, V. B. Mountcastle, E. R. Kandel, and S. R. Gieger, eds.), pp. 945–968, American Physiological Society, Bethesda, Maryland.

Schwartz, J. C., Lampart, C., and Rose, C., 1970, Properties and regional distribution of histidine decarboxylase in rat brain, *J. Neurochem.* **17:** 1527–1534.

Tran, V. T., Uhl, G. R., Perry, D. C., Manning, D. C., Vale, W. W., Perrin, M. H., Rivier, J. E., Martin, J. B., and Snyder, S. H., 1984, Autoradiographic localization of somatostatin receptors in rat brain, *Eur. J. Pharmacol.* **101:** 307–309.

Unnerstall, J. R., and Wamsley, J. K., 1983, Autoradiographic localization of high-affinity [^{3}H]-kainic acid binding sites in the rat forebrain, *Eur. J. Pharmacol.* **86:** 361–371.

Unnerstall, J. R., Molliver, M. E., Kuhar, M. J., and Palacios, J. M., 1983, Ontogeny of opiate binding sites in hippocampus, olfactory bulb and other regions of rat forebrain by autoradiographic methods, *Dev. Brain Res.* **7:** 157–169.

Unnerstall, J. R., Niehoff, D. L., Kuhar, M. J., and Palacios, J. M., 1982, Quantitative receptor autoradiography using [^{3}H]-Ultrofilm: Application to multiple benzodiazepine receptors, *J. Neurosci. Methods* **6:** 59–73.

Wamsley, J. K., 1984, Autoradiographic localization of receptor sites in the cerebral cortex, in: *Cerebral Cortex,* Vol. 2 (E. G. Jones and A. Peters, eds.), pp. 173–202, Plenum, New York.

Wamsley, J. K., and Palacios, J. M., 1983, Apposition techniques of autoradiography for microscopic receptor localization, in: *Current Methods in Cellular Neurobiology* (J. L. Barker and J. F. McKelvy, eds.), pp. 241–268, Wiley, New York.

Wamsley, J. K., and Palacios, J. M., 1984*a*, Histaminergic receptors, in: *Handbook of Chemical Neuroanatomy,* Vol. 3 (A. Bjorklund, T. Hokfelt, and M. J. Kuhar, eds.), pp. 386–406, Elsevier, Amsterdam.

Wamsley, J. K., and Palacios, J. M., 1984*b*, Amino acid and benzodiazepine receptors, in: *Handbook of Chemical Neuroanatomy,* Vol. 3 (A. Bjorklund, T. Hokfelt, and M. J. Kuhar, eds.), pp. 352–385, Elsevier, Amsterdam.

Wamsley, J. K., Zarbin, M. A., Birdsall, N. J. M., and Kuhar, M. J., 1980, Muscarinic cholinergic receptors: Autoradiographic localization of high and low affinity agonist binding sites, *Brain Res.* **200:** 1–12.

Wamsley, J. K., Zarbin, M. A., and Kuhar, M. J., 1983, Distribution of muscarinic cholinergic high and low affinity agonist binding sites: A light microscopic autoradiographic study, *Brain Res. Bull.* **12:** 233–243.

Wamsley, J. K., Gehlert, D. R., and Olsen, R. W., 1986, The benzodiazepine/barbiturate-sensitive convulsant/GABA receptor–chloride ionophore complex: Autoradiographic localization of individual components, in: *Benzodiazepine-GABA Receptors and Chloride Channels: Structural and Functional Properties* (R. W. Olsen and J. C. Venter, eds.), pp. 299–313, Liss, New York.

Watanabe, K., and Kawana, E., 1983, Selective retrograde transport of tritiated D-aspartate from the olfactory bulb to the anterior olfactory nucleus, pyriform cortex, and nucleus of the lateral olfactory tract in the rat, *Brain Res.* **296:** 148–151.

Watson, M., Yamamura, H. I., and Roeske, W. R., 1983, A unique regulatory profile and regional distribution of [^{3}H]-pirenzepine binding in the rat provide evidence for distinct M_1 and M_2 muscarinic receptor subtypes, *Life Sci.* **32:** 3001–3011.

Young, W. S., and Kuhar, M. J., 1979, A new method for receptor autoradiography: [^{3}H]-opioid receptor labeling in mounted tissue sections, *Brain Res.* **179:** 255–270.

Young, W. S., and Kuhar, M. J., 1981, Neurotensin receptor localization by light microscopic autoradiography in rat brain, *Brain Res.* **206:** 273–285.

Young, W. S., Kuhar, M. J., Roth, J., and Brownstein, M. J., 1980, Radiohistochemical localization of insulin receptors in the adult and developing rat brain, *Neuropeptides* **1:** 15–22.
Zarbin, M. A., Wamsley, J. K., and Kuhar, M. J., 1981, Glycine receptor: Light microscopic autoradiographic localization with [^{3}H]strychnine, *J. Neurosci.* **1:** 532–547.
Zarbin, M. A., Innis, R. B., Wamsley, J. K., Snyder, S. H., and Kuhar, M. J., 1983, Autoradiographic localization of cholecystokinin receptors in rodent brains, *J. Neurosci.* **3:** 877–906.

II

Molecular Biophysics and Membrane Function

6

Membrane Probes in the Olfactory System

Biophysical Aspects of Initial Events

Randall B. Murphy

1. INTRODUCTION

The mechanisms involved in the perception of chemical substances by the olfactory epithelium have been a matter of both great interest and rampant speculation since the time of the early Greek philosophers. In about 320 BC, Theophrastus (Stratton, 1917) appears to have been the first to suggest that the odor of a substance was due to some innate property and not simply to the act of inhalation.

The major focus of investigation in olfaction during the late nineteenth century became the classification of odors and the concomitant elucidation of a unique perceptual mechanism. Elaborate analogies were constructed to audition with odors placed in quasimusical "scales" (Piesse, 1880). This effort historically reflects an outgrowth of the period's preoccupation with simplistic biological reductionism based on physical and chemical analogies (Mayr, 1982; Huxley, 1942). Cain (1978) reviewed this period of development of olfaction in detail. Historical consideration is of importance as more recent biophysical theories arose directly from these "vibrational" preconceptions. Unfortunately, these theories were dependent, as was pointed out by Davies (1971), on systematic correlations that were statistically of dubious validity. Such correlations with physicochemical parameters (Laffort and Dravnieks, 1973) do not necessarily illuminate the fundamental underlying mechanisms that underlie initial events.

2. THE ROLE OF BIOPHYSICAL MODELS WITHIN A RECEPTOR-MEDIATED MODEL OF THE INITIAL EVENTS IN OLFACTION

The evidence for specific macromolecular receptors in the initial events of the olfactory neuron is described in detail by Lancet (1986). Given the acceptance of such a

Randall B. Murphy • Department of Chemistry, New York University, New York, New York 10003; and Department of Psychiatry, Cornell University Medical College-Westchester Division-New York Hospital, White Plains, New York 10605.

schema for the transduction process, it is germane to question the role that remains for biophysical models of odorant interaction. What then is the role of membrane probes in studying the initial mechanisms of mammalian olfaction? The recognition of a receptor mechanism itself implies little as to the inherent selectivity or sensitivity of the transductive process. It has been suggested that biophysical processes may serve to define these aspects of the olfactory response. Theoretically, a variety of mechanisms could be involved within the framework of a receptor model in the delineation of olfactory selectivity and specificity. In the simplest model, a somewhat modernized view of the Davies (1971) puncturing idea is suggested. Specificity of response could be caused by differences in partitioning of the odorants into mucus, regulating access to receptor sites. Within the olfactory receptor cell itself, changes in physical environment mediated by selective partitioning of odorant could also mediate receptor macromolecule access by the odorant. These two mechanisms could result in a pattern that could be interpreted as a specific signature of an odorant.

2.1. Possible Mechanisms

It now remains to be stated by virtue of what detailed mechanisms might these effects be mediated. With regard to changes in the receptor cell membrane, we may enumerate some of the possibilities:

1. Hydrophobic odorant molecules may associate with membrane lipids in a manner that results in selective changes in the immediate vicinity of a receptor macromolecule, for example, in order parameter of boundary lipid. Effects would presumably also be evidenced on intermediate- and long-range lipid order in the membrane, in terms of diffusion coefficients and other bulk parameters of suitable reporter molecules.
2. Odorant molecules, upon inclusion in the lipid bilayer, perturb fluctuations in membrane area or membrane volume, which results in a restriction of the conformational dynamics of lipid molecules.
3. The odorant molecule remains principally in the region close to the surface of the membrane, where it may be partially solvated by the water layer present there. This would result in profound changes in the surface potential of the cell membrane, which could modulate actions of macromolecular receptors in the membrane by a variety of distinct mechanisms.
4. An overt hole is punctured in the membrane by a hydrophobic odorant molecule.
5. The presence of a small hydrophobic molecule in the membrane bilayer results in changes in membrane dynamic properties, such as electrostriction, which influence receptor macromolecules in the membrane by virtue of global changes in mechanical or electric properties not apparent in steady state.

In principle, these possible mechanisms could be distinguished by the use of appropriate membrane probes to examine in detail the nature of the biophysical events which are involved. However, for the most part such studies have not been carried out to the extent that sufficient data are immediately available for us to distinguish all of these possibilities. We shall therefore review what results are available and shall further suggest what further approaches would be particularly of value in addressing these possible mechanisms. We shall further discuss the relationship of these mechanisms to the receptor-mediated initial processes in the initial steps of olfaction, for which overwhelming evidence now exists (Lancet, 1986).

2.2. Evidence for Nonspecific Mechanisms

We first consider the possibility that hydrophobic odorant molecules may directly perturb boundary lipid in the immediate vicinity of a putative chemoreceptor macromolecule. By virtue of linkage to an ion channel or by other less direct mechanisms, perturbation of some lipid subpopulation could result in the modulation of ionic inflow in the olfactory receptor neuron. This concept is problematic as a distinct recognition mechanism for odorants, as there is little convincing evidence from model protein–lipid systems that the concept of boundary lipid is meaningful. There is, however, little question that global lipid order parameters as studied, for example, by nuclear magnetic resonance (NMR), will be perturbed by the inclusion of hydrophobic molecules in the cell membrane. The locus of this effect is probably on lipid–protein interactions rather than on the lipid itself, as in systems that do not contain protein, even large amounts of hydrophobic molecules (hexane) included in the bilayer perturb lipid motional characteristics relatively little (Jacobs and White, 1984). In the erythrocyte membrane, fluorescence studies of lateral diffusion suggest that there is little if any evidence for discrete lipid domains (Golan *et al.*, 1984). In artificial, NMR and X-ray diffraction data indicate that the dependence of the order parameter on the lipid-to-protein ratio for a variety of very different proteins (Ca/Mg-ATPase, cytochrome C oxidase, Mellitin) is the same (Jahnig *et al.*, 1982). These investigators furthermore showed that the boundary tilt angle of these proteins varied in precisely the same manner in all cases with coherence length. The presence of discrete and specialized boundary lipid critical to the function of a given membrane protein would not be supported, albeit indirectly, by these data.

It should be stated, however, that these models bear little direct relationship to ionic channel systems; studies of order parameter with hexane concentration have not been performed, for example, in solvent-containing bilayers containing reconstituted purified acetylcholine receptor (AChR). However, there is no evidence that processes described under (1) play a primary role in odorant recognition.

A more interesting possibility is that described under (2), i.e., that volume fluctuations in membrane area or membrane volume are significantly perturbed by the inclusion of hydrophobic molecules in the bilayer. This presupposes that these fluctuations must normally be important in defining lipid–protein interaction, assuming that the protein–lipid lattice can be described in statistical–mechanical terms (Pink and Chapman, 1979). Little experimental evidence exists to support this hypothesis directly, although several theoretical treatments have suggested that this could be the case; without question, such rapid (picosecond) dynamical fluctuations in proteins themselves would seem to be important in defining conformational flexibility. It is clear that the activation volume of some membrane transport processes is large (Canfield and Macey, 1984) which is consistent with, but does not prove, this is how odorants exert their nonspecific effects.

2.3. Further Experimental Evidence

It is of interest to consider mechanisms of nonspecific interaction of organic molecules with the olfactory epithelium by analogy with biophysically better characterized model systems. For example, in connection with studies on the nonspecific mechanisms by which small organic molecules at high concentration can act as anesthetic agents, a considerable amount of work has been performed on model planar bilayers, micelles, and isolated nerve preparations. Although it can be disputed that any of these model systems necessarily corresponds closely to the olfactory epithelium, it is rational that some degree of parallelism can be claimed.

2.3.1. Odorous Isomers

It has long been known that certain terpene natural products, which are optical isomers (enantiomers), have different odors. However, a number of incorrect claims had been made in the past for such pairs of enantiomers, and it was not until Friedman and Miller (1971) interconverted the two compounds stereochemically that differing odors of the two members of an enantiomeric pair was rigorously demonstrated. Other enantiomeric pairs of terpenes have since been demonstrated to possess differing odors, such as the (R) and (S) *p*-mentha-1,8-dien-4-ols (Delay and Ohloff, 1979).

Structurally similar stereoisomers sometimes have very different odors, as is well known in the series of aromatic nitromusks or in other structurally related compounds (Polak *et al.*, 1978). In some of these cases, however, the construction of spacefilling models indicates that the overall shapes of the molecular van der Waals volumes can be quite different.

The differing odors of enantiomers have been used to substantiate the idea that receptor-specific interactions are necessary in order to account for such differences. However, this does not necessarily need to be the case. Chromatographic separation of chiral compounds on chiral high-pressure liquid chromatographic (HPLC) liquid phases is possible precisely because the thermodynamics of interaction of a chiral small molecule with a fixed chiral phase depends strongly on the chirality of the small molecule. An example of the differing thermochemical interactions is also found in the completely different packing arrangement observed in crystals of brucine or strychnine with benzoyl-L-alanine or benzoyl-D-alanine, respectively (Gould and Walkinshaw, 1984). Thus, it is probable that nonspecific interactions of small chiral hydrophobic molecules with cellular membranes would be influenced substantially by the chirality of the small molecule. This effect could be explored to advantage using bilayers prepared from chiral lipids, which in themselves possess many unusual properties (Kunitake *et al.*, 1979; Nakashima *et al.*, 1985).

2.3.2. Effect of Hydrophobic Organic Molecules on Nerve Conduction

The anesthetic potency of homologous series of alcohols and hydrocarbons correlates to some degree with the relative partition coefficients of the members of the series into lipid domains, as was first described more than 80 years ago in the Meyer–Overton theory (Holmstedt and Liljestrand, 1981; Meyer, 1899; Overton, 1901). The *n*-alkanols are perhaps the most thoroughly examined anesthetic series. Such alkanols have been shown to block the action potential reversibly in lobster and crayfish giant axons, squid giant axons, and frog sciatic nerve (Narahashi, 1964; Houck, 1969; Haydon and Urban, 1983; Oyama *et al.*, 1986). Although specific blockade of sodium channels has been claimed (Swenson and Oxford, 1980; Haydon and Urban, 1983), a qualitatively similar inhibition of the peak calcium current in *Helix aspersa* neurons has been reported. It would seem probable that nonspecific membrane changes induced by concentrations of these *n*-alkanols in the range 10–200 mM in the bathing medium might be responsible for these anesthetic effects, rather than a specific hydrophobic blockade of the Na^+ channel, as has been suggested. However, a problem remains in that long-chain alcohols (greater than C_{13}) exhibit little biological activity, yet it would appear that they would partition equally well into lipid bilayers as their smaller-chain homologues. A recent study (Franks and Lieb, 1986) in which partition coefficients of a homologous series of *n*-alkanols were carefully measured confirms that these long-chain alcohols can partition well into membranes; these investigators conclude that *n*-alkanol anesthesia must involve some specific protein target sites. Relatively complex structure–activity relationships for neuronal blockade are evidenced, as seen, for example, in the complete inactivity of tridecan-1-ol, but the considerable activity of tridecan-5-ol and tridecan-7-ol (Requena *et al.*, 1985). By contrast, these isomers have very different molecular volumes and, based on the study of the intercalation between lipid chains of hydrophobic molecules in model lipid bilayers as probed with proton NMR, such striking structure–activity relationships might be predicted (Pope and Dubro, 1986). However, it is not clear whether the anesthetic molecules intercalate with the lipid chains in all cases. While this is certainly rational with the steroid anesthetics, that is, to assume that they might behave crudely in a qualitatively similar manner to cholesterol, it would appear that benzyl alcohol does not penetrate deeply into the bilayer (Kaneshina *et al.*, 1984).

It is further germane in the present context to consider the phenomenon of the pressure-induced reversal of anesthesia. High hydrostatic pressures will generally reverse the anesthetic effect of gaseous anesthetics, *n*-alkanols, and so forth. The pressure-reversal phenomenon has been suggested to have a locus at the Na^+ channel, although the effect may be mediated indirectly. The latter is suggested in that Conti *et al.* (1982) observed activation volumes for activation of the squid giant axon Na^+ channel in the range of 32–58 A^3, depending on how they interpreted their data, whereas for the alamethicin channel Bruner and Hall (1983) interpret their pressure-dependence data to suggest an activation volume that is roughly double this value. Qualitatively, the electrophysiological effect of pressure on the latter model system resembled that upon the actual system, in that single-state conductance state times were lengthened in the latter, although the magnitudes of the open states were not much effected, whereas in the former biological system the overall kinetics as reflected in both Na^+ and K^+ were slowed. These effects of pressure, as well as those involving pressure reversal, would not appear to be mediated by changes in membrane thickness (Vodyanoy and Hall, 1984), as membrane capacitance with pressure is relatively small (Aldridge and Bruner, 1985).

Canfield and Macey (1984) demonstrated that the erythrocyte anion transporter has an activation volume in the vicinity of 150 $cm^3 \cdot mol^{-1}$, which, if the transport processes are in any way sterically similar to other ionic transport processes, is not in accord with the results on potassium and sodium conductances in neuronal systems.

Further confounding the issue of the pressure reversal of anesthesia is the suggestion that specific interactions at glycine receptors may be involved, as in the freshwater shrimp *Gammarus pulex,* which lacks these receptors, the phenomena of pressure reversal of anesthesia from a variety of agents is not observed (Smith *et al.*, 1984).

Accordingly, it would appear that the phenomena of pressure reversal may suggest that interactions of small hydrophobic molecules such as the *n*-alkanols with membranes has a selective locus in allowing conformational changes to occur in ionic channels that are necessary for channel opening by virtue of direct interactions with either (1) the channel itself, or (2) a specific boundary lipid population of some type, rather than by virtue of relatively nonspecific interactions with the plasma membrane environment as a whole. The second possibility would in our view be highly unlikely, based on the relative pausity of evidence in support of such specific boundary lipid; thus, the former hypothesis would be strongly favored.

With regard to molecular mechanisms of olfaction, the relevancy of these studies of anesthesia and pressure reversal are considerable. Initially, it is clear that specific interactions of small molecules with ionic channels are possible, possibly via the formation of hydrogen-bonded clatharates between these small molecules and the walls of the ionic channel. Conceptually, in the simplest sense, this would involve physical plugging of the channel. However, attenuation of gating currents by physically restricting conformational flexibility of a portion of the channel macromolecule that was involved in gating could equally be invoked, and this latter explanation is perhaps more in accord with the experimental data. In the case of some of the gaseous anesthetics, such as xenon, the structure of such clatharates has been extensively investigated. It would be presumed that molecules such as the *n*-alkanols, which are capable of hydrogen bonding yet contain a large excluded van der Waals volume, also can form a highly organized system in an aqueous environment. As in proteins, the change in solvent entropy on the formation of such complexes would represent the principal driving force to stabilize their formation. The fact that compounds such as benzyl alcohol apparently remain in a relatively superficial region of the bilayer would support the contention that it is not necessary for these small molecules to influence lipid order parameters in order to exert their effects. In other words, the study of the effects of anesthetic molecules on parameters of lipid phase transitions may not be productive, both from the standpoint of theories of anesthesia and from interpretation of nonspecific mechanisms of olfaction. Furthermore, as such phase transitions do not occur in most organisms under physiologically relevant conditions of pressure and temperature, they are of doubtful relevance to the problem at hand, unless we accept that a controlling mechanism of small-molecule interaction with the plasma membrane involves effects upon membrane volume fluctuations, as has been suggested (Pink and Chapman, 1979).

It is clear, however, that small organic molecules could interact in the olfactory system directly with single-ion channels, regardless of whether we invoke mechanisms that are specific in that direct interaction with a putative ion channel is required, as evidenced in terms of, for example, its gating kinetics, or that the interaction is indirect, with some portion of the lipid environment, which produces essentially the same effect. Thus, in one sense we have a concordance of such direct and indirect actions of small hydrophobic molecules. Experimental data from electrophysiological studies in the olfactory system could not possibly be expected to distinguish these two possibilities, that is, even if they can be considered distinct at a molecular level.

Unfortunately, few experimental data in the olfactory system are relevant to these questions. Pressure reversal has not been studied in the olfactory system; its demonstration, if present, would lend credence to a common modality of mechanism in odorant and anesthetic action. Such studies could be performed either in model membranes or at the patch-clamp level; a number of different workers have successfully performed such electrophysiological studies in a high-pressure cell.

3. SURFACE POTENTIAL AND SINGLE IONIC CHANNELS

For some time, experiments have been reported in which chemosensitive responses could be produced from simple systems that either did not contain membrane proteins or did not contain membrane proteins that could conceivably serve as selective chemoreceptors. Initially, Cherry *et al.* (1970) employed such a model in support of their idea that olfactory response resulted from puncturing of the membrane. A wide range of chemically unrelated odorants caused various complex changes in the transmembrane resistance of a Muller–Rudin bilayer, but these responses were usually observed at high concentrations of the odorant substances. Unfortunately, for a homologous series of alcohols in this study, biphasic responses were obtained across the series. It is not easy to simply interpret these or similar data of other investigators, which have since appeared. The puncturing notion is somewhat outmoded, as there is no evidence to suggest that small hydrophobic molecules produce such holes in membranes. For example, Jacobs and White (1984) found that hexane present in bilayers of phosphatidylcholine behaves as a relatively simple two-component system thermodynamically, the bilayer stays quite intact even at high mole fractions of the hexane, and the motional characteristics of the lipid chains are remarkably unperturbed by the presence of the alkane. These results are not consistent with the idea of puncturing or hole formation. On the other hand, under conditions of phase separation an amphipathic molecule such as a detergent or chlorpromazine could cause the formation of defect-like structures (Muller *et al.*, 1986). The physiological function of such defectlike structures in real membranes under conditions in which phase separation does not occur is entirely conjectural.

To continue with the consideration of chemosensitive model systems that presumably operate through nonselective mechanisms, Ueda (1975), for example, demonstrated that olfactory-like responses could be elicited by adding various odorants to the bathing medium of a protoplasmic droplet from *Nitella*. Koyama and Kurihara (1972) demonstrated odorant specific responses on surface properties of lipid monolayers. Kurihara *et al.* (1978) interpreted these results in terms of changes in surface potential forming the dominant contribution. Miyake and Kurihara (1983) and Miyake *et al.* (1983) subsequently studied the neuroblastoma cell as a model system for chemoselective effects due to changes in surface potential of the cell plasma membrane, using potential-sensitive dyes. Kashiwayanagi and Kurihara (1984) used this model system for olfactory responses, and Kumazawa *et al.* (1985) used it as a model for taste-selective responses.

While changes in surface potential may accompany olfactory responses at olfactory receptor cells, it is difficult to understand why surface potential change alone must play a seminal role in olfactory responses. Surface potential changes could markedly influence leakage through nonspecific ionic pores in the membrane by virtue of changing image potentials in the dielectric region. It is more a debatable matter as to whether such surface potential changes are of a magnitude as to modify the physiological function of ion transporters of a specific nature substantially. In these systems, the localized charge density in the central cavity through which (conceptually) the transported material travels is presumably so great that weaker surface potential changes will not much effect it. However, Haydon and Myers (1973) demonstrated that changes in surface charge of a membrane can modify the functional behavior of an antibiotic ion transporter in the membrane. Much indirect evidence can be brought to bear on the possibility that changes in surface charge at a plasma membrane could affect the function of an ion transporter in a real membrane, but there is little direct evidence. Coronado and Affolter (1986) reconstituted calcium channels from skeletal muscle transverse tubules into planar bilayers formed from lipids of widely differing surface charge. In this study, it was observed that barium open-channel currents were independent of the mole fraction of positively charged lipid currents; sodium conductances of the channel did depend to some degree on lipid surface charge, but the maximal sodium conductance did not. At least in this ionic channel system, it would appear that the conduction pathway of the channel is well separated from the lipid surface, and surface charge may here play a minor role. As we do not know the molecular nature of the ionic channels whose activation represents a primary step in the olfactory transduction pathway, this approach is highly speculative.

Much more indirectly, biochemical changes can be produced by apparently nonspecific effects of polyelectrolytes. Wolff and Cook (1977) showed that in a cultured tumor cell line polylysine will stimulate steroidogenesis, stimulate adenylate cyclase, and inhibit adrenocorticotropic hormone (ACTH) stimulation of adenylate cyclase. Under certain circumstances, however, interactions can occur between polyglutamic acid and polytyrosine with phosphatidylcholine vesicles, which appear relatively specific as probed by the use of spin-labeled stearic acids (Yu *et al.*, 1974). Thus, the specific or nonspecific nature of this interaction or its mediation solely by changes in surface potential remains somewhat unclear. Polyethylene glycol, which is well known to stimulate cell–cell fusion, causes large decreases in surface potential of phosphatidylcholine and phosphatidylethanolamine (Maggio *et al.*, 1976), suggesting that surface potential changes could be involved in some of the changes that this polymer effects.

An additional contribution to what might be termed the phase-boundary potential, would include not only the Gouy–Chapman surface potential, but long-range repulsive forces. Such forces have been shown to exist, but their importance in the physiology of excitable cells is unclear. In an elegant experiment, Gingell and Fornes (1976) measured such long-range repulsive forces between fixed erythrocyte membranes and a polarized metal electrode. Using a quite different approach, Parsegian *et al.* (1979) demonstrated such forces to exist between phosphatidylcholine bilayers.

4. PHYSICAL METHODS OF PROBING INITIAL CHEMORECEPTIVE MECHANISMS

The possible mechanisms by which mammalian chemoreceptive processes operate are reviewed elsewhere in this volume. Assuming that a discrete ligand–receptor interaction is primarily responsible for the generation of receptor cell potentials, our purpose here is to describe in what manner biophysical approaches can be helpful in discerning the details of such mechanisms. Unfortunately, these methods have only seldomly been applied to advantage in the olfactory system.

A major difficulty with many of these methods is that they can be only applied to disaggregated cells. As cell disaggregation can frequently cause significant alterations in the cell-surface properties (see, e.g., Carlsen *et al.*, 1981), this represents a real problem.

4.1. General Considerations

In principle, it is possible to use either electron-spin resonance (ESR) or various photophysical techniques, such as fluorescence photobleaching recovery to obtain detailed information as to the dynamics of individually labeled proteins in the cell membrane, provided that it is possible to specifically label such proteins. Such dynamics consist of the quantification of either transverse diffusion or the measurement of lateral motion. The latter is frequently, albeit somewhat incorrectly, termed lipid fluidity.

The relationship of lateral mobility of characteristic cell surface proteins to generation of receptor cell potentials is unclear, except that substantial lateral mobility of some membrane components is required in order to invoke adenylate cyclase-dependent mechanisms or receptor function. Measurement of lateral mobility of appropriate components of the receptor system could illuminate the question of the putative role of this coupling system. Furthermore, it is clear that alterations in lipid fluidity effect a variety of membrane–receptor-associated processes in a complex manner. For example, Heron *et al.* (1980) showed that in brain serotonin receptors, increases of membrane fluidity produced by lineoleic acid caused a decrease in binding of tritiated serotonin, while the decreased fluidity produced by incubation with cholesterol hemisuccinate caused increased radioligand binding. Axelrod (1983) reviewed the role of lateral fluidity in processes as cholinergic synaptogenesis, rhodopin function, viral infection of cells, and cell–cell contact specificity, all of which depend on the ready lateral motion of certain specific membrane proteins associated with these biological systems.

Unfortunately, measurements of lateral fluidity in the olfactory system have not been performed. This is probably a consequence of the relative complexity of this system with regard to others that have been more extensively investigated. For example, since a pharmacology of the olfactory receptor site in large part remains to be developed, specific labeling must depend on the application of appropriate monoclonal antibodies for characteristic cellular surface proteins or of sialoproteins (e.g., lectins) that associate with certain abundant varieties of cell-surface proteins with some degree of specificity (Chen and Lancet, 1984). These markers could themselves be expected to significantly perturb membrane–lipid interactions, and such possible perturbations would need to be quantified in appropriate model systems prior to actually performing experiments in isolated olfactory receptor cells.

4.2. Photophysical Methods

The primary method of importance in the measurement of lateral diffusion in biological membranes is fluorescence photobleaching recovery. A probe–protein complex in one patch of a cell is photobleached with a directed pulse from a laser of appropriate wavelength directed onto the cell patch. The return of photobleached probe to a ground state probe is slow compared with the lateral diffusional time in the membrane; thus, the return of the probe–protein complex to the bleached region (as quantified by spectrophotometric measurement) represents a measure of the time required for lateral diffusion of the probe–protein complex (Golan *et al.*, 1984). Using this technique, lateral diffusion coefficients of membrane proteins generally lie within the range of 10^{-9}–10^{-12} cm^2-sec^{-1} (Koppel *et al.*, 1981; Cherry, 1979). A recent modification of this technique in which linearly polarized light is used to measure the fluorescence anisotropy during probe recovery can also be used to quantify rotational diffusion coefficients (Yoshida and Barisas, 1986).

An alternative technique that is experimentally much simpler and that yields complementary information is the quantification of pyrene eximer formation. Eximers are excited-state dimers formed after the collision of an excited-state monomer with a monomer in the ground state. As this process is diffusion controlled, the measurement of the ratio of the amount of eximer to monomer present at any given time is a measure of the ability of the pyrene molecule to laterally diffuse within the membrane. This is a simple matter to quantify as the monomer band for pyrene is centered at 395 nm and the eximer band is broadly centered at 478 nm (Galla and Hartmann, 1981). Thus, the ratio of the intensities of the eximer to the monomer band can be directly related to the lateral fluidity of the membrane (Ellerbe *et al.*, 1986; Muller *et al.*, 1986). However, as the transition probabilities of the pyrene are influenced by the anisotropic nature of the cell plasma membrane, different absolute values of lateral fluidity can be obtained by the use of different pyrene probes or in different membranes (Melnick *et al.*, 1981). It is thus difficult to compare lateral diffusion coefficients obtained with this technique with those obtained by complementary biophysical approaches.

It is also possible to obtain information as to "lipid fluidity" somewhat more indirectly by examining the rotational diffusion coefficient of probe molecules such as 1,6-diphenyl-1,3,5-hexatriene in cell membranes (Lakowicz *et al.*, 1979*a,b;* Gilmore *et al.*, 1979; Bouchy *et al.*, 1981; Cherksey *et al.*, 1981). This is quantified by measuring the depolarization of fluorescence of the probe in the membrane environment; experimentally, this can be carried out using either steady-state fluorescence depolarization measurements, measurement of lifetime anisotropies by photon-correlation methods, or the so-called phase method. The latter depends upon measurement of the differential tangent between the phase angles of the parallel and perpendicular polarized emission components of the sample when excited by a sinusoidially modulated source (Lakowicz *et al.*, 1979*a*). All three methods yield similar information, although the latter two are less ambiguous with regard to interpretation as a rotational diffusion coefficient. It is important to realize that when using cell membrane suspensions or whole cells, scattering can contribute significantly to depolarization, confounding such measurements, as appropriate corrections are not easily made (Kutchal *et al.*, 1982). Although appropriate corrections are often not made, it is possible to extrapolate results to limiting dilution, but this approach is not rigorously accurate. Of equal importance is the inner filter effect, which results in loss of excitation intensity due to self-absorption by the sample. This is readily discerned by a nonlinearity if fluorophore concentration is plotted against fluorescence intensity; by working within only the linear region, this effect can be avoided, or appropriate corrections outside this region can sometimes be made (Lutz and Luisi, 1983).

4.3. Electron Spin Resonance

Studies of ESR of appropriate spin labels (stable free radicals) in biological membranes can in principle provide detailed information as to the geometry of the surrounding environment. However, using simple measurement of ESR spectra alone, such as lineshape analysis, the available information is limited except for systems of rigidly defined geometry, such as transition–metal complexes. The rotational correlational times that can be measured with the aid of simple lineshape analysis also generally do not fall within the range of those observed in most biological membranes. However, line-shape analysis alone has been used successfully in some systems, such as to measure the rotational and lateral diffusion coefficients of thyroxine in lipid bilayers, although the absolute value of such numbers is questionable (Lai and Cheng, 1982). The only straightforward ESR technique that can measure rotational diffusion coefficients directly is the saturation transfer approach. In this method, the rate of diffusion of saturation, produced by an out-of-phase pulse applied to the continuously irriadiated sample can be related to a rotational diffusion coefficient (Cherry, 1979). The major problem with the technique with regard to the study of low concentrations of probe–protein complex in biological membranes is that it is not very sensitive; micromolar concentrations or greater of probe are generally required. Recent methodological improvements may make this approach more practicable (Squier and Thomas, 1986).

The photobleaching recovery technique has also been adapted to ESR methodology, although it is instrumentally complex (Sheats and McConnell, 1978).

5. BIOPHYSICAL STUDIES OF OLFACTORY EPITHELIUM

5.1. ESR Studies

Bannister (1975) and Clark *et al.* (1980) described ESR studies of various spin-label probes in whole mouse olfactory epithelium. In the first study, significant differences were claimed in line shape between spin labels that were odorous and those that were not. The odorless spin labels elicited line shapes that indicated that they were in an aqueous environment, while the odorous spin labels appeared to be constrained. In the second study, however, these same differences were also seen in control tissue, mouse intestinal mucosa. Thus, it was suggested that odorous potential correlated with ability to associate with membranes in some manner. This is in itself, however, not surprising, as it has long been known that it is possible to correlate physicochemical parameters empirically, such as partition coefficients in the *n*-octanol–water system, with odorous thresholds (Laffort and Dravnieks, 1973). It is therefore difficult to accept the model of the second paper (Clark *et al.*, 1980), as experimental data are not available to validate it directly.

The application of various odorous substances to the spin-labeled olfactory epithelium was not reported to cause any discernible changes in the ESR line shapes (Clark *et al.*, 1981).

In summary, as these ESR studies did not use spin labels with the ability to associate with known cell-surface components of olfactory receptor cells (Chen and Lancet, 1984), nor were changes in line shape observed in the presence of high concentrations of odor-

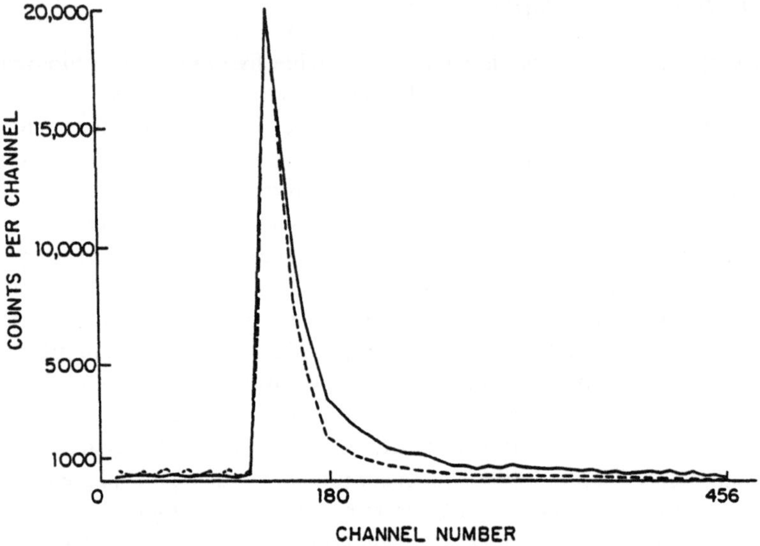

Figure 1. Fluorescence lifetime decay of 2-ethoxynapthalene saturated rat olfactory epithelium (—) and respiratory epithelium (---). Lifetime may be resolved into two components: a short component, which is about 3.9 nsec in both preparations, and a long lifetime component, which is 20–24 nsec (respiratory and olfactory, respectively).

ants, the information that they provide as to the nature of the initial mechanisms in the olfactory process is necessarily limited.

5.2. Fluorescence Studies

We have performed studies of the association of the fluorescent odorant 2-ethoxynaphthalene with rat olfactory epithelium (Murphy, 1982). Changes in fluorescence lifetime and steady-state fluorescence emission of the probe were evident between the rat olfactory and respiratory epithelium, the latter being used as a control tissue. These data are illustrated in Fig. 1. The fluorescence emission spectra with increasing concentration of the probe are illustrated in Fig. 2. Extrapolation of these data into a Scatchard plot format yields a biphasic plot with a high-affinity component, with an approximate K_d value of 25 nM and a low-affinity component with a K_d value in excess of 1 μM, which is only poorly saturable.

Thus, although it can be seen that this result is potentially of interest in that specific associations of the probe with the olfactory epithelium are observed, the nature of these associations or their relationship to the primary transduction mechanisms in olfaction can not readily be elucidated.

5.3. Raman and Infrared Spectroscopy

Given the historical interest in vibrational hypotheses of olfaction (Davies, 1971), it is surprising that raman spectroscopy apparently has not been carried out on odorants associated with the olfactory epithelium. The use of odorants that absorbed at a wavelength coincident with excitation could permit meaningful resonance–Raman studies to be carried out at relatively low probe concentrations. The use of conventional Raman techniques would not appear to be promising, based on the results of Savoie *et al.* (1986). These workers observed that both the lipid and carotenoid Raman bands did not change significantly in the olfactory nerve of the pike (*Esox lucius*) during nerve excitation. Interestingly, in other systems, such as the erythrocyte membrane, Raman bands of carotenoids seem quite sensitive to relatively perturbation, such as temperature changes, addition of fluidizers such as lysolecithin, or brief trypsin treatment (Verma and Wallach, 1975). Infared absorption studies using photoacoustic methodology applicable to water-containing biological systems have not yet been reported in the olfactory epithelium.

5.4. Difference Absorption Spectrophotometry

Absorption spectroscopy, aside from the use of membrane-potential sensitive indicators has the potential to discern membrane or membrane-putative receptor-site interactions. Although early work in this area initially appeared promising, the interpretation of the data was somewhat naive in light of more modern understanding of receptor–ligand associations. Ash (1968) initially presented data demonstrating that the addition of

Figure 2. Steady-state fluorescence emission spectrum (uncorrected) of the effect of increasing concentration of 2-ethyoxynaphthalene on its spectrum, in the presence of a constant concentration of olfactory epithelium (rat) homogenate. Medium: tris-maleate, pH 7.40, with 0.8% NaCl added. Concentrations (total) of ethoxynapthalene in the cuvette are, going upward from the bottom of graph, 1.5, 7.4, 14.3, 57.9, 96.9, 135, 166, 197, 229, 252, 276, and 301 nM.

odorants to crude homogenates of olfactory epithelium from rabbit caused small absorbance changes in the absorption spectrum of the homogenate. However, responses were not specific to olfactory epithelium homogenates, as crude brain homogenates evidenced similar absorbance changes. Although it was claimed that the wavelength at which the maximal effect was observed (267 nm) provided evidence that the substance responsible for the effect was a nucleoprotein, later studies suggested that it could be an ascorbate–protein complex, as this absorbs broadly in the same wavelength region. Further work (Ash and Skogen, 1970) suggested that a protein complex of some sort was involved; the effect could be abolished by thermal denaturation, or in the presence of 8 M urea. Thus, although the interpretation of the effect in terms of some specific receptor interaction was not correct, the observation itself was reproducible, and free ascorbate alone could not be responsible for it. A speculative interpretation of the effect, although it can not be wholly substantiated from the available data, is that cytochrome P-450 interactions with substrate could be partly responsible. Studies using more highly purified, solubilized membrane protein fractions from cultured receptor cell neurons would be necessary. We performed some initial experiments in which a partially purified solubilized fraction could be obtained from rabbit olfactory epithelium that evidenced an effect qualitatively similar to that reported by Ash and Skogen (1970). Figure 3 plots the change in absorbance versus molarity of odorant (in this case dibutyl disulfide). Clearly, an apparently saturable interaction of this odorant with the crude protein fraction was observed, in addition to a lower-affinity nonsaturable component. Although we did not obtain similar results using control preparations of respiratory epithelium, substantiating the specificity of these observations, we abandoned these efforts because it did not appear possible to demonstrate the relationship of such absorption changes, if any, to the initial chemoreceptive steps in the olfactory receptor neuron.

Although such methods as optical difference assays appear archaic in reference to pharmacologically specific, receptor–ligand interactions they may continue to be useful in examining lower affinity nonspecific interactions that may contribute, in part, to more specific mechanisms involved in the initial steps of olfactory transduction. For example, the use of ultraviolet difference spectra has been useful in examining the binding of steroids to serum progesterone binding proteins (Stroupe and Westphal, 1978).

5.5. Biophysical Studies Using Artificial Membranes

Most other biophysical studies of the olfactory epithelium that are not inherently electrophysiological in nature have examined the specificity of interactions of a variety of protein-modifying reagents with the intact olfactory epithelium. These were done in an attempt to discern patterns that would implicate a specific cell-surface receptor or that would at least begin to define a functional pharmacology of the compounds that might interact by virtue of structure–activity relationships. These studies are not reviewed here. However, another approach has been to attempt to simplify the biophysics of the olfactory receptor cell by functionally transferring some parts of it to an artificial membrane system. Other workers have simply employed bilayer membranes without proteins, as in the study conducted by Cherry *et al.* (1970), which attempted to validate the hypotheses underlying the puncturing theory. More recently, Dodd and his group have employed such membranes for chemosensing, and have attempted to develop theories for their operation

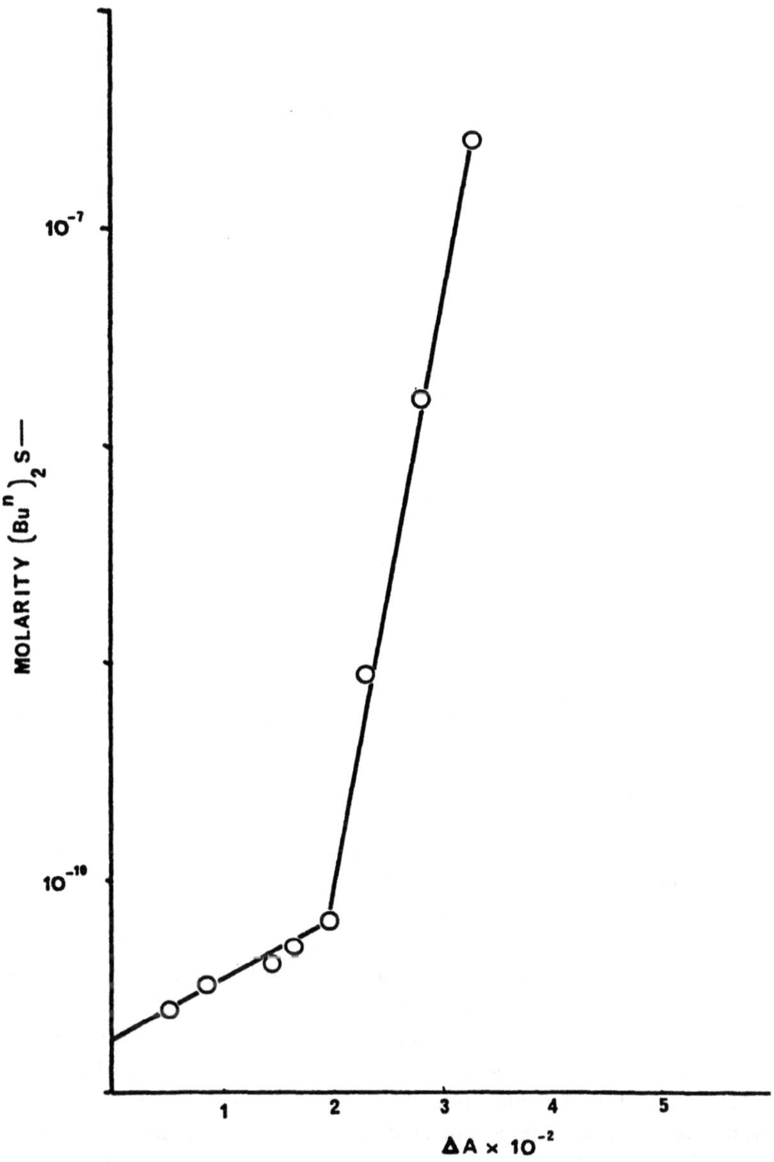

Figure 3. Difference spectrum illustrating changes observed at 215 nm in a soluble preparation obtained from rabbit olfactory epithelium upon the addition of diethyl sulfide. The correction has been made for increasing absorbance due to increased diethyl sulfide concentration. Rabbit olfactory epithelium was homogenized in MOPS–sucrose buffer, pH 7.40 (50 mM MOPS, 100 mM sucrose), and was sedimented at 100,000 × *g* for 1 hr. Homogenization was effected by a Teflon-in-glass pestle homogenizer with a tight clearance at 800 rpm for 1 min. The supernate was applied to a 90 × 2.6 cm Sephadex–G-200 column, and a void volume fraction was obtained. The column was eluted with 50 mM Tris–HCl, pH 7.4 (at 4°C). This material (protein concentration about 300 μg-ml^{-1} by Lowry) was then used for the optical difference study. Points represent means from 3–6 separate determinations.

which suggest that the initial olfactory receptor is essentially a nonspecific chemosensor network, where each element is equipped with nothing more than a level detector.

With regard to such model membrane systems, which do not contain proteins, it is unquestionable that they represent potentially analytically useful devices for specific chemosensing. It is rational to assume that a network of nonspecific chemosensors could acquire considerable selectivity if each element in a large matrix (of some defined geometry) were allowed to differ from every other element in some minor manner only in terms of latency of response, as governed by restricting access. For example, such an ensemble of membranes could have a graded coating of some polymer film that gradually became thicker across the assembly. If this polymer film only permitted access to the detector network relatively slowly, based on partition coefficients of compounds into it, it could model well the type of selectivities observed in the intact olfactory system, as Laffort and Dravnieks (1973) showed that thresholds can be reproduced by what amounts to a linear-free energy treatment of a few simple thermochemical parameters. A further improvement would consist of equipping every element in the detector network with the ability to amplify its response as well as mix with the outputs of its neighbors in a controllable but dynamically nonlinear manner. The network then begins to have the properties of a neural net, in that localized regions of excitability could be produced.

Although a highly analytically useful chemosensor might be built along these lines, it is not clear what, if any, relationship this has to function in the initial events of olfaction, as experimental data are not available to suggest the magnitudes of the many variable parameters that would have to be adjusted to produce any given result with such a model.

We therefore do not further consider such approaches that do not use material obtained from olfactory epithelium, rather, we concentrate on those studies that employ functional transfer techniques. Fesenko *et al.* (1977) first demonstrated in relatively simple experiments that some discrete material, which was suggested to be a protein, could be added to the bathing medium of a bilayer membrane; this would cause the membrane to respond by means of resistance change to the addition of relatively high concentrations of odorants, in particular camphor, as well as some other terpenes. In more recent studies, this group has partially purified a protein that has been hypothesized to be responsible for the effect, but this subsequent work has for the most part employed a radiolabeled-camphor binding assay rather than fallen back on the biophysical approach.

In the studies conducted by Fesenko *et al.* (1977), the basis for the putative effect was not clear, nor was the relationship to olfaction, but it did appear that some association of a protein which bound odorous molecules did seem to occur. We further investigated this phenomenon and found using large bilayer membranes (Vodyanoy and Murphy, 1982; Vodyanoy *et al.*, 1982) that a sonicated crude ciliary preparation from rat olfactory epithelium contained large lipid vesicles that could be fused with such bilayers. Proteins formerly present in such ciliary membranes could then be functionally transferred into the large bilayers and characterized biophysically. We found that the addition of nanomolar concentrations of odorants, particularly alkyl sulfides such as diethyl sulfide, caused large changes in membrane DC conductivity without significant perturbation of the membrane *in toto* as quantified by variations in membrane capacitance (Murphy, 1983). These data are illustrated in a typical experiment in Fig. 4. We observed a linear correlation between the relative stationary conductance evoked by diethyl sulfide with the logarithm of its concentration over three orders of magnitude (0.01–10 nM).

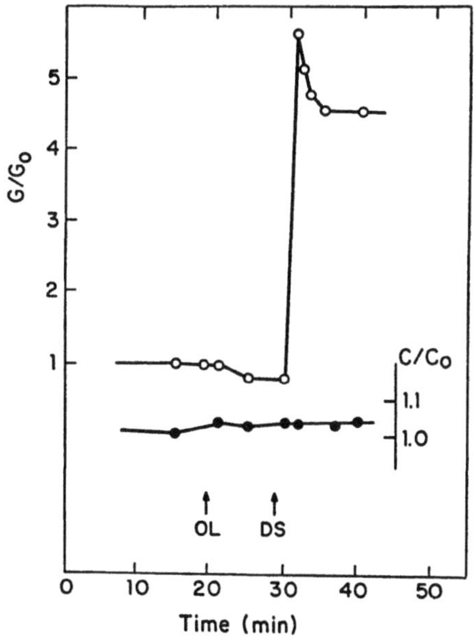

Figure 4. Variation of the relative DC conductivity (○) and the relative capacitance (●) with time for a planar lipid bimolecular membrane (solvent-free, egg PC; bathing medium 20 mM KCl, 2 mM $CaCl_2$, pH 7.4). The initial conductivity was 0.22×10^{-8} S/cm^2. The addition of rat olfactory epithelial homogenate (30 μg/ml) is denoted by OL; the addition of diethyl sulfide (final concentration 4 nM) is denoted by DS. The initial capacitance was observed to be 0.76 μF/cm^2.

In subsequent studies (Vodyanoy and Murphy, 1983), we investigated in detail the ionic selectivity of the effect and its single ionic channel basis. Briefly, we found that the channels were strongly K^+ selective, had a unitary conductance of about 56 pS, and were characterized by very long open times.

We did not examine in detail the influence of varying surface charge or other parameters on the preparation, as this also would have changed the extent of fusion dramatically and therefore would substantially change the results of the experiment. However, if a different experimental design was employed (e.g., prior fusion of the vesicles with a monolayer, from which the bilayer is then subsequently formed), such measurements would, in principle, be possible. Such an approach might permit evaluation of the influence of some of the parameters we have discussed on this model system.

6. SUMMARY

This chapter has examined some biophysical aspects hypothesized to be involved in the initial steps of vertebrate olfactory reception. These effects would appear to operate in addition to, and in no sense in place of, other putative receptor–ligand interactions at the olfactory receptor neurons. However, such nonspecific interactions may play a critical role in topographical patterning of response over the epithelium, removal of odorant from the epithelium, and selective modification of the more specific biological signal achieved by occupancy of a specific cell-surface receptor macromolecule.

ACKNOWLEDGMENTS. This work was supported in part by the National Science Foundation and the National Bureau of Standards. I thank Dr. Vitaly Vodyanoy, Dr. James

DeVoe, Dr. Martin Pope, Dr. Charles Swenberg, Dr. Joan Cho, Dr. Joel Laxer, Dr. Theodore Misailides, and Dr. Daniel Licht for their assistance in these studies.

REFERENCES

Adamek, G. D., Gesteland, R. C., Mair, R. G., and Oakley, B., 1984, Transduction physiology of olfactory receptor cells, *Brain Res.* **310:** 87–97.

Aldrige, B. E., and Bruner, L. J., 1985, Pressure effects on mechanisms of charge transport across bilayer membranes, *Biochim. Biophys. Acta* **817:** 343–354.

Arito, H., and Takagi, S. F., 1979, Response patterns of olfactory bulb neurons to stimulation of distilled water and odorous solutions, *Jpn. J. Physiol.* **29:** 645–649.

Arito, H., Iino, M., and Takagi, S. F., 1978, Water response of the frog olfactory epithelium as observed from the olfactory bulb, *J. Physiol. (Lond.)* **279:** 605–619.

Armstrong, C. M., and Binstock, L., 1964, The effects of several alcohols on the properties of squid axon, *J. Gen. Physiol.* **48:** 265–277.

Ash, K. O., 1968, Chemical sensing: An approach to biological molecular mechanisms using difference spectroscopy, *Science* **162:** 452–454.

Ash, K. O., and Skogen, J. D., 1970, Chemosensing: Selectivity, sensitivity, and additive effects on a stimulant-induced activity of olfactory preparations, *J. Neurochem.* **17:** 1143–1153.

Axelrod, D., 1983, Lateral motion of membrane proteins and biological function, *J. Membr. Biol.* **75:** 1–10.

Bannister, L. H., Clark, A. D., Dunne, L. J., and Wyard, S. J., 1975, Electron spin resonance measurements of molecular interactions in mouse olfactory epithelium, *Nature (Lond.)* **256:** 517–518.

Bohem, G., Hanke, W., Barrantes, F. J., Eibl, H., Sakmann, B., Fels, G., and Maelicke, A., 1981, Agonist-activated ionic channels in acetylcholine receptor reconstituted into planar lipid bilayers, *Proc. Natl. Acad. Sci. USA* **78:** 3586–3590.

Bouchy, M., Donner, M., and Andre, J. C., 1981, Evolution of fluorescence polarization of 1,6-diphenyl-1, 3,5-hexatriene (DPH) during the labeling of living cells, *Exp. Cell. Res.* **133:** 39–46.

Braganza, L. F., and Worcester, D. L., 1986, Hydrostatic pressure induces hydrocarbon chain interdigitation in single-component phospholipid bilayers, *Biochemistry* **25:** 2591–2596.

Bruner, L. J., and Hall, J. E., 1983, Pressure effects on alamethicin conductance in bilayer membranes, *Biophys. J.* **44:** 39–47.

Cain, W., 1978, History of research on smell, in: *Handbook of perception,* Vol. 6A (E. C. Carterette and M. P. Friedman, eds.), pp. 197–232, Academic, New York.

Canfield, V. A., and Macey, R. I., 1984, Anion exchange in human erythrocytes has a large activation volume, *Biochim. Biophys. Acta* **778:** 379–384.

Carlsen, S., Schimell, E., Weigel, P. H., and Roseman, S., 1981, The effect of the method of isolation on the surface properties of isolated rat hepatocytes, *J. Biol. Chem.* **256:** 8058–8062.

Carrier, D., and Pezolet, M., 1986, Investigation of polylysine-dipalmitoylphosphatidylglycerol interactions in model membranes, *Biochemistry* **25:** 4167–4174.

Chen, Z., and Lancet, D., 1984, Membrane proteins unique to vertebrate olfactory cilia: Candidates for sensory receptor molecules, *Proc. Natl. Acad. Sci. USA* **81:** 1859–1863.

Cherksey, B. D., Murphy, R. B.,and Zadunaisky, J. A., 1981, Fluorescence studies of the beta-adrenergic receptor topology, *Biochemistry* **20:** 4278–4283.

Cherry, R. J., 1979, Rotational and lateral diffusion of membrane proteins, *Biochim. Biophys. Acta* **559:** 289–327.

Cherry, R. J., Dodd, G. H., and Chapman, D., 1970, Small molecule–lipid membrane interactions and the puncturing theory of olfaction, *Biochim. Biophys. Acta* **21:** 409–416.

Clark, A. D., Dunne, L. J., and Wyard, S. J., 1980, Electron spin resonance spin-label studies of mose olfactory epithelium, *Physiol. Chem. Phys.* **12:** 139–151.

Conti, F., Fioravanti, R., Segal, J. R.,and Stuhmer, W., 1982, Pressure dependence of the sodium currents of squid giant axon, *J. Membr. Biol.* **69:** 23–34.

Coronado, R., and Affolter, H., 1986, Insulation of the conduction pathway of muscle transverse tubule calcium channels from the surface charge of bilayer lipid, *J. Gen. Physiol.* **87:** 933–953.

Darszon, A., Vandenberg, C. A., Schonfeld, M., Ellisman, M. H., Spitzer, N. C., and Montal, M., 1980, Reassembly of protein-lipid complexes into large bilayer vesicles: Perspectives for membrane reconstitution, *Proc. Natl. Acad. Sci. USA* **77:** 239–243.

Davies, J. T., 1971, Olfactory theories, in: *Handbook of Sensory Physiology,* Vol. 4 (L. M. Beidler, ed.), pp. 323–350, Springer-Verlag, New York.

Deckmann, M., Haimovitz, R., and Shinitzky, M., 1985, Selective release of integral proteins from human erythrocyte membranes by hydrostatic pressure, *Biochim Biophys. Acta* **821:** 334–340.

Delay, F., and Ohloff, G., 1979, Synthesis of (R) and (S)-p-mentha-1,8-dien-4-ols from (R)-limonene, *Helv. Chim. Acta* **62:** 2168–2173.

Duchamp, A., and Sicard, G., 1984, Influence of stimulus intensity on odour discrimination by olfactory bulb neurons as compared with receptor cells, *Chem. Senses* **8:** 355–366.

Ellerbe, P., Murphy, R. B., and Rose, H. G., 1986, Microenvironment effects of substrate upon lecithin-cholesterol acyltransferase; lateral fluidity studies using pyrene, *J. Lipid Res.* **57:** 232–237.

Fesenko, E. E., Novoselov, V. I., Pervukhin, G. Ya., and Fesenko, N. K., 1977, Molecular mechanisms of odor-sensing. 2. Studies of fractions from olfactory tissue scrapings capable of sensitizing artificial lipid membranes to action of odorants, *Biochim. Biophys. Acta* **466:** 347–356.

Franks, N. P., and Lieb, W. R., 1986, Partitioning of long-chain alcohols into lipid bilayers: Implications for mechanisms of general anesthesia, *Proc. Natl. Acad. Sci. USA* **83:** 5116–5120.

Freeman, W. J., 1981, A physiological hypothesis of perception, *Perspect. Biol. Med.* **8:** 561–592.

Friedman, L., and Miller, J. G., 1971, Odor incongruity and chirality, *Science* **172:** 1044–1046.

Galla, H.-J., and Trudell, J. R., 1980, Pressure-induced changes in the molecular organization of a lipid–peptide complex–polymyxin binding to phosphatidic acid membranes, *Biochim. Biophys. Acta* **602:** 522–530.

Galla, H.-J., and Hartmann, W., 1981, Pyrenedecanoic acid and pyrene lecithin, *Methods Enzymol.* **72:** 471–479.

Getchell, T. V., and Getchell, M. L., 1977, Early events in vertebrate olfaction, *Chem. Senses* **2:** 313–326.

Getchell, T. V., Heck, G. L., DeSimone, J. A., and Price, S., 1980, The location of olfactory receptor sites—Inferences from latency measurements, *Biophys. J.* **29:** 397–411.

Gingell, D., and Fornes, J. A., 1976, Interaction of red blood cells with a polarized electrode—Evidence of long-range intermolecular forces, *Biophys. J.* **16:** 1131–1152.

Getchell, T. V., Margolis, F. L., and Getchell, M. L., 1984, Perireceptor and receptor events in vertebrate olfaction, *Prog. Neurobiol.* **23:** 317–345.

Golan, D. E., Alecio, M. R., Veatch, W. R., and Rando, R. R., 1984, Lateral mobility of phospholipid and cholesterol in the human erythrocyte membrane: Effects of protein–lipid interactions, *Biochemistry* **23:** 332–339.

Gould, R. O., and Walkinshaw, M. D., 1984, Molecular recognition in model crystal complexes: The resolution of D and L amino acids, *J. Am. Chem. Soc.* **106:** 7840–7842.

Haydon, D. A., and Urban, B. W., 1983, The action of alcohols and other non-ionic surface active substances on the sodium current of squid giant axon, *J. Physiol. (Lond.)* **341:** 411–427.

Heron, D. S., Shinitzky, M., Hershkowitz, M., and Samuel, D., 1980, Lipid fluidity markedly modulates the binding of serotonin to mouse brain membranes, *Proc. Natl. Acad. Sci. USA* **77:** 7463–7467.

Holmstedt, B., and Liljestrand, G., 1981, *Readings in Pharmacology,* Raven, New York.

Houck, D. J., 1969, Effect of alcohols on potentials of lobster axons, *Am. J. Physiol.* **216:** 364–367.

Huxley, J., 1942, *Evolution: The Modern Synthesis,* Harper, New York.

Jacobs, R. E., and White, S. H., 1984, Behavior of hexane dissolved in dioleylphosphatidylcholine bilayers. An NMR and calorimetric study, *J. Am. Chem. Soc.* **106:** 6909–6912.

Jahnig, F., Vogel, H., and Best, L., 1982, Unifying description of the effect of membrane proteins on lipid order. Verification for the melittin/dimyristoylphosphatidylcholine system, *Biochemistry* **21:** 6790–6798.

Juge, A., Holley, A., and Rajon, D., 1979, Olfactory receptor cell activity under electrical polarization of the nasal mucuosa in the frog. 2. Responses to odour stimulation, *J. Physiol. (Paris)* **75:** 929–938.

Kaneshina, S., Kamaya, H., and Ueda, I., 1984, Benzyl alcohol penetration into micelles, dielectric constant of the binding site, partition coefficient and high-pressure squeeze-out, *Biochim. Biophys. Acta* **777:** 75–83.

Koppel, D. E., Sheetz, M. P., and Schindler, M., 1981, Matrix control of protein diffusion in biological membranes, *Proc. Natl. Acad. Sci. USA* **78:** 3576–3580.

Kumazawa, T., Kashiwaynagi, M., and Kurihara, K., 1985, Neuroblastoma cell as a model for a taste cell: Mechanism of depolarization in response to various bitter substances, *Brain Res.* **333:** 27–33.

Kunitake, T., Nakashima, N., Hayashida, S., and Yonemori, K., 1979, Chiral, synthetic bilayer membranes, *Chem. Lett.* 1413–1416.

Kutchal, H., Huxley, V. H., and Chandler, L. H., 1982, Determination of fluorescence polarization of membrane probes in intact erythrocytes—Possible scattering artifacts, *Biophys. J.* **39:** 229–232.

Laffort, P., and Dravnieks, A., 1973, An approach to a physicochemical model of olfactory stimulation in vertebrates by single compounds, *J. Theor. Biol.* **38:** 335–345.

Lakowicz, J. R., Prendergast, F. G., and Hogen, D., 1979*a*, Differential polarized phase fluorometric investigations of diphenylhexatriene in lipid bilayers. Quantitation of hindered depolarizing rotations, *Biochemistry* **18:** 508–519.

Lakowicz, J. R., Prendergast, F. R., and Hogen, D., 1979*b*, Fluorescence anisotropy measurements under oxygen quenching conditions as a method to quantify the depolarizing rotations of fluorophores. Application to diphenylhexatriene in isotropic solvents and in lipid bilayers, *Biochemistry* **18:** 520–527.

Lai, C.-S., and Cheng, S.-Y., 1982, Rotational and lateral diffusions of L-thyroxine in phospholipid bilayers, *Biochim. Biophys. Acta* **692:** 27–32.

Lancet, D., 1986, Vertebrate olfactory reception, *Annu. Rev. Neurosci.* **9:** 329–355.

Lutz, H.-P., and Luisi, P. L., 1983, Correction for inner filter effects in fluorescence spectroscopy, *Helv. Chim. Acta* **66:** 1929–1935.

McIntosh, T. J., and Simon, S. A., 1986, Hydration force and bilayer deformation: A reevaluation, *Biochemistry* **25:** 4058–4066.

Mackay-Sim, A., Shaman, P., and Moulton, D. G., 1982, Topographic coding of olfactory quality: Odorant specific patterns of epithelial responsivity in the salamander, *J. Neurophysiol.* **48:** 584–596.

Mackay-Sim, A., and Shaman, P., 1984, Topographic coding of odorant quality is maintained at different concentrations in the salamander olfactory epithelium, *Brain Res.* **297:** 207–216.

Maggio, B., Ahkong, Q. F., and Lucy, J. A., 1976, Poly(ethylene glycol), surface potential, and cell fusion, *Biochem. J.* **158:** 647–650.

Mayr, E., 1982, *The Growth of Biological Thought: Diversity, Evolution, and Inheritance,* Belknap/Harvard University Press, Cambridge.

Melnick, R. L., Haspel, H. C., Goldenberg, M., Greenbaum, L., and Weinstein, S., 1981, Use of fluorescent probes that form intramolecular excimers to monitor structural changes in model and biological membranes, *Biophys. J.* **34:** 499–515.

Meyer, H. H., 1899, Zur theorie der alkoholnarkose: Erste mitteilung, *Arch. Exp. Pathol. Pharmakol.* **42:** 109–118.

Middelkoop, E., Lubin, B. H., Op den Kamp, J. A. F., and Roelofsen, B., 1986, Flip-flop rates of individual molecular species of phosphatidylcholine in the human red cell membrane, *Biochim. Biophys. Acta* **855:** 421–424.

Mozell, M. M., 1964, Evidence for sorption as a mechanism in the olfactory analysis of vapors, *Nature (Lond.)* **203:** 578–593.

Mozell, M. M., 1970, Evidence for a gas chromatographic model of olfaction, *J. Gen. Physiol.* **83:** 233–267.

Muhlenbach, T., and Cherry, R. J., 1985, Rotational diffusion and self-association of band 3 in reconstituted lipid vesicles, *Biochemistry* **24:** 975–983.

Muller, H.-J., Luxnat, M., and Galla, H.-J., 1986, Lateral diffusion of small solutes and partition of amphipaths in defect structures of lipid bilayers, *Biochim. Biophys. Acta* **856:** 283–289.

Murphy, R. B., 1982, *Molecular Biophysics of Olfaction—Report of Progress,* NBS-GCR-82-378, National Bureau of Standards, Washington, D.C.

Murphy, R. B., 1983, *Molecular Biophysics of Olfaction—Progress Report 2,* NBS-GCR-83-442, National Bureau of Standards, Washington, D.C.

Mustparta, H., 1971, Spatial distribution of receptor-responses to stimulation with different odors, *Acta. Physiol. Scand.* **82:** 154–166.

Nakashima, N., Asakuma, S., and Kunitake, T., 1985, Optical microscopic study of helical superstructures of chiral bilayer membranes, *J. Am. Chem. Soc.* **107:** 509–510.

Overton, E., 1901, *Studien uber die Narkose zugleich ein Beitrag zur Allgemeinen Pharmakologie,* Fischer-Verlag, Jena.

Oxford, G. S., and Swenson, R. P., 1979, n-alkanols potentiate sodium channel inactivation in squid giant axons, *Biophys. J.* **26:** 585–590.

Oyama, Y., Akaike, N., and Nishi, K., 1986, Effects of n-alkanols on the calcium current of intracellularly perfused neurons of *Hexix aspersa, Brain Res.* **376:** 280–284.

Parsegian, V. A., Fuller, N., and Rand, R. P., 1979, Measured work of deformation and repulsion of lecithicin bilayers, *Proc. Natl. Acad. Sci. USA* **76:** 2750–2754.

Piesse, G. W. S., 1880, *The Art of Perfumery,* 4th ed., Presley Blakiston, Philadelphia.

Pink, D. A., and Chapman, D., 1979, Protein–lipid interactions in bilayer membranes: A lattice model, *Proc. Natl. Acad. Sci. USA* **76:** 1542–1546.

Pjura, W. J., Kleinfeld, A. M., and Karnovsky, M. J., 1984, Partition of fatty acids and fluorescent fatty acids into membranes, *Biochemistry* **23:** 2039–2043.

Polak, E., Trotier, G. D., and Baliguet, E., 1978, Odor similarities in structurally related odorants, *Chem. Senses* **3:** 369–380.

Pope, J. M., and Dubro, D. W., 1986, The interaction of n-alkanes and *n*-alcohols with lipid bilayer membranes: A proton–NMR study, *Biochim. Biophys. Acta* **858:** 243–253.

Price, S., 1984, Mechanisms of stimulation of olfactory neurons: An essay, *Chem. Senses* **8:** 341–354.

Requena, J., Vedlaz, M. E., Guerrero, J. R., and Medina, J. D., 1985, Isomers of long-chain alkane derivatives and nervous impulse blockade, *J. Membr. Biol.* **84:** 229–238.

Rienacker, R., and Ohloff, G., 1961, Optisch aktives beta-citronellol aus (+)-oder-(−)-pinan, *Angew. Chem.* **73:** 240.

Richards, C. D., Keightley, C. A., Hesketh, T. R., and Metcalfe, J. C., 1980, A critical evaluation of the lipid hypotheses of anesthetic action, *Prog. Anesthesiol.* **2:** 337–351.

Rosenberg, P. H., 1980, Synaptosomal studies of fluidity changes caused by anesthetics, *Prog. Anesthesiol.* **2:** 325–335.

Savoie, R., Pigeon-Gosselin, M., Pezolet, M., and Georgescauld, D., 1986, Effect of the action potential on the raman spectrum of the pike olfactory nerve, *Biochim. Biophys. Acta* **834:** 329–333.

Sheats, J. R., and McConnell, H. M., 1978, A photochemical technique for measuring lateral diffusion of spin-labeled phospholipids in membranes, *Proc. Nat. Acad. Sci. (U.S.A.)* **75:** 4661–4663.

Sicard, G., and Holley, A., 1984, Receptor cell responses to odorants: similarities and differences among odorants, *Brain. Res.* **292:** 283–296.

Silver, W. L., Mason, J. R., Adans, M. A., and Smeraski, C. A., 1986, Nasal trigeminal chemoreception: responses to n-aliphatic alcohols, *Brain Res.* **376:** 221–229.

Smith, E. B., Bowser-Riley, F., Daniels, S., Dunbar, I. T., Harrison, C. B., and Paton, W. D. M., 1984, Species variation and the mechanism of pressure-anaesthetic interactions, *Nature* **311:** 56–57.

Squier, T. C., and Thomas, D. D., 1986, Methodology for increased precision in saturation transfer electron paramagnetic resonance studies of rotational dynamics, *Biophys. J.* **49:** 921–935.

Stratton, G. M., 1917, *Theophrastus and the Greek Physiological Psychology Before Aristotle,* Macmillan, New York.

Stroupe, S. D. and Westphal, U., 1978, Alterations in the ultraviolet absorption spectra of steroids upon binding to serum proteins, *Biochemistry* **17:** 882–887.

Swenson, R. P., and Oxford, G. S., 1980, Modification of sodium channel gating by long-chain alcohols: Ionic and gating current measurements, *Prog. Anesthesiol.* **2:** 7–16.

Takagi, S. F., 1978, Biophysics of smell, in: *Handbook of Perception,* Vol. 6A (E. C. Carterette and M. P. Friedman, eds.), pp. 233–244, Academic, New York.

Thommesen, G., and Doving, K. B., 1977, Spatial distribution of the EOG in the rat: A variation with odour quality, *Acta Physiol. Scand.* **99:** 270–280.

Tonosaki, K., and Funakoshi, M., 1984, Effect of polarization of mouse taste cells, *Chem. Senses* **9:** 381–388.

Ueda, T., Muratsugu, M., Kurihara, K., and Kobatake, Y., 1975, Olfactory response in excitable protoplasmic droplet and internodal cell of *Nitella, Nature (Lond.)* **253:** 629–631.

van Drongelen, W., Holley, A., and Doving, K. B., 1978, Convergence in the olfactory system: Quantitative aspects of odour sensitivity, *J. Theor. Biol.* **71:** 39–48.

Verma, S. P., and Wallach, D. F. H., 1975, Carotenoids as raman-active probes of erythrocyte membrane structure, *Biochim. Biophys. Acta* **401:** 168–176.

Vodyanoy, I., and Hall, J. E., 1984, Thickness dependence of monoglyceride bilayer membrane conductance, *Biophys. J.* **46:** 187–193.

Vodyanoy, V., and Murphy, R. B., 1982, Solvent-free lipid bimolecular membranes of large surface area, *Biochim. Biophys. Acta* **687:** 189–194.

Vodyanoy, V., and Murphy, R. B., 1983, Single-channel fluctuations in bimolecular lipid membranes induced by rat olfactory epithelial homogenates, *Science* **220:** 717–719.

Vodyanoy, V., Halverson, P., and Murphy, R. B., 1982, Hydrostatic stabilization of solvent-free lipid bimolecular membranes, *J. Coll. Interf. Sci.* **88:** 247–257.

White, S. H., 1977, Studies of the physical chemistry of planar bilayer membranes using high-precision measurements of specific capacitance, in: *Electrical Properties of Biological Polymers, Water, and Membranes,* Vol. 303 (S. Takashima and H. Fishman, eds.), Annals of the New York Academy of Science Vol. 303 pp. 243–265, New York.

Wolff, J., and Cook, G. H., 1977, Simulation of hormone effects by polycations, *Endocrinology* **101:** 1767–1775.

Wu, K.-L., Baldassare, J. J., and Ho, C., 1974, Physical-Chemical studies of phospholipids and poly (amino acids) interactions, *Biochemistry* **13:** 4375–4381.

Yoshida, T. M., and Barisas, B. G., 1986, Protein rotational motion in solution measured by polarized fluorescence depletion, *Biophys. J.* **50:** 41–53.

7

Membrane Properties of Isolated Olfactory Receptor Neurons

Robert A. Maue and Vincent E. Dionne

1. INTRODUCTION

1.1. Basis of Interest in Olfactory Receptor Neurons

Olfactory receptor neurons from the nasal epithelium exhibit several unique features that have made them the subject of extensive research. Most notably, these neurons play a key role in the olfactory process, being responsible for both the detection and initial discrimination of odorants. Developmentally, the receptor neurons are of interest because they experience continual neurogenesis, even in adult animals. Concomitant with this are changes in neuronal morphology, biochemistry, and selectivity in the response to odorants associated with synaptogenesis in the olfactory bulb. Mammalian olfactory neurons also undergo a change in the selectivity of their response to odorants just prior to birth of the animal, from that of general irritability to the selective responsiveness found in the adult (Gesteland *et al.*, 1980, 1982). Comparing the conductance mechanisms present in olfactory neurons prior to and after this change occurs might provide insight into the mechanisms of discrimination. Because of these unique features, considerable effort has focused on understanding the membrane properties of olfactory receptor neurons.

1.2. Obstacles to Studying Receptor Neuron Membrane Currents

Despite the interest in olfactory neurons, information about their membrane currents has been difficult to obtain due to the limitations of conventional electrophysiological techniques and their cellular morphology. Intracellular recordings are technically difficult because the cells are usually very small and their somata are located out of sight beneath the surface of the epithelium. Penetration with intracellular electrodes often damages the cells, and the duration of stable recordings can be brief. In addition, the necessity of using high-resistance electrodes with low-current-passing capabilities limits control of the membrane potential.

Robert A. Maue • Division of Molecular Medicine, New England Medical Center, Boston, Massachusetts 02111. *Vincent E. Dionne* • Division of Pharmacology, Department of Medicine, University of California, San Diego, La Jolla, California 92093.

Apart from these methodological problems, there are other obstacles to studying receptor neurons *in situ* or in tissue explants. Unambiguous identification of the cells usually requires additional techniques and experiments. Knowledge of and control over the ionic milieu bathing the neurons is limited by the close apposition of cells in the epithelium and the presence of tight junctions between support cells and receptor neurons. These anatomical features also affect the application of pharmacological compounds, including odorants, by restricting their distribution to individual cells.

1.3. Advantages to Patch-Clamping Isolated Receptor Neurons

We have tried to overcome many of these technical problems by using the patch-clamp technique (Hamill *et al.*, 1981) to study isolated receptor neurons. Well suited to working with small cells, the patch-clamp technique is often less damaging to the cell and permits stable recordings to be made for long periods of time. The technique facilitates measurement of both macroscopic whole-cell currents and currents through individual ion channels with a high degree of temporal resolution. In other cells it has been used to examine the effects of transmitters, hormones, and drugs and is well suited to investigate the actions of intracellular second messengers. The use of isolated receptor neurons enables experiments to be carried out in defined media on identified cells and simplifies application of compounds to single cells.

This chapter surveys our progress in studying the membrane properties of receptor neurons isolated from mice. We were able to record from the somata of the receptor cells and from the terminal dendritic knob as well (Maue and Dionne, 1984, 1986). In addition, we have examined receptor neurons isolated from embryonic and neonatal mice using this procedure (Maue and Dionne, 1986).

1.4. Focus of Initial Patch-Clamp Studies

At the initiation of these studies, there was a scarcity of information concerning ionic conductances in the membranes of receptor neurons, while in other excitable cells a complexity of channel types and regulatory mechanisms was being described. In many excitable cells, especially those that were spontaneously or rhythmically active, a diversity of ionic conductances had been observed. The overall pattern of electrical activity in cells appeared to be the interplay of a variety of ionic currents, any or all of which were a possible means to modulate activity of the cell. Consideration of these findings seemed especially warranted, since there were suggestions that receptor neurons exhibited low levels of spontaneous activity (Mathews, 1972; Getchell 1974; Trotier and MacLeod, 1983), and that upon stimulation by odorants the activity became more regular, the odorants having perhaps modulatory-like actions (Trotier and MacLeod, 1983). These observations and others seemed to justify a careful examination of the ionic conductances in these cells. Guided by these considerations, our initial patch-clamp studies focused on the following specific questions:

1. What ionic conductances are present in olfactory receptor neurons?
2. Are there any properties of these conductances that might implicate them in the odorant response (i.e., ionic selectivity, activation properties, localization in

specific regions of the cell, or developmental changes correlated with the prenatal change in the selectivity)?
3. Do odorants affect any of these conductances?
4. What are the roles of these conductances in the electrical activity of the cells?

2. EXPERIMENTAL SYSTEM

2.1. Preparation of Isolated Olfactory Receptor Neurons

Olfactory receptor neurons were isolated in large numbers from mouse nasal epithelia using enzyme treatment, divalent-free media and mechanical disruption (Maue and Dionne, 1987). The isolated neurons retained their morphology (Fig. 1) and so could be easily distinguished from the other cell types of the epithelium. They comprised the largest fraction of identifiable cells in the preparation and remained largely intact, except for the axon, which presumably was severed from the soma during the dissociation procedure. Somata were spherical or ovoid, 5–8 μm in diameter. A dendrite approximately 1 μm in diameter extended from the soma and terminated in a swelling or knob bearing several fine cilia. Dendrites were usually 5–15 μm long, although dendrites as long as 30–35 μm were observed. Most cells appeared to have 5–10 cilia, with 15–20 in some cases; cilia were generally 10–25 μm long, occasionally extending 40–45 μm. As expected for mammalian olfactory neurons (Lidow and Menco, 1984), the cilia were immotile.

Several distinctive non-neuronal cell types were also dissociated from the epithelia, including respiratory epithelial cells, sustentacular or support cells, and flat fibroblast-like cells that would often extend processes onto the coverslip to which the cells were applied.

A variety of criteria suggested that the receptor neurons were viable and intact:

1. When assayed 1 hr after the cells were plated onto a coverslip, 75–80% excluded trypan blue dye. This percentage remained essentially unchanged for several hours then gradually declined; 5–6 hr after plating greater than 50% of the neurons still excluded dye.
2. Membrane potentials of −35 to −45 mV were measured in the isolated receptor neurons using intracellular techniques. In patch-clamp experiments, the reversal potential of K^+ channels determined in the cell-attached configuration gave estimates of membrane potential in the range of −30 to −80 mV and averaged −52 mV.
3. The indication of a large Ca^{2+} gradient across the membrane also suggested that the cells were viable. Among the ion channels we have studied are Ca-activated K^+ channels (Maue and Dionne, 1984). The activity of these channels in the cell-attached configuration was infrequent, suggesting that the $[Ca^{2+}]$ inside the cell was low.
4. Isolated neurons could fire action potentials. Currents associated with spontaneous action potentials were occasionally recorded from cell-attached membrane patches (Fig. 1).
5. We have also observed that the opening of single ion channels can trigger action

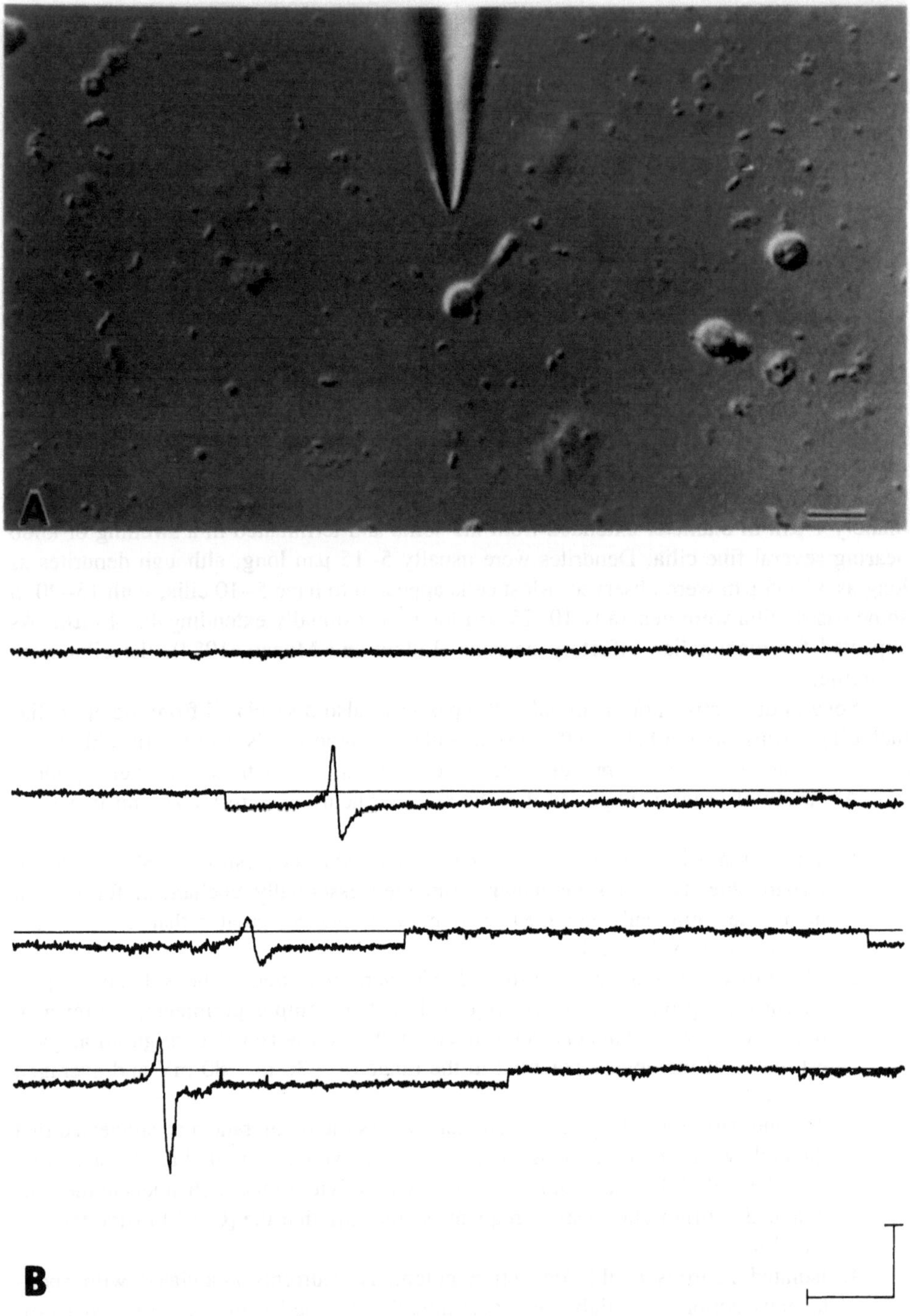
A
B

potentials in these cells (Fig. 1), suggesting that the isolated receptor neurons had high input resistances.

The isolated neurons contained olfactory marker protein (OMP). This small cytosolic protein is unique to the receptor neurons of the nasal epithelium and is a widely used biochemical marker for these cells (Keller and Margolis, 1975; Farbman and Margolis, 1980; Monti-Graziadei *et al.*, 1980). Most of the isolated neurons showed specific labeling when exposed to goat antiserum to OMP. No specific labeling was observed in any of the other cell types or in control experiments with nonimmune serum.

2.2. Patch-Clamp Technique and Data Analysis

Standard methods were used to record and analyse single-channel data from receptor neurons (Hamill *et al.*, 1981). Membrane currents were recorded at room temperature from neurons viewed at 500× magnification with differential interference contrast (Nomarski) optics. Patch electrodes were fire-polished to resistances of 3–5 MΩ and manually positioned on the cell surface with a Leitz manipulator (Fig. 1). High-resistance seals (1–10 GΩ) between the electrode and the cell surface were formed with gentle suction on the lumen of the pipette, and currents flowing through the small patch of membrane circumscribed by the electrode tip were measured with a patch-clamp amplifier built in the laboratory.

The current signal was low pass filtered using an 8-pole Bessel filter and digitally recorded with a laboratory computer system (Digital Equipment Corp., Lynwood, Massachusetts) equipped with a Cheshire Data interface and BASIC-23 software. The signal was typically filtered at 2 kHz (−3 dB) and digitally sampled at 100-μsec intervals. The computer was also used to apply voltage pulses of specified duration and amplitude to the patch pipette via the clamp electronics.

Digitized records of single-channel currents were analyzed using an operator-assisted computer program. The amplitude and duration of the single-channel events, as well as the length of time between events, were measured and stored for later compilation. In addition to allowing information about the conductance and ion selectivity of the channels to be obtained, this procedure also permitted examination of the kinetic behavior of a channel. The pattern of openings and closings can give insight into the mechanisms by which the channel protein functions by providing estimates of the number of conformational states it can adopt and the relationship between them. It may also provide an estimate of the relative rates with which the channel moves between states and indicate

Figure 1. Recording electrical activity from an isolated olfactory receptor neuron with the patch-clamp technique. (A) An isolated mouse olfactory receptor neuron is shown together with a patch-clamp electrode for comparison. Calibration bar: 10 μm. (B) Current recording made in the cell-attached configuration from a receptor neuron. Four contiguous current traces are shown with a baseline drawn at the current level where all the channels in the patch are closed. Openings of individual channels in this patch pass inward currents, depicted by downward deflections from the baseline. Currents associated with spontaneous action potentials in the cell are also evident. In this case, it appeared that action potentials could be triggered by the currents flowing through single channels; the action potential currents were biphasic, usually occurring at infrequent and irregular intervals, and correlated with single-channel activity. The recording electrode contained an elevated K^+ saline; pipette potential = 37 mV. Calibration bars: 10 pA, 20 msec.

which transitions are most influenced by changes in voltage, for example. As with conformational changes in other proteins, the rates with which a channel changes from one conformation to another, and the time it spends in an open or closed state, are random variables. The distributions of the open and closed durations were obtained by compiling histograms of many measurements; by examining and fitting these, estimates of the transition rates and state lifetimes were made. Exponential probability distribution functions fitted to the histograms were optimized using a maximum likelihood method.

2.3. Perfusion System and Odorant Application

Dissociated olfactory cells were plated onto glass coverslips previously coated with concanavalin A (Con A) (Maue, 1986) to hold the cells in place. For the electrical recordings fragments of coverslip containing the cells were placed in a drop of saline (~0.2 ml) between a 40× water-immersion objective and a glass plate. This small volume could be continuously perfused through pieces of stainless steel tubing suspended from the lens. The deadspace of the perfusion system was approximately 0.8 ml.

Heptane, isobutyl alcohol, ethyl acetate, and benzaldehyde were chosen as standard odorants and applied together by microperfusion from a pressure-driven pipet at concentrations of 10 and 100 μM. These compounds were selected because they had different molecular conformations and functional groups, were at least slightly soluble in water, and had been used as effective stimuli in studies of olfactory neurons from several species including rats (Gesteland *et al.*, 1982). Although thresholds of individual cells for some substances appear to be in the nanomolar to micromolar range (Getchell, 1974), odorant concentrations of 10–100 μM have often been found to be effective stimuli; these were chosen in an attempt to maximize the number of cells that would respond and the magnitude of their response. Mass spectrometry was used to verify the presence and concentration of the odorants in the solutions.

3. RESULTS

3.1. Ion Channels in Olfactory Receptor Neurons

Several distinct types of ion channels were active in the membranes of olfactory receptor neurons. Often more than one type of channel was detected in the same membrane patch (Fig. 2). Some of the characteristics of these ion channels are presented below; a more complete description is in preparation.

3.1.1. K^+ Channels

Four different types of K^+ channels were observed in the membranes of olfactory neurons. The most frequently observed channel type was a large conductance, Ca-activated K^+ channel (Fig. 3). Such channels were detected in excised patches of membrane and cell-attached patches from both the neuronal soma and the dendritic knob of mature and neonatal mice. Their single-channel conductance was 133 ± 14 pS (mean ±SD) in excised patches bathed on both sides with saline containing 145 mM K^+. The mean open

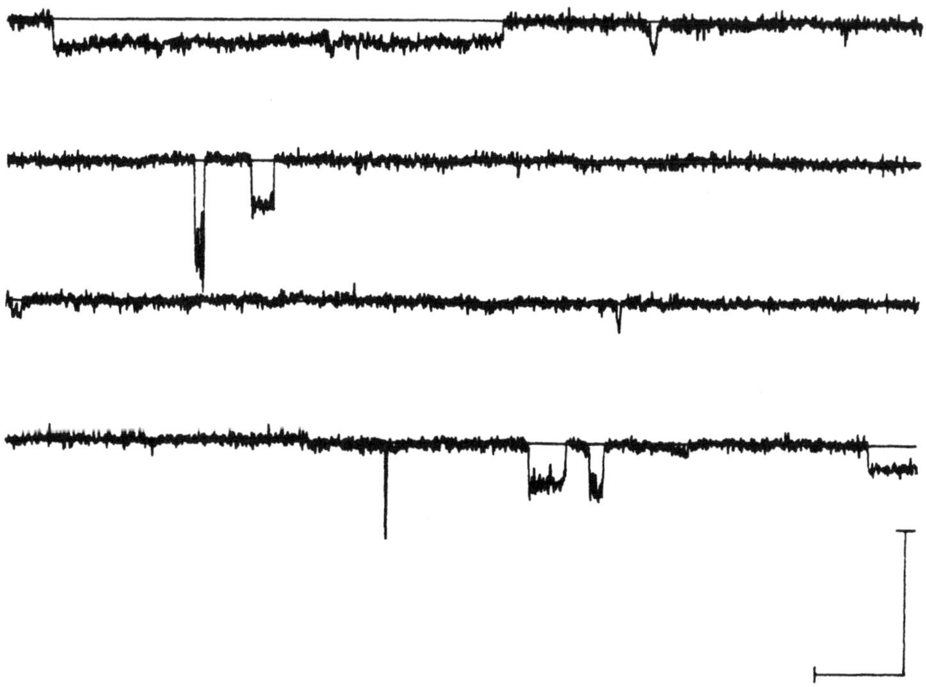

Figure 2. Multiple channel types. Four contiguous current traces illustrate the variety and complexity of single-channel currents found in isolated patches of membrane from olfactory receptor neurons. Downward deflections represent channel openings. Calibration bars: 10 pA, 20 msec.

time of this channel was several milliseconds, sensitive to both the membrane potential and the intracellular [Ca^{2+}]. The open time decreased with depolarization and increased from <10% to >90% when intracellular [Ca^{2+}] was increased from 0.1 to 1.0 μM. The 130-pS channel was reversibly blocked by intracellular Cs^+ (50 mM) but not by tetraethylamonium (30 mM) or 4-aminopyridine (10 mM).

A second type of Ca-activated K^+ channel was observed in patches of somal and dendritic membrane from receptor neurons of adult mice (Fig. 4). This channel was characterized by rapid flickery kinetic behavior and a lower mean conductance than the channel just described. It had a normal conductance of 82 ± 10 pS and was blocked in the presence of 50 mM intracellular Cs^+, similar to the 130-pS channel. Although the activity of this channel increased as intracellular Ca^{2+} was raised (Fig. 4), it was less sensitive to [Ca^{2+}] than was the 130-pS channel. In further contrast, the mean channel open time (typically ~1 msec) was independent of voltage and of intracellular [Ca^{2+}].

Openings of extremely long duration (LD), often exceeding 100 msec, were characteristic of a third type of K-selective ion channel observed in patches of membrane from the soma and the dendritic knob of olfactory neurons (Figs. 1 and 2). The activity of the LD channel was most often seen at hyperpolarized membrane potentials. It was not affected by intracellular Ca^{2+}. The single-channel current-voltage relationship for the channel was nonlinear, showing rectification at depolarized potentials. At hyperpolarized

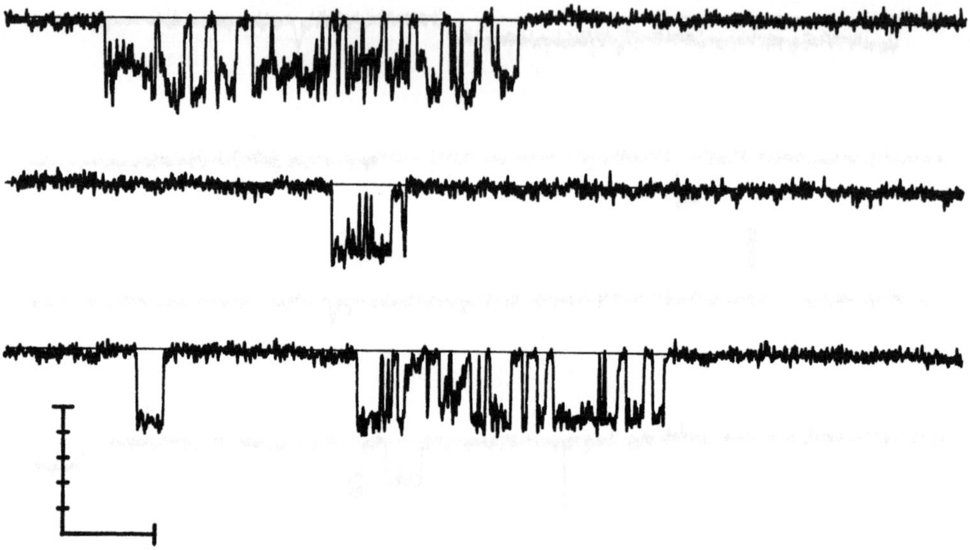

Figure 3. Activity of the 130-pS K^+ channel. Details in the text; downward deflections represent channel openings. Calibration bars: 10 pA, 50 msec.

potentials in cell-attached patches its conductance was 21 ± 9 pS. The channel was not blocked by Cs^+.

A fourth K-selective channel was identified which activated in response to rapid membrane depolarizations (Fig. 5); it was observed in cell-attached membrane patches on receptor neurons from adult and embryonic (E19) mice. The single-channel conductance

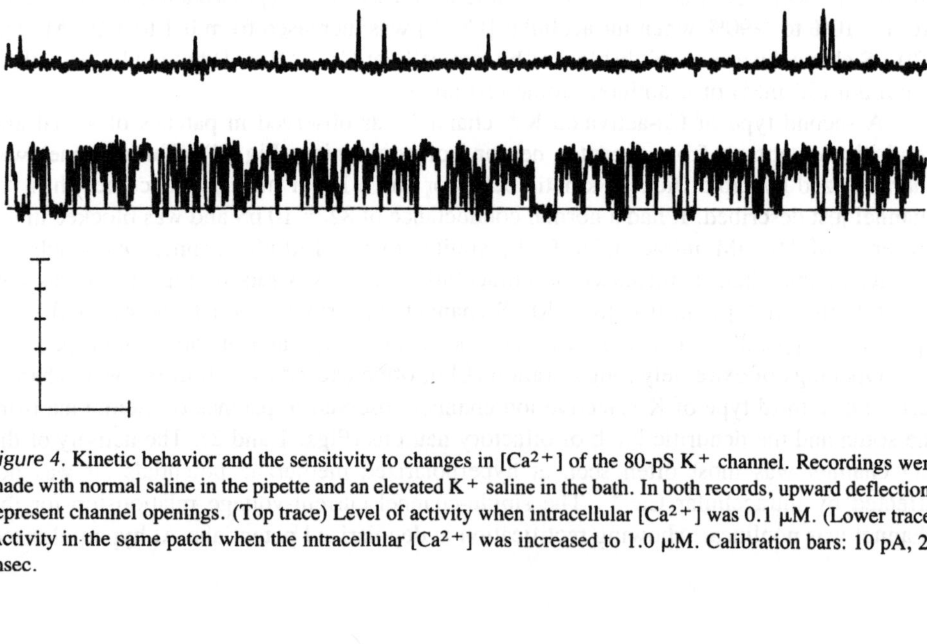

Figure 4. Kinetic behavior and the sensitivity to changes in $[Ca^{2+}]$ of the 80-pS K^+ channel. Recordings were made with normal saline in the pipette and an elevated K^+ saline in the bath. In both records, upward deflections represent channel openings. (Top trace) Level of activity when intracellular $[Ca^{2+}]$ was 0.1 μM. (Lower trace) Activity in the same patch when the intracellular $[Ca^{2+}]$ was increased to 1.0 μM. Calibration bars: 10 pA, 20 msec.

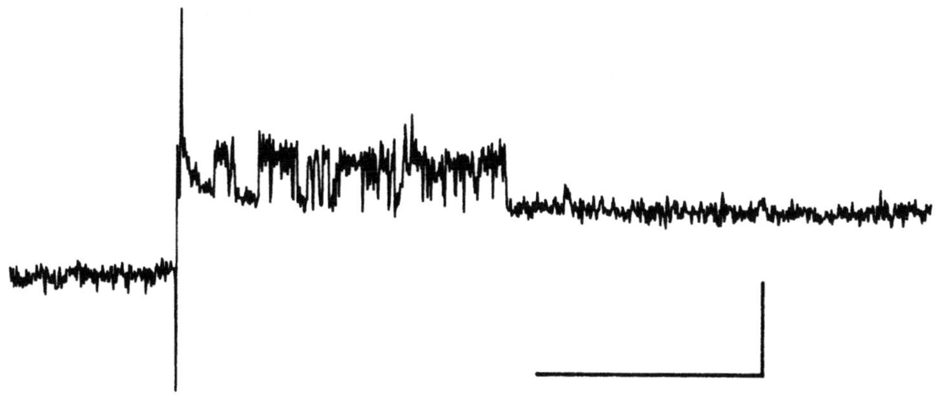

Figure 5. Activation and inactivation of a voltage-sensitive channel. Shown is a current recording made in the cell-attached configuration. Prior to a depolarizing step of 56 mV, the channel was not active. The channel rapidly activated in response to the voltage-step and then inactivated approximately 500 msec later. Inactivation time courses typically ranged from 500 to 900 msec. Calibration bars: 5 pA, 500 msec.

was ~45 pS. Both the mean open time of several milliseconds and the fraction of time open increased with depolarizaton, but at all voltages the channel eventually inactivated.

3.1.2. Other Channel Types

Two types of ion channels selective for ions other than K^+ were detected in membranes of olfactory neurons. Calcium channels, activated by depolarizing voltage steps, were found in patches from the neuronal soma and dendritic ending with pipettes containing isotonic $BaCl_2$. The activity was infrequent and, in both cell-attached and excised patches, short lived. The estimated single-channel conductance was ~16 pS, similar to the value obtained for Ca^{2+} channels in snail neurons using elevated $BaCl_2$ solutions (Brown *et al.*, 1982).

In several experiments using membrane patches from the neuronal soma and dendritic knob, Cl^--selective ion channels were observed. As with the other channel types, the ion selectivity was evaluated by ion substitution and determination of the unit current reversal potential. Chloride-selective channels were unaffected by cation substitution, but the reversal potential shifted toward the Cl^- equilibrium potential when glutamate was substituted for Cl^-. The single channel conductance in normal saline was 211 ± 21 pS.

3.2. Response of Receptor Neurons to Odorants

Odorant detection is signaled centrally with a change in the frequency of action potentials conducted by the axons of olfactory receptor neurons (Gesteland *et al.*, 1965). Presumably this change is due to odorant-induced changes in the neuronal membrane conductance and may appear as an alteration of the characteristics of the membrane ion channels. To evaluate this possibility, the response of the receptor neurons to odorants was examined by monitoring changes in single-channel activity and holding current in cell-attached patches of membrane. One advantage of this method was that the cell remained intact with its cytoplasmic constituents undisturbed.

In 30% of trials, the conductance of cell-attached membrane patches on isolated olfactory neurons changed when a mixture of four odorants was applied. Odorant concentrations of 1–50 μM rarely produced a detectable effect (1 of 17 cells), while 100-μM odorants induced responses in one-third of the trials (14 of 43 cells) (Fig. 6). Responses were not detected in excised patches of neuronal membrane, and perfusion of each neuron studied with normal saline alone had no effect.

Several qualitatively different types of odor-induced responses were observed:

1. An inward current (nominally 10 pA) was observed from patches on the soma and the dendritic knob of neurons from adult mice and in somal patches on neurons from 2-day-old mice. The currents were observed with both elevated K^+ and normal saline in the pipette.
2. An increase in the peak-to-peak noise of the membrane current that showed a time course similar to that of the induced inward current.
3. An increase in the activity of single channels.

Although these odor-induced responses were often repeatable in a cell, they usually could be elicited only two to three times before the cell became unresponsive or the recording configuration was lost. Furthermore, sensitive patches did not show all three types of response. This diversity together with the low frequency of successful trials made it difficult to study in detail these olfactory responses.

We examined whether the odor-induced responses could arise from membrane depolarization by perfusing the neurons with 145 mM KCl. In 9 of 15 cells, KCl induced effects opposite to those observed with odorants: a small outward current accompanied by a decrease in the peak-to-peak noise of the holding current (Fig. 6). The implication was that hyperpolarization rather than depolarization may underlie some of the odorant-induced responses that were observed.

4. DISCUSSION

4.1. Roles of Ion Channels in Receptor Neuron Activity

Only recently has information about the specific ionic conductances present in olfactory receptor neurons begun to accumulate, providing a basis for interpreting the excitability of these cells and for speculating on their response to odorants. Ca-dependent K^+ currents that were blocked by internal Cs^+ have been reported in olfactory receptor neurons from salamander (Firestein and Werblin, 1985). Prolonged afterhyperpolariza-

Figure 6. Odorant application to an isolated receptor neuron. (A) After a seal was formed on the surface of the cell with a patch-clamp electrode (on the right), a multibarrelled perfusion pipette (on the left) containing the four odorants was positioned near the cell. Calibration bar: 10 μm. (B) Current recordings made in the cell-attached configuration. (Top trace) Response to the standard mixture of odorants (100 μM). (Bottom trace) Response when 140 mM KCl was applied by perfusion. In each trace, the leftmost bar under the record designates the duration of control saline application followed several seconds later by the test application. Calibration bars: 10 pA, 4.6 sec.

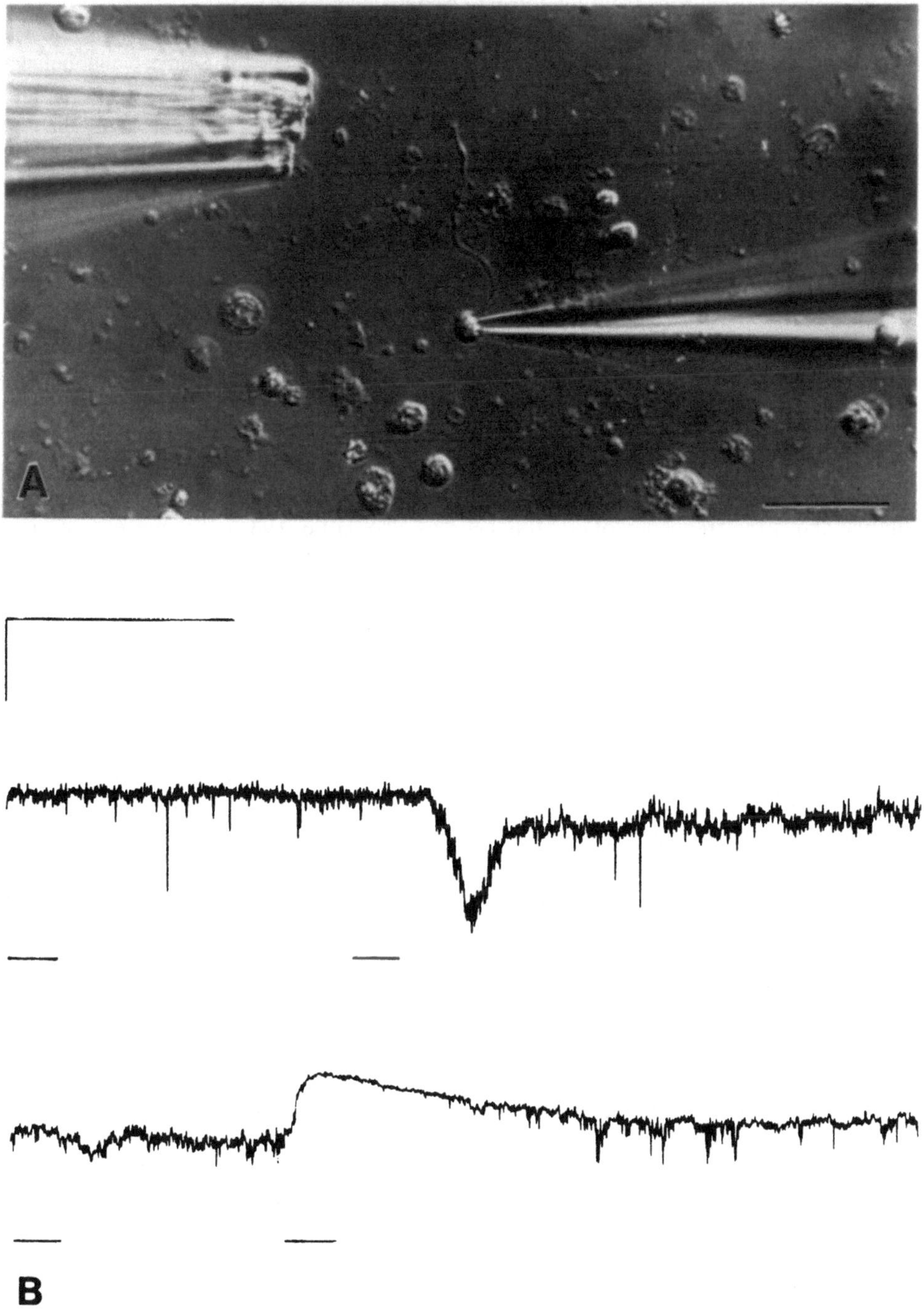
A
B

tions, believed to be generated by Ca-dependent K^+ currents, have also been observed in olfactory receptor neurons from the lobster (Anderson and Ache, 1985). The two types of channels most frequently observed in our studies of mouse olfactory neurons were Ca-dependent K^+ channels with conductances of 130 and 80 pS. The larger channel was more sensitive to $[Ca^{2+}]$ and was also sensitive to voltage, in contrast to the smaller channel. Both Ca-activated channels were blocked by internal Cs^+. The 130-pS channel, the 80-pS channel, or both, may represent the unitary conductance mechanisms underlying the currents in salamander and lobster olfactory neurons.

The properties of the Ca-activated K^+ channels in the mouse receptor neurons, especially the sensitivity of the 130-pS channel to voltage and $[Ca^{2+}]$ and its kinetic behavior, are similar to those of large conductance Ca-activated K^+ channels reported in a variety of cell types (Wong *et al.*, 1982; Magleby and Pallotta, 1983; Gallin, 1984; Findlay *et al.*, 1985). In most excitable cells, the role postulated for Ca-activated K^+ currents is the repolarization of the cell following a depolarization and influx of Ca^{2+}, as during an action potential. That this is the role of these channels in mouse receptor neurons is suggested by the correlation between openings of these channels and the repolarizing phase of action potentials observed in cell-attached patches of membrane (Fig. 1). Ca-activated K^+ channels have been implicated in control of the interspike interval and firing frequency of bursting neurons also (Hille, 1984). The presence of more than one type of Ca-activated channel in the same cell has been reported by others (Marty *et al.*, 1984; Peterson and Maruyama, 1984). The channels often have different Ca^{2+} sensitivities and are thought to provide a method of fine-tuning the activity of the cell (Peterson and Maruyama, 1984).

The mouse olfactory receptor neurons also contained a K^+ channel, most often observed at hyperpolarized membrane potentials, with characteristically long open durations. This is the first report of a conductance mechanism active at hyperpolarized membrane potentials in olfactory neurons, and the role of these channels is unclear. The characteristics of this channel were similar to those described for inwardly rectifying K^+ channels in tunicate eggs (Ohmori, 1978), and cardiac tissues (Bechem *et al.*, 1983; Noma, 1983; Sakmann and Trube, 1984; Trube and Heschler, 1984). Inwardly rectifying K^+ conductances have also been reported in Aplysia neurons (Kandel and Tauc, 1966) and cat spinal cord motorneurons (Nelson and Frank, 1967). In these cells, the channels have been implicated in the maintenance of the resting potential.

Potassium channels activated by rapid depolarizations of 30–40 mV and inactivated at maintained depolarization were also found in the membranes of the mouse receptor neurons. The properties of this channel suggest it may underlie a portion of the voltage-activated K^+ current in salamander receptor neurons described by Firestein (1985). The single-channel conductance, activation range, and kinetics of activation and inactivation of this voltage-sensitive channel in mouse receptor neurons is similar to that of delayed rectifier K^+ channels in other cell types (Quandt and Narahashi, 1982; Ebihara and Speers, 1984; Standen *et al.*, 1985). In excitable cells, delayed rectifiers, like Ca-activated K^+ currents, act to repolarize the cell following an action potential.

Calcium channels similar to those observed in the mouse receptor neurons may be the underlying mechanism of the transient inward current in salamander receptor neurons that activates between -40 to -30 mV, inactivates within 20 msec, and is blocked by external

Co^{2+} (Firestein, 1985). Although their role in olfactory neurons has yet to be determined, Ca^{2+} channel activity appears to be fundamental in a variety of biological responses, including electrical excitability, contraction, and secretion (Hille, 1984).

Chloride currents have not been previously described in olfactory receptor neurons and the role of the Cl^- channels observed in these patch-clamp studies of mouse receptor neurons is not clear. Large-conductance Cl^- channels, insensitive to intracellular [Ca^{2+}] and open more at depolarized membrane potentials, have also been described in rat skeletal muscle (Blatz and Magleby, 1985), where they have been proposed to stabilize the membrane potential.

The mechanism underlying the odorant response and the role that the channels described above have in it are unknown at this time. However, the results described here and those from other sources may provide clues to the mechanisms by which these cells respond to odorants. Experiments using intracellular techniques (Trotier and MacLeod, 1983) and those measuring impedance changes at the surface of the mucosa (Gesteland *et al.*, 1965) have suggested that the response of odorants involves an increase in membrane conductance. If so, excitatory responses may involve an increased permeability to ions such as Na^+ that depolarize the cell and lead to an action potential, while inhibitory responses may involve increased permeabilities to ions such as K^+ and Cl^- that prevent the cell from reaching threshold. Preliminary estimates of the reversal potential of the odorant-evoked response, made using intracellular techniques (Trotier and MacLeod, 1983) and whole-cell patch-clamp recordings (Anderson and Ache, 1985), range from 0 mV to +40 mV, suggesting an increased nonselective cation permeability or one selective for Na^+ or Ca^{2+}. In lobster neurons, odorant application elicited small currents of approximately 5–6 pA (Anderson and Ache, 1985). This is consistent with other intracellular experiments, where depolarizations induced by picoampere currents were sufficient to elicit action potentials (Trotier and MacLeod, 1983; Masukawa *et al.*, 1985). In experiments where olfactory homogenates were incorporated into bilayers (Vodyanoy and Murphy, 1983), increases in K^+-channel activity were observed in response to odorants; the single channels had a conductance of 60 ± 10 pS in 20 mM KCl and were blocked by 0.125 μM 4-aminopyridine. However, the mean channel open time (29 ± 8 sec) was very long in comparison to that commonly observed for channels in cell membranes.

In these patch-clamp studies of mouse receptor neurons, membrane hyperpolarizaton and an increase in membrane conductance were induced by the application of odorants. This is consistent with the results obtained in the bilayer experiments and implicates an inhibitory response mechanism as at least one element elicited by odorants. Using the cell-attached recording configuration, it was not possible to determine whether odor-evoked changes were elicited secondarily or were a component of the primary response. If the hyperpolarization was a secondary response to odorants, it is possible that the initial response involved membrane depolarization or an increase in intracellular [Ca^{2+}]. If the primary response is mediated through an intracellular second messenger such as cyclic adenosine monophosphate (cAMP), it may be that K^+ channels are among a spectrum of targets in the cell and that a hyperpolarizing conductance change results from kinase activation. Potassium channels in other active cells have been shown to be the targets through which cellular activity and responsiveness are modulated (Levitan, 1985; Siegelbaum and Tsien, 1983).

4.2. Future Directions

These studies demonstrate that the patch-clamp technique is a powerful tool with which to examine the membrane properties of olfactory receptor neurons. Two important topics that need to be resolved next are the nature of the neuronal response to odorants and the effects of intracellular second messengers on the membrane conductances of these cells. Along these lines, the evidence that olfactory neurons have receptors for GABA and muscarinic agonists (Hedlund and Shepard, 1983; Anholt *et al.*, 1984) invites further scrutiny. The ability to conduct such studies in cultured olfactory neurons (Noble *et al.*, 1984; Schubert *et al.*, 1985) would permit possible protein degradation during tissue dissociation to be avoided. The study of tissue-cultured cells holds promise, especially with the availability of monoclonal antibodies to the receptor neurons (Allen and Akeson, 1985; Hempstead and Morgan, 1985) and their potential use for cell identification. Finally, the combined application of molecular biology and the patch clamp technique that has been advantageous for studying ion channels in other systems may prove useful in the study of olfaction. Irrespective of its final utilization, it is apparent that the patch-clamp technique will contribute a great deal to understanding the olfactory system on both the cellular and molecular levels.

ACKNOWLEDGMENTS. This work was supported by the Office of Naval Research, The University of California, and a grant from the National Institute of Health.

REFERENCES

Allen, W. K., and Akeson, R., 1985, Identification of a cell surface glycoprotein family of olfactory receptor neurons with a monoclonal antibody, *J. Neurosci.* **5**:284–296.

Anderson, P. A. V., and Ache, B. W., 1985, Voltage- and current-clamp recordings of the receptor potential in olfactory cells in situ, *Brain Res.* **338**:273–280.

Anholt, R. R. H., Murphy, K. M. M., Mack, G. E., and Snyder, S. H., 1984, Peripheral-type benzodiazepine receptors in the central nervous system: localization to olfactory nerves, *J. Neurosci.* **4**:593–603.

Bechem, M., Glitsch, H. G., and Pott, L., 1983, Properties of an inward rectifying K^+ channel in the membrane of guinea pig atrial cardioballs, *Pflugers Arch.* **399**:186–193.

Blatz, A., and Magleby, K. L., 1985, Single chloride selective channels active at resting membrane potentials in cultured rat skeletal muscle, *Biophys. J.* **47**:119–123.

Brown, A. M., Camerer, H., Kunze, D. L., and Lux, H. D., 1982, Similarity of unitary Ca^{++} currents in three different species, *Nature (Lond.)* **299**:156–158.

Ebihara, L., and Speers, W. C., 1984, Ionic channels in a line of embryonal carcinoma cells induced to undergo neuronal differentiation, *Biophys. J.* **46**:827–830.

Farbman, A. I., and Margolis, F. L., 1980, Olfactory marker protein during ontogeny: Immunohistochemical localization, *Dev. Biol.* **74**:205–215.

Findlay, I., Dunne, M. J., and Peterson, O., 1985, High-conductance K^+ channel in pancreatic islet cells can be activated and inactivated by internal calcium, *J. Membr. Biol.* **83**:169–175.

Firestein, S., and Werblin, F., 1985, Electrical properties of olfactory cells isolated from the epithelium of the salamander, *Soc. Neurosci. Abst.* **11**:970.

Gallin, E. K., 1984, Calcium- and voltage-activated potassium channels in human macrophages, *Biophys. J.* **46**:821–825.

Gesteland, R. C., Lettvin, J. Y., and Pitts, W. H., 1965, Chemical transmission in the nose of the frog, *J. Physiol. (Lond.)* **181**:525–559.

Gesteland, R. C., Yancey, R. A., Mair, R. G., Adamek, G. D., and Farbman, A. I., 1980, Ontogeny of olfactory receptor specificity, in: *Olfaction and Taste*, Vol. VII (H. Van derStarre, ed.), pp. 143–146, I. R. L., London.

Gesteland, R. C., Yancey, R. A., and Farbman, A. I., 1982, Development of olfactory receptor neuron selectivity in the rat fetus, *Neuroscience* **7**:3127–3136.

Getchell, T. V., 1974, Unitary responses in frog olfactory epithelium to sterically related molecules at low concentrations, *J. Gen. Physiol.* **64**:241–261.

Hamill, O. P., Marty, A., Neher, E., Sakmann, B., and Sigworth, F. J., 1981, Improved patch clamp techniques for high resolution current recording from cells and cell free patches, *Pflugers Arch.* **391**:85–100.

Hedlund, B., and Shepard, G. M., 1983, Biochemical studies on muscarinic receptors in the salamander olfactory epithelium, *FEBS Lett.* **162**:428–431.

Hempstead, J. L., and Morgan, J. I., 1985, A panel of monoclonal antibodies to the rat olfactory epithelium, *J. Neurosci.* **5**:436–449.

Hille, B., 1984, *Ionic Channels of Excitable Membranes*, 1st ed., Sinauer, Sunderland, Massachusetts.

Kandel, E., and Tauc, L., 1966, Anomalous rectification in the metacerebral giant cells and its consequences for synaptic transmission, *J. Physiol. (Lond.)* **183**:287–304.

Keller, A., and Margolis, F. L., 1975, Immunological studies of the rat olfactory marker protein, *J. Neurochem.* **24**:1101–1106.

Levitan, I. B., 1985, Phosphorylation of ion channels, *J. Membr. Biol.* **87**:177–190.

Lidow, M. S., and Menco, B. P. M., 1984, Observations on axonemes and membranes of olfactory and respiratory cilia in frogs and rats using tannic acid-supplemented fixation and photographic rotation, *J. Ultrastruct. Res.* **86**:18–30.

Magleby, K. L., and Pallotta, B. S., 1983, Calcium dependence of open and shut interval distributions from calcium-activated potassium channels in cultured rat muscle, *J. Physiol. (Lond.)* **344**:585–604.

Marty, A., Tan, Y. P., and Trautmann, A., 1984, Three types of calcium-dependent channels in rat lacrimal glands, *J. Physiol. (Lond.)* **357**:293–325.

Masukawa, L. M., Hedlund, B., and Shepard, G. M., 1985, Electrophysiological properties of identified cells in the in vitro olfactory epithelium of the tiger salamander, *J. Neurosci.* **5**:128–135.

Mathews, D. F., 1972, Response patterns of single neurons in the tortoise olfactory epithelium and olfactory bulb, *J. Gen. Physiol.* **60**:166–180.

Maue, R. A., 1986, Patch clamp studies of isolated mouse olfactory receptor neurons, Ph.D. thesis, University of California, San Diego, LaJolla, California.

Maue, R. A., and Dionne, V. E., 1984, Ion channel activity in isolated murine olfactory receptor neurons, *Soc. Neurosci. Abst.* **10**:655.

Maue, R. A., and Dionne, V. E., 1986, Membrane conductance mechanisms in neonatal and embryonic mouse olfactory neurons, *Biophys. Soc. Abst.* **49**:556a.

Maue, R. A., and Dionne, V. E., 1987, Preparation of isolated mouse olfactory receptor neurons, *Pflugers Arch.* **409**:244–250.

Monti-Graziadei, G. A., Stanley, R. S., and Graziadei, P. P. C., 1980, The olfactory marker protein in the olfactory system of the mouse during development, *Neuroscience* **5**:1239–1252.

Nelson, P. G., and Frank, K., 1967, Anomalous rectification in cat spinal motorneurons and effect of polarizing currents on excitatory postsynaptic potential, *J. Neurophysiol.* **30**:1097–1113.

Noble, M., Mallaburn, P. S., and Klein, N., 1984, The growth of olfactory neurons in short-term cultures of rat olfactory epithelium, *Neurosci. Lett.* **45**:193–198.

Noma, A., 1983, ATP-regulated K^+ channels in cardiac muscle, *Nature (Lond.)* **305**:147–148.

Ohmori, H., 1978, Inactivation kinetics and steady state current noise in the anomalous rectifier of tunicate egg cell membranes, *J. Physiol. (Lond.)* **281**:77–99.

Peterson, O., and Maruyama, Y., 1984, Calcium-activated potassium channels and their role in secretion, *Nature (Lond.)* **307**:693–696.

Quandt, F. N., and Narahashi, T., 1982, Properties of delayed rectifier K^+ channels in neuroblastoma cells, 1982, *Soc. Neurosci. Abst.* **8**:124.

Sakmann, B., and Trube, G., 1984, Conductance properties of single inwardly rectifying potassium channels in ventricular cells from guinea pig heart, *J. Physiol. (Lond.)* **347**:641–657.

Schubert, D., Stallcup, W., LaCorbiere, M., Kiddokoro, Y., and Orgel, L., 1985, Ontogeny of electrically excitable cells in cultured olfactory epithelium, *Proc. Natl. Acad. Sci. USA* **82**:7782–7786.

Siegelbaum, S., and Tsien, R. W., 1983, Modulation of gated ion channels as a mode of transmitter action, *Trends Neurosci.* **6**:307–313.

Standen, N. B., Stanfield, P. R., and Ward, T. A., 1985, Properties of single potassium channels in vesicles formed from the sarcolemma of frog skeletal muscle, *J. Physiol. (Lond.)* **364**:339–359.

Trotier, D., and MacLeod, P., 1983, Intracellular recordings from salamander olfactory receptor cells, *Brain Res.* **268**:225–237.

Trube, G., and Heschler, J., 1984, Inward-rectifying channels in isolated patches of the heart cell membrane: ATP-dependence and comparison with cell-attached patches, *Pflugers Arch.* **401**:178–184.

Wong, B. S., Lecar, H., and Adler, M., 1982, Single calcium-dependent potassium channels in clonal anterior pituitary cells, *Biophys. J.* **39**:313–317.

8

Voltage-Clamp Studies of the Isolated Olfactory Mucosa

John A. DeSimone, Krishna C. Persaud, and Gerard L. Heck

1. INTRODUCTION

1.1. Chemoreception: A Property of Every Cell

All cells recognize chemicals in their immediate environment. They detect and respond specifically to substances as simple and ubiquitous as sodium ions to substances as complex as protein antigens. In many cases the response involves the translocation of the ligand or the activation of an ion channel.

Although many membrane proteins possess specific receptor sites for ligands, not every cell is regarded as a chemoreceptor cell. These are distinguished by their ability to transmit information regarding their present status, modulating the function of other systems. This information relay is accomplished by transducing the peripheral chemical or electrochemical event into the secretion of neurotransmitters or hormones. Internal chemoreception relies on both modes of signal relay. The carotid bodies are examples of the former chemoreceptor type, responding specifically to changes in arterial P_{O_2}, P_{CO_2}, or pH with afferent impulses to the respiratory center (Acker *et al.*, 1983). The gastrointestinal (GI) hormone-releasing cells of the stomach and intestines are examples of hormone signaling. I cells found in the duodenum respond vigorously to the presence of L-phenylalanine (among other stimuli) releasing cholecystokinin into the blood (Meyer and Grossman, 1972). We are seldom aware of these chemoreceptors or of the consequences of their action. However, we are conscious of chemoreception when it directly affects behavior, as with olfaction and taste. It is on this basis that these two senses are commonly grouped with vision and audition as special senses. From the standpoint of the initial events of chemoreception, this classification is not exact. For many purposes better

Abbreviations used in this chapter: ADP, adenosine diphosphate; ATP, adenosine triphosphate; cAMP, adenosine-3′,5′-monophosphate; cGMP, guanosine-3′,5′-monophosphate; EOG, electro-olfactogram; GDP, guanosine diphosphate; GDPβS, 5′-*O*-(2-thio)diphosphate; GI, gastrointestinal; G_i, inhibitory guanine nucleotide binding protein; G_s, stimulatory guanine nucleotide binding protein; GTP, guanosine triphosphate; GTPγS, 5′-*O*-(3-thio)triphosphate.

John A. DeSimone, Krishna C. Persaud, and Gerard L. Heck • Department of Physiology, Virginia Commonwealth University, Richmond, Virginia 23298.

analogs of the early ligand–receptor events can be found among the chemoreceptor cells of the gastrointestinal tract and also the transporting epithelia. Applying the electrophysiological methods of epithelial transport to the mammalian dorsal lingual epithelium has led to the identification of two types of sodium taste receptors, at least one of which is a channel (DeSimone *et al.*, 1981, 1984; Schiffman *et al.*, 1983). The secretory and the absorptive properties of the olfactory mucosa have also been studied using *in vitro* electrophysiological methods (Heck *et al.*, 1984*a*). Recent research shows that olfactory receptor function can also be probed by these techniques when pharmacological agents are used (DeSimone and Getchell, 1985; Persaud *et al.*, 1986). This approach has revealed a high degree of active ion transport in the olfactory mucosa. This is measured as a steady-state short-circuit current across the mucosa. Odorant stimulation introduces a perturbation current superimposed on the steady-state baseline. The odorant-evoked current has many of the properties expected of a generator current. This current can be measured accurately under voltage clamp.

1.2. Chapter Organization

Sections 2–4 review the methods and theory of ion transport across epithelia. We focus mainly on methods that have already been applied to olfaction and taste. We have, however, included several methods that should be useful in future studies. Section 5 deals with the lingual epithelium as a transporting epithelium and relates these results to salt taste transduction. Sections 6 and 7 extend the methodology to include the olfactory mucosa. Here electrophysiological results are presented bearing on the role of second messengers in olfactory transduction. Section 8 summarizes the results to date.

2. ION-TRANSPORTING EPITHELIA

2.1. Asymmetrical Structure: Consequences for Function

Epithelia typically form an interface between extracellular fluid and a compartment that is functionally outside the body. This is literally true in the case of sodium-transporting anuran skins. In the case of the GI tract, the respiratory tract, and the renal system, the contents of the respective lumina are in a milieu distinct from the homeostatically regulated internal fluids, and in that sense are functionally outside the body. The two fluid compartments separated by the epithelium are different, so the epithelium functions as a passive barrier. However epithelial cells, through their specific transport functions, also modify compartment compositions. Therefore, the asymmetry in compartment composition is in part created by the cells themselves through transport of ions, water, and organic substances. The vectorial character of these processes implies a functional asymmetry at the cellular level. Membrane vesicle fractions enriched in either apical or basolateral membrane proteins can be prepared (Baumann and Kinne, 1972; Hopfer *et al.*, 1973), demonstrating the asymmetrical localization of transport function. Studies confirm the topological arrangement of transport elements that permit transport functions at individual loci to be coupled. The simplest conceptual decomposition of a transporting epithelium consists of a series arrangement of apical and basolateral membranes in parallel with a paracellular shunt. This concept has gained universal acceptance and is the starting point of most analyses (e.g., Ussing and Zerahn, 1951; Diamond and Bossert, 1967; DiBona and Civan, 1973; Mikulecky, 1979; Essig, 1982; DeSimone *et al.*, 1984; Simon and Garvin, 1985; Okada *et al.*, 1986).

2.2. Special Membrane-Transport Systems

Many of the ion transport systems involving the major constituents of extracellular fluid are not unique to either the apical or basolateral membrane but can be found in one or the other depending on the specific tissue. This is true of the electroneutral ion-exchange systems, such as the Na^+/H^+ and Cl^-/HCO_3^-, and the cotransport system Na^+/Cl^-. Thus, in pancreatic duct cells the Na^+/H^+ exchanger is indicated for the basolateral membrane (Swanson and Solomon, 1975), while in the renal proximal tubule it is found in the apical membrane (Murer *et al.*, 1976). A Cl^-/HCO_3^- exchanger has been proposed for the basolateral membrane of oxyntic cells (Rehm, 1967) and has also been isolated in the brush-border membranes of enterocytes (Liedtke and Hopfer, 1982). The Na^+/Cl^- cotransport system appears to be present in the apical membranes of gallbladder epithelial cells (Frizzell *et al.*, 1975), in the cells in the olfactory mucosa of bullfrog (Heck *et al.*, 1984*a*), and in the basolateral membranes of canine tracheal epithelial cells (Widdicombe and Welsh, 1980).

Transport systems are located in either the apical or basolateral membranes of transporting cells. Among these are the Na,K-electrogenic exchanger found exclusively in basolateral membranes of virtually all transporting cells and the Na-glucose and Na-amino acid transporters found in the brush border membranes of enterocytes and proximal tubule cells (Kinne and Kinne-Saffran, 1978). With hormone-regulated ion transport in which cyclic adenosine monophosphate (cAMP) is a second messenger, if the site of the hormone receptor is on the basolateral membrane, and the site of final action on the apical membrane, adenylate cyclase activity is found in the basolateral membrane and a protein kinase for which cAMP is a substrate is found in the apical membrane (Kinne and Kinne-Saffran, 1978). In blowfly taste receptors, supporting cells actively secrete potassium across their apical membranes (Wieczorek, 1982). This electrogenic process is thought to provide the driving force for receptor function.

2.3. Paracellular Shunts

Paracellular shunts exist in parallel with transcellular routes of ion transport. These are believed to be across the tight junctional complexes along the apical pole of the cells. Epithelia are considered physiologically tight or leaky, depending on the resistance of the shunt pathways. Typical tight epithelia, are the frog skin or the toad urinary bladder, i.e., those that function primarily to conserve sodium (Civan, 1983; Lewis and Wills, 1981). Transepithelial resistances are usually several thousand $\Omega\text{-cm}^2$. Leaky epithelia usually function to absorb or secrete ions and water. Examples are the gallbladder, small intestine, and canine trachea where resistances range from approximately 50 $\Omega\text{-cm}^2$ to a few hundred $\Omega\text{-cm}^2$. The bullfrog olfactory mucosa has a resistance ranging from 50–150 $\Omega\text{-cm}^2$ (Heck *et al.*, 1984*a*), placing it among the class of leaky epithelia. Thus, the shunt pathways are important routes of electrodiffusion across the olfactory mucosa. The existence of these shunts and their possible role in olfactory transduction has been previously discussed (Gesteland, 1971; Getchell, 1974; Juge *et al.*, 1979; Heck *et al.*, 1984*a*).

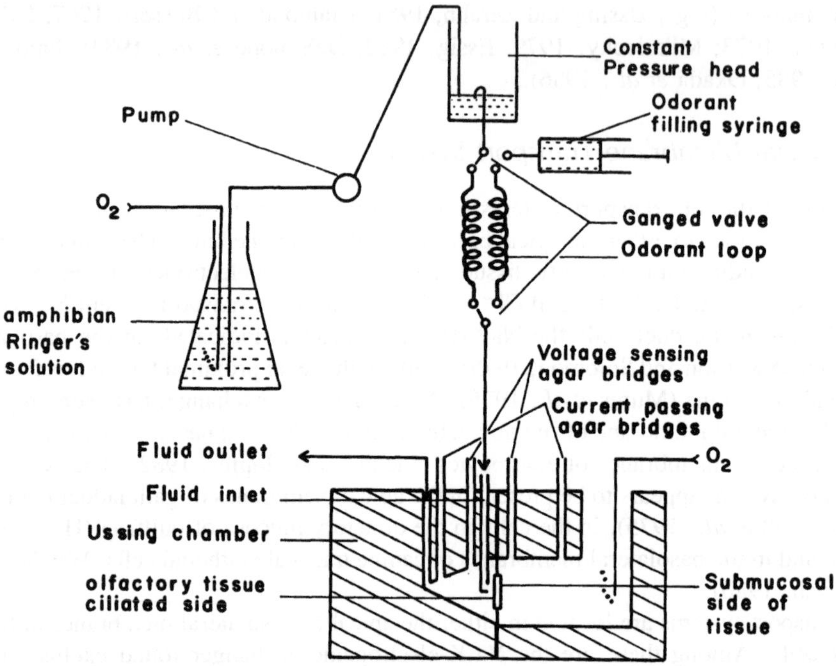

Figure 1. Modified Ussing chamber apparatus used to record transmucosal currents from the olfactory mucosa under voltage clamp. The olfactory tissue was mounted between two Ussing chambers. Oxygenated amphibian Ringer's solution was pumped up to a constant pressure head reservoir. The solution then flowed at 0.2 ml/sec through either a 4-ml sample loop containing odorant solution or a dummy loop of identical volume entering the chamber facing the ciliated side of the tissue. After leaving the chamber, the solution was discarded. The loops were switched by a 2-gang 3-way valve shown in simplified form. While Ringer's solution flowed through the dummy loop, the sample loop could be filled with odorant solution via a syringe. When the valve was switched, a bolus of odorant solution then entered the Ussing chamber. Agar bridges connected to both chambers were connected electrically to the input headstage of a voltage-clamp apparatus.

3. ELECTROPHYSIOLOGICAL METHODS

3.1. The Short-Circuit Method (Ussing Method)

The driving forces for ion transport across an epithelium can be both passive and active. In his pioneering work on frog skin, Huf (1936) used these terms to distinguish passive electrochemical differences as driving forces, from cellular mechanisms that convert chemical free energy to translocate ions. One major focus of epithelial electrophysiology has been the unequivocal identification of those ions actively transported. This can be accomplished *in vitro* by placing the tissue between two reservoirs (cf. Fig. 1). Passive transepithelial fluxes are nullified by eliminating all passive driving forces for transport across the tissue. For nonelectrolytes this is achieved by placing the same solution on both sides of the tissue. This is not sufficient in the case of ions because most active transport processes are electrogenic; i.e., they create a transmural electrical potential that in turn acts as a passive driving force for ion flow. Ussing and Zerahn (1951) solved this problem by short-circuiting the potential using a transmural voltage clamp. The current (i.e., the short-circuit current) is equal to the sum of the net fluxes of all the actively transported ions. This is an extremely useful result. When only one ion is actively transported, the short-circuit current and the net flux are identical. When more than one ion is actively transported the short-circuit current serves as a constraining relationship such that of the n ion fluxes only $n-1$ need be determined directly.

The identity of the short-circuit current and the transepithelial ion fluxes holds only in the steady state. Passive ion flow from one reservoir into the epithelial cells may be the cause of fast current transients. The odorant induced perturbations in the short-circuit current discussed in Section 6 appear to be associated with a transient passive sodium influx across the apical membranes of receptor cells. Transient changes in either the flux of an ion across the apical barrier of an epithelium or in the short-circuit current, resulting from a variation in the composition of the apical reservoir, often demonstrate properties characteristic of that particular barrier. Biber (1971) and Fuchs *et al.* (1977) studied the sodium permeability of the outer barrier of the frog skin in this way. DeSimone *et al.* (1984) analyzed sodium-induced current transients originating at the apical barrier of the lingual epithelium in their probe of taste receptor function.

3.2. Radioisotopes and Active Ion Transport

The short-circuit current gives only the sum of the actively transported ions. The individual ion fluxes can be measured using radioisotopes. When the short-circuit method is used in conjunction with flux measurements, a complete profile of each actively transported species is possible. The net current contributed by each ion is the difference between an influx and an efflux. These are measured separately. If two radioisotopes of one ion are available, the two unidirectional fluxes can be measured with a single epithelial preparation. This is often impractical, particularly when the fluxes of more than one ion are to be determined. An alternative technique is to measure the influx and efflux using tissue pairs. This method requires that the pairs be matched to within 10–15% with respect to their electrical parameters. Flux studies in the steady state under short-circuit conditions are considered basic in the analysis of all transporting epithelia. The special sodium transport system of the lingual epithelium was in part characterized through radioisotopic flux determinations as a function of the sodium concentration in the apical medium (Mierson *et al.*, 1985).

3.3. Other Methods

Transporting epithelia can be regarded as tesselations of basic resistive elements. The functional unit in its simplest form consists of a series arrangement of apical and basolateral membranes and a parallel shunt. Microelectrode impalement of individual cells under voltage clamp permits the electrical properties of the apical, basolateral and shunt pathways to be separated. The potential change across the individual barriers can be obtained, and the resistances associated with each barrier can be measured at different transepithelial voltages. Analysis of both tight and leaky epithelia is discussed extensively in a symposium volume edited by Schultz (1981).

Fluctuation or noise analysis of apical sodium channel opening/closing kinetics in tight epithelia has given a better understanding of sodium self-inhibition of sodium channel permeability. Current fluctuations around short circuit are analyzed in the frequency domain. For Lorentzian noise (common for membrane noise sources), it is possible to relate the frequency corresponding to one half the plateau value of the current power-density spectrum to the relaxation times for ion binding to the channel sites (Li and Lindemann, 1983). These in turn allow for the computation of microscopic rate constants and channel densities. Fluctuation analysis has now been joined by patch clamp methods as a means of characterizing single-channel kinetics (Frizzell *et al.*, 1986). Patch-clamp techniques have been introduced recently to olfactory receptor research (cf. Maue and Dionne, 1986). It is possible to form membrane vesicles from either apical or basolateral membranes (Kinne and Kinne-Saffran, 1978). Specific transport or receptor sites can thereby be studied in an enriched preparation. Enriched olfactory ciliary preparations have recently enabled Pace *et al.* (1985) to propose a G-protein-mediated cAMP-stimulated olfactory transduction mechanism.

Various methods are used to estimate intracellular ion activity and the extent to which ions are in the bound state. Nuclear magnetic resonance (NMR) has been used to show that very little Na or K is bound inside cells and that intracellular concentrations of adenosine diphosphate (ADP) may be far less than chemical determinations would suggest (Civan, 1983). Ion-selective microelectrodes are available for Na, K, and Cl. These are now commonly used to estimate intracellular ion activities and have played major role in investigating cell energetics (Armstrong *et al.*, 1981). They have also been used to investigate possible interactions between apical and basolateral membranes (Lewis and Wills, 1981) and the role of intracellular potassium in the control of apical membrane sodium permeability in tight epithelia (Civan, 1983). Electron-probe X-ray microanalysis also allows for an estimate of intracellular ion concentrations and can be used in cases in which microelectrode impalement is technically unfeasible (Civan *et al.*, 1980). It can be anticipated that many of these methods will ultimately be useful in probing the olfactory epithelium.

4. THEORETICAL METHODS

4.1. Nonequilibrium Thermodynamics

Classic thermodynamics treats systems at equilibrium or systems evolving toward equilibrium. However, this is the single steady state of a functioning epithelium that is intrinsically least interesting. The epithelium performs its ligand recognition and ion translocation functions about steady states remote from equilibrium. The energetics of nonequilibrium steady states is best approached from the standpoint of nonequilibrium thermodynamics. We can consider the flow of any substance as a dissipative process when it occurs in the same direction as its conjugate driving force. In the case of the flux of a cation, J_+, the conjugate driving force for transepithelial transport is its electrochemical potential difference, X_+. In general, the rate of entropy production in a system is given as the sum of products of all conjugate forces and flows. Let us consider the case of two force-flow pairs, one of which is J_+ and X_+, while the other is J_r and X_r. The *r*-

subscripted variables refer to the flow (or rate) of a metabolic reaction, J_r, driven by its chemical affinity (Gibbs free energy at constants T and P), X_r. When both processes occur simultaneously, the dissipation function, Φ (rate of entropy production times the temperature) is given by

$$\Phi = J_+X_+ + J_rX_r \tag{1}$$

The dissipation function is a positive definite quantity. This is a constraint imposed by the second law of thermodynamics. However, this requirement applies only to the sum of the individual contributions of the force-flow products. If the two processes are coupled, only one process (say the metabolic reaction) need contribute positive entropy production. The other flux may be driven against its conjugate driving force. The coupled system can often be represented formally by a system of coupled linear equations that express the fact that a flux such as J_+ can also be driven by the chemical affinity X_r. This is exactly the case for the short-circuited frog skin, where sodium is absorbed in the absence of its electrochemical driving force because it is driven by the hydrolysis of adenosine triphosphate (ATP) through the sodium pump. Vieira *et al.* (1972) used suprabasal oxygen consumption as a measure of the metabolic flux, and the linear phenomenological equations to compute the affinity of the reactions coupled to sodium transport.

The strength of nonequilibrium thermodynamics is its ability to treat coupled processes where the degree of coupling may be incomplete, i.e., nonstoichiometric. Its usefulness has been recently extended by Mikulecky (1982) through network theory, which allows complex arrays of coupled flows to be analyzed for a given topology of structures.

4.2. Kinetic Approaches

The kinetic approaches applied to epithelial transport vary considerably. Compartmental analysis has been used extensively in the treatment of sodium transport (Huf and Mikulecky, 1985, 1986). Electrodiffusion equations have been solved numerically to model ion flow across epithelia (Lew *et al.*, 1979). Electrodiffusion coupled with network thermodynamic methods allows very complicated topologies to be treated using electrical circuit simulation computer programs (Fidelman and Mikulecky, 1986). Recently, Biber *et al.* (1986) modeled chloride transport through the paracellular shunts of frog skin as electrodiffusion through a region with fixed charge on the apical surface. This accounted for the rectification with transepithelial potential in the unidirectional chloride fluxes. One of the first kinetic approaches dealing with criteria for active transport is due to Ussing (1949). He showed that the ratio of the unidirectional fluxes of an ion subject only to passive forces of electrodiffusion across a single barrier was given by

$$\frac{J_{\mathrm{ms}}}{J_{\mathrm{sm}}} = \frac{a_{\mathrm{m}}}{a_{\mathrm{s}}} \exp\left[-\frac{zF(V_{\mathrm{s}} - V_{\mathrm{m}})}{RT} \right] \tag{2}$$

Here J_{ms} is the ion flux from mucosa to serosa; J_{sm} is the ion flux from serosa to mucosa; a_{m} and a_{s} are the ion activities on the mucosal and serosal side, respectively; z is the ionic valance; F, Faraday's constant; R, the gas constant; T, the temperature, and $V_s - V_m$, the potential difference. When both reservoirs are identical and the system is short-circuited and the ion transport is driven by processes other than simple electrodiffusion, including active transport, the flux ratio will deviate from unity. While originally derived for steady state conditions only, Sten-Knudsen and Ussing (1981) showed it also to be valid during transients for membranes in which paracellular shunt pathways are negligible. These principles are now being applied to the olfactory mucosa (Heck *et al.*, 1984*a*).

5. ION TRANSPORT AND CHEMORECEPTION

A large range of compounds are capable of being detected by vertebrate chemoreceptory systems. It is likely that more than one transduction system operates in dealing with chemical stimuli. One mechanism in chemoreception may be the interaction of a chemical with a proteinaceous receptor, triggering changes in cyclic nucleotide levels, and leading to the opening of ion channels in the membrane (Dodd and Persaud, 1981). In the detection of Na^+ ions in gustation, the channel itself appears to be the taste receptor.

5.1. The Sodium Taste Receptor

DeSimone and co-workers discovered that sodium and other ions could cross the lingual epithelium and that sodium was actively transported. In addition, there was evidence for a special sodium transport system extending into the hyperosmotic salt concentration range. This coincided with the normal range of gustatory sensitivity to salt, and it was hypothesized that this was part of the sodium taste transduction system (DeSimone *et al.*, 1981). Confirming this, it was found that the diuretic amiloride, a Na^+ channel blocker in other systems, was a specific blocker of sodium chloride taste as determined both neurophysiologically and psychophysically (Schiffman *et al.*, 1983; Heck *et al.*, 1984*b*; Brand *et al.*, 1985). Amiloride was more effective as a blocker of sodium chloride taste in the hyperosmotic range and this was specific for sodium chloride and lithium chloride. It was much less effective against potassium chloride taste. The results suggest that Na^+ ions stimulate the cells by way of two independent pathways. A part of the response is not inhibited by amiloride and this implies the existence of a second pathway (DeSimone *et al.*, 1984; DeSimone and Ferrell, 1985). The results imply that one type of sodium taste receptor is an apical sodium ion channel; in this case the ligand (Na^+) is the direct agent of the receptor cell depolarization. Recent measurements of intracellular potentials in frog taste cells by Okada *et al.* (1986) are consistent with this conclusion, in that amiloride hyperpolarizes the cells in the face of a sodium stimulus.

5.2. Other Ions

Ion substitution experiments using the lingual epithelium (Mierson *et al.*, 1985) indicate that some large ions may also be permeable although not to the extent of Na^+ and Cl^-. Choline chloride and arginine hydrochloride, which have a bitter to salty taste, as well as the large cations Tris and tetramethylammonium, support a short-circuit current.

Barium partially blocks potassium chloride current across the canine lingual epithelium (DeSimone, Persaud, and Heck, unpublished observations), suggesting the presence of passive potassium ion channels (Hille, 1984).

6. THE OLFACTORY MUCOSA

The resting membrane potential of any cell depends on the passive permeability of its membranes to certain ions such as sodium, potassium and chloride, and the presence of certain nonconductive, but electrogenic elements such as the Na^+/K^+ ion pump.

A simple circuit model of the passive electrical properties of a membrane is a resistance and capacitance in parallel. Any change in resting membrane potential implies a current across the membrane. Consider the probable events that occur when olfactory receptor cells transmit information to second-order sensory neurons:

1. The stimulus directly or indirectly alters the properties of a specialized area of the receptor cell membrane, such that the ionic permeability changes and a change in conductance occurs.
2. The change in conductance, together with the electrochemical gradient across the membrane, cause an inward current (receptor or transduction current).
3. The receptor current produces a change in membrane potential (receptor potential).
4. The receptor potential spreads to an area of membrane that has a regenerative current-voltage relation. This is the spike-initiating region located in the axon hillock of the olfactory neuron. The electrotonic spread of the receptor potential is called the generator potential.

When an electrode is placed on the surface of the olfactory epithelium and an odor stimulus is applied, a slow potential change of up to several millivolts is observed. This phenomenon, first recorded by Hosoya and Yoshida (1937) in canine olfactory epithelium, has been found in many other animals and was called the electro-olfactogram (EOG) by Ottoson (1956). This potential change is generally regarded as a summated receptor potential.

The most thorough studies of the ionic basis of the odorant-evoked electrical transients are due to Takagi and co-workers. These investigators used an *in vitro* preparation of the bullfrog olfactory mucosa, replacing sodium and potassium ions in the solution bathing both sides of the mucosa and recording the EOGs in response to odor stimuli. On the basis of these experiments, Takagi *et al.* (1969) and Okano and Takagi (1974) suggest that the EOG consists of three components:

1. A negative receptor potential, due to influx of Na^+ with a corresponding increase in membrane permeability to K^+ ions.
2. A hyperpolarizing positive component mainly due to the movement of Cl^- ions, with a contribution from K^+ ions.
3. A slow positive potential, accompanied by vigorous secretory activity in the supporting cells.

However, EOG measurements do not prove this conclusively. Transmembrane potentials can also change because of nonspecific changes in whole tissue resistance. A more direct means of establishing the presence of ion fluxes involved with transduction is to measure generator currents rather than potentials. The current across the olfactory mucosa can be measured under voltage clamp and, with care, those current components associated with transduction can be identified. Results using the voltage-clamp technique are discussed in Section 6.3.

Several cell types in the olfactory mucosa may support transmural ion transport. The superficial olfactory epithelium contains the receptor neurons and the sustentacular cells that may contribute to both the standing current and the odorant-stimulated current. The underlying layer, the lamina propria, contains the acinar cells of Bowman's glands, which produce mucopolysaccarides (M. L. Getchell *et al.*, 1984). Since the secretion or absorption of aqueous solution in many glands and epithelia is driven osmotically by electrolyte transport (Macknight and Leaf, 1977), this may contribute to the observed steady-state current.

The initial events in olfactory transduction are associated with the spread of current in the microenvironment of the stimulated receptor neurons (Getchell, 1974; Suzuki, 1977). The nature of the current pathways, identification of current carriers, and location of current sources and sinks have until now only been modeled on the basis of the limited experimental data available. The models indicate that the current associated with olfactory transduction, spreads inward from a locus on the apical membrane of the olfactory receptor neuron. The current shunts include the paracellular spaces and the sustentacular cells. The paracellular spaces are probably the lowest resistance pathways for the return currents. Secretion or absorption within the mucosa will contribute to the total current measured. The sustentacular cells are postulated to be the source of various secretory products (Getchell and Getchell, 1977; Getchell *et al.*, 1984). If the active transport processes are transmucosal in character, they could influence the ion composition in the microenvironment of the olfactory receptor neurons. Therefore events originating in the sustentacular cells could affect indirectly the olfactory receptor neurons. We have used the Ussing chamber technique described in Section 3.1 to examine ion transport across frog olfactory mucosa *in vitro*, both in the steady-state condition and when stimulated by an odorant.

6.1. Materials and Methods

Adult bullfrogs (*Rana catesbeiana*) were anesthetized by being placed in 0.4% ethyl-*m*-aminobenzoate for 30–40 min. The dorsal olfactory mucosa was removed as a single sheet of tissue. This was placed on a drop of amphibian Ringer's solution and floated in position on a nylon mesh-backed silicone rubber washer so that the submucosal side of the tissue was supported mechanically. The edges of the mucosa were attached to the nylon mesh with a trace of cyanoacrylate adhesive; a Lucite washer lightly coated with adhesive was then placed on the ciliated side of the mucosa so that a watertight seal was formed between the two washers and the tissue. The area of mucosa then available for access to odorants was 0.075 cm^2. The mounted tissue was then placed between two Ussing chambers held against the washers by silicone rubber gaskets, as shown in Fig. 1.

The entire assembly was secured by a screw brace. The chamber volumes were 0.5 ml on the ciliated side and 1.5 ml on the submucosal side. The tissue was maintained in oxygenated amphibian Ringer's solution consisting of 100 mM NaCl, 5 mM $MgCl_2$, 2.5 mM KCl, 2.5 mM $CaCl_2$, 1.1 mM Na_2HPO_4, 0.4 mM NaH_2PO_4 and 10 mM glucose. In most experiments, the solution on the ciliated surface was replaced continuously at a flow rate of 0.2 ml/sec via a constant-pressure head flow system. The solution on the submucosal side was kept oxygenated.

Odorants were presented in aqueous solution to the ciliated side. A bolus of odorant solution was introduced into the stream of amphibian Ringer's by switching an odorant charged loop of Ringer's solution into the flow system via a 2-gang-3-way valve. The stimulus profile was monitored at intervals using a photodarlington transistor to follow the light adsorption profile when a blue dye (5,5′-indigosulfonic acid) was used instead of an odorant.

The transmucosal potential difference and short circuit current were recorded as described by Heck *et al.* (1984*a*). Potentials were recorded with saturated calomel electrodes connected to each reservoir with 3% agar/0.15 M NaCl salt bridges. Asymmetry potentials (<0.5 mV) were nulled to zero, and the solution resistance was automatically compensated for. Current was passed via Ag/AgCl electrodes that were connected to the Ussing chamber with NaCl/agar salt bridges. A Physiologic Instruments VCC 600 voltage (or current) clamp was connected to the electrodes and the output was recorded on a strip-chart recorder with a potentiometer backoff for the necessary scale expansion.

6.2. Active Ion Transport in the Steady State

When bathed symmetrically with oxygenated Ringer's solution, the steady-state open-circuit potential, short-circuit current and transmucosal resistance were respectively -3.8 mV $\pm$ 0.6 mV, 56.0 $\pm$ 6.3 $\mu A/cm^2$, and 73 $\pm$ 15 Ω-cm^2 (mean values $\pm$SEM, N=7). The ciliated side of the mucosa was electronegative, indicating that a major fraction of the current is due to cation transport from the ciliated side to the submucosal side or anion secretion in the opposite direction. Both dorsal and ventral olfactory mucosa gave similar evidence of electrogenic ion transport. Confirming this, the short-circuit current was abolished by ouabain, a potent blocker of Na^+-K^+-ATPase activity. Cation transport from the ciliated side to the submucosal side appears to contribute the major part of the steady-state short-circuit current. Symmetrical replacement of Na^+ with *N*-methyl-D-glucammonium caused the short circuit current to decline to zero; this could be reversed by restoring normal sodium concentrations. In addition, the diuretic furosemide caused a decrease in the steady-state short-circuit current. It is likely that the steady-state short-circuit current is a reflection of processes that maintain water flow and the renewal of electrolytes bathing the olfactory mucosa.

When Na^+ ions in the bathing medium were replaced with K^+ ions, the steady-state current increased by up to 50% of the original standing current. This can be attributed directly to K^+ movement into the cells via apical K^+ ion channels. The steady state current returned to normal when the Na^+ ions were restored in the bathing medium. Symmetrical replacement with K^+ on both sides of the tissue caused the short circuit current to drop to zero.

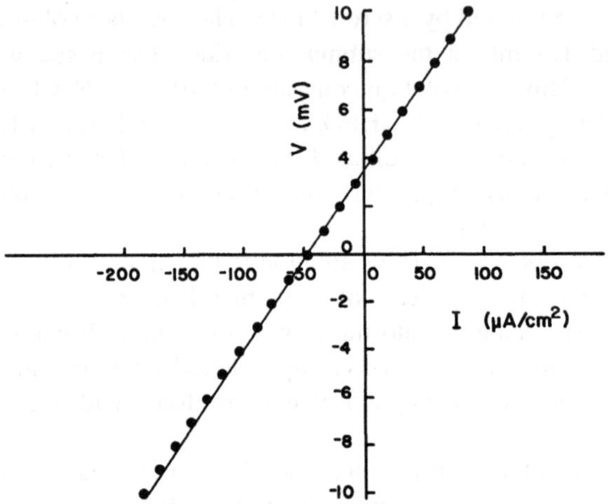

Figure 2. Current – voltage curve recorded from frog olfactory mucosa. The mucosa was bathed on both sides with amphibian Ringer's solution and voltage clamped. The current excursions produced when external voltage pulses were applied across the epithelium were then recorded. A representative current–voltage curve is shown, the electrical parameters in this preparation being 3.8 mV open-circuit potential, 48 μA/cm² short-circuit current, and 79 Ω-cm².

6.3. The Current–Voltage Relationship

Figure 2 shows a current–voltage curve recorded under symmetrical conditions wherein both sides of the mucosa were in contact with Ringer's solution. The specific resistance (79 Ω-cm² in this preparation) measured from the *I–V* curve is quite low, suggesting a leaky epithelial structure. The olfactory mucosa is very similar to the epithelia of the gut, the oral cavity and respiratory epithelia in this respect (Heck *et al.*, 1984*a*). The current–voltage relationship was linear over the range ±10 mV. This may imply that the passive transmucosal resistance is determined mainly by low impedance pathways, such as the paracellular shunts, and not by current paths through the cells. However, the active surface area of the tissue is greater than the geometrical area of the mucosa, since the olfactory mucosa contains both olfactory cilia and microvilli on the sustentacular cells; thus, it is difficult to make an estimate of the actual specific transmucosal resistance. The low transmucosal resistance suggests that paracellular shunts may be the main pathways for the applied external current. This is operationally advantageous because the cells are left virtually unperturbed.

6.4. Odorant-Evoked Current Transients under Voltage Clamp

One advantage of the voltage-clamped olfactory mucosa preparation is that known concentrations of odorants or pharmacological agents can be applied directly in aqueous solution to the ciliated side of the epithelium. Most investigations of the EOG have been limited by the fact that the actual concentration of odorant in contact with the receptors is very difficult to know with certainty. Stimuli are applied in the vapour phase, and only

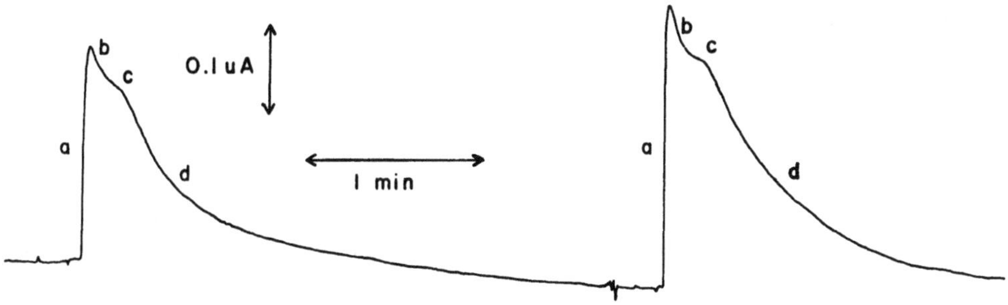

Figure 3. Odorant-induced current transients under voltage clamp. Shown from left to right, respectively, current transients recorded when a bolus of 1.2×10^{-4} M 1,8-cineole or 2.34×10^{-4} M 2-isobutyl-3-methoxypyrazine was presented to the ciliated side of the olfactory mucosa. The short-circuit responses are characterized by (a) a steep rise in the short-circuit current, (b) a sharp decline toward a plateau, (c), followed by (d) a slow decline of the current toward baseline as the odorant is washed out of the system.

after partitioning into the aqueous phase of the mucus overlying the olfactory mucosa are they accessible to the olfactory receptors.

Typical current transients to 1,8-cineole, and 2-isobutyl-3-methoxy pyrazine are shown in Fig. 3. Referring to Fig. 3, the short-circuit current responses are characterized by (a) a steep rise in the short-circuit current, reaching a maximum within 5 sec, (b) a sharp decline in current toward a plateau (c), although the odorant stimulus is still present at constant concentration, followed by (d) a slow decline of the current toward baseline as the bolus of odorant is washed out of the system. We have found that with high concentrations of odorants ($>10^{-4}$ M) or with repeated stimulations over a period of time, a slow rising current component becomes apparent (arrowhead, Fig. 4), which may under some conditions be of equal or greater amplitude than the initial fast component. The final relaxation of the slower second component may take much longer than the clearance of the odorant from the system. In some preparations, the first and second components are merged and appear as one (see Fig. 7). The early rapid component of the olfactory current response transient is probably consistent with the EOG measurements of an odorant induced summated receptor generator potential and represent a summated inward current

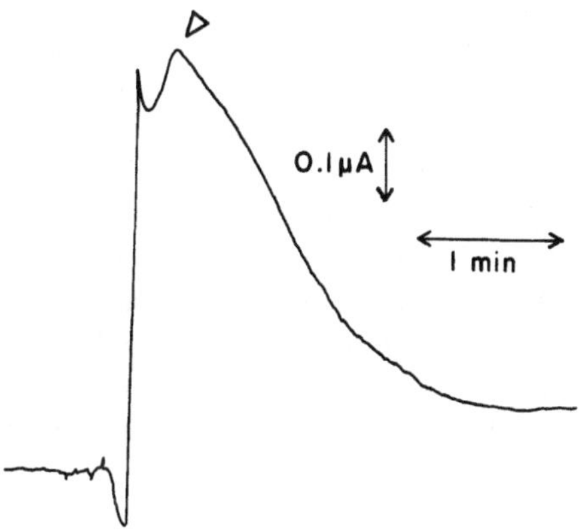

Figure 4. Response to 1.18×10^{-4} 1,8-cineole showing a slow rising current component (arrowhead) after the initial steep rise in current. This type of response is found in some preparations after several repeated stimulations with odorants. The second slower component of the response becomes more pronounced with each subsequent stimulation. A small switching artifact is present in this trace at the onset of the odorant response.

originating in the cilia and apical dendritic knob of olfactory receptor neurons. The sharp decline in current observed in step (c) of the odorant-induced current transient while the stimulus is still present gives evidence for a restoring current. The restoring current is clearly controlled by processes originating in the receptor neurons and not the rate of odorant clearance because the response declines during the interval when the odorant concentration is at a plateau. We have evidence that Na^+ transport is chiefly responsible for the inward current transient, since 10^{-4} M amiloride, when applied simultaneously with an odorant stimulus, causes a large decrease in the current transient, and this effect is fully reversible.

When the Na^+ ions in the bathing medium on the ciliated side were replaced by K^+ ions the steady-state short-circuit current increased. We found that the odorant responses under these conditions were largely inhibited, and only traces of odorant-evoked current could be seen. This indicates that K^+ may completely depolarize the olfactory receptor neurons such that they are no longer able to responsd to odorants.

When the Na^+ ions in the bathing medium on the ciliated side of the mucosa were replaced by *N*-methyl-D-glucammonium ions, we found that although the short-circuit current dropped through zero and became a positive outward current (ciliated side positive), the tissue still responded with an inward positive current response when stimulated with an odorant. This response was still amiloride sensitive indicating that a sodium current was still involved although the bathing medium on the ciliated side of the mucosa contained no Na^+ ions. We attribute this effect to the secretion of Na^+ ions into the microenvironment at the ciliated surface of the mucosa via glands and the paracellular shunts. The opening of odorant gated Na channels permitted Na ions to carry a generator current into the receptor cells utilizing the energy made available, in part, by the clamping current, which under these conditions was directed inward. This result emphasizes the importance of the paracellular leak pathways in maintaining sodium levels on the ciliated surface. Symmetrical replacement of Na^+ ions with *N*-methyl-D-glucammonium ions eliminated the odorant response. This shows that the submucosal side was the source of the ions carried into the cells upon stimulation with odorant.

6.5. Dose–Response Relationship

Figure 5 shows the concentration response curve observed with 2-isobutyl-3-methoxypyrazine over the range 12 μM to 1 mM. A nonlinear monotonic dose response relationship is observed over the entire concentration range. The minimum detectable concentration is expected to be less than micromolar, but this was not investigated, as it is dependent mainly on the tissue signal to noise ratio. Saturation of the response begins to be apparent at concentrations above 1 mM. However, such high concentrations may cause degradative changes in the tissue. The results show that the bullfrog olfactory mucosa responds to odorants over at least three log units. We found that the sensitivity to different odorants varied. When stimulated with 10^{-4} M odorant solutions, the amplitude of the current responses were in the order 1,8-cineole > amyl acetate > 2-isobutyl-3-methoxy-pyrazine > citral. The slower second component of the olfactory response was also concentration dependent and showed a monotonic increase in amplitude with odorant concentration.

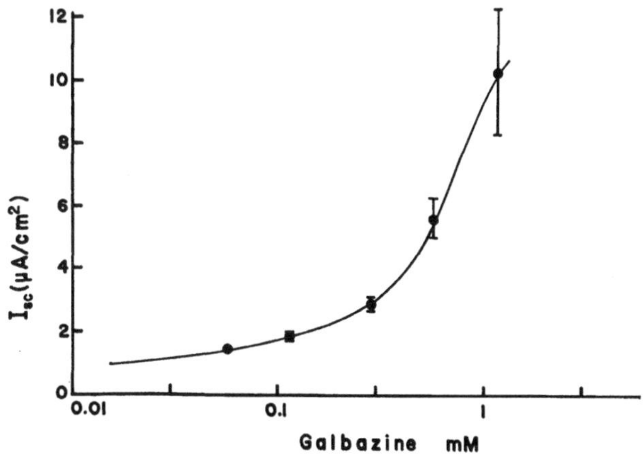

Figure 5. Concentration–response curve recorded to 2-isobutyl-3-methoxypyrazine. The olfactory mucosa was voltage clamped and the current responses measured when it was presented with varying concentrations of odorant. The amplitude of the initial fast component of the response was measured. This was corrected for decrease in tissue response over the course of the experiment by reference to a control stimulus of constant concentration presented before each new odorant concentration was used.

7. CYCLIC NUCLEOTIDE MODULATION OF OLFACTORY TRANSDUCTION

Substantial progress is being made in understanding the transduction mechanisms involved in initiating current transients across the olfactory mucosa when receptors are stimulated. Kurihara and Koyama (1972) and Bitensky *et al.* (1972) found a substantial amount of adenylate cyclase activity present in homogenates of rabbit olfactory epithelium (4.7–5.7 nmoles cAMP/mg protein/10 min). Minor and Sakina (1973) investigated the effects of cyclic nucleotide phosphodiesterase inhibitors on the EOG. The phosphodiesterase inhibitors theophylline and papaverine were found to reduce the peak amplitude and to increase the duration of the EOG responses. Imidazole, which stimulates phosphodiesterase activity, produced a small increase in the amplitude of the EOG but reduced the duration of the response. Menevse (1976), Menevse *et al.* (1977) and Squirrell (1978) further studied these effects. Reagents were applied in solution to the olfactory mucosa of the frog and EOGs were recorded to vapour phase pulses of odorant. Phosphodiesterase inhibitors, in this case, caused a reversible concentration-dependent reduction of the EOG response amplitudes. Membrane-permeable derivatives of cAMP, such as dibutyryl and 8-bromo-3′,5′-cAMP, caused a reduction in the EOG response independent of the type of odorants used. Squirrell (1978) observed that dibutyryl cAMP did not affect the EOG response from sheep olfactory epithelium, while dibutyryl-3′,5′-cGMP and 8-bromo-3′,5′-cGMP caused stimulation. The results can be interpreted as specific involvement of cyclic nucleotides in the transduction process (Dodd and Persaud, 1981). However, the experimental evidence on the effects on the EOG are not clear. More recently, an odorant stimulated adenylate cyclase activity in isolated dendritic membranes of the olfactory receptor neurons has been identified (Pace *et al.*, 1985). This group also electrophoretically identified polypeptides from olfactory cilia that show some properties similar to the signal transducing guanine nucleotide-binding proteins (G proteins) that modulate adenylate cyclase activity in other sensory and hormonal systems (Birnbaumer *et al.*, 1985). These data provide further evidence for cyclic nucleotide involvement in the transduction process.

7.1. cAMP Evokes an Inward Current Transient

We have used the modified Ussing preparation of the bullfrog olfactory mucosa to probe directly the effects of cyclic nucleotides on the apical sources of transductory current.

We monitored the short-circuit current after bathing the ciliated surface with 8-bromo-3′,5′-cAMP. Minor and Sakina (1973) found that some cAMP analogues cause a slow prolonged change in the transmucosal potential. We found that a bolus of 1 μM 8-bromo-3′,5′-cAMP produced a rapid inward current transient resembling an odorant response (arrowhead, Fig. 6). However, 2′,3′-cAMP was not stimulatory even at 10^{-3} M. 8-Bromo-3′,5′-cGMP caused no change in the short circuit current at micromolar concentrations, but 10^{-4} M caused a slow rise in the current over 5 min. When the ciliated side of the mucosa was bathed in 10^{-4} M 8-bromo-3′,5′-cAMP, the initial transient increase in inward positive current returned to baseline values in about 5 min, although the cyclic nucleotide was still present. No changes in current occurred when 10^{-4} M 8-bromo-3′,5′-cAMP was presented to the submucosal side of the tissue. However, this may only reflect that a longer incubation time may be necessary before the compound reaches the site of action. The data suggest that cAMP analogues can increase the conductance of some portion of the membranes on the ciliated side of the mucosa. The rapidity of the response suggests that a high concentration of cyclic nucleotide binding sites are available and that adequate amounts of cAMP or 8-bromo-cAMP diffuse through the membranes. The olfactory cilia have a large surface area and contain the highest activity of adenylate cyclase in the olfactory epithelium. It is likely that cAMP can gain

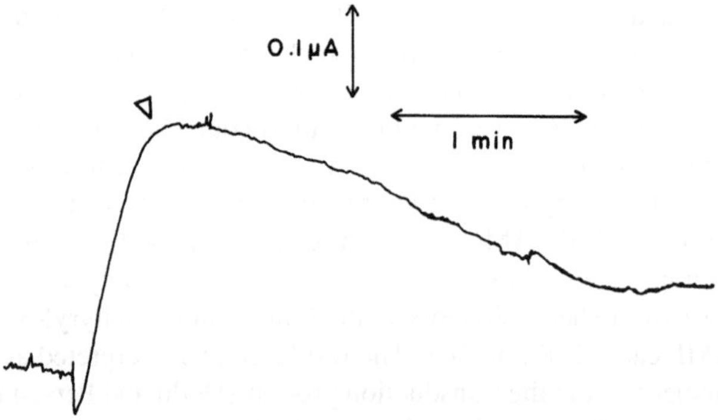

Figure 6. Current transient (arrowhead) recorded when the ciliated side of the tissue was presented with 2.4 μM 8-bromo-3′,5′-cAMP. A small switching artifact is present at the onset of the current response.

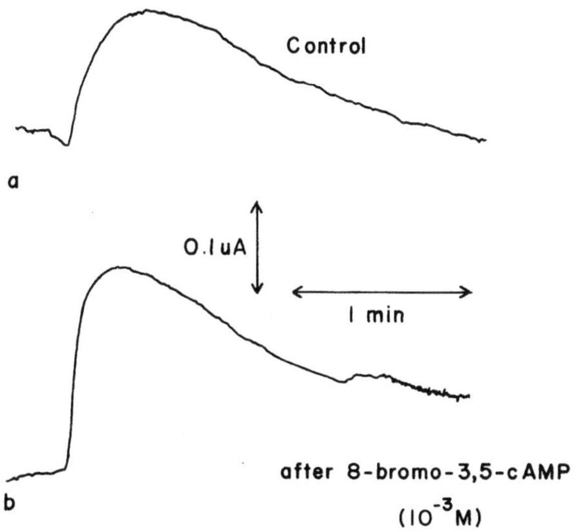

Figure 7. Effect of 8-bromo-3′,5′-cAMP on the odorant response, showing current responses to 1.2×10^{-4} M 1,8-cineole: (a) control, and (b) after bathing the ciliated side of the tissue with 10^{-3} M 8-bromo-3′,5′-cAMP for 5 min. The amplitude in this case was increased by 51%. With subsequent stimulations the response returned to normal amplitudes. Higher concentrations of 8-bromo-3′,5′-cAMP caused elevation of the standing current and only small odorant responses could be measured.

ready access to the cell interior by diffusing across the ciliary membranes, and activate apical ion channels.

7.2. cAMP Enhances Odorant-Evoked Current Transients

When the ciliated side of the mucosa was bathed with 10^{-3} M 8-bromo-3′,5′-cAMP for 5 min, we observed that the odorant induced current transient was increased by up to 51% (Fig. 7). This effect was reversible with time. Increasing the endogenous cAMP concentration with the phosphodiesterase inhibitors theophylline (5 mM) or isobutylmethyl xanthine (10^{-4} M) caused increases in the odorant evoked currents by 43% and 39%, respectively. Forskolin, a stimulant of adenylate cyclase activity, amplified the response by 53% at 10-μM concentrations. The data suggest that cAMP levels modulate ion-channel opening on the apical membranes of the olfactory mucosa. High concentrations of cAMP analogues (mM) or compounds that increase endogenous cAMP, such as phosphodiesterase inhibitors or forskolin, increased the standing short-circuit current by up to 30%. The current remained high for up to 30 min after the compounds had been washed away from the bathing solution. During this period, the tissue responded poorly to odorants, recovering only when the baseline current returned to control conditions. In some tissue preparations this effect was observed at 10^{-4} M 8-bromo-3′,5′-cAMP. The rise in the standing short-circuit current suggests that the higher levels of cAMP caused the opening of the normally odorant-gated ion channels, hence depolarization of the cells. Being fixed in an open configuration, the ion channels can no longer be modulated by odorous stimuli, and no olfactory response is seen. We believe that this may account for some of the previous conflicting observations of the effects of cyclic nucleotides on the EOG where either amplification or inhibition was observed.

7.3. Evidence for a Stimulatory G Protein

There are many examples of receptors that affect cAMP formation. These can be classified into two types: R_s receptors increase cAMP levels by stimulating adenylate cyclase, while R_i receptors decrease cAMP levels by inhibition of the enzyme. β-Adrenergic, glucagon, secretin, and luteinizing hormone receptors are of type R_s, while acetylcholine (Ach) of muscarinic type and opioid peptide receptors are examples of R_i receptors (Birnbaumer *et al.*, 1985). The results described above would indicate that the olfactory receptors are of the R_s type. The mechanism by which R_s and R_i receptors couple to modulate adenylate cyclase activity involves two oligomeric coupling proteins called G proteins. These proteins can bind as well as hydrolyze GTP and can regulate ligand affinity for receptors as well as the catalytic activity of adenylate cyclase. The activity of these proteins requires Mg^{2+} ions. There are both stimulatory G proteins (G_s) as well as inhibitory G proteins (G_i). The adenylate cyclase can be regarded as a three-component system containing a catalytic unit that forms cAMP from MgATP and the two types of G proteins that are in turn coupled in some way to receptors. Olfactory G proteins have been identified in the cilia of the receptor neurons by labelling with bacterial toxins and electrophoretic separation (Pace *et al.*, 1985). A cholera toxin labelled polypeptide of 42,000 M_r had properties similar to those of a G_s protein, while a pertussis toxin-labeled polypeptide of 40,000 M_r had properties similar to those of G_i proteins. *In vitro*, it was found that GTP and its nonhydrolyzable analogues guanosine 5′-*O*-(3-thio)triphosphate (GTPγS) stimulated the odorant-sensitive adenylate cyclase activity while the GDP analog, guanosine 5′-*O*-(2-thio)diphosphate (GDPβS) was inhibitory. This suggests that olfactory transduction may be mediated by a system similar to hormone and neurotransmitter stimulated enzyme systems.

We have investigated the effects of GTP and GDP analogues on the intact olfactory mucosa using the voltage-clamped preparation. GTPγS (10 μM) caused an amplification of the odorant response similar to that seen with cAMP analogues (Fig. 8a,b). GDPβS (10 μM), on the other hand, proved to be a good inhibitor of the odorant response (Fig. 8c,d). These findings provide additional evidence that apical ion channel opening is mediated through adenylate cyclase activity which is regulated by a G_s protein.

7.4. cGMP Inhibits Odorant-Evoked Current Transients

8-Bromo-3′,5′-cGMP proved to be an inhibitor of the odorant response at the concentrations tested. Bathing the ciliated side of the olfactory mucosa in 1 μM BrcGMP reduced the odorant evoked current by 58%. This was only partially reversible, and application of 8-bromo-3′,5′-cAMP after 8-bromo-3′5′-cGMP had no effect in stimulating the odorant response. Bathing the ciliated side of the tissue with 10^{-4} M 8-bromo-3′,5′-cGMP reduced subsequent odorant responses to less than 10% of control values (Fig. 9). Trotier and MacLeod (1986) showed from patch-clamp studies of ion channels in the cilia of isolated olfactory receptor cells that both cAMP and cGMP result

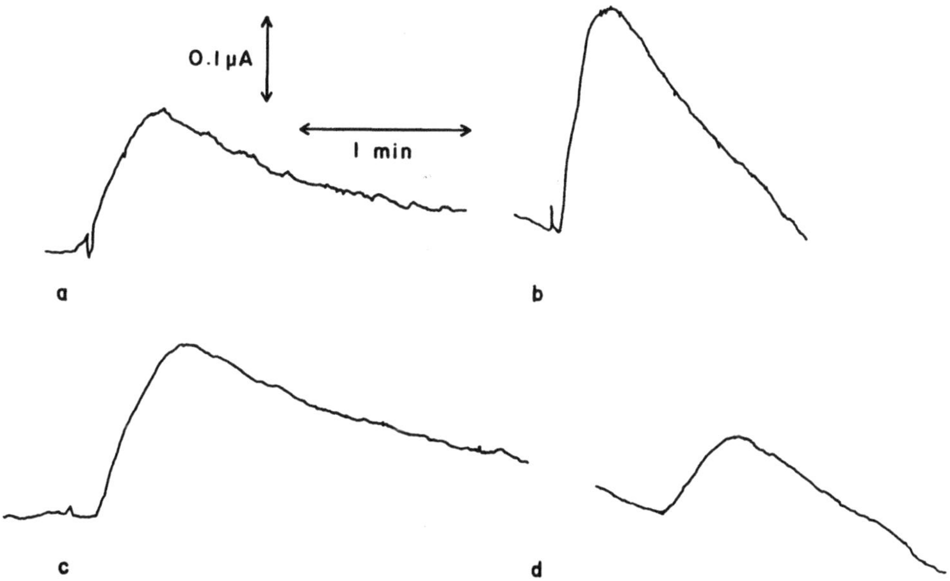

Figure 8. Effect of GTPγS and GDPβS on the odorant response. (a) Control response to 1.2×10^{-4} M 1,8-cineole. (b) Odorant response after bathing the tissue with 8.8 μM GTPγS for 5 minutes. (c) Control response. (d) Odorant response after bathing the tissue with 8.8 μM GDPβS for 5 min.

in increased conductance. This suggests that the inhibitory effect of cGMP on the odorant response occurs by way of a decoupling of the odorant–receptor interaction from subsequent increases in channel conductance. Since cAMP had no effect in reversing the inhibition, this may suggest that cGMP may operate on a different part of the transductory pathway from cAMP or that cGMP has a much higher binding affinity for a site which also binds cAMP. The precise role of cGMP in olfactory transduction remains to be investigated.

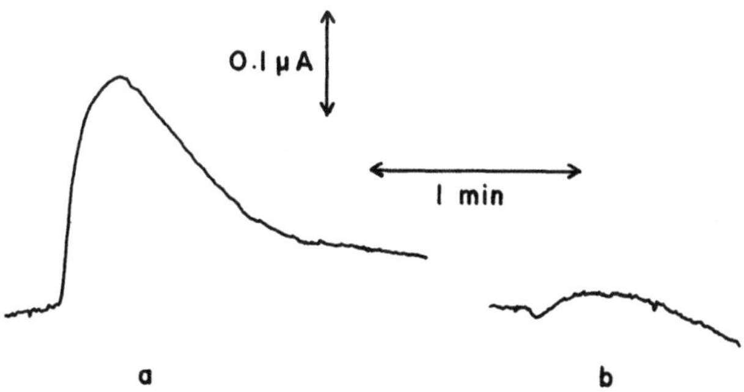

Figure 9. Effect of 8-bromo-cGMP on the odorant response. (a) Control response to 1.18×10^{-4} M 1,8-cineole. (b) Odorant response after incubating the tissue with 10^{-4} M cGMP for 5 min.

8. SUMMARY AND CONCLUSIONS

In vitro preparations of epithelia have been used widely to obtain information on absorption and secretion. When the lingual and olfactory epithelia are subjected to analysis by voltage clamp, the ion-translocation aspects of chemosensory transduction can also be studied. The preparation is well suited to pharmacological dissection, and there is little question about the viability of the preparation. The stability of the short-circuit current and the odorant-induced response are reliable monitors. The preparation of olfactory mucosa gives evidence for electrogenic ion transport, which in the unstimulated state may involve Na^+/Cl^- cotransport across the apical membranes. The results obtained thus far suggest that the odorant gated channel is an amiloride sensitive sodium channel. The permeability of this channel may be modulated by both cAMP and cGMP. These cyclic nucleotides may act either independently or in concert on processes that are triggered when an odorant becomes bound to a receptor, resulting in changes in membrane permeability. The appearance of a slow second component in the odorant-evoked current transient suggests that apart from transductory events, odorants may initiate other ion transport processes. These may be secretory in nature and may thus be involved in odorant clearance from the surface of the mucosa. Much more will be learned about both the secretory and sensory aspects of the olfactory mucosa using the methods discussed in this chapter.

ACKNOWLEDGMENTS. Supported by National Institutes of Health grant NS-13767 and the Campbell Institute for Research and Technology.

REFERENCES

Acker, H., Delpiano, M., and Degner, F., 1983, The meaning of the pO_2 field in the carotid body for the chemoreceptive process, in: *Physiology of the Peripheral Arterial Chemoreceptors* (H. Acker and R. G. O'Regan, eds.), pp. 89–115, Elsevier Science Publishers, Amsterdam.

Armstrong, W. McD., Garcia-Diaz, J. F., and Diez de los Rios, A., 1981, Energetics of coupled sodium chloride entry in absorbing cells of leaky epithelia, in: *Ion Transport by Epithelia* (S. G. Schultz, ed.), pp. 151–162, Raven, New York.

Baumann, K., and Kinne, R., 1972, The effect of sodium on the transtubular transport of D-glucose in rat kidney and on the D-glucose binding to isolated brush border membranes, in: *Na-Linked Transport of Organic Solutes* (E. Heinz, ed.), pp. 130–133, Springer-Verlag, Berlin.

Biber, T. U. L., 1971, Effects of changes in transepithelial transport on the uptake of sodium across the outer surface of frog skin, *J. Gen. Physiol.* **58**:131–144.

Biber, T. U. L., DeSimone, J. A., and Drewnowska, K., 1986, Potential dependence of unidirectional chloride fluxes across isolated frog skin, *Biochim. Biophys. Acta* **862**:27–38.

Birnbaumer, L., Codina, J., Mattera, R., Cerione, R. A., Hildebrandt, J. D., Sunyer, T., Rojas, F. J., Caron, M. G., Lefkowitz, R. J., and Iyengar, R., 1985, Structural basis of adenylate cyclase stimulation and inhibition by distinct guanine nucleotide regulatory proteins, in: *Molecular mechanisms of transmembrane signalling* (P. Cohen and M. D. Houslay, eds.), pp. 131–182, Elsevier Science Publishers, Amsterdam.

Bitensky, M. W., Miller, W. H., Gorman, R. E., Neufield, A. H., and Robinson, R., 1972, The role of cyclic AMP in visual excitation, *Adv. Cyclic Nucleotide Res.* **1**:317–335.

Brand, J. G., Teeter, J. H., and Silver, W. L., 1985, Inhibition by amiloride of chorda tympani responses evoked by monovalent salts, *Brain Res.* **334**:207–214.

Civan, M. M., 1983, *Epithelial Ions and Transport,* Wiley, New York.

Civan, M. M., Hall, T. A., and Gupta, B. J., 1980, Microprobe study of toad urinary bladder in absence of serosal K^+, *J. Membr. Biol.* **55**:187–202.

DeSimone, J. A., and Ferrell, F., 1985, Analysis of amiloride inhibition of chorda tympani taste response of rat to NaCl, *Am. J. Physiol.* **249**:R52–R61.

DeSimone, J. A., and Getchell, T. V., 1985, Odorant-stimulated current transients across the frog olfactory mucosa *in vitro*, in: *Proceedings of the Nineteenth Japanese Symposium on Taste and Smell* (S. Kimura, A. Miyoshi, and I. Shimada, eds.), pp. 46–48, Asahi University, Gifu, Japan.

DeSimone, J. A., Heck, G. L., and DeSimone, S. K., 1981, Active ion transport in dog tongue: A possible role in taste, *Science* **214**:1039–1041.

DeSimone, J. A., Heck, G. L., Mierson, S., and DeSimone, S. K., 1984, The active ion transport properties of canine lingual epithelia *in vitro, J. Gen. Physiol.* **83**:633–656.

Diamond, J. M., and Bossert, W. H., 1967, Standing-gradient osmotic flow: A mechanism for coupling of water and solute transport in epithelia, *J. Gen. Physiol.* **50**:2061–2083.

DiBona, D. R., and Civan, M. M., 1973, Pathways for movement of ions and water across toad urinary bladder. I. Anatomic site of transepithelial shunt pathways, *J. Membr. Biol.* **12**:101–128.

Dodd, G., and Persaud, K., 1981, Biochemical mechanisms in vertebrate primary neurons, in: *Biochemistry of taste and olfaction* (R. Cagan and M. Kare, eds.), pp. 333–357, Academic, New York.

Essig, A., 1982, Influence of cellular and paracellular conductance patterns on epithelial transport and metabolism, *Biophys. J.* **38**:143–152.

Fidelman, M. L., and Mikulecky, D. C., 1986, Network thermodynamic modeling of hormone regulation of active Na^+ transport in cultured renal epithelium (A6), *Am. J. Physiol.* **250**:C978–C991.

Frizzell, R. A., Dugas, M. C., and Schultz, S. G., 1975, Sodium chloride transport by rabbit gallbladder: Direct evidence for a coupled NaCl influx process, *J. Gen. Physiol.* **65**:769–795.

Frizzell, R. A., Rechkemmer, G., and Shoemaker, R. L., 1986, Altered regulation of airway epithelial cell chloride channels in cystic fibrosis, *Science* **233**:558–560.

Fuchs, W., Hviid Larson, E., and Lindemann, B., 1977, Current-voltage curve of sodium channels and concentration dependence of sodium permeability in frog skin, *J. Physiol. (Lond.)* **267**:137–166.

Gesteland, R. C., 1971, Neural coding in olfactory receptor cells, in: *Handbook of Sensory Physiology*. Vol. IV: *Chemical Senses. Part I, Olfaction* (L. M. Beidler, ed.), pp. 132–150, Springer-Verlag, Berlin.

Getchell, M. L., Rafols, J. A., and Getchell, T. V., 1984, Histological and histochemical studies of the secretory components of the salamander olfactory mucosa: Effects of isoproterenol and olfactory nerve section, *Anat. Rec.* **208**:553–565.

Getchell, T. V., 1974, Electrogenic sources of slow voltage transients recorded from frog olfactory epithelium, *J. Neurophysiol.* **37**:1115–1130.

Getchell, T. V., and Getchell, M. L., 1977, Histochemical localization and identification of secretory products in salamander olfactory epithelium, in: *Olfaction and Taste*. Vol. VII (J. LeMagnen and P. Macleod, eds.), pp. 105–112, IRL Press, London.

Getchell, T. V., Margolis, F. L., and Getchell, M. L., 1984, Perireceptor and receptor events in vertebrate olfaction. *Prog. Neurobiol.* **23**:317–345.

Heck, G. L., DeSimone, J. A., and Getchell, T. V., 1984, Evidence for electrogenic active ion transport across the frog olfactory mucosa *in vitro, Chem. Senses* **9**:273–283.

Heck, G. L., Mierson, S., and DeSimone, J. A., 1984b, Salt taste transduction occurs through an amiloride-sensitive transport pathway, *Science* **223**:403–405.

Hille, B., 1984, Ionic channels of excitable membranes, Sinauer, Sunderland, Massachusetts.

Hopfer, N., Nelson, K., Perrotto, J., and Isselbacher, K. J., 1973, Glucose transport in isolated brush border membrane from rat small intestine, *J. Biol. Chem.* **248**:25–32.

Hosoya, Y., and Yoshida, H., 1937, Uber die bioelektrische Erscheinungen an der Reichschleimhaut, *Jpn. J. Med. Sci. Biophys.* III **5**:2223.

Huf, E. G., 1936, Ueber aktiven Wasser-und Salztransport durch die Froschhaut, *Pflugers Arch. Ges. Physiol.* **237**:143–166.

Huf, E. G., and Mikulecky, D. C., 1985, Compartmental analysis of the Na^+ flux ratio with application to data on the frog skin epidermis, *J. Theor. Biol.* **112**:193–220.

Huf, E. G., and Mikulecky, D. C., 1986, Role of topology in bioenergetics of sodium transport in complex epithelia, *Am. J. Physiol.* **250**:F1107–F1118.

Juge, A., Holley, A., and Rajon, D., 1979, Olfactory receptor cell activity under electrical polarization of the nasal mucosa of the frog. II. Responses to odour stimulation, *J. Physiol. (Paris)* **75:**929–938.

Kurihara, K., and Koyama, K., 1972, High activity of adenylate cyclase in olfactory and gustatory organs, *Biochem. Biophys. Res. Commun.* **48:**30–34.

Kinne, R., and Kinne-Saffran, E., 1978, Differentiation of cell faces in epithelia, in: *Molecular Specialization and Symmetry in Membrane Function* (A. K. Solomon and M. Karnovsky, eds.), pp. 272–293, Harvard University Press, Cambridge.

Li, J. H.-Y., and Lindemann, B., 1983, Competitive blocking of epithelial sodium channels by organic cations: The relationship between macroscopic and microscopic inhibition constants, *J. Membr. Biol.* **76:**235–251.

Liedtke, C. M., and Hopfer, U., 1982, Mechanism of Cl^- translocation across small intestinal brush-border membrane, *Am. J. Physiol.* **242:**G272–G280.

Lew, V. L., Ferreira, H. G., and Moura, T., 1979, The behaviour of transporting epithelial cells. I. Computer analysis of a basic model, *Proc. R. Soc. Lond. [Biol.]* **206:**53–83.

Lewis, S. A., and Wills, N. K., 1981, Interaction between apical and basolateral membranes during sodium transport across tight epithelia, in: *Ion Transport by Epithelia* (S. Schultz, ed.), pp. 93–107, Raven, New York.

Macknight, A. D. C., and Leaf, A., 1977, Regulation of cellular volume, *Physiol. Rev.* **57:**510–573.

Maue, R. A., and Dionne, V. E., 1986, Membrane conductance changes in neonatal and embryonic mouse neurons, *Biophys. J.* **49:**556a.

Menevse, A., 1976, Biochemical aspects of olfactory mechanisms, Ph.D. thesis, University of Warwick, United Kingdom.

Menevse, A., Dodd, G., and Poynder, T. M., 1977, Evidence for the specific involvement of cyclic AMP in the olfactory transduction mechanism, *Biochem. Soc. Trans.* **5:**191–194.

Meyer, J. H., and Grossman, M. I., 1972, Comparison of *d*- and *l*-phenylalanine as pancreatic stimulants, *Am. J. Physiol.* **222:**1058–1063.

Mierson, S., Heck, G. L., DeSimone, S. K., Biber, T. U. L., and DeSimone, J. A., 1985, The identity of the current carriers in canine lingual epithelium *in vitro, Biochim. Biophys. Acta* **816:**283–293.

Mikulecky, D. C., 1979, A network thermodynamic two-port element to represent coupled flow of salt and current: An improvement on the equivalent circuit approach, *Biophys. J.* **25:**323–340.

Mikulecky, D. C., 1982, Network thermodynamic simulation of biological systems: An overview, *Math. Comp. Simul.* **24:**437–441.

Minor, A. V., and Sakina, N. L., 1973, The role of cyclic adenosine-3′,5′-monophosphate in olfactory reception, *Neirofiziologiya* **5:**415–422.

Murer, H., Hopfer, U., and Kinne, R., 1976, Sodium proton antiport in brush border membrane vesicles isolated from rat small intestine, *Biochem. J.* **154:**597–604.

Okada, Y., Miyamoto, T., and Sato, T., 1986, Contribution of the receptor and basolateral membranes to the resting potential of a frog taste cell, *Jpn. J. Physiol.* **36:**139–150.

Okano, M., and Takagi, S. F., 1974, Secretion and electrogenesis of the supporting cells in the olfactory epithelium, *J. Physiol.* **242:**353–370.

Ottoson, D., 1956, Analysis of the electrical activity of the olfactory epithelium, *Acta Physiol. Scand.* **35:**1–83.

Pace, U., Hanski, E., Salomon, Y., and Lancet, D., 1985, Odorant-sensitive adenylate cyclase may mediate olfactory reception, *Nature (Lond.)* **316:**255–258.

Persaud, K. C., Heck, G. L., DeSimone, S. K., Getchell, T. V., and DeSimone, J. A., 1988, Ion transport across the frog olfactory mucosa: The action of cyclic nucleotides on the basal and odorant-stimulated states, (in preparation).

Rehm, W. S., 1967, Ion permeability and electrical resistance of the frog's gastric mucosa, *Fed. Proc.* **26:**1303–1313.

Schiffman, S. S., Lockhead, E., and Maes, F. W., 1983, Amiloride reduces the taste intensity of Na^+ and Li^+ salts and sweeteners, *Proc. Natl. Acad. Sci. USA* **80:**6136–6140.

Schultz, S. G., 1981, *Ion Transport Across Epithelia,* Raven, New York.

Simon, S. A., and Garvin, J. L., 1985, Salt and acid studies on canine lingual epithelium, *Am. J. Physiol.* **249:**C398–C408.

Squirrell, 1978, A study of olfactory mechanisms Ph.D thesis, University of Warwick, United Kingdom.

Sten-Knudsen, O., and Ussing, H. H., 1981, The flux ratio equation under nonstationary conditions, *J. Membr. Biol.* **63:**233–242.

Suzuki, N., 1977, Intracellular responses of lamprey olfactory receptors to current and chemical stimulation, in: *Food Intake and the Chemical Senses* (Y. Katsuki, M. Sato, S. F. Takagi, and Y. Oomura, eds.), pp. 13–22, Tokyo University Press, Tokyo.

Swanson, C. H., and Solomon, A. K., 1975, Micropuncture analysis of the cellular mechanisms of electrolyte secretion by the *in vitro* pancreas, *J. Gen. Physiol.* **65:**22–45.

Takagi, S. F., Kitamura, H., Imai, K., and Takeuchi, H., 1969, Further studies on the roles of sodium and potassium in the generation of the electro-olfactogram: Effects of mono-, di- and trivalent cations, *J. Gen. Physiol.* **53:**115–130.

Trotier, D., and MacLeod, P., 1986, cAMP and cGMP open channels and depolarize olfactory receptor cells, *Abstr. ISOT/ACHEMS Chem. Senses* **11:**674.

Ussing, H. H., 1949, The distinction by means of tracers between active transport and diffusion, *Acta Physiol. Scand.* **19:**43–56.

Ussing, H. H., and Zerahn, K., 1951, Active transport of sodium as the source of the electrical current in the short-circuited isolated frog skin, *Acta Physiol. Scand.* **23:**110–127.

Vieira, F. L., Caplan, S. R., and Essig, A., 1972, Energetics of sodium transport in frog skin. I. Oxygen consumption in the short-circuited state, *J. Gen. Physiol.* **59:**60–76.

Wieczorek, H., 1982, A biochemical approach to the electrogenic potassium pump of insect sensills: Potassium sensitive ATPases in the labellum of the fly, *J. Comp. Physiol. A* **148:**303–311.

Widdicombe, J. H., and Welsh, M. J., 1980, Ion transport by dog tracheal epithelium, *Fed. Proc.* **39:**3062–3066.

III

Biochemical–Molecular Biological Studies

9

Neurotransmitter Plasticity in the Juxtaglomerular Cells of the Olfactory Bulb

Harriet Baker

1. INTRODUCTION

1.1. Overview

The existence of trophic interactions between neurons both in adults and during development had been established by a number of investigators using a variety of experimental paradigms and neuronal systems. For example, the catecholamine neurons of the superior cervical ganglion exhibit developmental anomalies following decentralization (deafferentation) (Black *et al.*, 1971). In addition, the levels of the catecholamine synthesizing enzymes in both the adult superior cervical ganglion and adrenal were altered by changes in presynaptic input and/or activity (Thoenen, 1974). Recent experiments of the author and others also suggested that such interactions might occur between olfactory receptor neurons and their targets in the olfactory bulb. A number of approaches, briefly outlined here and described in more detail in the discussion that follows, have been utilized to study these interactions. Lesions, including both chemical and surgical destruction of the olfactory epithelium and nerves, as well as neonatal closure of the nares produced alterations in catecholamine and/or substance P expression in the olfactory bulb (Margolis *et al.*, 1974; Harding *et al.*, 1978; Nadi *et al.*, 1981; Kawano and Margolis, 1982; Baker *et al.*, 1983*b*, 1984; Kream *et al.*, 1984; and Brunjes *et al.*, 1984). Developmental studies suggested that transmitter expression in bulbar neurons coincided with the time some received receptor cell input (Specht *et al.*, 1981). In addition the expression of olfactory

Abbreviations used in this chapter: AC, anterior commissure; CC, corpus callosum; CI, internal capsule; CP, caudate putamen; ep, external plexiform layer; GABA, gamma amino butyric acid; gl, glomerular layer; GP, globus pallidus; HRP, horseradish peroxidase; IC, islands of Callejae; ig, internal granule cell layer; ip, internal plexiform layer; LV, lateral ventricle; m, mitral cell layer; MS, medial septal nucleus; OMP, olfactory marker protein; NE, norepinephrine; on, olfactory nerve layer; PIR, piriform cortex; TH, tyrosine hydroxylase; TOL, lateral olfactory tract; TUO, olfactory tubercle; III, third ventricle.

Harriet Baker • Laboratory of Molecular Neurobiology, The Burke Rehabilitation Center, Cornell University Medical College, White Plains, New York 10605.

marker protein (Margolis, 1978), a molecule specific to olfactory receptor cells, was temporally correlated with the arrival of receptor cell axons in the olfactory bulb and at least partially dependent on interactions with the bulb (Farbman and Margolis, 1980; Chuah and Farbman, 1983). The strong inductive capacity of receptor neurons was illustrated in both amphibians and mammals, in which they produced bulb like structures when placed in contact with a variety of brain tissues (Graziadei and Kaplan, 1980; Graziadei and Samanen, 1980; Stout and Graziadei, 1980). Finally, the olfactory receptor cells appear to be capable of transporting a whole spectrum of inorganic and organic materials. A number of investigators have demonstrated that viruses enter the brain through the olfactory epithelium (Jackson *et al.*, 1979; Monath *et al.*, 1983; Eseri and Tomlinson, 1983). In addition, transport of amino acids (Weiss and Holland, 1967; Burd *et al.*, 1982), gold (de Lorenzo, 1970), HRP (Kristensson and Olsson, 1971), and dyes (Holl, 1980) was observed into the olfactory glomeruli after intranasal application. Transneuronal transport, meaning transport from the axon of one neuron to the axon or dendrite of a second neuron, also has been observed. Specifically, the lectin wheat germ agglutinin conjugated to horseradish peroxidase (WGA/HRP) could be demonstrated in a number of brain regions following application to the olfactory epithelium (Baker and Spencer, 1986; Shipley, 1985; Broadwell and Balin, 1985). This transport phenomenon may reflect the mechanisms by which trophic material is transferred between neuronal populations.

The interactions mentioned above and described in more detail in the discussion that follows also may reflect the unique properties of olfactory receptor cells and their relationship with the olfactory bulb (Graziadei and Monti-Graziadei, 1978, 1980). First, the receptor neurons can be replaced from a population of stem cells present in the olfactory epithelium. The stem cells are capable of cell division even in adult animals; either in response to lesion or on a cyclical basis (these may be the same phenomena, since receptor cells may respond to even mild trauma or continuous odorant stimulation by degeneration and replacement), these cells mature, send out axons, and reinnervate the olfactory bulb. Thus, denervation and reinnervation are a constant phenomenon in this system, producing a unique environment for and responses in the target neurons of the olfactory bulb. The alterations in olfactory bulb produced by lesions of the olfactory epithelium are discussed with emphasis on the biochemical and anatomical evidence for trophic interactions.

1.2. Anatomy

The following brief outline of the anatomy of olfactory neuronal circuits, especially with respect to those elements that have glomerular contacts and their neurotransmitter phenotypes, is provided to facilitate analysis of the experiments described. The material presented was adapted primarily from the review of Halasz and Shepherd (1983), which should be consulted for both references and details. Only references not found in this review are cited. Olfactory receptor axons originating from receptor cells in the epithelium traverse the cribriform plate to terminate in the glomeruli of the main olfactory bulb. Within the glomeruli, they form synaptic contacts with the dendrites of mitral cells, the primary output neurons of the olfactory bulb. Contacts are also found with dendrites of periglomerular and some tufted cells. Periglomerular cells in turn form reciprocal dendrodendritic synapses with mitral and tufted cell dendrites in the glomerulus. There do not appear to be direct contacts of receptor cell axons with the granule cells of the internal granule cell layer.

Only some of the transmitters in the above neuronal pathways are known. For example, there is some evidence to suggest that carnosine may be the transmitter of the olfactory receptor neurons. The identity of the transmitter of the mitral cells also has not been established definitively, but aspartate, glutamate or more recently *N*-acetylaspartylglutamate (Ffrench-Mullen *et al.*, 1985) have been postulated for this pathway. Periglomerular and tufted cells containing a number of transmitter candidates have been identified, including dopamine, GABA, substance P (in some species), enkephalin, and CCK. Granule cells are thought to be primarily gabaergic. Lastly, a number of centrifugal afferents terminate in the olfactory bulb. In the glomerular region cholinergic afferents, arising from the neurons of the basal forebrain, serotonergic fibers from the raphe dorsalis, and noradrenergic terminals originating in the locus ceruleus have been described. Also, centrifugal afferents containing substance P, possibly originating from the raphe, have been demonstrated (Kream *et al.*, 1984).

2. ALTERATIONS IN OLFACTORY BULB STRUCTURE AND FUNCTION FOLLOWING RECEPTOR AFFERENT LESIONS

2.1. Changes in the Size of the Olfactory Bulb

Surgical or chemical lesions of the rodent olfactory epithelium result in a number of morphological and biochemical alterations in the olfactory bulb. Most prominent is a decrease in the size of the olfactory bulb (Nadi *et al.*, 1981). The shrinkage is thought to result primarily from a loss of the receptor cell axonal arborization. Cell loss has not been reported in the rat or mouse (Baker *et al.*, 1983). Cell loss was reported, however, in the rabbit suggesting species or strain specificity may exist in this phenomenon (Pinching and Powell, 1971). Interestingly, neonatal closure of the nares also resulted in a decrease in the size of the olfactory bulb suggesting that receptor cell stimulation, producing the release of trophic substances, transmitters, ions, and/or electrical activity, may be necessary for the maintenance of normal olfactory structure and thus function (Meisami, 1976; Meisami and Satari, 1981; Benson *et al.*, 1984; Brunjes *et al.*, 1985). These effects are reciprocal; i.e., removal of the bulb also resulted in a decrease in the thickness of the epithelium, indicating that trophic interactions may be bidirectional in this system as in others (Costanzo and Graziadei, 1985; Samanen and Forbes, 1985).

2.2. Biochemical Alterations in the Olfactory Bulb

Accompanying the alterations in olfactory bulb size are profound changes in a number of biochemical parameters. Chemically produced deafferentation using a number of substances, such as $ZnSO_4$, Triton X-100, and vinblastine, resulted in a decrease in carnosine and OMP levels, produced by loss of afferents, and in a profound reduction in parameters reflecting dopamine synthesis and release (Nadi *et al.*, 1981; Kawano and Margolis, 1982; Baker *et al.*, 1984). Dopamine levels, the concentration of its major metabolite (DOPAC), and tyrosine hydroxylase activity, the rate-limiting enzyme in dopamine synthesis, were all reduced significantly (Tables 1 and 2). In both mice and rats, TH activity and dopamine levels exhibited a parallel decrease suggesting that the alterations in enzyme activity was restricted to the intrinsic dopamine neurons. Nor-

Table 1. Influence of Chemical and Surgical Deafferenting Lesions on Catecholamines in Rat and Mouse Olfactory Bulb[a,b]

Species and treatment	Time after lesions	*N*	NE	DA	DOPAC	Tissue weight	TH staining
Mouse							
Saline	21 days	6	1531 ± 113	1180 ± 78	359 ± 32	23.1 ± 1.9	3
$ZnSO_4$	21 days	3	2388 ± 222[c]	381 ± 85[c]	96 ± 16[c]	15.1 ± 0.7[c]	±
Triton X-100	21 days	3	1674 ± 171	485 ± 51[c]	139 ± 16[c]	19.2 ± 0.6[e]	1
Triton X-100	49 days	3	1971 ± 239[d]	1227 ± 248	515 ± 114[e]	21.3 ± 1.4	3
Rat							
Unlesioned side	28 days	3	1976 ± 239	330 ± 5	218 ± 28	37.2 ± 1.4	3
Lesioned side	28 days	3	2558 ± 294[e]	101 ± 33[c]	55 ± 22[c]	24.1 ± 1.1[c]	±

[a]Values are means ±SD, NE, DA, and DOPAC levels are in pmoles/g tissue, tissue weight in mg, and TH staining on a scale of 0–3, with ± indicating occasional patches of staining.
[b]All p values expressed as compared with saline control or unlesioned side by Student's t-test (two-tailed).
[c]$p < 0.05$.
[d]$p < 0.01$.
[e]$p < 0.001$.

epinephrine concentrations, representing the centrifugal afferents from the locus ceruleus which also contain TH, were actually increased when expressed per milligram tissue, but unchanged when expressed per bulb. The increase in norepinephrine (NE) concentration when expressed per milligram tissue was proportional to the decrease in bulb size.

2.3. Immunocytochemical Observations in the Deafferented Olfactory Bulb

The question remained, however: Did the loss of TH activity reflect a decrease in metabolism in all cells or in only some cell types of the juxtaglomerular dopamine cell population, e.g., tufted versus periglomerular cells? Alternatively, was there a general reduction or loss of TH protein in all cell types as had been previously described in other systems? Or was it possible that the remaining activity represented TH in noradrenergic afferents and/or nondeafferented intrinsic dopamine neurons and that most dopaminergic neurons completely lost the capacity to express TH?

2.3.1. Changes in Tyrosine Hydroxylase Expression in Olfactory Bulb

To distinguish between the above hypotheses, it was necessary to identify the neuronal elements containing the TH enzyme. This end was accomplished using immunocytochemical techniques and a specific antibody directed against TH enzyme. The experiments were performed in mice deafferented by bilateral intranasal irrigation with $ZnSO_4$ (0.17 M) or Triton X-100 (0.7%) and in rats with unilateral surgical deafferentation. The peroxidase antiperoxidase technique described by Sternberger *et al.* (1970) was applied to frozen sections from control and lesioned animals, in the case of mice, or from control and lesioned bulbs in rats. In mouse (Fig. 1) 21 days postlesion, cell staining in all

Table 2. Influence of Chemical and Surgical Deafferenting Lesions on Tyrosine Hydroxylase Activity in Rat and Mouse Olfactory Bulb

Species and treatment	TH activity
Mouse	
Control	60.6 ± 12.9[a]
Lesion ($ZnSO_4$)	4.5 ± 1.3[c]
Ratio-L/C	0.07
Rat	
Control bulb	671 ± 131[b]
Lesioned bulb	181 ± 90[c]
Ratio-L/C	0.27

[a]Data present as $\bar{x}$ ±SD of four mice (in pmoles/mg protein/20 min) 3 months following intranasal irrigation with $ZnSO_4$ (0.17 M).
[b]Data presented $\bar{x}$ ±SD for five rats at 1 month following unilateral surgical deafferentation.
[c]$p < 0.001$.

juxtaglomerular neurons was eliminated in those animals with most extensive $ZnSO_4$ lesions. Deafferentation produced by Triton and analyzed 21 days postlesion also produced significant loss in the number of TH-stained neurons (Fig. 1). Unilateral surgical lesions in rat produced similar results (Fig. 2). There was no change in TH staining in regions containing the noradrenergic centrifugal afferents. The biochemical data from a series of mice and rats deafferented at the same time as those analyzed immunohistochemically confirmed previous data and were in complete agreement with TH-staining density (Table 1). Thus, staining was eliminated from all juxtaglomerular cell populations including periglomerular and tufted cells. The residual TH activity could be attributed to the TH enzyme in the noradrenergic terminals and to those few neurons that presumably were not deafferented or that survived the treatment.

2.3.2. *Presence of Aromatic Amino Acid Decarboxylase in Deafferented Olfactory Bulb*

In rat and mouse previous investigators suggested that cell loss in the olfactory bulb was not associated with peripheral deafferentation. Our data support the postulate that cell death does not accompany the absence of TH staining. First, the density of juxtaglomerular neurons in lesioned animals, observed in Nissl stained sections, appeared increased (Figs. 2 and 3). Second, dopamine synthesis following exogenous administration of L-DOPA was still observed, albeit at a reduced level (Baker *et al.*, 1983*b*). In addition, TH staining returned in mice treated with Triton X-100 (Fig. 1). Application of Triton presumably destroyed only mature receptor cells but not stem cells which then matured to reinnervate the olfactory bulb. To confirm the continued presence of the presumptive

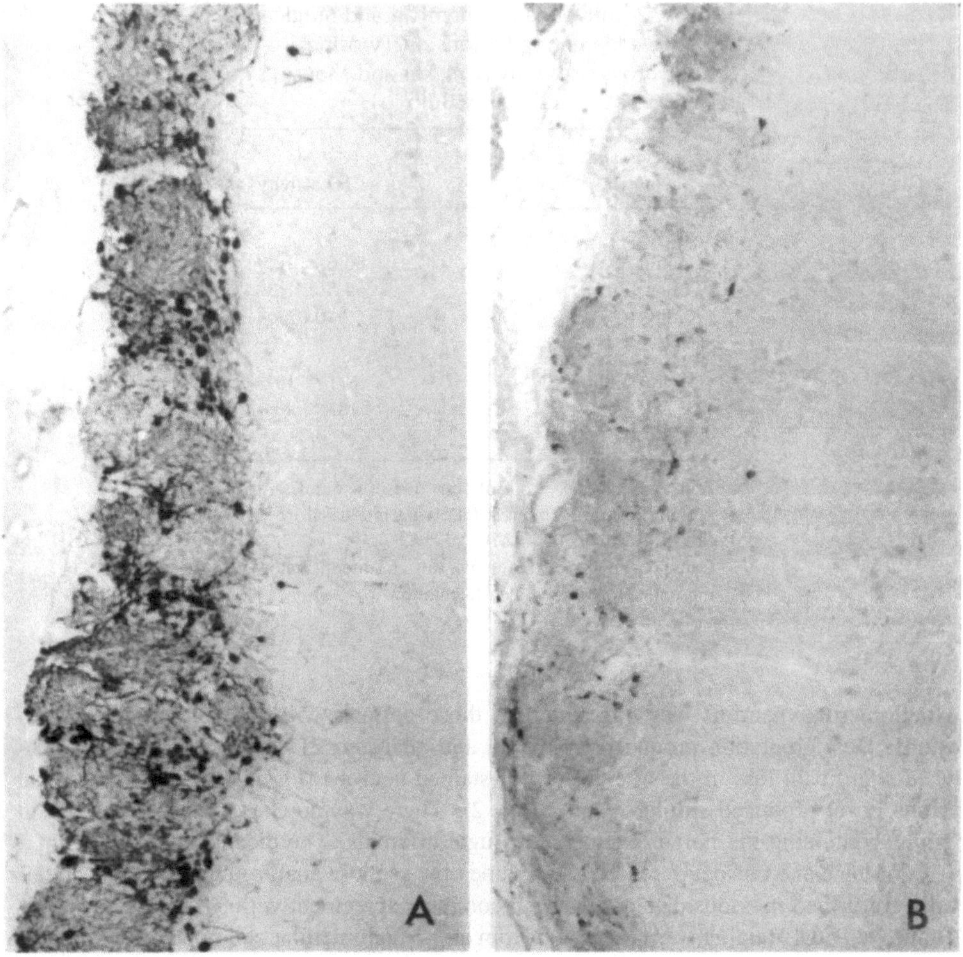

Figure 1. Alterations in number of neurons exhibiting tyrosine hydroxylase (TH) immunoreactivity in mouse main olfactory bulb following intranasal irrigation with (A) Saline; (B) $ZnSO_4$; (C) Triton X-100 (21 days post-lesion); (D) Triton X-100 (49 days post-lesion). In $ZnSO_4$-lesioned mice TH immunoreactivity is almost completely eliminated. Twenty-one days following treatment with Triton X-100 only a limited number of neurons retain TH-staining. At 49 days post-Triton X-100 treatment TH-staining is at or above control levels. (Bar = 100 μm.)

dopamine neurons, an antibody to L-aromatic amino acid decarboxylase (AADC) was used as a marker. This enzyme (AADC also known as dopa-decarboxylase or DDC) is not specific to dopamine systems, and its activity was not altered under other experimental conditions that reduced TH activity (Reis *et al.*, 1975). In adjacent sections from control bulbs stained with TH and AADC (Fig. 4), similar numbers of neurons were labeled (TH, 1,020 ± 47; AADC, 994 ± 63 neurons/section ±SE). In the lesioned bulb (Fig. 4), only a few neurons stained with TH (255 ± 38 neurons/section), while many were stained with AADC (679 ± 41). The same results were obtained in sections stained first with anti-

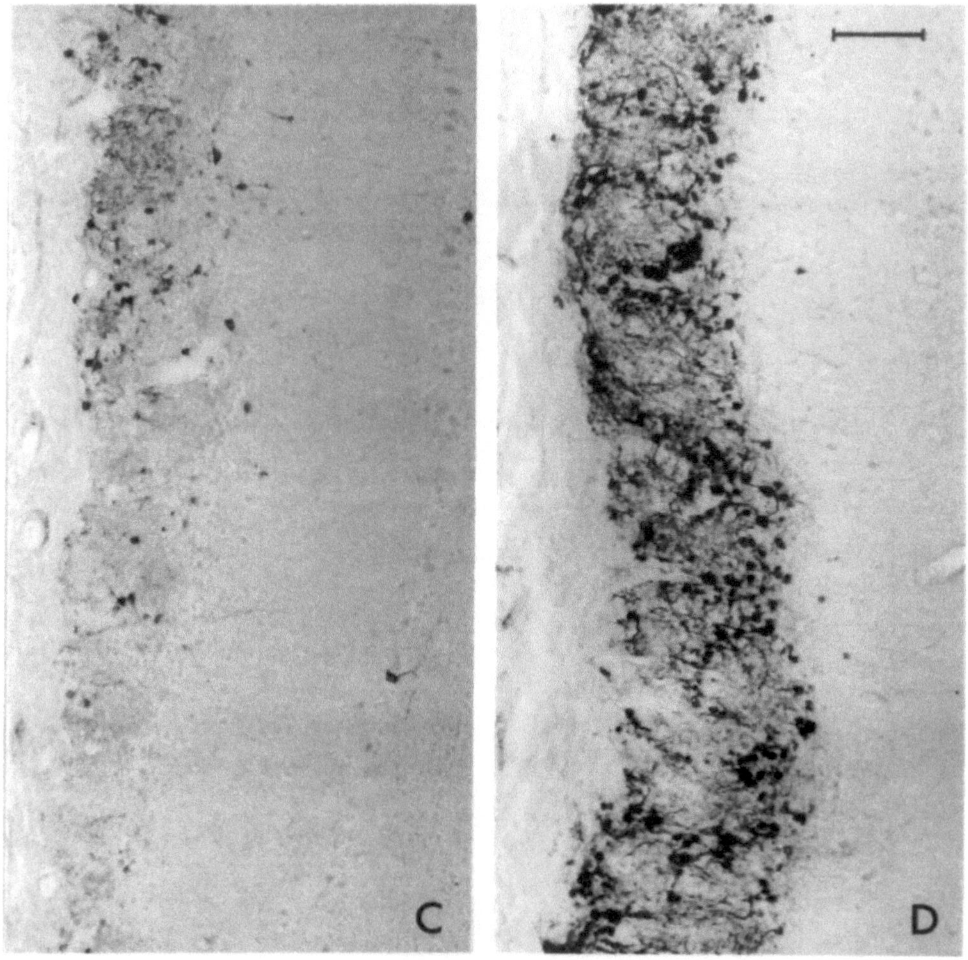

Figure 1. (continued)

bodies to AADC; the antibody and chromagen were removed and the sections restained with TH antibodies. The procedure followed, a modification of the method of Tramu *et al.* (1978), was described previously (Baker *et al.*, 1984). In sections from unlesioned (normal) bulbs, cells labeled with AADC also stained with TH (Fig. 5). In lesioned bulbs, there was a population of AADC-stained neurons that did not stain with TH antibodies (Fig. 6). Biochemical analysis of olfactory bulbs from animals lesioned at the same time indicated that parallel decreases in enzyme activity occurred. TH activity in the lesioned bulb was 25% and AADC 65% of that in control bulbs. These data indicated that a population of neurons, presumably those that formerly stained with TH, can be demonstrated with AADC antibodies. Thus, a majority of the presumptive dopamine neurons were still present, even though they no longer synthesized TH and thus do not make dopamine.

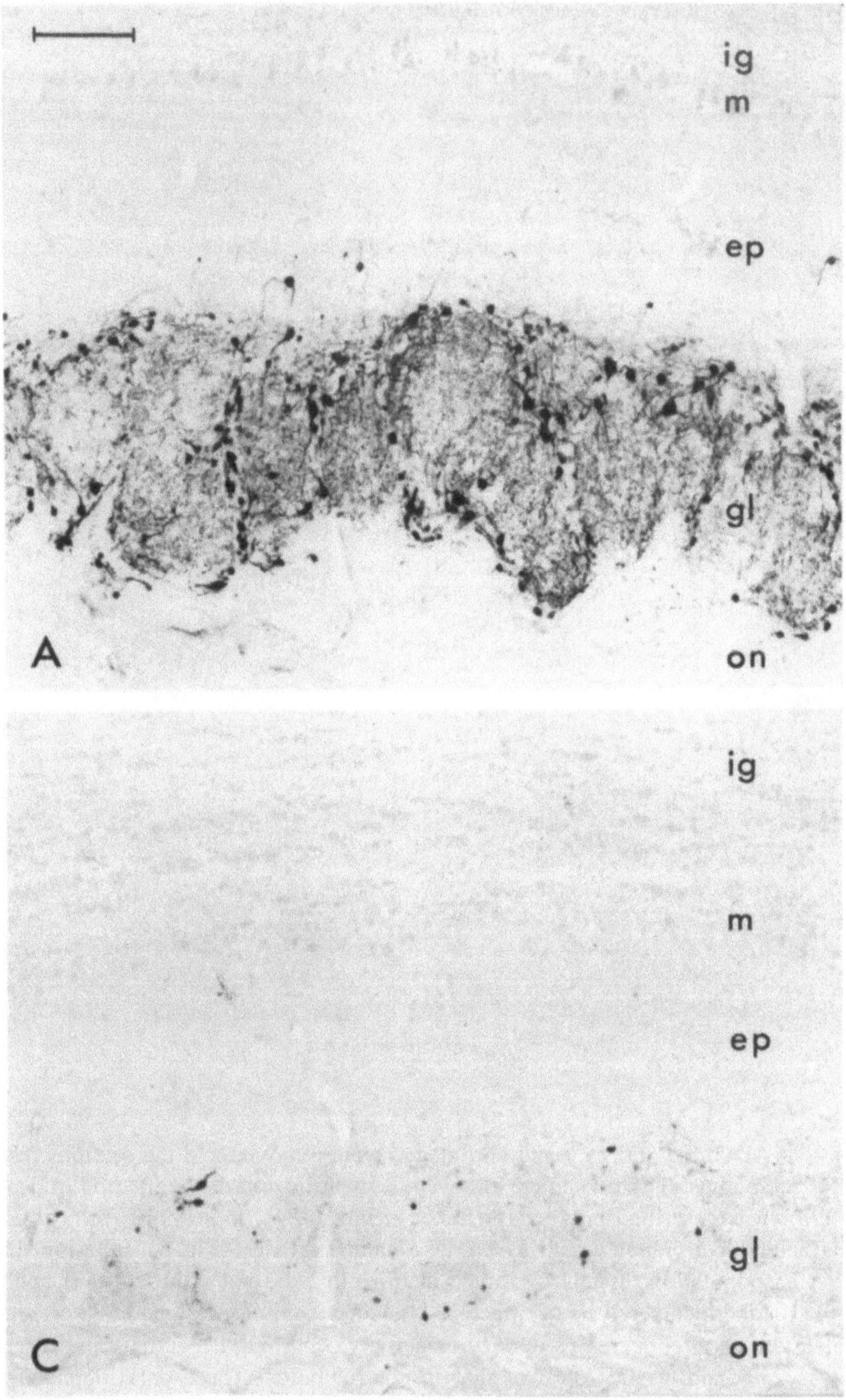

Figure 2. Tyrosine hydroxylase (TH) and Nissl-stained sections of olfactory bulbs from a rat which was unilaterally surgically deafferented. As in mouse (see Fig. 1) the number of neurons exhibiting TH-immunoreactivity is dramatically reduced in the lesioned (C) as compared to the unlesioned (A) bulb. The loss of afferent input results in a reduction in the size of the olfactory bulb and a concomitant increase in the density of juxtaglomerular neurons (compare B and D). (Bar = 100 μm.)

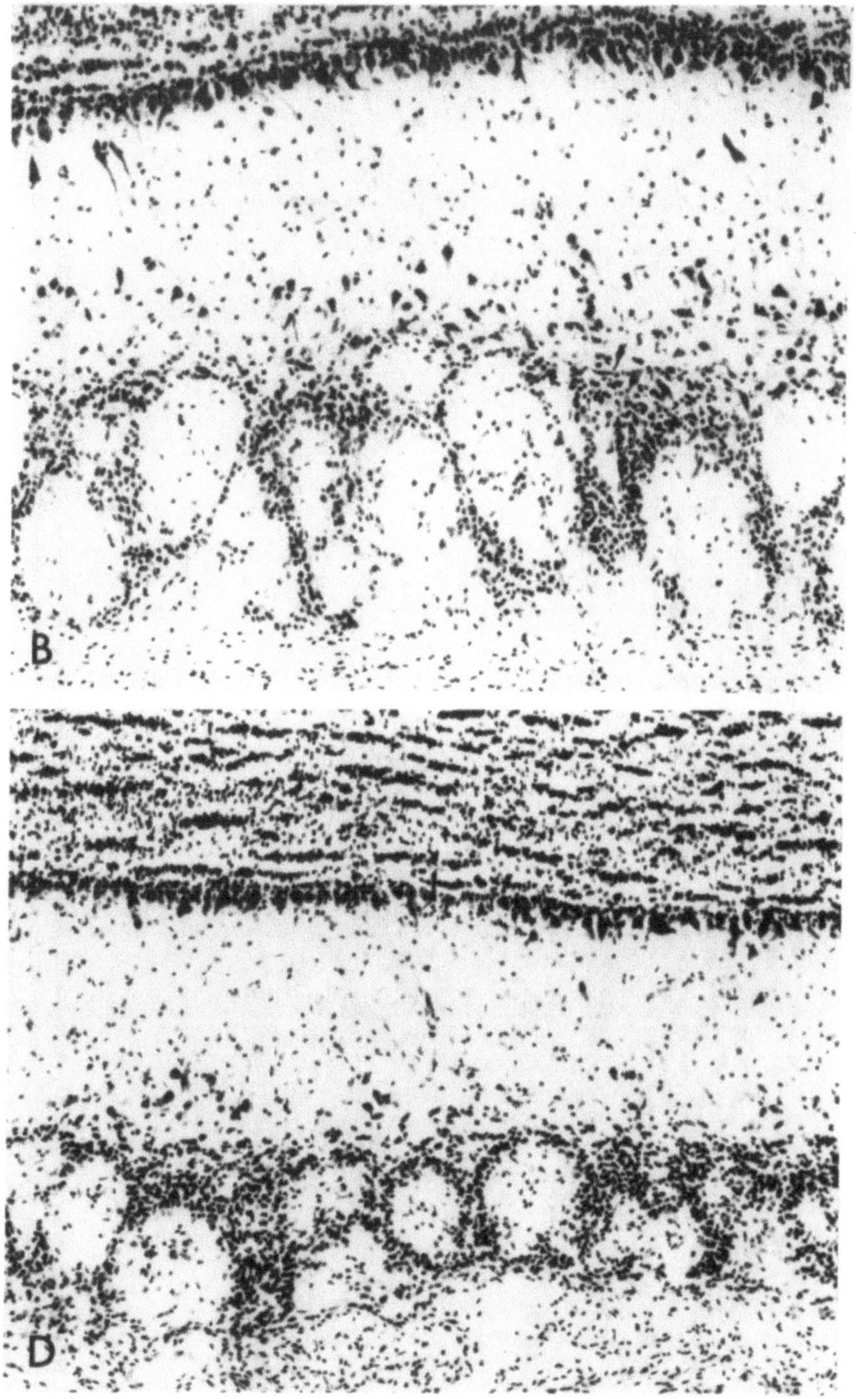

Figure 2. (*continued*)

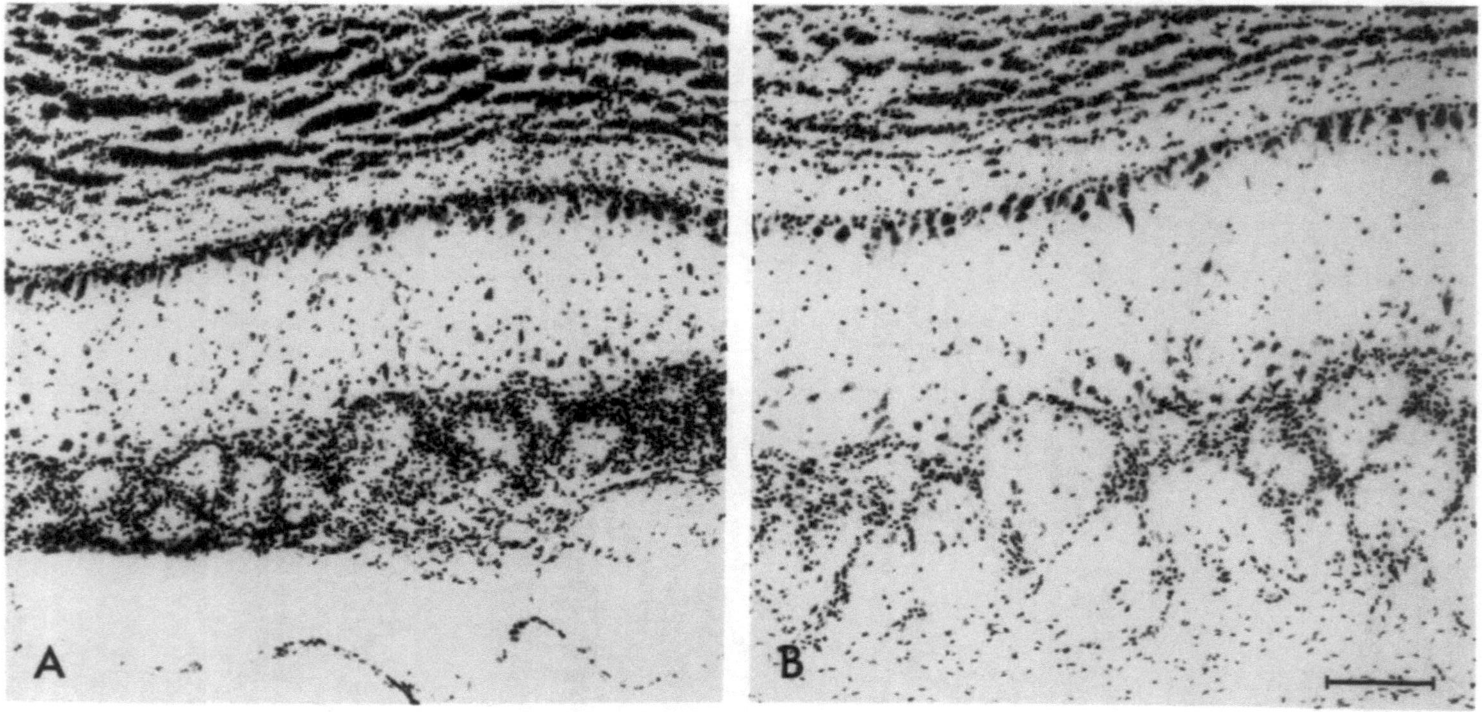

Figure 3. Nissl-stained sections from $ZnSO_4$ (A) and saline (B) treated mice. The olfactory bulb is reduced in size following intranasal $ZnSO_4$ treatment. The glomeruli shrink as a result of the loss of receptor cell afferents. An increase in the density of juxtaglomerular cells also is observed supporting the hypothesis that neuronal loss does not contribute to the decrease in the number of neurons containing TH-immunoreactivity. (Bar = 100 μm.)

2.3.3. Alterations in Substance P Expression in Deafferented Olfactory Bulb

These experiments support the hypothesis that neurons that lose the ability to synthesize TH survive and that following reafferentation reexpress the dopaminergic phenotype. The combination of immunocytochemical and biochemical analysis has allowed us to dissect the neuronal response to deafferentation of the dopamine system. The loss of TH was found to be in all types of dopamine neurons and not in centrifugal afferents of the noradrenergic terminals which also contain TH. The use of immunocytochemical techniques to dissect alterations in intrinsic versus extrinsic neuronal elements also was fundamental to the analysis of the effects of deafferentation on the substance P neurons in the hamster olfactory bulb (Kream *et al.*, 1984). The decrease in substance P staining following intranasal irrigation with $ZnSO_4$ was localized to a population of juxtaglomerular neurons in the main olfactory bulb. No alterations were observed in centrifugal afferents containing substance P. Only a modest decrease in the levels of substance P, analyzed by radioimmunoassay (RIA), was observed in the same group of animals in which staining in the juxtaglomerular neurons was abolished. In fact, the decrease in substance P levels was only apparent when the data were expressed per bulb and not per milligram protein as the decrease (about 30%) was equivalent to the reduction in size of the main olfactory bulb. Interestingly, using similar immunocytochemical techniques, substance P could not be demonstrated in the rat or mouse main olfactory bulb (Baker, 1986*a*), although there are several reports of significant levels of substance P measured by RIA. Substance P is formed from a propeptide (Kream *et al.*, 1984) and, since in other brain regions neuronal staining can be observed only in colchicine-treated animals, it is possible that either the levels of peptide or precursor processing in these neurons exhibit species differences. Taken together, these experiments demonstrate afferent regulation of transmitter expression in the dopaminergic and substance P neuronal systems. They also indicate that these effects occur without cell loss and, at least, for the dopaminergic system are reversible.

2.3.4. Are There Alterations in Other Transmitter Systems?

These studies also raise a number of questions. First, does deafferentation produce the same alterations in transmitter synthesis in all systems intrinsic to the olfactory bulb. For example, GABA (Mugnaini *et al.*, 1984), enkephalin (Bogan *et al.*, 1982), and CCK (Gall *et al.*, 1985) all have been localized to juxtaglomerular neurons in the main olfactory bulb. Transmitter expression following deafferentation in these neuronal systems has not been examined. It is also possible that more than one transmitter can be expressed in the same juxtaglomerular neurons and that deafferentation produces an alteration in the predominant transmitter or even induces synthesis of a new transmitter. The coexistence of additional transmitters in dopamine neurons has been demonstrated in other systems, e.g., the substantia nigra (Hokfelt *et al.*, 1980). In the hamster, however, recent studies indicated that dopamine (using TH as a marker) and substance P are not found in the same neurons (Baker, 1986*b*). The data in rat for GABA and dopamine (DA) are contradictory, but it is possible that a population of GABA neurons also may synthesize dopamine (Mugnaini *et al.*, 1984; Kosaka *et al.*, 1985; Gall *et al.*, 1985).

The mechanisms underlying the deafferentation-induced alterations in transmitter

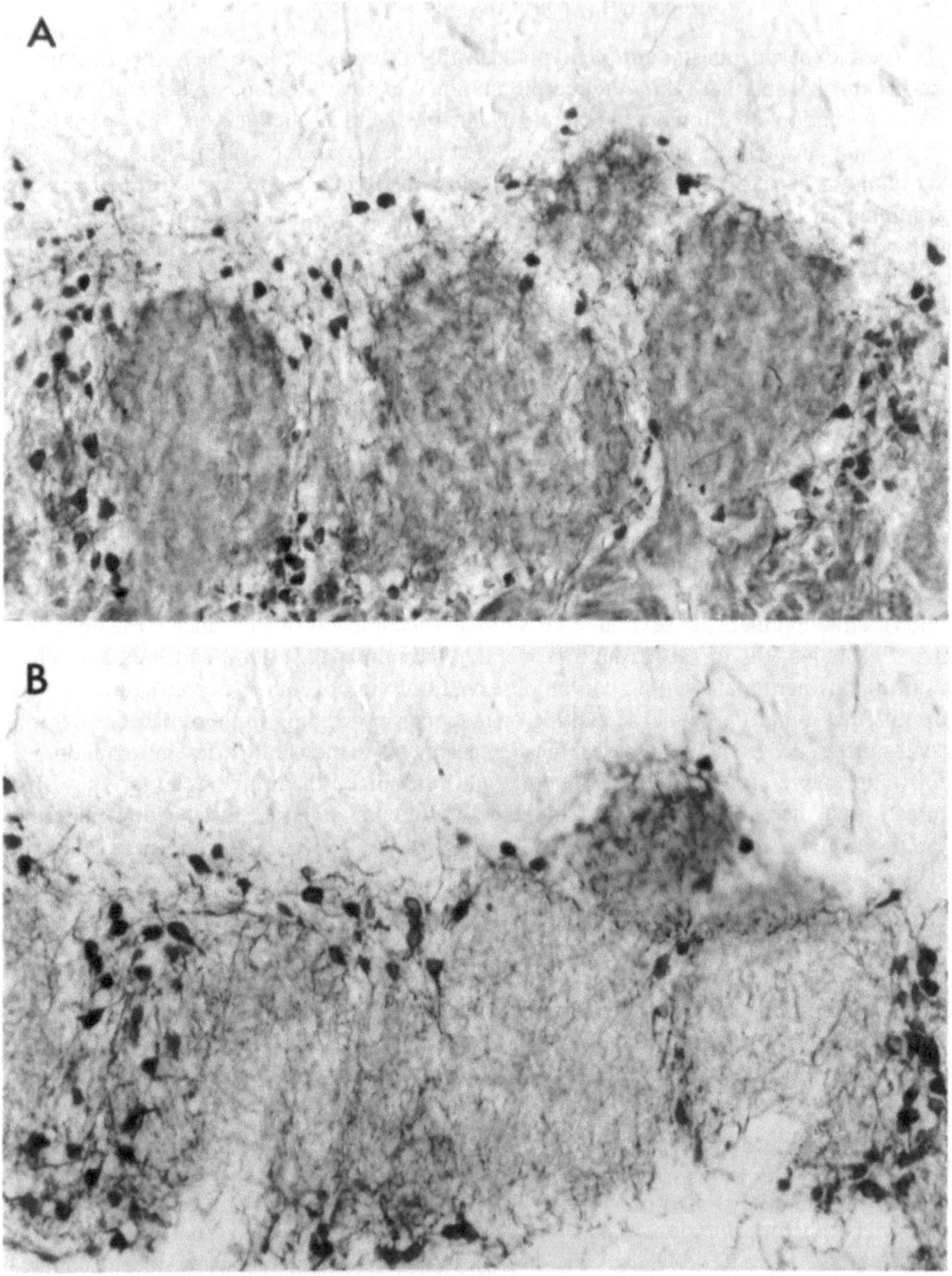

Figure 4. Adjacent sections from control and deafferented rat olfactory bulbs stained with antibodies to tyrosine hydroxylase (TH) and aromatic amino acid decarboxylase (AADC). The distribution and number of cells immunoreactive with the two antibodies are the same in control bulbs [(A) AADC, (B) TH] indicating that the antigens are contained in the same neurons. Many more neurons stain with antibodies to AADC (C) than with TH (D) demonstrating that although these neurons are unable to synthesize TH they still express the monoamine phenotype as exemplified by their synthesis of AADC. b: blood vessel. (Bar = 50 μm.)

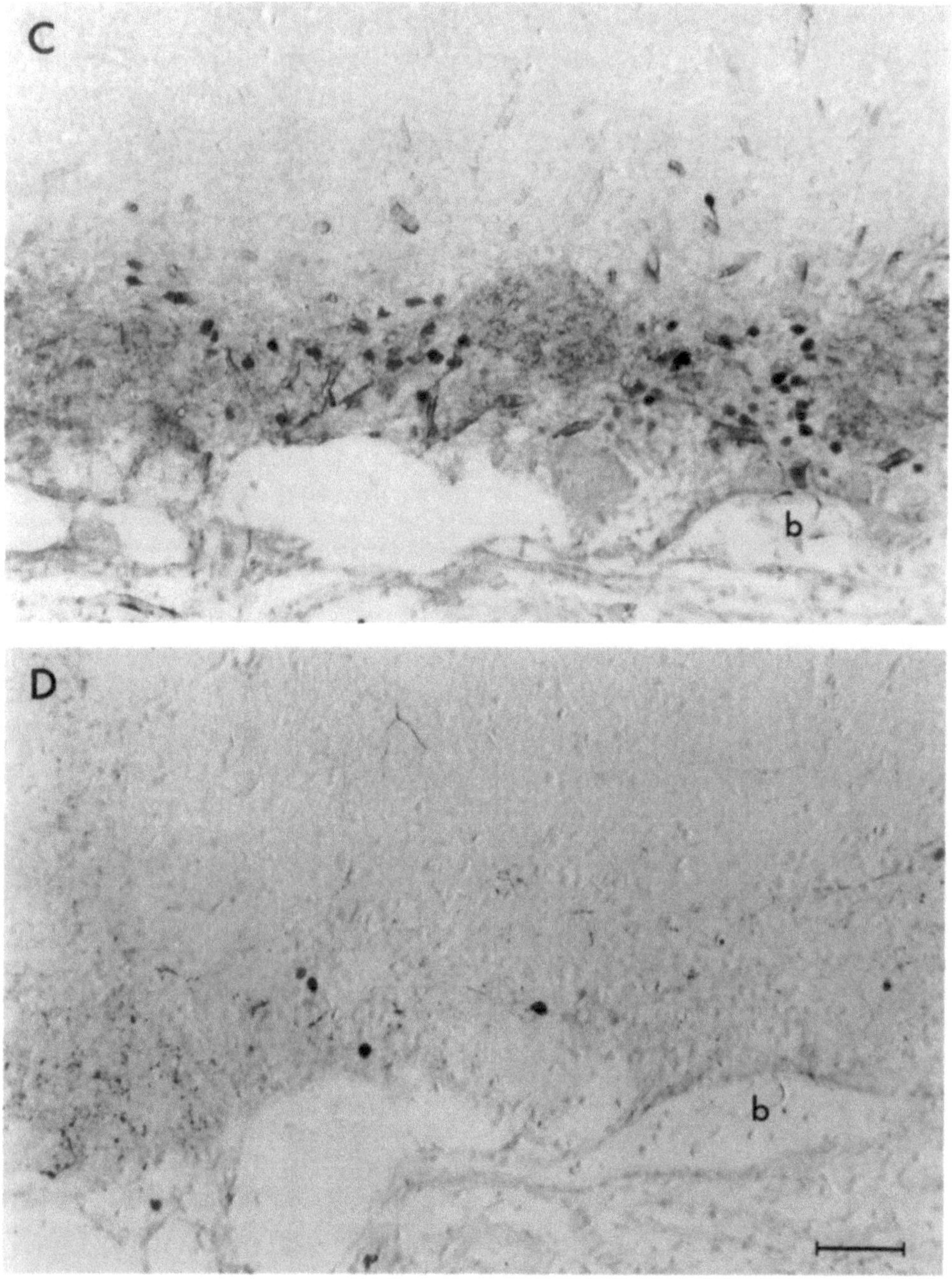

Figure 4. (*continued*)

synthesis have yet to be elucidated. Experiments investigating the changes in TH message produced by receptor cell lesions will be helpful in indicating how these affects are genomically controlled. Studies in the superior cervical ganglion (Black *et al.*, 1985) indicate that the levels of TH message can be increased by afferent stimulation. Comparison of the time course of the effect of deafferentation to that which occurs during development are especially important in this system, in which the afferent neuron is capable of replacement even in adults. The olfactory bulb may have developed unique responses to deafferentation because of this continuous turnover of afferent input, which allows the newly formed fibers to correctly find their synaptic targets in the glomerulus.

3. DEVELOPMENT AND PLASTICITY

3.1. Inductive Capacity of the Olfactory Epithelium

As demonstrated by Graziadei, Monti-Graziadei, and co-workers, the olfactory epithelium has enormous inductive capacity in the frog and salamander, producing olfactory bulblike structures in many brain regions. These investigators have also demonstrated that the epithelium has inductive effects in the mammal producing ectopic glomeruli even in the frontal cortex (Graziadei and Samanen, 1980). These studies suggest that the olfactory epithelium can exert strong trophic actions. The developmental sequence of tyrosine hydroxylase activity and receptor cell input to the olfactory bulb also suggests that the epithelium may play a role in the determination of transmitter phenotype during ontogenesis. Although the time of arrival of the afferents in the olfactory bulb is known, the precise time at which synapses are functionally mature in a trophic sense, i.e., is capable of influencing transmitter expression, is unknown. Since cell division in the olfactory bulb continues well into the neonatal period (Hinds *et al.*, 1982), it is possible that neurons that leave the cell cycle later find a different environment with respect to the amount and functional maturity of receptor cell input. For example, the level of activity in or biochemical maturity of the axons may change at different developmental stages. In fact, Mair and Gesteland (1982) showed that modulation of mitral cell activity does not occur during the first postnatal week suggesting that periglomerular and/or granule cell activity is not functional at birth.

3.2. Strain Differences in Transmitter Expression in the Olfactory Bulb

The phenotype expressed by the juxtaglomerular neurons may be altered by developmental anomalies in receptor cell synaptogenesis. For example, the literature suggests that juxtaglomerular neurons in the BALB/cJ strain of mice do not receive direct synaptic input from receptor cell afferents (White, 1973). This strain also has very high tyrosine hydroxylase activity and/or dopamine levels in all brain regions (Baker *et al.*, 1980, 1983*a*) compared with other mouse strains. These differences are reflected in a proportional increase in the number of TH-stained neurons demonstrable immunocytochemically in the olfactory bulb of normal BALB/cJ mice (Fig. 7). Following receptor cell lesions, BALB/cJ mice continue to express TH. The levels are reduced but are still much higher than in control mice whether analyzed immunohistochemically (Fig. 8) or biochemically

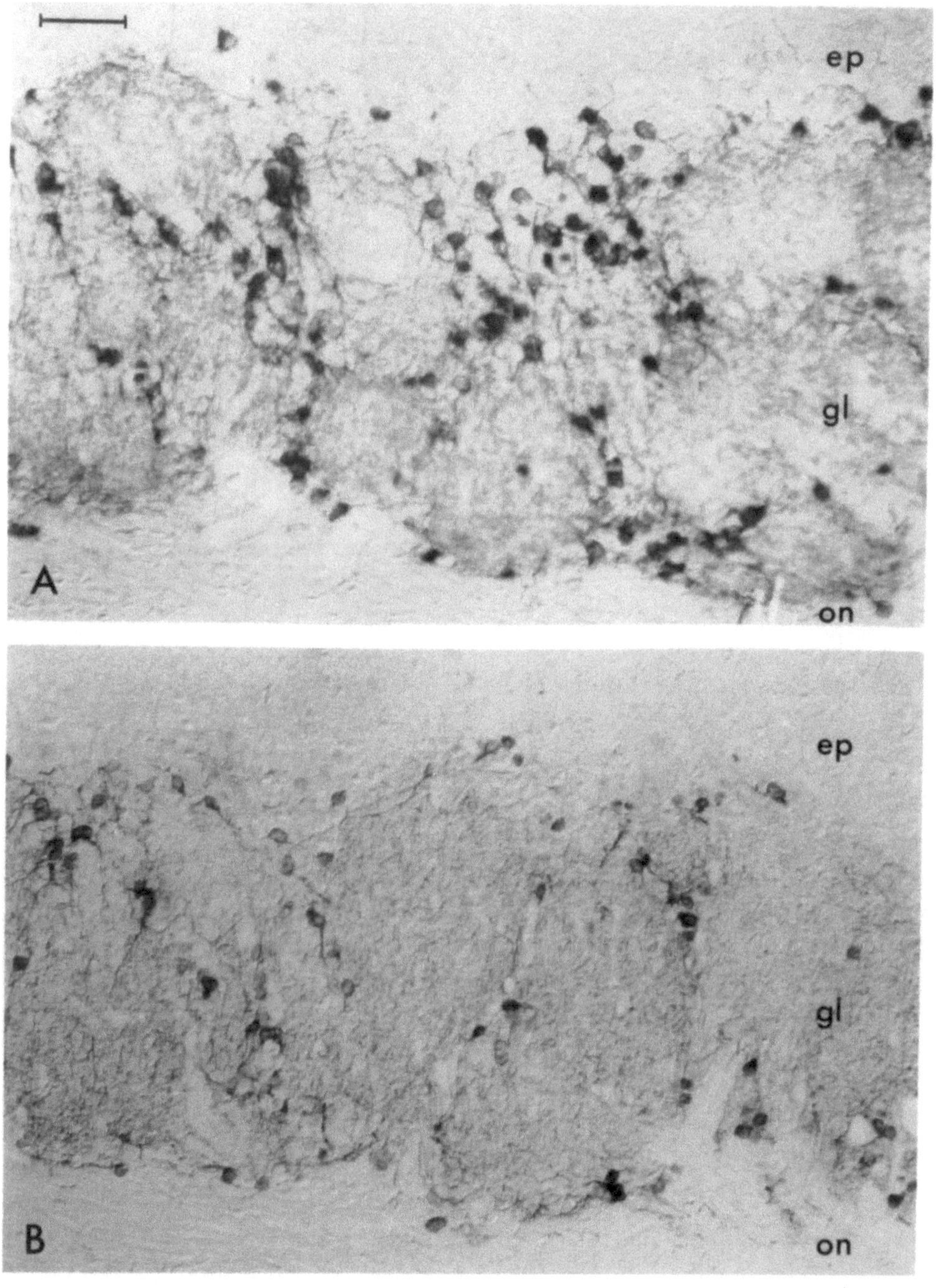

Figure 7. Strain differences in the number of TH-immunoreacitive neurons in the mouse main olfactory bulb. Many more neurons contain TH in normal BALB/cJ (A) than CD-1 (B) mice. However, the distribution of stained cells appears similar in the two strains. (Bar = 50 μm.)

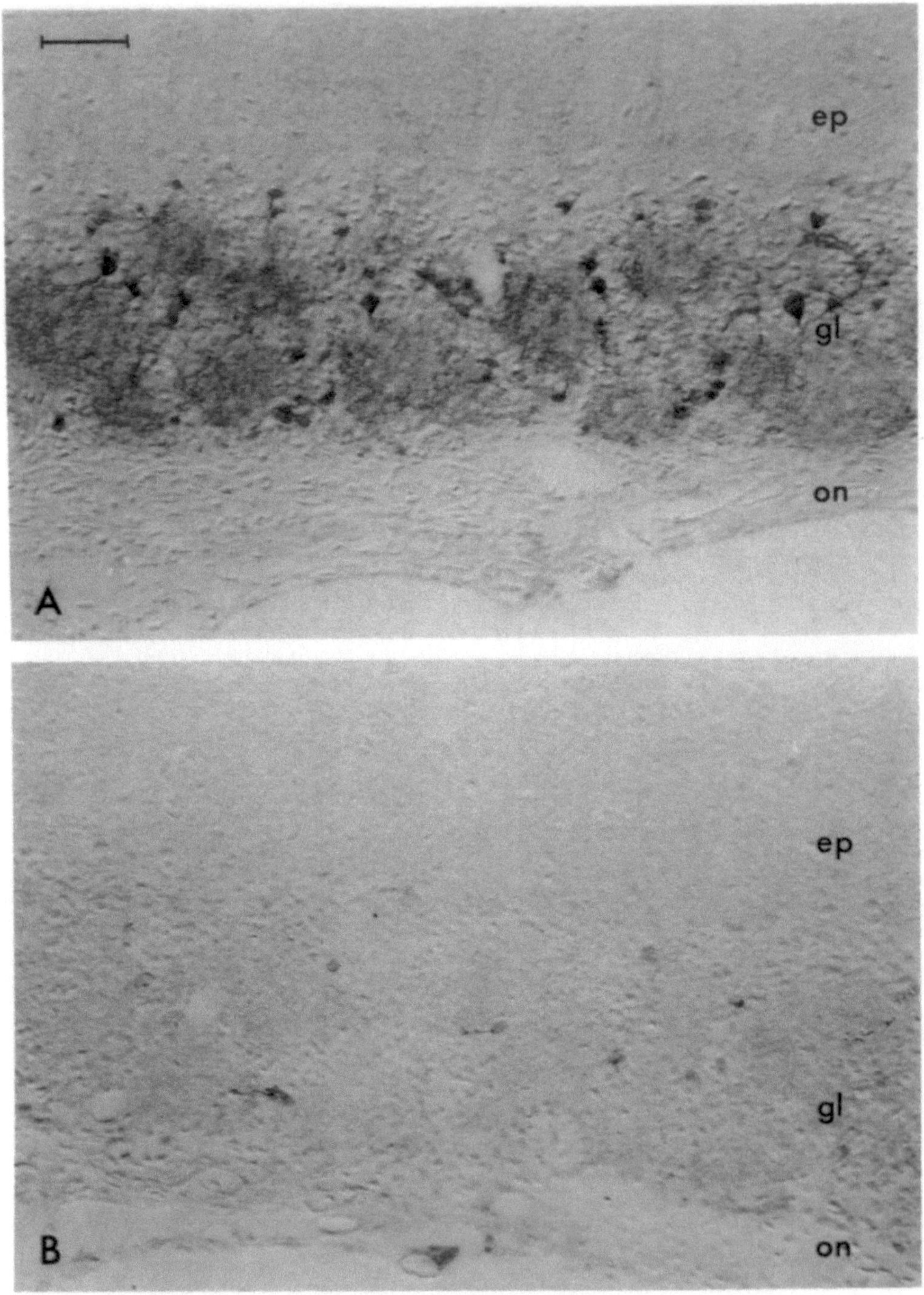

Figure 8. Strain difference in the number of TH-immunoreactive neurons following chemical deafferentation by intranasal irrigation with $ZnSO_4$ in the mouse main olfactory bulb. The number of neurons remaining in the BALB/cJ (A) is greater than in the CD-1 (B) mouse. Strain differences in afferent innervation in the BALB/cJ mice may produce this alteration in the response to deafferentation. (Bar = 50 μm.)

Table 3. Strain Differences in Dopamine Levels in Olfactory Bulbs from Normal and Deafferented Mice

Species and treatment	Dopamine levels	Weight of bulb
CD-1		
Control	1180 ± 109[a]	23.0 ± 0.95[b]
$ZnSO_4$	336 ± 60	14.5 ± 0.80
BALB/cJ		
Control	2207 ± 169	18.9 ± 0.75
$ZnSO_4$	674 ± 110	12.0 ± 1.00

[a]Dopamine levels are expressed as $\bar{x}$ pmoles/g tissue ±SD.
[b]Tissue weight is expressed in mg.

(Table 3). These strain differences in the normal levels of TH and in the response to deafferentation suggest that genetic variations in afferent synaptic input and connections can alter phenotypic expression in the olfactory system. Species differences in the expression of transmitters in both the main and accessory olfactory bulb were found that may also reflect variations in afferent input (see Section 2.3.3).

4. TRANSNEURONAL TRANSPORT OF EXOGENOUS MATERIALS FROM THE OLFACTORY EPITHELIUM TO BRAIN

The olfactory sensory epithelium, by virtue of its function as an odor detector, is in constant contact with the constituents of inspired air. Intuitively, and by analogy with the blood–brain barrier, which limits access to the central nervous system (CNS), most investigators, myself included, have assumed that the tight junctions found in the olfactory epithelium would prevent exogenous environmental substances from reaching the CNS. From a historical perspective, however, this assumption was challenged frequently. Both organic and inorganic substances were shown to be transported from the epithelium to the olfactory bulb in a number of species. The list includes dyes, amino acids, gold, and HRP. In addition, several viruses have been shown to enter the brain through this route. The studies described below demonstrate that transneuronal transport of lectins occurs in the olfactory system.

4.1. *Transport of Lectins in the Olfactory System*

Recently, a number of studies demonstrated that a lectin, wheat germ agglutinin (WGA) or its conjugate to HRP (WGA–HRP), was transneuronally transported (i.e., from one neuron to another) within the visual system following intraocular injection (Itaya and Van Hoesen, 1982; Ruda and Coulter, 1982; Trojanowski, 1983). In the experiments described below WGA–HRP (50 μl, 0.1%, Sigma Chemical Co.) was applied dropwise through the external nares of a rat, a form of intranasal irrigation. The application was unilateral. After 2–4 days survival, transport demonstrated by TMB histochemistry (Baker and Spencer, 1986) was observed unilaterally in the olfactory nerve and glomerular layers of the main olfactory bulb (Fig. 9). In addition, periglomerular, tufted, and mitral cells were labeled. Only a few cells in the most superficial aspects of the internal granule cell layer contained reaction product. Electron microscopic analysis demonstrated that transport from the nasal epithelium was intraneuronal, since peroxidase labeled vesicles were found in olfactory receptor cell axons (Fig. 10). Transneuronal transport within the glomeruli was verified by the presence of apposed labeled axonal and dendritic profiles. Periglomerular, tufted, and mitral cell bodies with reaction product also were observed, confirming the transneuronal transport phenomenon.

4.2. Transneuronal Transport of Lectin to the Forebrain

The presence of the labeled mitral and tufted cells suggested that transport of WGA-HRP through their axons to the forebrain might have occurred as well. Following 4–6-day survival, labeled terminals were observed in several forebrain regions including anterior aspects of the piriform cortex, olfactory tubercle and dorsal to the lateral olfactory tract (Fig. 11). Thus, mitral and/or tufted cells incorporated the lectin conjugate and transported it to some terminal fields in the forebrain. In addition to terminals, retrogradely labeled neurons were observed in basal forebrain. A few cells with reaction product were observed in the vertical limb of the diagonal band of Broca; however, the largest number of labeled neurons occurred in the lateral nucleus of the horizontal limb of the diagonal band at the level of the decussation of the anterior commissure (Fig. 11). In caudal aspects of the preoptic area, labeled neurons decreased in number and disappeared in the retrochiasmatic area. In mouse, a few neurons were observed in the raphe dorsalis. Using another mode of application, a few labeled neurons were reported in the locus ceruleus (Shipley, 1985). The ability to label these latter two regions may depend on the amount of conjugate applied or the duration and mode of application. The basal forebrain, raphe dorsalis, and locus ceruleus all have terminal arborizations in the glomerular region of the olfactory bulb, supporting the hypothesis that WGA–HRP transported by receptor cell axons is released into the synaptic space and internalized not only by dendrites of mitral, tufted, and periglomerular cells, but by the terminals of centrifugal axons located within or near glomeruli as well. Brain regions with centrifugal afferents to the olfactory bulb that do not terminate predominantly within the glomerular layer, for example, those of the anterior olfactory nucleus, were not labeled.

In addition lectin-labeled glial cells were observed in the olfactory bulb. Following intraocular injection of WGA–HRP glial elements with reaction product were not observed in adult animals. In young animals, however, labeled glia were observed (R. Spencer, personal communication). Thus, the olfactory system in some ways displays characteristics consistent with continuous development where under normal conditions deinnervation and reinnervation occur continuously following differentiation of stem cells to form new olfactory receptor neurons. The presence of HRP-reaction product in glia may indicate the relative immaturity of some elements in the olfactory system.

4.3. Implications of Transneuronal Transport for Disease Processes

The localization of retrogradely labeled neurons to the basal forebrain has significant implications for disease processes. The lateral nucleus of the horizontal limb of the diagonal band is the rat homologue of the nucleus basalis of Meynert (Fibiger, 1982, Mesulam *et al.*, 1982), which is altered in degenerative brain syndromes, such as Alzheimer disease (Coyle *et al.*, 1983; McGeer *et al.*, 1984; Rosser *et al.*, 1983). It has been postulated that the loss of these neurons leads to the memory loss syndrome associated with the disease. Interestingly, deficits in two other brain regions are seen in Alzheimer disease. Transmitter content and synthesis decrease in the locus ceruleus and raphe, both of which have projections to the glomerular region of the olfactory bulb (Bondareff *et al.*, 1981; Bowen *et al.*, 1983; Gottfries, 1983; Iversen *et al.*, 1983; Tomlinson *et al.*, 1981; Yates *et al.*, 1983). In addition, a number of recent reports indicate that even in normal aging, deficits are found in olfaction that are greater than those observed in other sensory systems such as taste (Stevens *et al.*, 1984). These data taken together suggest that the olfactory system is peculiarly sensitive to environmental influence and may, through transport of viruses and toxins, be more susceptible to injury than previously suspected. It also may be a route of entry for substances or pathogens which alter its function as well as that of other brain regions.

5. CONCLUSIONS

The olfactory system exhibits reciprocal neuronal interactions. The expression of some transmitters in the olfactory bulb is dependent on receptor cell input as is the integrity of the epithelium influenced by the presence of the olfactory bulb. The changes in transmitter synthesis in the olfactory bulb are produced without apparent cell loss, at least in rat and mouse. The question remains, however, as to the process that permits cell survival with apparent loss of phenotype. Juxtaglomerular neurons, both periglomerular and tufted cells, express several transmitters. Is it possible that these neurons can alter their phenotype, expressing dopamine when afferent input is present and another transmitter following deafferentation. Alternatively, another transmitter, e.g., GABA, may be synthesized normally by dopamine neurons, as suggested by several investigators (Gall *et al.*, 1985; Kosaka *et al.*, 1985). The expression of this second transmitter may not be sensitive to deafferentation, permitting survival of the dopamine neuron.

All these possibilities may be a reflection of the role of the receptor afferent in the determination of phenotype during development. It is curious that the alterations in transmitter expression are observed only in the juxtaglomerular neurons. These neurons, some of which are formed postnatally, do not influence mitral cell activity in the neonate. Alterations in mitral cell transmitter expression have not been studied for two reasons. First, the phenotype has not been definitively established, and second, there are no specific markers, as yet, for the proposed transmitters. The mitral cell, like the juxtaglomerular, not only does not degenerate following deafferentation, but forms new synapses with periglomerular and tufted cell dendrites. The earlier neurogenesis of the mitral cell, that is, prior to the arrival of receptor cell afferents may be a factor in its presumed resistance to deafferentation.

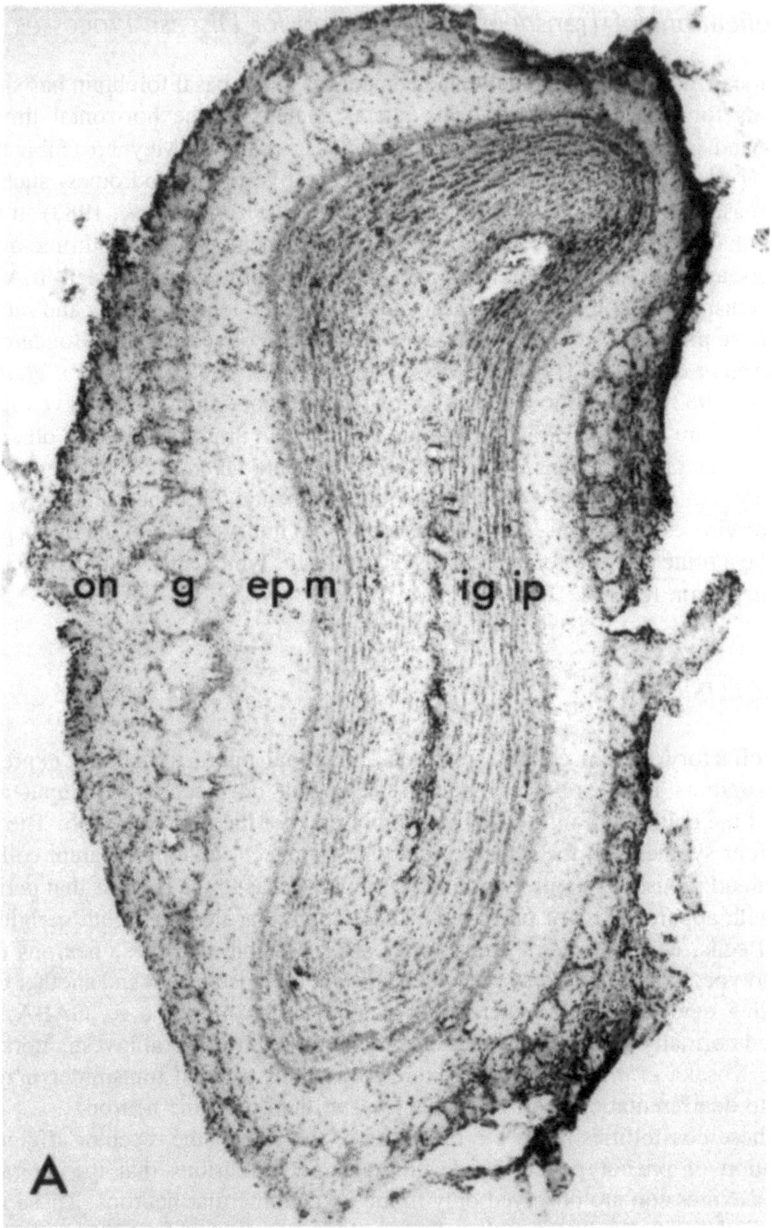

Figure 9. Brightfield photomicrographs of olfactory bulbs contra- (A) and ipsi- (B) lateral to the external nares of a rat intranasally irrigated with wheat germ agglutinin conjugated to horseradish peroxidase (WGA-HRP). Only the main olfactory bulb ipsilateral to the irrigated nares contains reaction product. Label is distributed evenly in axonal terminals throughout all glomeruli indicating that the lectin-conjugate was transported in the anterograde direction from olfactory receptor cells located in widely distributed regions of nasal epithelium. In addition, a number of neuronal populations are labeled including periglomerular, tufted, mitral, and a few internal granule cells demonstrating the occurrence of retrograde transneuronal transport from olfactory receptor cell axons to the dendritic processes of the labeled cell populations. (Bar = 200 μm.)

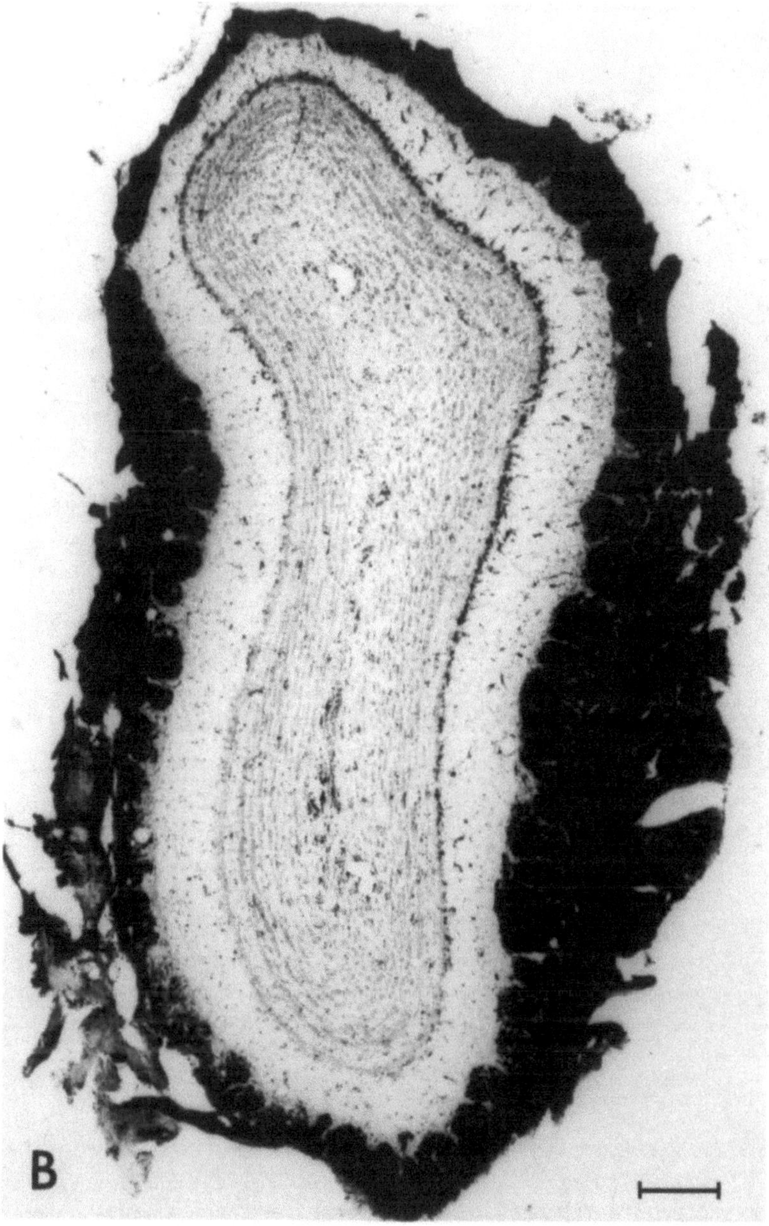

Figure 9. (*continued*)

The differences in normal transmitter expression as well as the altered response to deafferentation in the BALB/cJ mice suggest that intrabulbar connections, in addition to receptor afferent input, may be important factors in the maintenance of phenotype, especially in the dopamine system. Since the periglomerular neuron in this mouse strain does not synapse directly with receptor cell axons, is there an increase, for example, in the

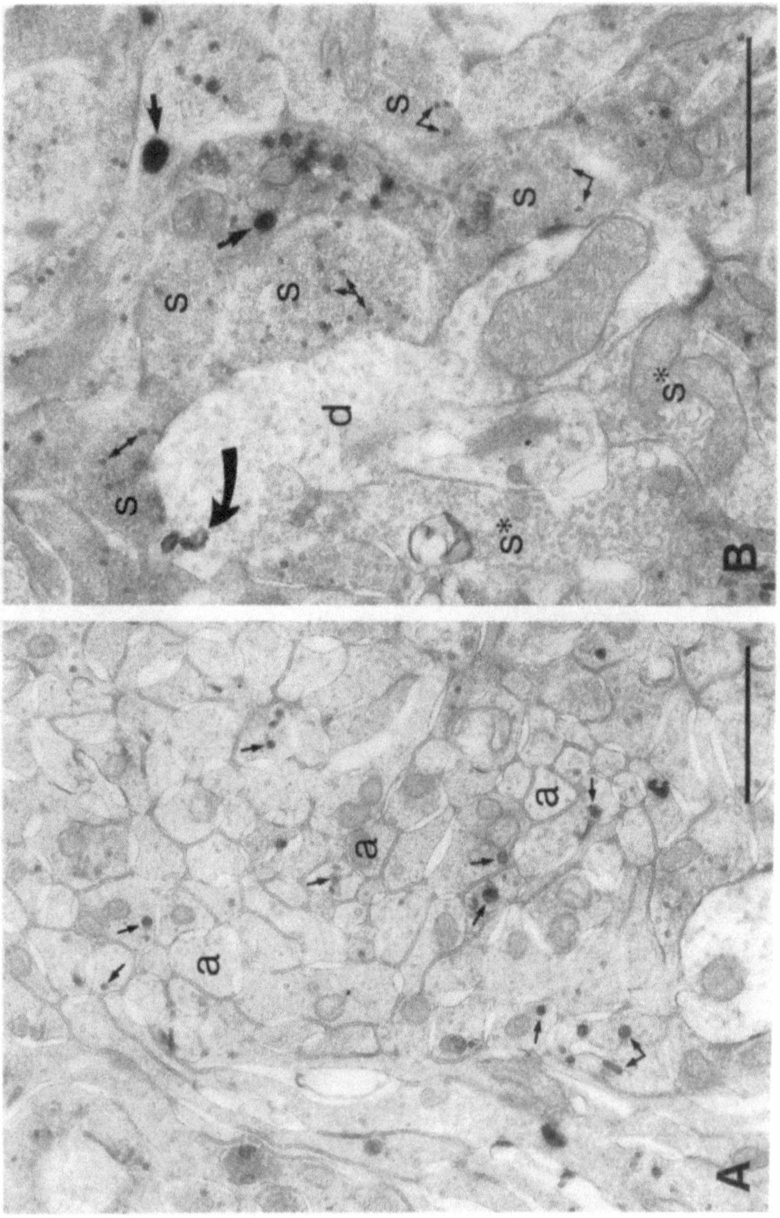
A
a
B
s
d
s*

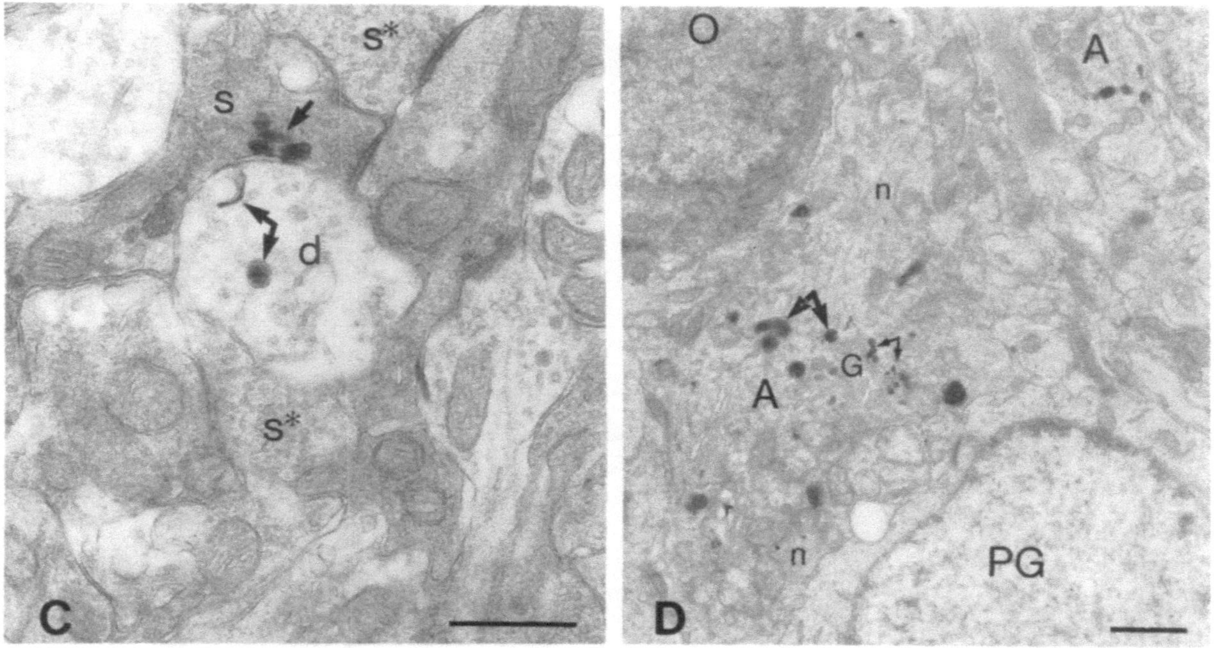

Figure 10. Electron microscopic evidence for intraaxonal transport of WGA-HRP in olfactory receptor cell axons. (A) Electron micrograph of unmyelinated primary olfactory axons (a) with WGA-HRP reaction product localized intraaxonally in 60–70 nm diameter tubulo-vesicular profiles (arrows). (Bar = 1 μm.) (B) Electron micrograph of primary olfactory synaptic endings (S) in a glomerulus, exhibiting typical electron dense axoplasmic matrix. WGA-HRP reaction product is localized in 75–150 nm diameter dense bodies (large arrows) and in 30 nm diameter vesicular profiles (small arrows). Transneuronal transport of WGA-HRP is indicated by reaction product located in dense bodies (large curved arrow) in a presynaptic dendrite (d). Note also unlabeled synaptic endings (S*) with electron lucent matrix in same vicinity. (Bar = 1 μm.) (C) Electron micrograph of primary olfactory synaptic endings (S) with WGA-HRP reaction product (large arrows) localized in 70 nm diameter tubulo-vesicular profiles presynaptically and cisternal and lysosomal profiles postsynaptically in a dendrite (d). Note unlabeled synaptic endings (S*) with electron lucent matrix in close proximity. (Bar = 5 μm.) (D) Electron micrograph of an astrocyte (A), characterized by bundles of neurofilaments (n), with WGA-HRP reaction product localized in vesicular (small arrows) and lysosomal (large arrows) profiles in the vicinity of Golgi apparatus (G). O, satellite oligodendrocyte; PG, periglomerular cell. (Bar = 1 μm.) (Taken by permission from Baker and Spencer, 1986.)

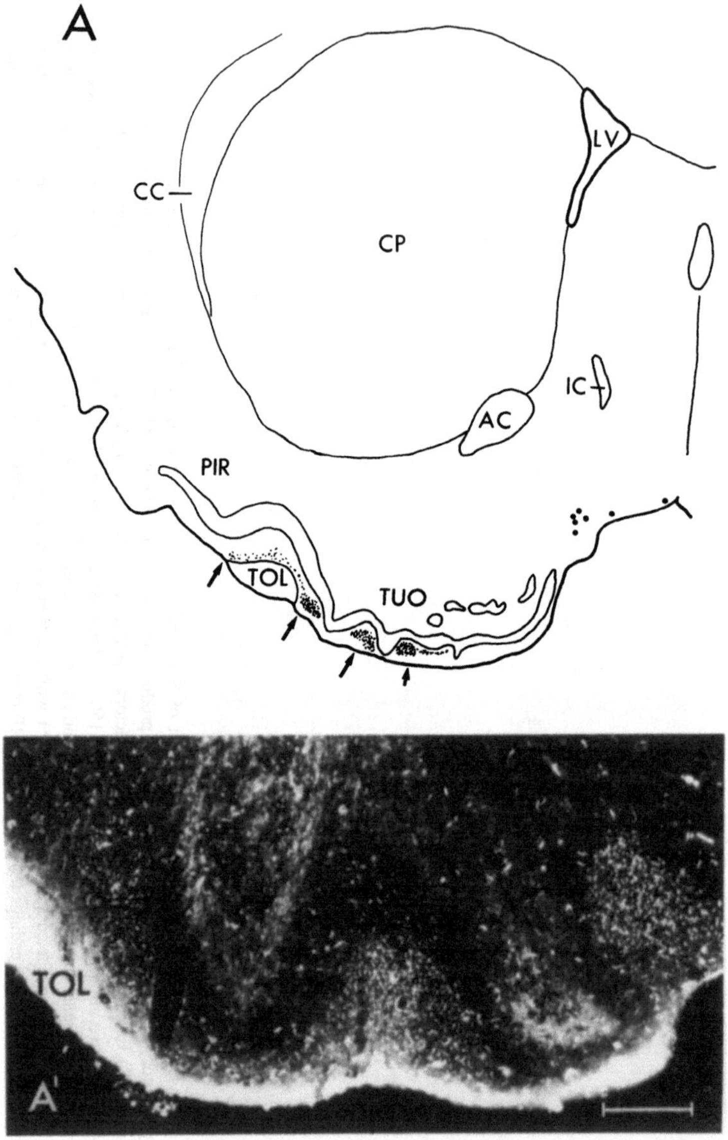

Figure 11. Anterograde and retrograde transneuronal transport of WGA-HRP to the forebrain of the rat. A and B are diagrammatic representations of the forebrain regions containing the largest density of labeled terminals and cells respectively. A′ is a darkfield photomicrographs of anterograde transneuronal label in terminals in the olfactory tubercle. B′ illustrates in darkfield the large number of neurons in the lateral nucleus of the horizontal limb of the diagonal band of Brocca at the level of the decussation of the anterior commissure (see B above) containing lectin-conjugate following its retrograde transneuronal transport from the main olfactory bulb. This forebrain region contains cholinergic neurons and is the rat homologue of the primate nucleus basalis. Bars A′ = 100 μm; and B′ = 50 μm.

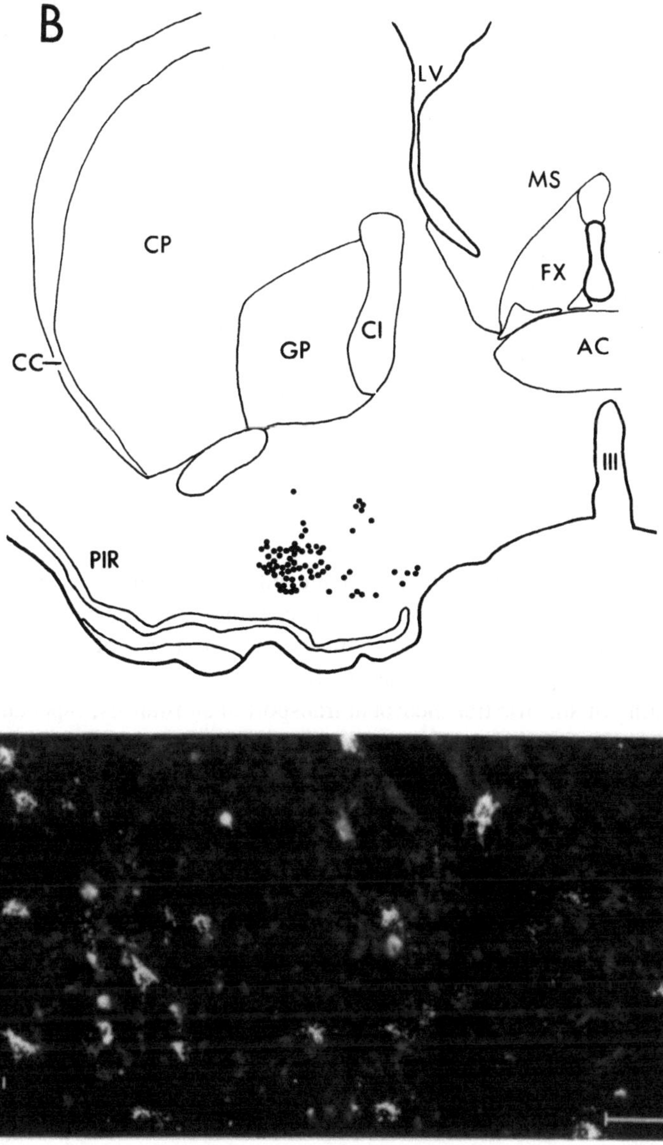

Figure 11. (continued)

reciprocal synapses with mitral cells? Or is the dopamine phenotype in the BALB/cJ mouse expressed in more neurons as opposed to a second transmitter? Many of these questions require the identification of another transmitter in the dopamine neuron and the demonstration that its expression is not altered by deafferentation, experiments currently in progress.

An understanding of the mechanisms controlling the expression of phenotype during development and regeneration, as well as its maintenance in the adult, is the focus of numerous investigations. The studies outlined above indicate that transneuronal events may be most important in both expression and maintenance of phenotype. Thus, the receptor cell interactions with the olfactory bulb can be considered an ideal model for examining trophism in the CNS. In fact, trophic factors, such as nerve growth factor, are found in very high concentrations in the adult olfactory system including the hippocampus, cortex, and olfactory bulb. In addition, the proto-oncogene, c-myc, a nuclear DNA-binding protein, correlated with cell proliferation, is found in high levels in adult mouse olfactory bulb but not in other brain regions in which it is expressed only during a specific period of ontogenesis (Whittemore *et al.*, 1986). These data suggest that adult olfactory bulb as well as the olfactory epithelium express characteristics of continuous development. Therefore, parallels between the ontogeny of TH and its expression during denervation and reinnervation should be sought. The role of nerve growth factor and/or other trophic factors in these phenomena must be investigated. Since it is known that environment plays a large role in phenotypic expression (Le Douarin, 1982) and specifically in TH expression (Teitelmen *et al.*, 1981), the olfactory dopamine system may be ideal for the determination of factors mediating transneuronal trophic interactions both during development and in response to lesion.

The molecular genetics of these trophic interactions also can be addressed using newly developed cDNA probes to the specific enzymes in the catecholamine biosynthetic pathway. One can determine whether changes in transmitter synthesis produced by deafferentation are accompanied by alterations in gene expression.

The finding of specific transneuronal transport of substances, especially lectin conjugates, suggests that trophic substances might be transferred between cells in a similar manner. In addition, studies of specific transport from the epithelium to the brain may be important in elucidating the mechanisms by which viruses or toxins can be introduced into the brain. The latter possibility takes on special import, since the regions labeled following intranasal application of WGA–HRP, for example, are implicated in the etiology of Alzheimer disease. Thus, the deafferentation-induced changes in olfactory bulb TH activity can serve as a model system for assessing both the toxicity of transneuronally transported substances as well as a means of understanding trophism in general in the CNS.

Acknowledgment. The research described in this chapter was supported in part by National Institutes of Health Grant No. NS-23103.

REFERENCES

Baker, H., 1986*a*, Species differences in the distribution of substance P and tyrosine hydroxylase immunoreactivity in the olfactory bulb, *J. Comp. Neurol.* **252:**206–226.

Baker, H., 1986*b*, Substance P and tyrosine hydroxylase are found in different neurons of the hamster olfactory bulb, *Exp. Brain Res.* **65:**245–249.

Baker, H., and Spencer, R. F., 1986, Transneuronal transport of peroxidase conjugated wheat germ agglutinin (WGA–HRP) from the olfactory epithelium to the brain of adult rat, *Exp. Brain Res.* **63:**461–473.

Baker, H., Joh, T. H., Reis, D. J., 1980, Genetic control of the number of midbrain dopaminergic neurons in inbred strains of mice: relationship to size and neuronal density of the striatum, *Proc. Natl. Acad. Sci. USA* **77**:4369–4373.

Baker, H., Joh, T. H., Ruggiero, D. A., and Reis, D. J., 1983*a*, Variations in the number of dopamine neurons and tyrosine hydroxylase activity in hypothalamus of two mouse strains, *J. Neurosci.* **3**:832–843.

Baker, H., Kawano, T., Margolis, F. L., and Joh, T. H., 1983*b*, Transneuronal regulation of tyrosine hydroxylase expression in olfactory bulb of mouse and rat, *J. Neurosci.* **3**:69–78.

Baker, H., Kawano, T., Albert, V., Joh, T. H., Reis, D. J., and Margolis, F. L., 1984, Olfactory bulb dopamine neurons survive deafferentation induced loss of tyrosine hydroxylase, *Neuroscience* **11**:605–615.

Baker, H., Ruggiero, D. A., Alden, S., Anwar, M., and Reis, D. J., 1986, Anatomical evidence for interactions between catecholamine and ACTH containing neurons, *Neuroscience* **17**:469–484.

Benson, T. E., Ryugo, D. K., and Hinds, J. W., 1984, Effects of sensory deprivation on the developing mouse olfactory system: A light and electron microscopy, morphometric analysis, *J. Neurosci.* **4**:638–653.

Black, I. B., and Geen, S. C., 1975, Inhibition of the biochemical and morphological maturation of adrenergic neurons by nicotinic receptor blockade, *J. Neurochem.* **22**:301–306.

Black, I. B., Hendry, I. A., and Iversen, L. L., 1971, Transsynaptic regulation of growth and development of adrenergic neurons in a mouse sympathetic ganglion, *Brain Res.* **34**:229–240.

Black, I. B., Chikaraishi, D. M., and Lewis, E. J., 1985, Trans-synaptic increase in RNA coding for tyrosine hydroxylase in a rat sympathetic ganglion, *Brain Res.* **339**:151–153.

Bogan, N., Brecha, N., Gall, C., and Karten, H. J., 1982, Distribution of enkephaline-like immunoreactivity in the rat main olfactory bulb, *Neuroscience* **7**:895–906.

Bondareff, W., Mountjoy, C. A., and Roth, M., 1981, Selective loss of neurons of origin of adrenergic projections to cerebral cortex (nucleus locus coeruleus) in senile dementia, *Lancet* **1**:783–784.

Bowen, D. M., Allen, S. J., Benton, J. S., Goodhardt, M. J., Haan, E. A., Palmer, A. M., Sims, N. R., Smith, C. C. J., Spillane, J. A., Esiri, M. M., Neary, D., Snowdon, J. S., Wilcock, G. K., and Davison, A. N., 1983, Biochemical assessment of serotonergic and cholinergic dysfunction and cerebral atrophy in Alzheimer's disease, *J. Neurochem.* **41**:266–272.

Broadwell, R. D., 1975, Olfactory relationships of the telencephalon and diencephalon in the rabbit. I. An autoradiographic study of the efferent connections of the main and accessory olfactory bulb, *J. Comp. Neurol.* **163**:329–346.

Broadwell, R. D., and Balin, B. J., 1985, Endocytic and exocytic pathways of the neuronal secretory process and transsynaptic transfer of wheat germ agglutinin–horseradish peroxidase in vivo, *J. Comp. Neurol.* **242**:632–650.

Brunjes, P. C., Smith-Crafts, L. K., and McCarty, R., 1985, Unilateral odor deprivation: Effects on development of olfactory bulb catecholamines and behavior, *Dev. Brain Res.* **22**:1–6.

Burd, G. D., Davis, B. J., Macrides, F., Grillo, M., and Margolis, F. L., 1982, Carnosine in primary afferents of the olfactory system: An autoradiographic study, *J. Neurosci.* **2**:244–255.

Chuah, M. I., and Farbman, A. I., 1984, Olfactory bulb increases marker protein in olfactory receptor cells, *J. Neurosci.* **3**:2197–2205.

Corwin, J., Serby, M., Conrad, P., and Rotrasen, J., 1985, Olfactory recognition deficit in Alzheimer's and Parkinsonian dementias, *IRCS Med. Sci.* **13**:260.

Costanza, R. M., and Graziadei, P. P. C., 1983, A quantitative analysis of changes in the olfactory epithelium following bulbectomy in hamster, *J. Comp. Neurol.* **215**:370–381.

Coyle, J. T., Price, D. L., DeLong, M. R., 1983, Alzheimer's disease: A disorder of cortical cholinergic innervation, *Science* **219**:1184–1190.

Davis, B. J., Burd, G. D., and Macrides, F., 1982, Localization of met-enkephalin, substance P and somatostatin immunoreactivities in the main olfactory bulb of the hamster, *J. Comp. Neurol.* **204**:377–383.

Davis, B. J., and Macrides, F., 1981, The organization of centrifugal projections from the anterior olfactory nucleus, ventral hippocampal rudiment and piriform cortex to the main olfactory bulb in the hamster: an autoradiographic study, *J. Comp. Neurol.* **203**:475–493.

de Lorenzo, A. J. D., 1970, The olfactory neuron and the blood brain barrier, in: *Taste and Smell in Vertebrates* (G. E. W. Wolstenholme and J. Knight, eds.), pp. 151–175, Churchill Livingstone, London.

de Olmos, J. Hardy, H., and Heimer, L., 1978, The afferent connections of the main and accessory olfactory bulb formation of the rat. An experimental HRP study, *J. Comp. Neurol.* **181**:213–244.

Eseri, M. M., and Tomlinson, A. H., 1984, Herpes simplex encephalitis, *J. Neurol. Sci.* **64**:213–217.

Eseri, M. M., and Wilcock, G. K., 1984, The olfactory bulbs in Alzheimer's disease, *J. Neurol. Neurosurg. Psychiatry* **47**:56–60.

Farbman, A. I., and Margolis, F. L., 1980, Olfactory marker protein during ontogeny: Immunohistochemical localization, *Dev. Biol.* **74**:207–215.

Ffrench-Mullen, J. M. H., Koller, K., Zaczek, R., Coyle, J. T., Hori, N., and Carpenter, D. O., 1985, N-Acetylaspartylglutamate: Possible role as the neurotransmitter of the lateral olfactory tract, *Proc. Natl. Acad. Sci. USA* **82**:3897–3900.

Fibiger, H. C., 1982, The organization and some projections of cholinergic neurons of the mammalian forebrain, *Brain Res. Rev.* **4**:327–388.

Gall, C. M., Hendry, S. H. C., Seroogy, K. B., and Jones, E. G., 1985, Co-localization of GABA- and tyrosine hydroxylase-like immunoreactivities in neurons of the rat olfactory bulb, *Soc. Neurosci. Abst.* **11**:89.

Gottfries, C. G., 1982, The metabolism of some neurotransmitters in ageing and dementia disorders, *Gerontology* **28**:11–19.

Graziadei, P. P. C., and Kaplan, M. S., 1980, Regrowth of olfactory sensory axons into transplanted neural tissue. I. Development of connections with occipital cortex, *Brain Res.* **201**:39–44.

Graziadei, P. P. C., and Monti-Graziadei, G. A., 1978, The olfactory system: A model for the study of neurogenesis and axon regeneration in mammals, in: *Neuronal Plasticity* (C. W. Cotman, ed.), pp. 131–153, Raven, New York.

Graziadei, P. P. C., and Monti-Graziadei, G. A., 1980, Neurogenesis and neuron regeneration in the olfactory system of mammals. III. Deafferentation and reinnervation of the olfactory bulb following section of the fila olfactoria in rat, *J. Neurocytol.* **9**:145–162.

Graziadei, P. P. C., and Samanen, D. W., 1980, Ectopic glomerular structures in the olfactory bulb of neonatal and adult mice, *Brain Res.* **187**:467–472.

Halasz, N., and Shepherd, G. M., 1983, Neurochemistry of the vertebrate olfactory bulb, *Neuroscience* **10**:579–619.

Harding, J. P., Graziadei, P. P. C., Monti-Graziadei, G. A., and Margolis, F. L., 1977, Denervation in the primary olfactory pathway of mice. IV. Biochemical and morphological evidence for neuronal replacement following nerve section, *Brain Res.* **132**:11–32.

Holl, A., 1980, Selective staining by Procion dyes of olfactory sensory neurons in the catfish Ictalurus nebulosus, *Z. Naturforsch.* **35**:526–528.

Hokfelt, R., Rehfeld, J. F., Skirboll, L., Ivemark, B., Goldstein, M., and Markey, K., 1980, Evidence for coexistence of dopamine and CCK in mesolimbic neurons, *Nature (Lond.)* **285**:476–477.

Itaya, S. K., and Van Hoesen, G. W., 1982, WGA HRP as a transneuronal marker in the visual pathways of monkey and rat, *Brain Res.* **236**:199–204.

Iversen, L. L., Rossor, M. N., Reynolds, G. P., Hills, R., Roth, M., Mountjoy, C. Q., Foote, S. L., Morrison, J. H., and Bloom, F. E., 1983, Loss of pigmented dopamine B-hydroxylase positive cells from locus coeruleus in senile dementia of Alzheimer's type, *Neurosci. Lett.* **39**:95–100.

Jackson, R. T., Tigges, J., and Arnold, W., 1979, Subarachnoid space of the CNS, nasal mucosa, and lymphatic system, *Arch. Otolaryngol.* **105**:180–184.

Hinds, J. W., 1978, Autoradiographic study of histogenesis in the mouse olfactory bulb. I. Time of origin of neurons and neuroglia, *J. Comp. Neurol.* **134**:287–304.

Kawano, T., and Margolis, F. L., 1982, Transsynaptic regulation of olfactory bulb catecholamines in mice and rats, *Neurochemistry* **39**:342–348.

Kosaka, T., Hataguchi, Y., Hama, K., Nagatsu, I., and Wu, J.-Y., 1985, Coexistence of immunoreactivities for glutamate decarboxylase and tyrosine hydroxylase in some neurons in the periglomerular region of the rat main olfactory bulb: possible coexistence of gamma-aminobutyric acid (GABA) and dopamine, *Brain Res.* **343**:166–171.

Kream, R. M., Davis, B. J., Kawano, T., Margolis, F. L., and Macrides, F., 1984, Substance P and catecholaminergic expression in neurons of the hamster main olfactory bulb, *J. Comp. Neurol.* **222**:140–154.

Kream, R. M., Schoenfeld, T. A., Mancuso, R., Clancy, A. N., El-Bermani, W., and Macrides, F., 1985, Precursor forms of substance P (SP) in nervous tissue: Detection with antisera to Sp, Sp-Gly, and Sp-Gly-Lys, *Proc. Natl. Acad. Sci. USA* **82**:4832–4836.

Kristensson, K., and Olsson, Y., 1971, Uptake of exogenous proteins in mouse olfactory cells, *Acta Neuropathol. (Berl.)* **19:**145–154.

Le Douarin, N. M., 1982, *The Neural Crest,* Cambridge University Press, London.

Mair, R. G., and Gesteland, R. C., 1982, Response properties of mitral cells in the olfactory bulb of the neonatal rat, *Neuroscience* **7:**3117–3125.

Margolis, F. L., 1980*a,* Carnosine: An olfactory neuropeptide, in: *Role of Peptides in Neuronal Function* (J. L. Baker and T. G. Smith, Jr., eds.) pp. 59–85, Dekker, New York.

Margolis, F. L., 1980*b,* A marker protein for the olfactory chemoreceptor neuron, in: *Proteins of the Nervous System* (R. A. Bradshaw and D. Schneider, eds.), pp. 59–84, Raven, New York.

Meisami, E., 1976, Effects of olfactory deprivation on postnatal growth of the rat olfactory bulb utilizing a new method for production of neonatal unilateral anosmia, *Brain Res.* **107:**437–444.

Meisami, E., and Satai, L., 1981, A quantitative study of the effects of early unilateral olfactory deprivation on the number and distribution of mitral and tufted cells and of glomeruli in the rat olfactory bulb, *Brain Res.* **221:**81–107.

Mesulam, M.-M., Mufson, E. J., Wainer, B. H., and Levey, A. I., 1983, Central cholinergic pathways in the rat: An overview based on an alternative nomenclature (CH1–CH6), *Neuroscience* **10:**1185–1201.

Monath, T. P., Cropp, C. B., and Harrison, A. K., 1983, Mode of entry of a neurotropic arbovirus into the central nervous system, *Lab. Invest.* **48:**399–410.

Mugnaini, E., Oertel, W. H., and Wouterlood, F. F., 1984, Immunocytochemical localization of GABA neurons and dopamine neurons in the rat main and accessory olfactory bulbs, *Neurosci. Lett.* **47:**221–226.

Mugnaini, E., Wouterlood, F. G., Dahl, A.-L., and Oertel, W. H., 1984, Immunocytochemical identification of gabaergic neurons in the main olfactory bulb of the rat, *Arch. Ital. Biol.* **122:**83–113.

Nadi, N. S., Head, R., Grillo, M., Hempstead, J., Granno-Reisfeld, N., and Margolis, F. L., 1981, Chemical deafferentation of the olfactory bulb: Plasticity of the levels of tyrosine hydroxylase, dopamine and norepinephrine, *Brain Res.* **213:**365–371.

Pinching, A. J., and Powell, T. P. S., 1971, Ultrastructural features of transneuronal cell degeneration in the olfactory system, *J. Cell Sci.* **8:**253–287.

Reis, D. J., Joh, T. H., and Ross, R. A., 1975, Effects of reserpine on activities and amounts of tyrosine hydroxylase and dopamine B-hydroxylase in catecholamine neuronal systems in rat brain, *J. Pharmacol. Exp. Ther.* **193:**775–784.

Rosser, M. N., Svendsen, C., Hunt, S. P., Mountjoy, C. Q., Roth, M., and Iversen, L. L., 1982, The substantia innominata in Alzheimer's disease; an histochemical and biochemical study of cholinergic marker enzymes, *Neurosci. Lett.* **28:**217–222.

Samanen, D. W., and Forbes, W. B., 1984, Replication and differentiation of olfactory receptor neurons following axotomy in the adult hamster: A morphometric analysis of postnatal neurogenesis, *J. Comp. Neurol.* **225:**201–211.

Ruda, M., and Coulter, J. D., 1982, Axonal and transneuronal transport of wheat germ agglutinin demonstrated by immunocytochemistry, *Brain Res.* **249:**237–246.

Shipley, M. T., 1985, Transport of molecules from nose to brain: Transneuronal anterograde and retrograde labeling in the rat olfactory system by wheat germ agglutinin–horseradish peroxidase applied to the nasal epithelium, *Brain Res. Bull.* **15:**129–142.

Shipley, M. T., and Adamek, G. D., 1984, The connections of the mouse olfactory bulb: A study using orthograde and retrograde transport of wheat germ agglutinin conjugated to horseradish peroxidase, *Brain Res. Bull.* **12:**669–688.

Specht, L. A., Pickel, V. M., Joh, T. H., and Reis, D. J., 1981, Light-microscopic immunocytochemical localization of tyrosine hydroxylase in prenatal rat brain. I. Early Ontogeny, *J. Comp. Neurol.* **199:**233–253.

Spencer, R. F., Baker, H., and Baker, R., 1982, Evaluation of wheat germ agglutinin immunohistochemistry as a neuroanatomical method for retrograde, anterograde and anterograde transsynaptic labeling in the cat visual occulomotor systems, *Soc. Neurosic. Abst.* **8:**785.

Sternberger, L. A., Hardy, P. H., Jr., Cuculis, J. J., and Meyer, H. B., 1970, The unlabeled antibody enzyme method of immunocytochemistry. Preparation and properties of soluble antigen-antibody complex (horseradish peroxidase–antihorseradish peroxidase) and its use in identification of spirochetes, *J. Histochem. Cytochem.* **18:**315–333.

Stevens, J. C., Bartoshuk, L. M., and Cain, W. S., 1984, Chemical senses and aging: Taste versus smell, *Chem. Senses* **9**:167–179.

Stout, R. P., and Graziadei, P. P. C., 1980, Influence of the olfactory placode on the development of the brain in Xenopus laevis (Daudin). I. Axonal growth and connections of the transplanted olfactory placode, *Neuroscience* **5**:2175–2186.

Stroop, W. G., Rock, D. L., and Fraser, N. W., 1984, Localization of herpes simplex virus in the trigeminal and olfactory systems of the mouse central nervous system during acute and latent infections by in situ hybridization, *Lab. Invest.* **51**:27–38.

Teitelman, G., Joh, T. H., and Reis, D. J., 1981, Transformation of catecholaminergic precursors into glucagon (A) cells in the mouse embryonic pancreas, *Proc. Natl. Acad. Sci. USA* **78**:5225–5229.

Thoenen, H., 1974, Trans-synaptic enzyme induction, *Life Sci.* **14**:223–235.

Tomlinson, B. E., Irving, D., Blessed, G., 1981, Cell loss in the locus coeruleus in senile dementia of Alzheimer type, *J. Neurol Sci.* **49**:419–428.

Tramu, G., Pillez, A., and Leonardelli, J., 1978, An effective method of antibody elution for the successive or simultaneous localization of two antigens by immunocytochemistry, *J. Histochem. Cytochem.* **26**:322–324.

Trojanowski, J. Q., 1983, Native and derivatized lectins for in vivo studies of neuronal connectivity and neuronal cell biology, *J. Neurosci. Methods* **9**:185–204.

Weiss, P., and Holland, Y., 1967, Neuronal dynamics and axonal flow in the olfactory nerve as model test object, *Proc. Natl. Acad. Sci. USA* **57**:258–264.

White, E. L., 1973, Synaptic organization of the mammalian olfactory glomerulus: New findings including an intraspecific variation, *Brain Res.* **60**:299–313.

Whittemore, S. R., Evendal, T., Larkfors, L., Olson, L., Seiger, A., Stromberg, I., and Persson, H., 1986, Developmental and regional expression of B-nerve growth factor messenger RNA and protein in the rat central nervous system, *Proc. Natl. Acad. Sci. USA* **83**:817–821.

Yates, C. M., Simpson, J., Gordon, A., Maloney, A. F. J., Allison, Y., Ritchie, I. M., and Urquhart, A., 1983, Catecholamines and cholinergic enzymes in pre-senile and senile Alzheimer-type dementia and Down's syndrome, *Brain Res.* **280**:119–126.

10

Axoplasmic Transport in Olfactory Receptor Neurons

Dieter G. Weiss and Klaus Buchner

1. INTRODUCTION

The function of a cell can only be fully understood through the knowledge of the internal machinery acting both in preserving the cell's shape and biochemical integrity and in maintaining the specific requirements of any highly specialized cell. Because nerve cells are made to convey information over long distances along their axons, knowledge of the internal mechanisms necessary to extend these long processes, to nourish them, and to establish regional differences in both macromolecular composition and metabolism is essential for the understanding of the function of the neuron in the physiological context. With this in mind, we believe that axoplasmic transport is one important facet to be studied for a thorough understanding of a specialized cell, such as the olfactory receptor neuron.

2. THE OLFACTORY RECEPTOR NEURON AS A SPECIALIZED NEURONAL SYSTEM

The location of the olfactory nerve in the animal and its morphology makes it exceptionally suitable for the study of axoplasmic transport. Its use was suggested about 20 years ago by Paul Weiss (P. Weiss and Holland, 1967); in the years to follow, several groups have indeed adopted this system for their studies.

The olfactory receptor neurons have relatively short axons in most mammals, but in many species of fish and amphibians they are of considerable length (Gasser, 1956; P. Weiss and Holland, 1967; Easton, 1971; Hara, 1975; Muralt *et al.*, 1976; Kreutzberg and Gross, 1977). Thus, most of the work reported here was performed using either the garfish, pike, or frog olfactory nerves. These systems offer the following advantages (see also Easton, 1971; Kreutzberg and Gross, 1977):

Dieter G. Weiss and Klaus Buchner • Institut für Zoologie, Technische Universität München, D-8046 Garching, Federal Republic of Germany.

1. The nerves are relatively long (4–5 cm in the pike, up to 30 cm in the garfish) and yield enough material to permit biochemical work.
2. The nerves are not branched but have a circumscribed region of origin containing the perikarya (olfactory mucosa), a well-defined axonal region (olfactory nerve) and a small region containing all synaptic terminals of these axons (olfactory bulb) (Fig. 1).
3. Marker substances and precursors can be applied to the nasal cavity without requiring surgery (Weiss and Holland, 1967; Gross and Beidler, 1973).
4. Quantitive dissection of all three neuronal regions (perikaryal, axonal, and synaptic) is possible, providing access to balanced biochemical analyses of synthesis, transport, and turnover.
5. Contrary to other vertebrate nerves, in the olfactory nerve, all axons belong to only one class of, very thin, axons (0.2–0.3 μm in diameter; C fibers) all of which have the same polarity.
6. The nerve consists of an extremely homogeneous population of several million unmyelinated axons, whereas only a minor fraction of the nerve volume is composed of glial cells and fibrocytes (Easton, 1971; Kreutzberg and Gross, 1977) (Fig. 2).
7. The ultrastructure of these axons is well characterized, having been the subject of several quantitative studies showing all the organelles, cytoskeletal elements, and their distributions in the axon (Easton, 1971; Kreutzberg and Gross, 1977; Burton and Paige, 1981; Burton, 1985; Burton and Laveri, 1985; Buchner *et al.*, in preparation).
8. Biochemical analysis of the axonal proteins shows a relatively high proportion of cytoskeletal proteins (Fig. 3). Therefore, this system is also well suited for the study of the transport of cytoskeletal elements (slow transport), as well as its role in the mechanism of rapid transport and in axonal growth and regeneration.

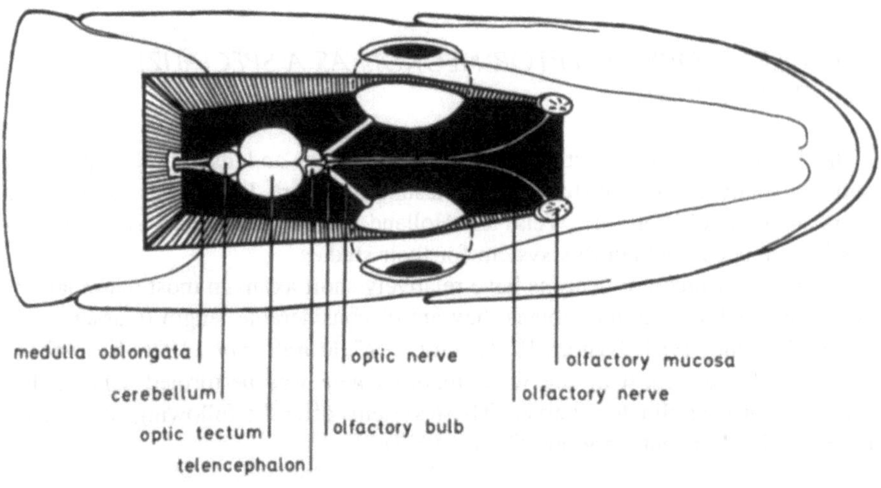

Figure 1. Schematic drawing of the pike brain showing the olfactory system.

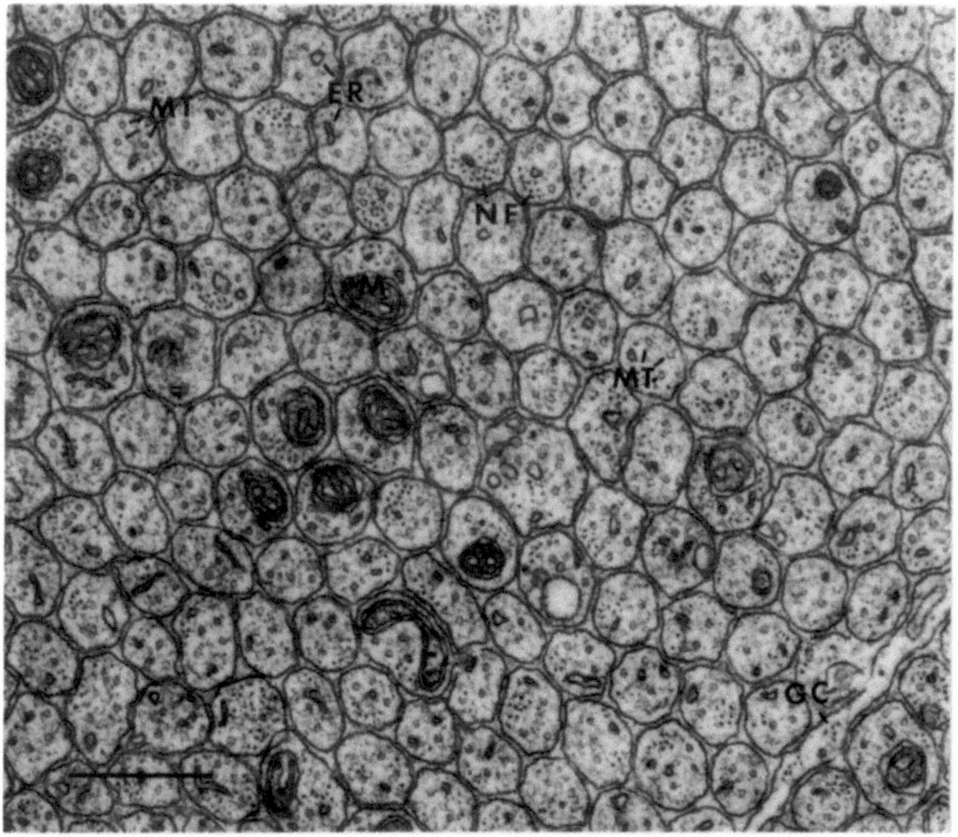

Figure 2. Detail from a cross section of pike olfactory nerve. The axons contain 3 ± 1 microtubules (MT), a variable number of neurofilaments (NF), which most often appear in clusters, and profiles of smooth endoplasmic reticulum (ER). In about 10% of fibers, mitochondria (M) are discernable, which usually cause the axons to bulge. GC, process of a glial cell. Bar: 0.5 μm. (This electron micrograph was kindly provided by D. Seitz-Tutter.)

9. The primary olfactory neurons undergo a continuous turnover (Graziadei and Monti Graziadei, 1978). They degenerate and are replaced by undifferentiated cells, which develop into receptor neurons and grow new axons to the bulb. This makes these cells exceptionally well suited for the study of degeneration, neurogenesis and regeneration (Graziadei and Monti Graziadei, 1978; Cancalon, 1985; Costanzo, 1985), as well as the role of axoplasmic transport in these states (Elam and Cancalon, 1984; Cancalon, 1985).

Taken together, the olfactory nerve is probably the simplest, most homogeneous and most easily accessible nervous tissue available for biochemical and axoplasmic transport studies. This turns out to be most advantageous for the understanding of both the olfactory receptor neuron and axoplasmic transport.

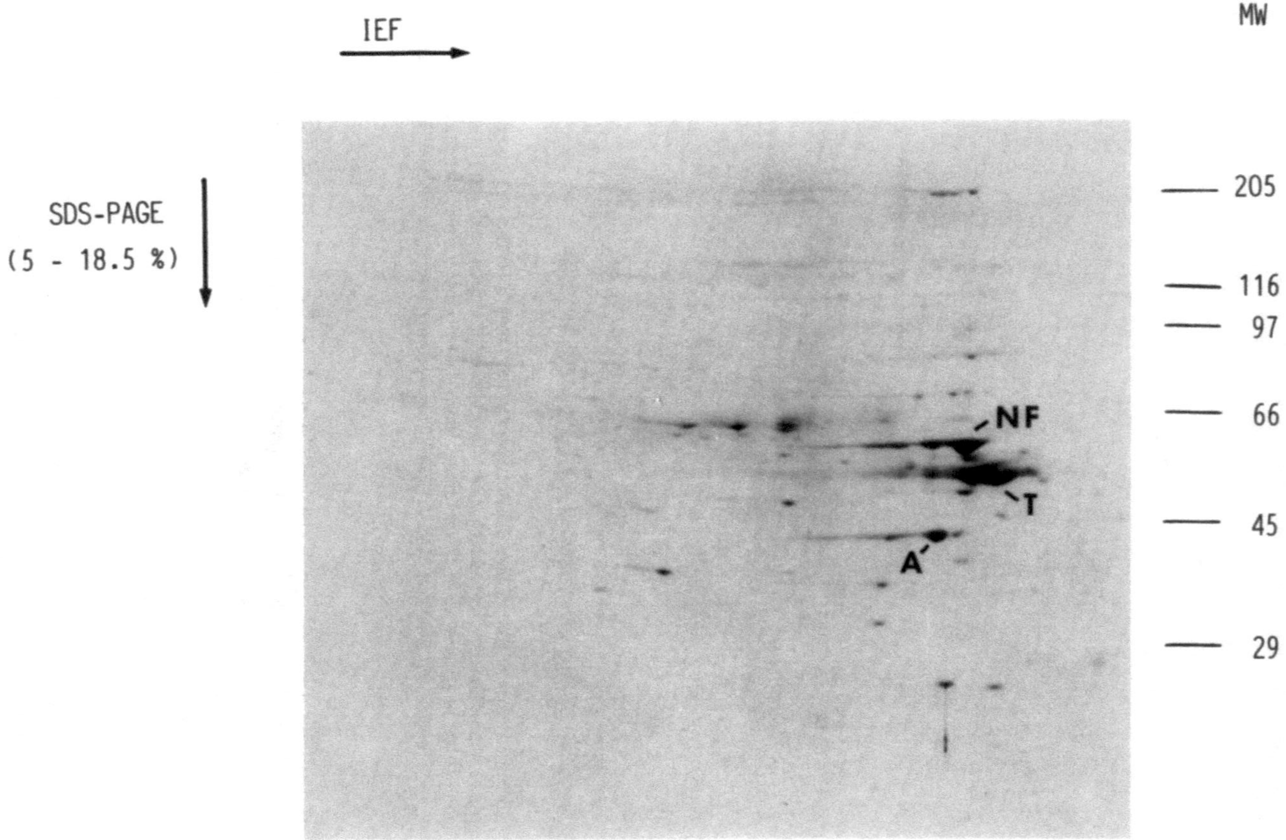

Figure 3. Pattern of the proteins of the pike olfactory nerve after two-dimensional gel electrophoresis according to O'Farrell (1975). The gel was stained with Coomassie brilliant blue. Cytoskeletal proteins are dominating the pattern. NF, neurofilament protein; T, tubulins; A, actin. Range of pI, 5.5–7.0; acidic to the right.

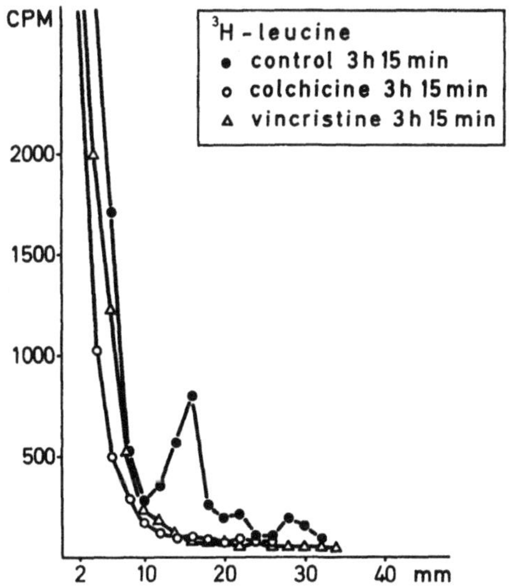

Figure 4. Radioactivity distribution profile along the nerve from an untreated animal (●) compared with profiles from animals treated with colchicine (○) and vincristine (△) 3 hr, 15 min after isotope application. The leading peak as well as part of the radioactivity of the main front fail to appear after treatment with the microtubule inhibitors. Radioactivity of the contralateral nerve (blood-borne radioactivity) is subtracted. The nerves were fixed with glutaraldehyde, which retains the radioactivity of both proteins, and free amino acids. (From Schmid *et al.*, 1983.)

3. TRANSPORT OF LOW-MOLECULAR-WEIGHT MATERIAL

Several studies have shown that a variety of low-molecular-weight compounds, such as leucine, proline, β-alanine, and 3-0-methyl-D-glucose, are transported anterogradely in the fish olfactory nerve (Cancalon and Beidler, 1975; Gross and Kreutzberg, 1978; Weiss, 1982; Schmid *et al.*, 1983). The transport of these compounds and that of proteins is similar, as both depend on oxidative phosphorylation (Ochs, 1971; Weiss, 1982; Schmid *et al.*, 1983), both classes of molecules are transported in the form of radioactivity peaks (Gross and Beidler, 1975; Gross and Kreutzberg, 1978; Weiss, 1982; Schmid *et al.*, 1983), and both can be inhibited by microtubule inhibitors (Price *et al.*, 1979; Schmid *et al.*, 1983) (Fig. 4). These substances are transported rapidly (Di Giamberardino, 1971; Gross and Kreutzberg, 1978; Weiss, 1982; Schmid *et al.*, 1983), and no somal delay time as seen with protein transport (Gross and Beidler, 1975) can be observed for these substances (Gross and Kreutzberg, 1978; Weiss, 1982; Schmid *et al.*, 1983). It has been shown that the maximal velocity is identical with that of protein transport (Gross and Kreutzberg, 1978; Weiss, 1982; Schmid *et al.*, 1983) (Fig. 5). Proteolysis is excluded as a source of the rapidly transported amino acids (Di Giamberardino, 1971; Gross and Kreutzberg, 1978; Schmid *et al.*, 1983).

It was also reported that the amount of low-molecular-weight material decreases during transport (Csanyi *et al.*, 1973; Weiss, 1982; Schmid *et al.*, 1983). In protein transport, a similar linear reduction (2–3% per cm) of the rapidly transported material has been shown (Gross and Beidler, 1975; Muñoz-Martínez *et al.*, 1981), but for amino acids the losses from the rapidly moving compartment are much more dramatic (Weiss, 1982; Schmid *et al.*, 1983). In the case of leucine transport, it is known that the rapidly moving peak loses 20% of its contents per hour (Schmid *et al.*, 1983). This leads to a situation in which a rapidly moving radioactive peak should no longer be detectable after 5 hr or a transport distance of 4–5 cm (19°C). Schmid *et al.* (1983) concluded that this may explain

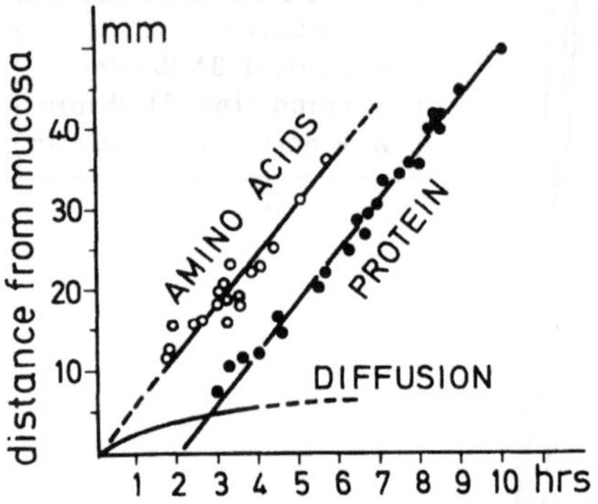

Figure 5. The wavefront loci of profiles such as those shown in Fig. 4. are plotted for amino acids and proteins. The slope representing the velocity remains unchanged with time and is identical for both substances. The velocity of amino acid diffusion was calculated and can be seen to be much slower than the transport observed, and to decrease with time. (From Weiss *et al.*, 1980.)

the frequent failures to detect transport of low-molecular-weight material to the terminals in other nerves (McEwen and Grafstein, 1968; Elam and Agranoff, 1971; Neale *et al.*, 1974; Karlsson, 1977). It is unknown whether the rapidly moving fraction of leucine is lost from the axon and perhaps even from the nerve to the circulation or degraded locally. A smaller amount of the radioactivity is deposited along the nerve and may have been fixed by glial cell and fibrocyte protein synthesis.

It was recently shown that diffusion cannot account for this movement (Schmid *et al.*, 1983). In this study, diffusion was calculated to be much slower than the observed displacement of radioactive amino acids. Furthermore, the velocity of diffusion should decrease with time, while the observed velocity remains constant. Also, microtubule inhibitors did block this movement and the same is true if the oxygen supply is interrupted. Finally, the finding that transport of various small molecules occurs in the form of moving peaks (Gross and Kreutzberg, 1978; Weiss, 1982; Schmid *et al.*, 1983) alone would exclude diffusion.

The state of this low-molecular-weight material during transport is unknown. It could move freely in solution, associated with macromolecules, or stored in organelles. Transport by carrier proteins appears unlikely, however, since free proteins are in general regarded to be transported only slowly (Grafstein and Forman, 1980; Baitinger *et al.*, 1982). Furthermore, the amount of all the soluble protein in axons would not be sufficient to carry the amount of small molecules transported (Weiss, 1982; Schmid *et al.*, 1983).

Transport of low-molecular-weight material stored in vesicles would be compatible with most proposed models for axoplasmic transport. This is not easily reconciled, however, with the fact that all low-molecular-weight substances studied so far have been shown to exhibit the same behavior. These substances include not only leucine and proline, but also exogenous substances that are not readily incorporated into macromolecules, such as β-alanine and 3-*O*-methyl-D-glucose (Gross and Kreutzberg, 1978; Weiss, 1982; Schmid *et al.*, 1983). The existence of mechanisms for the uptake of all these substances into vesicles seems rather unlikely. Transport in free solution is likewise unproved, but it has been shown that a considerable amount of low-molecular-weight compounds are present in free form in axons (Koike and Nagata, 1979).

Following injection of ^{3}H-labeled leucine into the nerve proper, bidirectional transport of free leucine can be observed (Fig. 6). The velocities (V_{max}) were similar for both directions (36 mm/day anterogradely and 46 mm/day, retrogradely) (Buchner, 1986), but

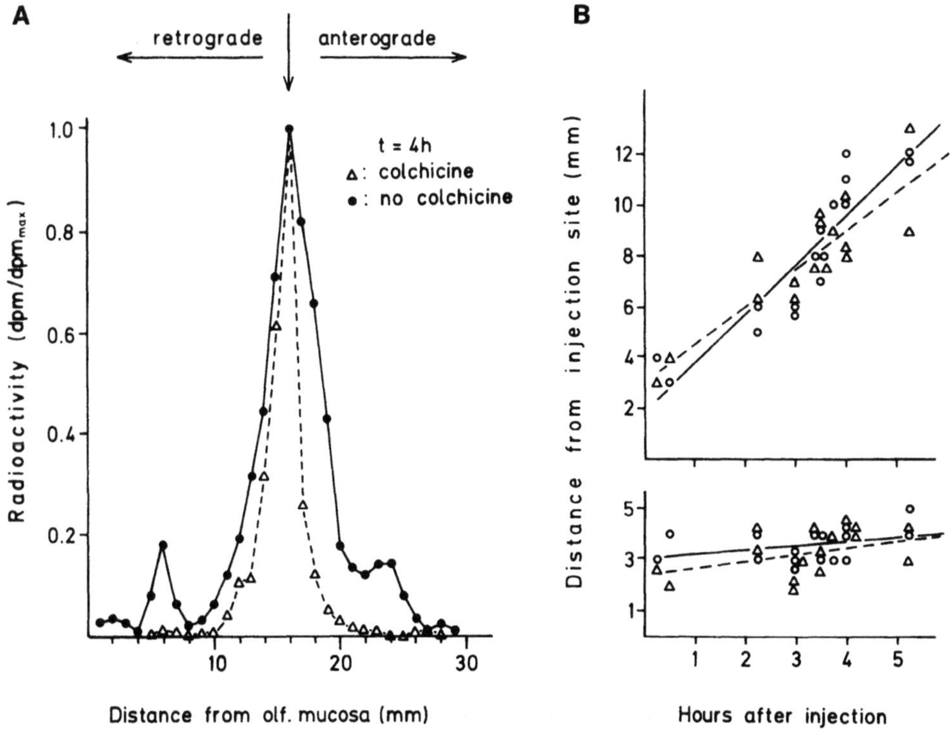

Figure 6. Anterograde and retrograde transport of free [^{3}H]leucine after injection into the pike olfactory nerve. (A) Distribution of radioactivity 4 hr after injection shows a rapidly moving peak in the anterograde as well as in the retrograde direction (-●——●-). This bidirectional transport was inhibited by colchicine; 4 hr after injection, the radioactivity in nerves treated with colchicine (-△- -△-) is confined to a narrow region around the injection site. To permit comparison of the profiles, the values are depicted as dpm versus maximal dpm (injection site). (B) (Top) Spread of the front of the rapidly moving peak (V_{max}). The regression lines are calculated for both the anterograde (- -△- -) and the retrograde directions (—○—). The velocities (V_{max}) derived from the slopes of the regression lines are 36 mm/day for the anterograde and 46 mm/day for the retrograde direction. These values are statistically not significantly different. (Bottom) Spread of the front of the main radioactivity pool around the injection site (main front). (- -△- -) anterograde direction. (—○—) retrograde direction. These velocities are slow for both directions and are not significantly different from zero.

considerably slower than the anterograde transport of free leucine taken up into the perikarya in the olfactory mucosa (cf. Table 1) (Gross and Kreutzberg, 1978; Schmid *et al.*, 1983). This difference in transport velocity is not understood. Since labeled inulin, a marker of the extracellular space, is not transported and colchicine inhibits free amino acid transport, it was concluded to be an intracellular phenomenon. Also retrograde transport of free leucine was found to exist and to be inhibited by colchicine (Fig. 6).

In fish and amphibian olfactory receptor neurons, the transmitter substance is not known and the axoplasmic transport studies using β-alanine as precursor of carnosine (β-alanyl-L-histidine) do not support the view that this may be the synaptic transmitter (Weiss, 1982). By contrast, mammalian systems studied show an enrichment of both carnosine and carnosine synthetase in the olfactory neurons, suggesting that carnosine

Table 1. Velocities of Axonal Transport in Pike Olfactory Nerve[a]

Substance	Anterograde (mm/day) Perikaryal region[h] (olf. mucosa)	Anterograde (mm/day) Nerve[h]	Retrograde (mm/day) Nerve[h]	Retrograde (mm/day) Synaptic region[h] (olf. bulb)
	Tracer experiments (population velocities, V_{max})			
[^{3}H]-WGA[b]	No transport	No transport	No transport	26
HRP[b]	25	No transport	20–30	25
[^{3}H]Leucine[c]	150	36	46	n.d.
[^{3}H]-β-Alanine[d]	150	n.d.	n.d.	n.d.
3-O-[^{3}H]-Methyl-D-glucose[d]	150	n.d.	n.d.	n.d.
Proteins labeled with [^{3}H]leucine[e]	150	No transport	No transport	n.d.
N-Succinimidyl[^{3}H]propionate[f]	? (Trace amounts)	No transport	No transport	27
	Organelles (velocity of individual organelles, mean values)			
Kind of organelles				
Lysosome-like[g]	1.45 ± 0.5 μm/sec (125 mm/day)		1.22 ± 0.5 μm/sec (105 mm/day)	
Mitochondria[g]	0.65 ± 0.28 μm/sec (56 mm/day)		1.15 ± 0.39 μm/sec (100 mm/day)	

[a]n.d., not determined.
[b]Buchner *et al.* (1987).
[c]Schmid *et al.* (1983).
[d]Weiss (1982).
[e]Gross and Kreutzberg (1978).
[f]Buchner (1986).
[g]Buchner *et al.* (in preparation).
[h]Site of marker or precursor application.

may be the olfactory transmitter in mammals (see Margolis, 1980, for review). This view is supported by the findings of Margolis and Grillo (1977) showing that in the mouse olfactory nerve β-alanine is not transported, but rather incorporated into carnosine, which is then axonally transported and accumulated in the synapse region, i.e., the olfactory bulb. This corresponds closely to what one would expect for the transmitter substance of the olfactory receptor neurons.

Further experiments are necessary to clarify the question of the physiological relevance of the rapid axoplasmic transport of low-molecular-weight substances. It may be a potent means to supply the synapses of specific neurons with amino acid transmitters and to supply the axon rapidly with small molecules necessary for metabolism or protein modification (e.g., Ludueña, 1979; Chakraborty *et al.*, 1986) and export to surrounding cells (e.g., Giorgi *et al.*, 1973; Giorgi, 1978; Schubert and Kreutzberg, 1982).

4. TRANSPORT OF BULK PROTEINS

4.1. Rapid Transport

Owing to its exceptional length and homogeneity, the garfish olfactory nerve was used in a number of studies of exceptional quality and detail. The exact velocity parameters, temperature dependence, and wave characteristics were determined for rapidly transported bulk proteins by Gross and Beidler (1973, 1975), and for slow transport by Cancalon (1979*a*). Gross found a maximum velocity for rapid transport of 160 mm/day at 19°C. This is almost exactly the same velocity as that found for the pike olfactory system (151 ± 5 mm/day) (Gross and Kreutzberg, 1978).

A review of the literature demonstrates that the transport V_{max} show a remarkable similarity in a variety of other nerves from phylogenetically different animals (Ochs, 1972; Edström and Hanson, 1973). Furthermore, the V_{max} in olfactory C fibers of fish has been shown to follow the following equation:

$$V\ (\text{cm/hr}) = 0.055T - 0.345$$

which predicts a V_{max} of 406 mm/day when 37°C is substituted for T into the equation (Gross and Beidler, 1975). This is identical to the V_{max} of 410 ± 30 mm/day reported by Ochs for sciatic nerves of several mammals (Ochs, 1972). It is therefore justified to suggest that if the temperature is controlled, and the physiological state of the nerve is not altered, a characteristic temperature-correlatable V_{max} may be found in all these nerves. This would mean that the force-generation process giving rise to the transport phenomenon is largely independent of axonal type, diameter, and function and that the force-generating enzymes have probably been held constant, at least in vertebrates, during evolution.

For fish olfactory nerves, we also know both the subcellular distribution of rapidly transported proteins and their molecular weight composition (Cancalon and Beidler, 1975; Cancalon *et al.*, 1976; Weiss *et al.*, 1978). The distribution of radioactivity among electrophoresed proteins from the pike olfactory nerve shows a high percentage of the transported label concentrated in only three protein bands of high molecular weight (Weiss *et al.*, 1978) (Fig. 7). Such simplicity is highly unusual. Since we know that only very little of the free leucine remains associated with the rapid transport system in such a way that it is transported efficiently over a long distance, one has to assume that labeled proteins have been transported as such and not that local proteins have been modified post-translationally by amino acylation. These results are probably a direct consequence of the morphological homogeneity and simplicity of this nerve. The anatomically similar olfactory nerve of the garfish also shows a major accumulation of radioactivity in the high-molecular-weight region (Cancalon *et al.*, 1976). In the garfish system, more than 60% of the radioactivity was found associated with proteins larger than 50,000 M_r. A 126,000-M_r peak was found to contain 12% of the total gel radioactivity while representing only 5% of the total protein. The garfish gels are far more complex, however, and show several additional bands in the lower-molecular-weight region, the identity of which has not yet been determined.

In the regenerating garfish olfactory system, both protein composition and transport velocity were determined (Cancalon and Elam, 1980; Cole and Elam, 1983; Elam, 1984). Whereas the velocity was not changed, the amount of protein transported was considerably increased and the protein pattern showed some, mainly qualitative, differences.

4.2. Slow Transport

In his elegant studies on slow transport in garfish olfactory nerve, Cancalon was able to show that although transport to the synaptic terminals may, depending on the tem-

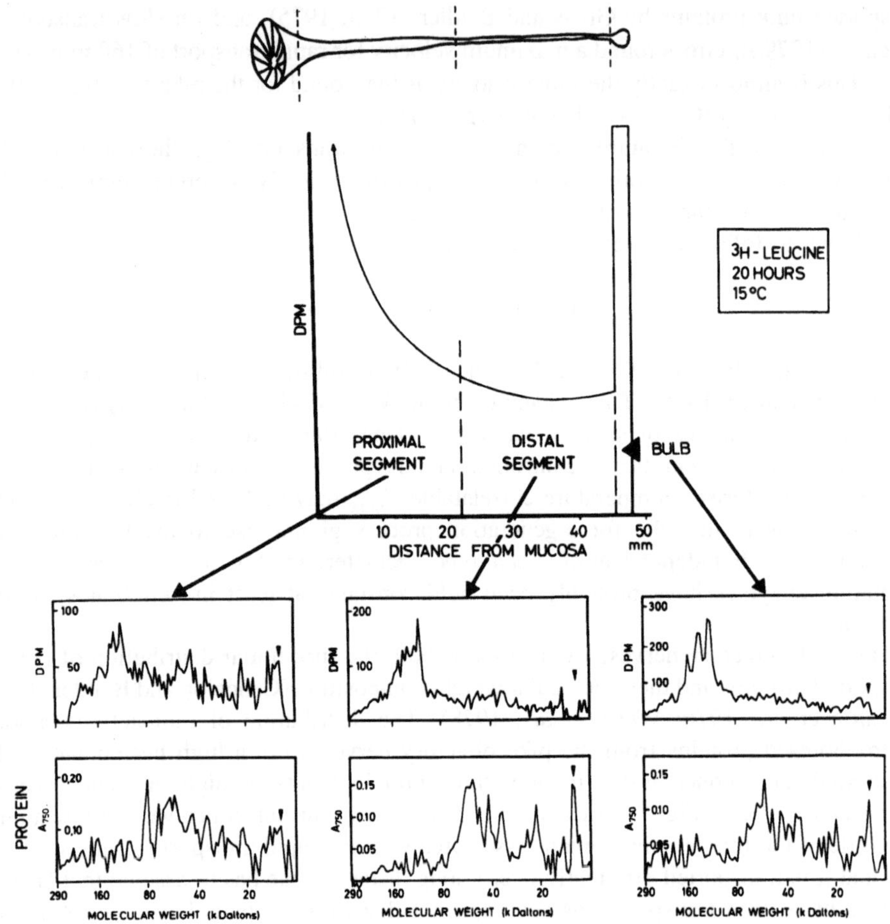

Figure 7. Distribution of radioactivity among transported proteins in three different segments of the pike olfactory system. The upper part of the figure shows (from left to right) the olfactory system with the mucosa, proximal and distal segments of the nerve and olfactory bulb, and a typical isotope distribution profile existing in the olfactory nerve 20 hr after isotope application (T = 19°C). Three dominant peaks with molecular weights of 118 ± 8, 137 ± 11, and 157 ± 11 kD contain up to 40% of the total radioactivity while representing only 10% of the total protein. The high radioactivity, fast transport, and small protein pool indicate rapid turnover for these proteins. The percentage of the total gel radioactivity in these peaks increases from 18 ± 4% in the proximal nerve segment to 33 ± 4% in the distal segment and 40 ± 5% in the synaptic region. Since the protein content of the high-molecular-weight peaks remains the same in all nerve regions, the transport system produces a separation of high- and low-molecular-weight proteins during transit from soma to synapse. The high-molecular-weight proteins are apparently favored to participate in rapid transport. (From Weiss *et al.*, 1978.)

perature, require more than 100 days, it was possible to use tracer techniques to follow the fate of these presumably cytoskeletal and soluble proteins (Cancalon, 1979*a,b*). He demonstrated that the catabolism of the proteins undergoing transport is very slow and matches the transport rate, so that these proteins are distributed all along the axon but are not accumulating at the terminals. Contrary to other nerves, there seems to be only one group of slowly transported proteins. Optic and other nerves show a slow component A and a somewhat faster component B (e.g., Baitinger *et al.*, 1982). Also contrary to most other systems studied, the olfactory system is capable of maintaining slow transport, even in axons detached from their perikarya (Cancalon, 1982, 1984). It is unclear why under these circumstances slow transport not only continues for several weeks but actually accelerates by a factor of 3.

5. TRANSPORT OF CHARACTERIZED PROTEINS

While the identity of virtually all the above-mentioned proteins remains unknown, it has already been possible to study the turnover and transport of one characterized protein. This 19,000-M_r protein specific for olfactory receptor cells was first described by Margolis (1972) and designated olfactory marker protein. It behaves as a typical slowly transported soluble protein (V_{max} = 2–4 mm/day) (Kream and Margolis, 1984). The characterization of proteins whose behavior can subsequently be studied in detail can be expected with the application of molecular biological techniques to the olfactory system (Margolis *et al.*, 1985).

6. RETROGRADE TRANSPORT

The olfactory nerve of the pike was recently used to study retrograde transport and to compare its properties with those of anterograde transport in the same system (Buchner and Weiss, 1983, 1986; Buchner, 1986; Buchner *et al.*, 1987). The first finding was that almost all materials examined were indeed found to be transported both anterogradely *and* retrogradely (see Table 1). Furthermore, in many cases transport in both directions was very similar (Table 1; see also Fig. 6). For [^{3}H]leucine, horseradish peroxidase (HRP), and wheat germ agglutinin (WGA) transport profiles for retrograde transport could be established for the first time, showing both peaks and saddle regions. This finding permitted the precise determination of maximal transport velocities for retrograde transport (Figs. 8 and 9; Table 1).

From the transport profiles of HRP, a maximal velocity of 25 mm/day (19°C) for the leading peak, and of about 7 mm/day for the slower component were measured. Both velocity components did not significantly differ for anterograde and retrograde transport.

In the case of WGA, only injection into the synaptic region resulted in typical transport profiles (retrograde transport) with peak and saddle regions. The maximum velocities of retrograde transport were the same as for HRP.

The electron microscopic distribution of HRP showed that after injection into the olfactory bulb, HRP was taken up into the neurons, where it was found mainly in multivesicular bodies (0.5-μm diameter). In longitudinal sections of the nerve, similar but slightly elongated organelles (diameter 0.25 μm, length 0.5 μm) were found in those segments in which the slowly moving bulk of the peroxidase activity was located (Fig. 10). The number of these organelles decreased with distance from the site of injection. The amount of HRP transported within the leading peak was so small that it could not be assigned to specific structures, although several electron microscopic histochemical methods were tried (Buchner *et al.*, 1987).

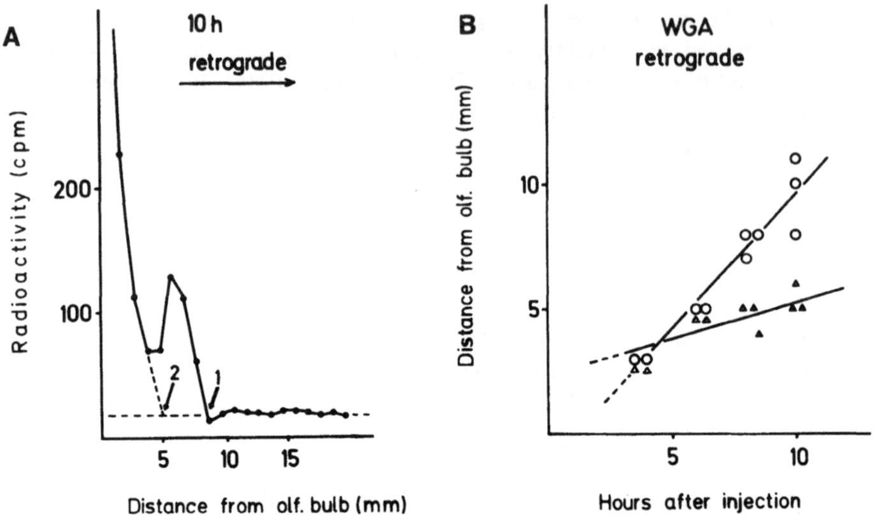

Figure 8. (A) Retrograde transport of [^{3}H]WGA 10 hr after injection into the olfactory bulb. For each experiment, the difference between radioactivities of experimental and contralateral (control) nerves in 1-mm nerve segments is plotted against the distance from the injection site. A clear division into peak (with front base, 1), saddle region, and main front (with front base, 2) can be seen. (B) Distances for peak front base and main front base plotted against survival times. The slope of the regression line for the peak front base (○) represents the maximum velocity (26 mm/day at 19°C). The slope of the regression line for the main front base (△) indicates a velocity of 7.2 mm/day at 19°C. (From Buchner *et al.*, 1987.)

Retrograde transport was reported to occur as well in the olfactory system of the rat (Jastreboff *et al.*, 1984). This study, which was designed to demonstrate topographical relationships between the olfactory epithelium and the olfactory bulb, confirmed earlier results (Land and Shepherd, 1974; Kauer, 1981), indicating that the olfactory receptors form a functional as well as an anatomical mosaic in the epithelium (Jastreboff *et al.*, 1984).

In order to study the retrograde transport of endogenous proteins, Buchner (1986) applied the *in vivo* covalent labeling technique (Fink and Gainer, 1980*a,b*). Proteins labeled with *N*-succinimidyl-[^{3}H]propionate were transported with a velocity of about 27 mm/day (19°C) (Table 1). When the transported proteins were characterized with the use of one- and two-dimensional gel electrophoresis and fluorography, it became evident that a complex set of proteins is transported retrogradely, from which the larger polypeptides (molecular weight higher than 50,000) were, however, largely excluded. Retrogradely transported proteins therefore differ markedly from the proteins in rapid anterograde transport (Fig. 7).

7. ORGANELLE MOVEMENT STUDIED WITH AVEC-DIC MICROSCOPY

There are several studies of axoplasmic transport of organelles by direct light microscopic observation (Cooper and Smith, 1974; Forman *et al.*, 1977, for review, see Smith,

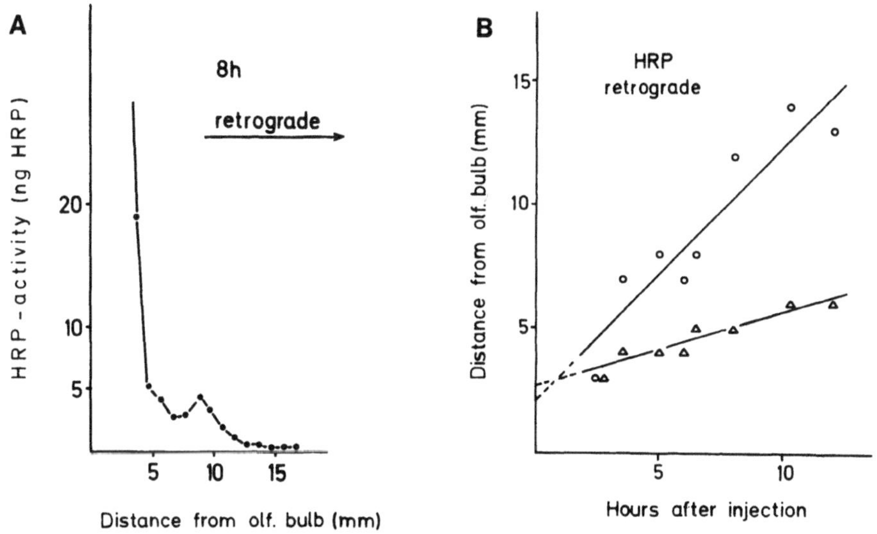

Figure 9. Retrograde and anterograde transport of horseradish peroxidase (HRP). (A) Retrograde transport of HRP 8 hr after injection into the bulb. For each experiment, the amount of enzymatically determined HRP within single nerve segments is plotted against the distance from the olfactory bulb. This results in transport profiles quite similar to those obtained for [^{3}H]WGA (see Fig. 8A), showing again a clear division into peak, saddle region, and bulk. (B) Distances from the injection site of the peak front base and main front base, respectively, plotted as a function of survival times. The slope of the regression line for the peak front base (○) represents the maximum velocity (25 mm/day at 19°C). The slope of the regression line for the main front base (△) represents a velocity of 7.4 mm/day at 19°C. (From Buchner *et al.*, 1987.)

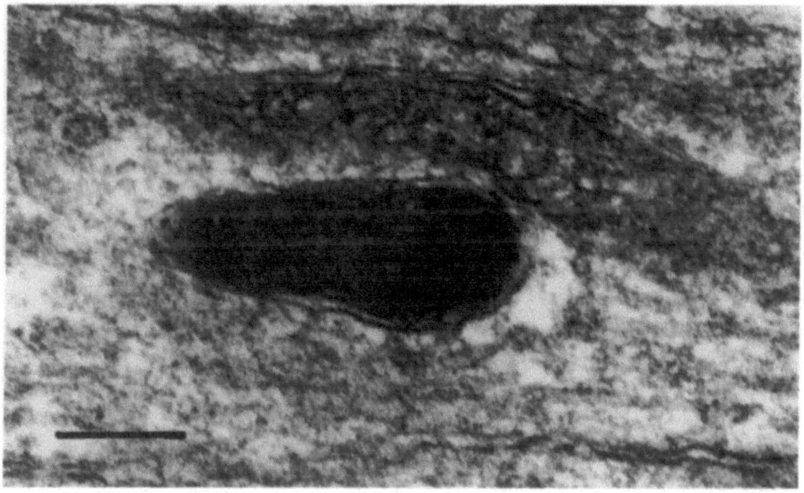

Figure 10. Longitudinal section of the pike olfactory nerve after injection of HRP into the olfactory bulb; 6–8 hr after injection, HRP-containing organelles, mainly elongated multivesicular bodies, are found in the first 3–4 mm of the nerve. These organelles, with a mean diameter of 0.25 ± 0.06 μm and a length of 0.5 ± 0.11 μm almost fill the axon and decrease in number with distance from the site of injection (electron microscopic histochemical method) Bar: 0.2 μm. (From Buchner *et al.*, 1987.)

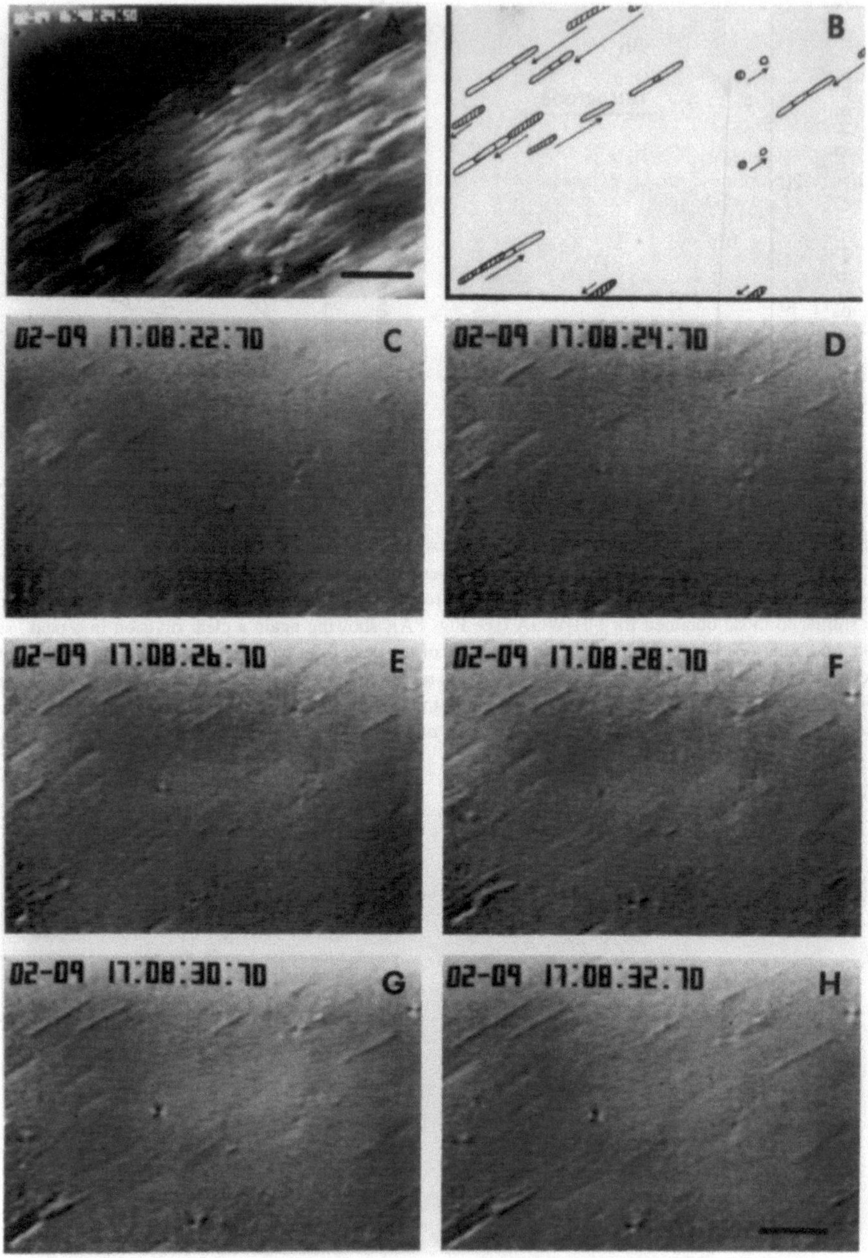

Figure 11. Movement of organelles in bundles of a few hundred axons isolated from the pike olfactory nerve and observed with AVEC-DIC microscopy. (A) Videomicrograph obtained with the AVEC-DIC method showing a bundle of axons. Longitudinally arranged linear elements which most probably represent mitochondria and not individual axons as well as round lysosome-like organelles are visible. Bar: 2 μm. (B) Schematic drawing summarizing the positions of the moving organelles of the sequence shown in (C)–(H). Hatched areas: Starting positions of the organelles. Arrows indicate directions of movement. The retrograde direction is to the upper right. (C–H) Sequence of videomicrographs showing the same bundle of axons as in a (different plane of focus).

1982; Weiss, 1987). These studies were only possible, however, in relatively large axons of a diameter of at least 10 μm. Since the olfactory nerve fibers are only 0.2 μm in diameter, which is about the limit of resolution of the optical microscope, it was considered impossible to observe transport inside them. However, since the development of the Allen video-enhanced contrast-differential interference contrast (AVEC-DIC) microscopy (Allen *et al.*, 1981*a,b*), it is possible to visualize much smaller objects, although they are not resolved optically (Allen *et al.*, 1982, 1985).

The olfactory nerve axons are among the first subjects of a detailed video-microscopic study (Buchner *et al.*, 1984, 1985; Buchner *et al.*, in preparation). Since it is well known that the prelysosomal organelles and mitochondria almost fill the axon (Kreutzberg and Gross, 1977; Seitz-Tutter *et al.*, 1985; Buchner *et al.*, in preparation), it was interesting to determine whether such a thin axon would allow for organelle transport in both directions.

Indeed, movement of organelles was found to occur in both anterograde (toward the olfactory bulb) and retrograde (toward the olfactory mucosa) directions. Occasionally an organelle displayed oscillatory motions, but no clear-cut reversal of directions was observed. The moving organelles could be divided into two categories: (1) lysosome-like organelles, i.e., round organelles with an apparent diameter of 0.5–0.7 μm; and (2) elongated organelles (length 1–10 μm, apparent width 0.5–0.8 μm), which can be considered mitochondria (Fig. 11). The lysosome-like organelles show average velocities of 1.45 ± 0.5 μm/sec when moving anterogradely and 1.22 ± 0.5 μm/sec when moving retrogradely. The difference between these velocities is not statistically different. The velocity of mitochondria in the retrograde direction was about twice that of anterograde movement: 1.15 ± 0.39 μm/sec versus 0.65 ± 0.28 μm/sec (Fig 12). No correlation was found between the length of the mitochondria and their velocities.

The movement of both types of organelles in the retrograde as well as in the anterograde direction was reversibly inhibited by 2 mM erythro-9-[3-(2-hydroxynonyl)]adenine (ENHA) (Buchner *et al.*, in preparation), an analogue of ATP which inhibits dynein and other ATPases (Schliwa *et al.*, 1984).

Since the velocities in these fine axons are very similar to the velocities found in very large axons (Forman *et al.*, 1983), we can conclude that the axon diameter does not significantly influence the velocity of organelle movement, although the larger organelles almost fill the axonal diameter and are transported as if moving in a sleeve. The simultaneous accomodation of transport systems for both directions in these narrow axons suggests that the axons are able to transiently enlarge their diameter considerably when two organelles pass each other (Buchner *et al.*, 1987).

The new technique of videomicroscopy permits direct assay of molecular mechanisms during normal function as well as in biochemical and pharmacological situations by the quantitative determination of the various parameters of organelle movement (Weiss *et al.*, 1986).

This sequence was obtained with the sequential subtraction mode of the image processor (Hamamatsu Photonic Microscope System, C 1966), which greatly facilitates the analysis of moving organelles. With this function a stored image is subtracted from the live image picked up by the TV camera with the result that only moving objects (and their starting positions) are displayed. All observations were made at 20°C. Intervals between the individual videomicrographs: 2 sec. Bar: 2 μm.

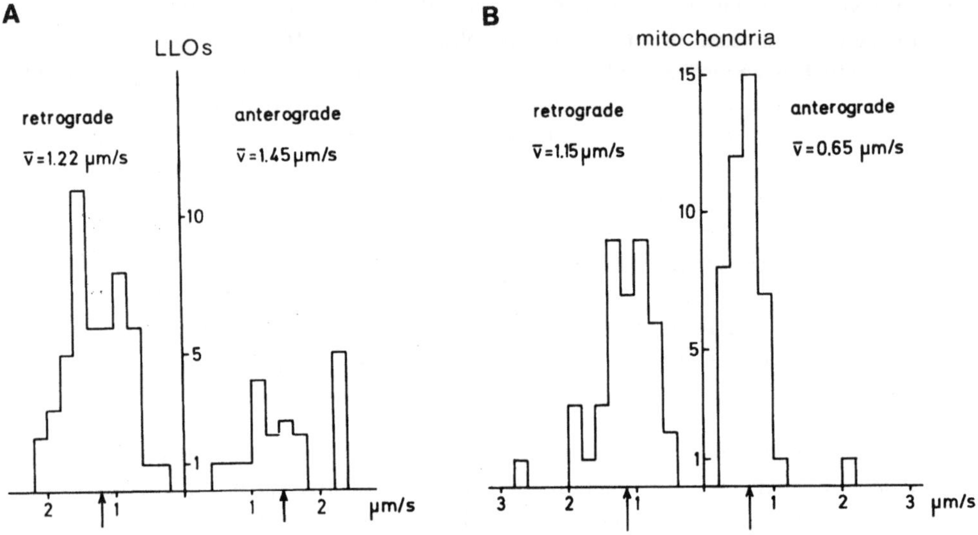

Figure 12. (A) Distribution of the velocities of lysosome-like organelles (LLOs). These organelles show average velocities (indicated by the arrows) of 1.45 ± 0.5 µm/sec when moving in the anterograde and 1.22 ± 0.5 µm/sec when moving in the retrograde direction. The difference between these two velocities is not statistically significant. T = 20°C. (B) Distribution of the velocities of mitochondria. Mitochondria show average velocities (indicated by the arrows) of 0.65 ± 0.28 µm/sec when moving in the anterograde and 1.15 ± 0.39 µm/sec when moving in the retrograde direction; i.e., the velocity of mitochondria in the retrograde direction is about twice that of anterograde movement. T = 20°C.

8. TRANSNEURONAL TRANSPORT AND THE SPREAD OF VIRUS

Several reports have shown metabolites or exogenous protein not only to be transported within a neuron but to be transmitted across synapses to neighboring cells as well (e.g., Schubert and Kreutzberg, 1982; La Vail, 1987). This mechanism was also studied in the olfactory system, especially of mammals, since its presence could mean that inhaled substances are taken up at the olfactory epithelium and transported to the brain. There they may spread, influence the function of higher level neurons (odor molecules, drugs, pheromones), or play a role in pathogenesis (toxins, viruses).

A sensitive tracer for receptor-mediated endocytosis (WGA coupled to HRP) was used to demonstrate anterograde transneuronal transport to first- and second-odor olfactory target neurons of the rat (Shipley, 1985; Baker and Spencer, 1986). Combined with retrograde transport in neurons projecting to the olfactory system, WGA–HRP reaches areas as distant as the raphe, the nucleus of the diagonal band, or the nucleus coeruleus.

In addition, the olfactory epithelium has been shown to be another entry port for neurotropic viruses long known to be transported along axons and to spread trans-synaptically in other cranial nerves (Kristensson, 1982). Herpes simplex virus type 1 (HSV-1) was shown to reach at least the olfactory bulb (Esiri and Tomlinson, 1984), while certain arboviruses appear not only in the olfactory bulb but later in the remainder of the brain as well (Monath *et al.*, 1983). Lundh *et al.* (1987) found that in mice Sendai virus after intranasal lavage infected the olfactory epithelium only to a limited extent while extensively affecting the adjacent respiratory epithelium. Vesicular stomatitis virus, on the other hand, infected exclusively the olfactory neurons and supporting cells and spread to the glomeruli in the bulb from where it propagated transneuronally into the rest of the brain (Lundh *et al.*, 1987).

9. THE STUDY OF AXOPLASMIC TRANSPORT USING THE OLFACTORY RECEPTOR NEURON

This chapter shows that the olfactory receptor neuron, especially that of certain fish and amphibians, is one of the most completely studied systems with respect of axoplasmic transport. There may be more studies on the optic nerve system (reviewed by Grafstein and Forman, 1980; Karlsson, 1983), but the olfactory system is superior with respect to the uniformity along its length and its anatomical and biochemical simplicity and homogeneity. Furthermore, it is better suited for the study of intracellular processes during degeneration, regeneration and growth. Its axonal transport is well characterized with respect of all transport classes and directions (see Table 1). Olfactory neurons can, therefore, be used as a valid and convenient model system for axoplasmic transport in general.

A number of basic problems of the molecular biology of the cell can be studied by analyzing axonal transport:

1. Distinction between neuronal and non-neuronal (e.g., glial) gene products (tracer studies)
2. Fate of synaptic vesicles and their contents
3. Processing of cellular components during transport (cytoskeleton, neuropeptides)
4. Routing of proteins to specialized regions of the plasma membrane (axolemma, synaptic membrane)
5. Mitochondrial synthesis versus import of proteins in mitochondrial biogenesis
6. Studies on the mechanisms of cytoplasmic transport

The work reported here shows that a number of steps toward this goal have already been successfully made, much to the mutual benefit of those interested in the basic problems of cell function and those interested in unraveling olfaction at a molecular level.

ACKNOWLEDGMENTS. We are most grateful to Dieter Seitz-Tutter for contributing the electron micrographs and to Wendy B. Green for assistance in the preparation of the manuscript. Our own research reported here was supported by grant We 790/1 to 7 from the Deutsche Forschungsgemeinschaft.

REFERENCES

Allen, R. D., Allen, N. S., and Travis, J. L., 1981*a*, Video-enhanced contrast, differential interference contrast (AVEC-DIC) microscopy: A new method capable of analyzing microtubule-related motility in the reticulopodial network of *Allogromia laticollaris*, *Cell Motil.* **1:**291–302.

Allen, R. D., Travis, H. L., Allen, N. S., and Yilmaz, H., 1981*b*, Video-enhanced contrast polarization (AVEC-POL) microscopy: A new method applied to detection of birefringence in the motile reticulopodial network of *Allogromia laticollaris*, *Cell Motil.* **1:**275–289.

Allen, R. D., Metuzals, J., Tasaki, I., Brady, S. T., and Gilbert, S. P., 1982, Fast axonal transport in squid giant axon, *Science* **218:**1127–1129.

Allen, R. D., Weiss, D. G., Hayden, J. H., Brown, D. T., Fujiwake, H., and Simpson, M., 1985, Gliding movement of and bidirectional transport along native microtubules from squid axoplasm: Evidence for an active role of microtubules in cytoplasmic transport, *J. Cell Biol.* **100**:1736–1752.

Baitinger, C., Levine, J., Lorenz, T., Simon, C., Skene, P., and Willard, M., 1982, Characteristics of axonally transported proteins, in: *Axoplasmic Transport* (D. G. Weiss, ed.), pp. 110–120, Springer-Verlag, Berlin.

Baker, H., and Spencer, R. F., 1986, Transneuronal transport of peroxidase-conjugated wheat germ agglutinin (WGA-HRP) from the olfactory epithelium to the brain of the adult rat, *Exp. Brain Res.* **63**:461–473.

Buchner, K., 1986, Vergleich von anterogradem und retrogradem axonalen Transport im Riechnerv des Hechtes, Ph.D. thesis, University Munich, Federal Republic of Germany.

Buchner, K., and Weiss, D. G., 1983, Anterograde and retrograde axoplasmic transport of exogenous proteins in C-fibers, *J. Neurochem.* **41**(Suppl.):S96.

Buchner, K., Gulden, J., and Weiss, D. G., 1984, Bidirectional cytoplasmic transport of organelles in olfactory nerve axons: An AVEC-DIC study, in: *International Cell Biology 1984–1985* (S. Seno, and Y. Okada, eds.), p. 491, Academic, Tokyo.

Buchner, K., Gulden, J., and Weiss, D. G., 1985, An AVEC-DIC study on the movement of organelles in olfactory nerve axons, *Eur. J. Cell Biol.* **36**(Suppl.)**7**:12.

Buchner, K., Seitz-Tutter, D., Schönitzer, K., and Weiss, D. G., 1987, A quantitative study of anterograde and retrograde axoplasmic transport of exogenous proteins in olfactory nerve C-fibers, *Neuroscience* **22**:697–707.

Buchner, K., Seitz-Tutter, D., and Weiss, D. G., Bidirectional movement of large organelles in very thin axons: An EM and AVEC-DIC study (in preparation).

Burton, P. R., 1985, Ultrastructure of the olfactory neuron of the bullfrog: The dendrite and its microtubules, *J. Comp. Neurol.* **242**:147–160.

Burton, P. R., and Laveri, L. A., 1985, The distribution, relationships to other organelles, and calcium-sequestering ability of smooth endoplasmic reticulum in frog olfactory axons, *J. Neurosci.* **5**:3047–3060.

Burton, P. R., and Paige, J. L., 1981, Polarity of axoplasmic microtubules in the olfactory nerve of the frog, *Proc. Natl. Acad. Sci. USA* **78**:3269–3273.

Cancalon, P., 1979*a*, Influence of temperature on the velocity and on the isotope profile of slowly transported labeled proteins, *J. Neurochem.* **32**:997–1007.

Cancalon, P., 1979*b*, Subcellular and polypeptide distributions of slowly transported proteins in the garfish olfactory nerve, *Brain Res.* **161**:115–130.

Cancalon, P., 1982, Slow flow in axons detached from their perikarya, *J. Cell. Biol.* **95**:989–992.

Cancalon, P., 1984, Role of slow flow in axonal degeneration and regeneration, in: *The Role of Axonal Transport in Neuronal Growth and Regeneration* (J. Elam and P. Cancalon, eds.), pp. 211–242, Plenum, New York.

Cancalon, P., 1985, Influence of temperature on various mechanisms associated with neuronal growth and nerve regeneration, *Prog. Neurobiol.* **25**:27–92.

Cancalon, P., and Beidler, L. M., 1975, Distribution along the axon and into various subcellular fractions of molecules labeled with ^{3}H leucine and rapidly transported in the garfish olfactory nerve, *Brain Res.* **89**:225–244.

Cancalon, P., and Elam, J. S., 1980, Rate and composition of rapidly transported proteins in regenerating olfactory nerves, *J. Neurochem.* **35**:889–897.

Cancalon, P., Elam, J. S., and Beidler, L. M., 1976, SDS gel electrophoresis of rapidly transported proteins in garfish olfactory nerve, *J. Neurochem.* **27**:687–693.

Chakraborty, G., Leach, T., Zanakis, M. F., and Ingoglia, N. A., 1986, Posttranslational protein modification by amino acid addition in regenerating optic nerves of goldfish, *J. Neurochem.* **46**:726–732.

Cole, G. J., and Elam, J. S., 1983, Characterization of axonally transported glycoproteins in regenerating garfish olfactory nerve, *J. Neurochem.* **41**:691–702.

Cooper, P. D., and Smith, R. S., 1974, The movement of optically detectable organelles in myelinated axons of *Xenopus laevis, J. Physiol. (Lond.)* **242**:77–97.

Costanzo, R. M., 1985, Neuronal regeneration and functional reconnection following olfactory nerve transection in hamster, *Brain Res.* **361**:258–266.

Csanyi, V., Gervai, J., and Lajtha, A., 1973, Axoplasmic transport of free amino acids, *Brain Res.* **56**:271–284.

Di Giamberardino, L., 1971, Independence of the rapid axonal transport of protein from the flow of free amino acids, *Acta Neuropathol.* Suppl. V: 132–135.

Easton, D. M., 1971, Garfish olfactory nerve: Easily accessible source of numerous long, homogeneous, non-myelinated axons, *Science* **172**:952–955.

Edström, A., and Hanson, M., 1973, Temperature effects on fast axonal transport of proteins *in vitro* in frog sciatic nerves, *Brain Res.* **58**:345–354.

Elam, J. S., 1984, Axonal transport of glycoproteins in regenerating nerve, in: *Axonal Transport in Neuronal Growth and Regeneration* (J. S. Elam, and P. Cancalon, eds.), pp. 87–104, Plenum, New York.

Elam, J. S., and Agranoff, B. W., 1971, Rapid transport of protein in the optic system of the goldfish, *J. Neurochem.* **18**:375–387.

Elam, J. S., and Cancalon, P. (eds.), 1984, *Axonal Transport in Neuronal Growth and Regeneration*, Plenum, New York.

Esiri, M. M., and Tomlinson, A. H., 1984, Herpes simplex encephalitis. Immunohistological demonstration of spread of virus via olfactory and trigeminal pathways after infection of facial skin in mice, *J. Neurol. Sci.* **64**:213–217.

Fink, D. J., and Gainer, H., 1980*a*, Axonal transport of proteins: A new view using *in vivo* covalent labeling, *J. Cell Biol.* **85**:175–186.

Fink, D. J., and Gainer, H., 1980*b*, Retrograde axonal transport of endogenous proteins in sciatic nerve demonstrated by covalent labeling *in vivo*, *Science* **208**:303–305.

Forman, D. S., Padjen, A. L., and Siggins, G. R., 1977, Axonal transport of organelles visualized by light microscopy: Cinematographic and computer analysis, *Brain Res.* **136**:197–213.

Gasser, H. S., 1956, Olfactory nerve fibers, *J. Gen. Physiol.* **39**:473–496.

Giorgi, P. P., 1978, Relationship between axonal transport and the Wolfgram proteins of myelin, *Proc. Eur. Soc. Neurochem.* **1**:118.

Giorgi, P. P., Karlsson, J.-O., Sjöstrand, J., and Field, E. J., 1973, Axonal flow and myelin protein in the optic pathway, *Nature* (New Biol.) **244**:121–124.

Grafstein, B., and Forman, D. S., 1980, Intracellular transport in neurons, *Physiol. Rev.* **60**:1167–1283.

Graziadei, P. P. C., and Monti Graziadei, G. A., 1978, The olfactory system: A model for the study of neurogenesis and axon regeneration in mammals, in: *Neuronal Plasticity* (C. W. Cotman, ed.), pp. 131–153, Raven, New York.

Gross, G. W., and Beidler, L. M., 1973, Fast axonal transport in the C-fibers of the garfish olfactory nerve, *J. Neurobiol.* **4**:413–428.

Gross, G. W., and Beidler, L. M., 1975, A quantitative analysis of isotope concentration profiles and rapid transport velocities in the C-fibers of the garfish olfactory nerve, *J. Neurobiol.* **6**:213–232.

Gross, G. W., and Kreutzberg, G. W., 1978, Rapid axoplasmic transport in the olfactory nerve of the pike. I. Basic transport parameters for proteins and amino acids, *Brain Res.* **139**:65–76.

Hara, T. J., 1975, Olfaction in fish, *Prog. Neurobiol.* **5**:217–335.

Jastreboff, P. J., Pederson, P. E., Greer, C. A., Stewart, W. B., Kauer, J. S., Benson, T. E., and Shepherd, G. M., 1984, Specific olfactory receptor populations projecting to identified glomeruli in the rat olfactory bulb, *Proc. Natl. Acad. Sci. USA* **81**:5250–5254.

Karlsson, J.-O., 1977, Is there an axonal transport of amino acids? *J. Neurochem.* **29**:615–617.

Karlsson, J.-O., 1983, Axonal transport in retinal ganglion cells, in: *Progress in Retinal Research*, Vol. 3 (N. N. Osborne and G. J. Chader, eds.), pp. 84–96, Pergamon, New York.

Kauer, J. S., 1981, Olfactory receptor cells staining using horseradish peroxidase, *Anat. Rec.* **200**:331–336.

Koike, H., and Nagata, Y., 1979, Intra-axonal diffusion of (^{3}H)-acetylcholine and (^{3}H)-γ-aminobutyric acid in a neurone of *Aplysia*, *J. Physiol. (Lond.)* **295**:397–417.

Kream, R. M., and Margolis, F. L., 1984, Olfactory marker protein: Turnover and transport in normal and regenerating neurons, *J. Neurosci.* **4**:868–879.

Kreutzberg, G. W., and Gross, G. W., 1977, General morphology and axonal ultrastructure of the olfactory nerve of the pike, *Esox lucius*, *Cell Tissue Res.* **181**:443–457.

Kristensson, K., 1982, Implications of axoplasmic transport for the spread of virus infections in the nervous system, in: *Axoplasmic Transport in Physiology and Pathology* (D. G. Weiss and A. Gorio, eds.), pp. 153–158, Springer-Verlag, Berlin.

Land, L. J., and Shepherd, G. M., 1974, Autoradiographic analysis of olfactory receptor projections in the rabbit, *Brain Res.* **70**:506–510.

La Vail, J. H., and Margolis, T. P., 1987, The anterograde axonal transport of wheat-germ agglutinin as a model for transcellular transport in neurons, in: *Axonal Transport* (R. S. Smith and M. A. Bisby, eds.), pp. 311–326, Liss, New York.

Ludueña, R. F., 1979, Biochemistry of tubulin, in: *Microtubules* (K. Roberts and J. S. Hyams, eds.), pp. 65–116, Academic, London.

Lundh, B., Kristensson, K., and Norrby, E., 1987, Selective infections of olfactory and respiratory epithelium by vesicular stomatitis and Sendai viruses, *Neuropathol. Appl. Neurobiol.* **13:**111–122.

Margolis, F., 1972, A brain protein unique to the olfactory bulb, *Proc. Natl. Acad. Sci. USA* **69:**1221–1224.

Margolis, F. L., 1980, Carnosine: An olfactory neuropeptide, in: *Role of Peptides in Neuronal Function* (J. L. Barker and T. Smith, eds.), pp. 545–572, Dekker, New York.

Margolis, F. L., and Grillo, M., 1977, Axoplasmic transport of carnosine (β-alanyl-L-histidine) in the mouse olfactory pathway, *Neurochem. Res.* **2:**507–519.

Margolis, F. L., Sydor, W., Teitelbaum, Z., Blacher, R., Grillo, M., Rogers, K., Sun, R., and Gubler, U., 1985, Molecular biological approaches to the olfactory system: Olfactory marker protein as a model, *Chem. Senses* **10:**163–174.

McEwen, B. S., and Grafstein, B., 1968, Fast and slow components in axonal transport of protein, *J. Cell. Biol.* **38:**494–508.

Monath, T. P., Cropp, C. B., and Harrison, A. K., 1983, Model of entry of a neurotropic arbovirus into the central nervous system, *Lab. Invest.* **48:**399–410.

Muñoz-Martínez, E. J., Núñez, R., and Sanderson, A., 1981, Axonal transport: A quantitative study of retained and transported protein fraction in the cat, *J. Neurobiol.* **12:**15–26.

Muralt, A. von, Weibel, E. R., and Howarth, J. V., 1976, The optical spike. Structure of the olfactory nerve of pike and rapid birefringence changes during excitation, *Pflügers Arch.* **367:**67–76.

Neale, J. H., Elam, J. S., Neale, E. A., and Agranoff, B. W., 1974, Axonal transport and turnover of proline- and leucine-labeled protein in the goldfish visual system, *J. Neurochem.* **23:**1045–1055.

Ochs, S., 1971, Local supply of energy to the fast axoplasmic transport mechanism, *Proc. Natl. Acad. Sci. USA* **68:**1279–1282.

Ochs, S., 1972, Rate of fast axoplasmic transport in mammalian nerve fibers, *J. Physiol. (Lond.)* **227:**627–645.

O'Farrell, P. H., 1975, High resolution two-dimensional electrophoresis of proteins, *J. Biol. Chem.* **250:**4007–4021.

Price, C. H., McAdoo, D. J., Farr, W., and Okuda, R., 1979, Bidirectional axonal transport of free glycine in identified neurons R3-R14 of *Aplysia, J. Neurobiol.* **10:**551–571.

Schliwa, M., Ezzel, R. M., and Euteneuer, U., 1984, Erythro-9-[3-(2-hydroxynonyl)]adenine is an effective inhibitor of cell motility and actin assembly, *Proc. Natl. Acad. Sci. USA* **81:**6044–6048.

Schmid, G., Wagner, L., and Weiss, D. G., 1983, Rapid axoplasmic transport of free leucine, *J. Neurobiol.* **14:**133–144.

Schubert, P., and Kreutzberg, G. W., 1982, Transneuronal transport: A way for the neuron to communicate with the environment, in: *Axoplasmic Transport in Physiology and Pathology* (D. G. Weiss and A. Gorio, eds.), pp. 32–43, Springer-Verlag, Berlin.

Seitz-Tutter, D., Buchner, K., and Weiss, D. G., 1985, Rapid bidirectional transport of horseradish peroxidase (HRP) in pike olfactory nerve, *Eur. J. Cell Biol.* **36**(Suppl. 7)**:**61.

Shipley, M. T., 1985, Transport of molecules from nose to brain: Transneuronal anterograde and retrograde labeling in the rat olfactory system by wheat germ agglutinin-horseradish peroxidase applied to the nasal epithelium, *Brain Res. Bull.* **15:**129–142.

Smith, R. S., 1982, Saltatory organelle movement and the mechanism of fast axonal transport, in: *Axoplasmic Transport* (D. G. Weiss, ed.), pp. 234–240, Springer-Verlag, Berlin.

Weiss, D. G., 1982, 3-0-methyl-D-glucose and β-alanine: Rapid axoplasmic transport of metabolically inert low molecular weight substances, *Neurosci. Lett.* **31:**241–246.

Weiss, D. G., 1987, The mechanism of axoplasmic transport, in: *Axoplasmic Transport* (Z. Iqbal, ed.), pp. 275–307, CRC Press, Boca Raton, Florida.

Weiss, D. G., Krygier-Brevart, V., Gross, G. W., and Kreutzberg, G. W., 1978, Rapid axoplasmic transport in the olfactory nerve of the pike. II. Analysis of transported proteins by SDS gel electrophoresis, *Brain Res.* **139:**77–87.

Weiss, D. G., Schmid, G., and Wagner, L., 1980, Influence of microtubule inhibitors on axoplasmic transport of free amino acids. Implications for the hypothetical transport mechanism, in: *Microtubules and Microtubule Inhibitors 1980* (M. De Brabander and J. De Mey, eds.), pp. 31–41, Elsevier, Amsterdam.

Weiss, D. G., Keller, F., Gulden, J., and Maile, W., 1986, Towards a new classification of intracellular particle movement based on quantitative analyses, *Cell Motil.* **6:**128–135.

Weiss, P., and Holland, Y., 1967, Neuronal dynamics and axonal flow. II. The olfactory nerve as model test object, *Proc. Natl. Acad. Sci. USA* **57:**258–264.

11

Molecular Cloning of Olfactory-Specific Gene Products

Frank L. Margolis

1. INTRODUCTION

The specificity of cell, tissue, and organismal development and differentiation derives from selective regulation of differential gene expression along spatial and temporal coordinates. Understanding the mechanisms controlling these events is among the key problems in contemporary biology. These questions are being addressed by many laboratories applying a wide range of techniques to a variety of organisms and neural systems. Our concern here is to consider those approaches that may help identify the genes that are specifically expressed and selectively regulated in the various cells of the olfactory system. Furthermore, it is our intention to identify the products of these various genes and to determine how they may be related to the diverse functional activities of the olfactory system. Thus, the major thrust of this chapter is to consider the use of biochemical and molecular biological approaches in the study of the regulation of specific gene expression in the peripheral olfactory system.

In contrast to most of the vertebrate nervous system, the olfactory pathway has the unusual ability to replace dying neurons from stem cells (Graziadei, 1971; Graziadei and Monti-Graziadei, 1978*a,b,* 1985; Simmons and Getchell, 1981; Wang and Halpern, 1982). This property can be viewed as a manifestation of continuing neural development in an adult animal. What can we learn from it about the regulation of normal neural ontogeny and plasticity? Furthermore, if we consider the full developmental spectrum, what can this system teach us about degenerative and aging-associated neural deficits and recovery from traumatic lesions that might apply to other areas of the nervous system.

The diverse biological properties of the olfactory system derive from selective tissue-specific expression of a subset of the total genome as well as of tissue-specific regulation of the expression of more common genes. Olfactory neurons express at least one unique developmentally regulated protein, the olfactory marker protein (OMP) (Margolis, 1972, 1980*a*). Additional examples have been identified by monoclonal antibody screening (Hempstead and Morgan, 1983, 1985*a,b*; Allen and Akeson, 1985; Fujita *et al.*, 1985;

Frank L. Margolis • Department of Neuroscience, Roche Institute of Molecular Biology, Roche Research Center, Nutley, New Jersey 07110.

Chen *et al.*, 1986*a*; Mori, 1987; Mori *et al.*, 1985) or directly as for novel G proteins (Anholt *et al.*, 1987; Jones and Reed, 1987) and the transmembrane ciliary glycoprotein (Chen *et al.*, 1986*b*). The enzymes carnosine synthetase (Horinishi *et al.*, 1978; Margolis *et al.*, 1985*a*, 1987) and adenylate cyclase (Pace *et al.*, 1985; Sklar *et al.*, 1986) and the peripheral type benzodiazepine receptor (Anholt *et al.*, 1986) are examples of gene products that are very abundantly, but not uniquely, expressed in these cells. However, the roles played by these components in the cellular activities of these neurons and the manner in which they are regulated are still largely unknown. Additional examples of restricted gene expression surely exist in these sensory cells (Khew-Goodall *et al.*, 1987; Wensley *et al.*, 1986).

Furthermore, the various properties of this tissue are not solely associated with its neuronal component. Activities of secretory cells (Getchell *et al.*, 1984*a,b*), their content of the P450 family of mixed function oxygenases (Dahl *et al.*, 1982; Voigt *et al.*, 1985; Reed *et al.*, 1986; Hadley and Dahl, 1982), of pyrazine-binding proteins (Cavaggioni *et al.*, 1987; Bignetti *et al.*, 1985; Pevsner *et al.*, 1985), as well as the presence of other novel cellular products in the adjacent tissue (Khew-Goodall *et al.*, 1987; Wensley *et al.*, 1986; Margolis *et al.*, 1983; Lee *et al.*, 1987) testify to the subtlety and complexity of function manifested here. Thus, it is necessary to address both the inherent complexity of a set of sensory neurons as well as the entire array of cells in the tissue associated with it. We need to understand how the numerous associated non-neuronal cells are involved to a greater or lesser extent in the regulation of neuron metabolism and turnover; in stimulus access, detection, and elimination; in transsynaptic target identification, regulation, and transmitter selection; and in excluding noxious environmental agents. This is further complicated, as we are dealing with a tissue that is extremely plastic and in constant biochemical and anatomical flux in both its central and peripheral compartments (Brunjes *et al.*, 1985; Baker *et al.*, 1983, 1984; Rochel and Margolis, 1980; Kream and Margolis, 1983; Kawano and Margolis, 1982; Nadi *et al.*, 1981).

Among the approaches that promise to give entree to the understanding of these questions are those associated with the methodology of recombinant DNA or molecular biology. Although applications of this technology to the vertebrate olfactory system are only beginning (Wensley *et al.*, 1985, 1987; Margolis *et al.*, 1985*b;* Lee *et al.*, 1987; Khew-Goodall *et al.*, 1987), it is the purpose of this chapter to illustrate some of the initial achievements, to comment on some of the pitfalls, and to consider some possible future directions. The anatomical organization of this tissue is considered initially, followed by an overview of olfactory tissue-specific gene expression. Specific approaches to some of these are then considered, followed by an indication of possible future directions.

2. ANATOMICAL ORGANIZATION

The olfactory mucosa of vertebrates (Graziadei, 1971; Menco, 1980; Kessel and Kardon, 1979) lies high in the nasal vault, separated from the entry to the nasal cavity by the intervening respiratory mucosa. It consists of two layers: a superficial avascular epithelium and a deeper lamina propria. The former consists of three cell types: the olfactory receptor neurons, the sustentacular cells, and the basal cells. The lamina propria contains fascicles of unmyelinated olfactory receptor cell axons, Bowman's glands, and blood vessels as well as connective tissue. Olfactory receptor cells are bipolar neurons along whose unmyelinated axons information about the quality and quantity of stimuli is transmitted centrally to the olfactory bulb of the brain. The olfactory bulb (Shepherd, 1972, 1977), the most rostral portion of the forebrain, is a laminated structure, of high cell density, receiving peripheral input from the olfactory receptor neurons as well as centrifugal input from several loci in the CNS. It is here that the afferent neurons synapse, in complex glomeruli, with the dendrites of mitral cells, the major output neurons of the bulb, as well as with dendrites of a biochemically diverse population of intrinsic neurons. The chemical nature of the transmitter used by the olfactory neuron synapses is unknown (Margolis, 1981; Macrides and Davis, 1983; Halasz and Shepherd, 1983), as is the mechanism by which functionally intact synapses at this site regulate neurotransmitter gene expression in their target neurons. The receptor neuron extends a process peripherally toward the mucosal surface, where it terminates in a dendritic knob from which several cilia project.

These cilia, which are believed to be the initial site at which detection and transduction of odorants occurs (Lancet, 1986; Getchell *et al.*, 1980), lie in the mucous layer covering the surface of the olfactory mucosa. The mucus derives from the secretions of the sustentacular cells and the Bowman's and submucosal glands (Getchell *et al.*, 1984*a*). The extent to which these secretions also derive from respiratory, nonolfactory mucosa is unknown. The basal cells in the neuroepithelium include a population of stem cells (Graziadei and Monti-Graziadei, 1978*a,b*). These are the source of olfactory receptor neurons during normal ontogeny and serve as precursors to replace olfactory neurons lost as a result of normal neuronal turnover in the mature animal or in response to experimentally induced chemical or surgical insult. Although this brief anatomical sketch derives from the vertebrate olfactory system, the general functional and organizational properties manifest phylogenetically striking similarities even to the insect olfactory system (Vinnikov, 1982), although neurochemical similarities are not yet evident.

3. BIOCHEMICAL PROPERTIES

The first biochemical example of olfactory tissue-specific gene expression was the isolation of the olfactory marker protein (OMP) from mouse olfactory bulb (Margolis, 1972). Subsequently, the dipeptide carnosine (*b*-alanyl L-histidine), a potential olfactory neurotransmitter or modulator (Margolis, 1974), and its biosynthetic enzyme carnosine synthetase (Margolis *et al.*, 1985*a*, 1987; Horinishi *et al.*, 1978) were characterized and shown to be present in olfactory neurons at very high levels relative to other tissues. The olfactory neurons have also been shown to be extremely rich in peripheral benzodiazepine receptors (Anholt *et al.*, 1984) and adenylate cyclase (Sklar *et al.*, 1986; Pace *et al.*, 1985) and G proteins (Anholt *et al.*, 1987; Jones and Reed, 1987). A unique transmembrane glycoprotein has been demonstrated in olfactory cilia that may be a candidate receptor molecule (Chen *et al.*, 1986*b*). In addition, a number of specific neuronal and non-neuronal antigens have been detected in this tissue by monoclonal antibody techniques (Allen and Akeson, 1985; Hempstead and Morgan, 1985*a,b;* Fujita *et al.*, 1985). Novel pyrazine-binding proteins have been identified in the non-neuronal Bowman's gland compartments of this tissue by Bignetti *et al.* (1985) and Pevsner *et al.* (1985). Lee *et al.* (1987) isolated a specific complementary DNA (cDNA) clone that predicts a protein sequence homologous to hepatic retinol-binding proteins and whose mRNA is located in the cells of Bowman's glands. These glands are also a rich source of the P450 mixed-function oxygenase (Reed *et al.*, 1986; Hadley and Dahl, 1982; Voigt *et al.*, 1985). Recently, a cDNA clone was isolated for an abundant messenger RNA (mRNA) present in the submucosal glands of adjacent nasal respiratory mucosa (Khew-Goodall *et al.*, 1987; Wensley *et al.*, 1986; E. Barbatos and R. Reed, personal communication). The olfactory neurons can regulate the expression of tyrosine hydroxylase (Baker *et al.*, 1983, 1984) and substance P (Kream *et al.*, 1984) in second-order target neurons in mammals and of sexually dimorphic glomerular structures in *Manduca* (Schneiderman *et al.*, 1982). These observations demonstrate that the olfactory mucosa manifests extensive cell- and tissue-specific gene expression as well as the ability to regulate gene expression of those target cells with which it interacts.

4. OLFACTORY MARKER PROTEIN

4.1. Overview

The archetypical example of olfactory neuron-specific gene expression is the olfactory marker protein (OMP). It was originally observed as a band on acrylamide gel electrophoresis of native mouse olfactory tissue cytoplasmic extracts (Margolis, 1972). The protein was subsequently purified and characterized from both mouse and rat olfactory tissue and specific antibodies prepared. OMP has a molecular mass of 19,000 M_r and an isoelectric point of about 5. The purified protein is devoid of carbohydrate and of the amino acid cysteine. OMP represents 0.1–1% of the total olfactory tissue cytoplasmic protein, and antisera directed against rat OMP cross-react with extracts of olfactory tissue from many mammalian species, including humans, as well as the garfish (for review, see Margolis, 1980*a*, 1985). Little or no cross-reactivity with OMP antisera is seen in extracts of frog and chicken. Attempts by several investigators to demonstrate cross reactivity of OMP antisera with insect and molluskan olfactory tissue have been unsuccessful.

By immunocytochemistry OMP has been observed only in olfactory neurons (Farbman and Margolis, 1980; Monti-Graziadei *et al.*, 1980; Miragall and Monti-Graziadei, 1982; Nakashima *et al.*, 1984). OMP is present in the cytoplasm of the cell body and its peripherally directed dendritic knob and in the proximal portion of the cilia as well as in the axon and synaptic terminals in the olfactory bulb. It is absent from every other cell type in the olfactory mucosa and disappears at the boundary between the olfactory and respiratory mucosa. Since OMP is absent from the neural precursor basal cells in the olfactory neuroepithelium, it has been used as a marker for the mature, functioning receptor cells whose axons comprise the first cranial nerve. In rodents it is also present in the olfactory neurons of the vomeronasal organ and those present in the organ of Masera in rats (W. Breipohl, personal communication) and hamsters (F. Macrides, personal communication).

Olfactory marker protein appears in olfactory neurons early in the last trimester of gestation at about the time when innervation of the bulb begins and when several other specific events associated with olfactory neuronal differentiation occur (Farbman, 1986).

OMP appears a few days in advance of the appearance of tyrosine hydroxylase in the target dopaminergic neuron in the bulb. OMP is synthesized in the mature olfactory receptor perikaryon from which it is transported to the entire cytoplasmic compartment of the neuron. Like many other cytoplasmic proteins, the transport of OMP is associated with the slow component of axoplasmic flow (Kream and Margolis, 1983). OMP turnover manifests a biphasic half-life with a rapid component of about 1 day and slower component of about 1 week. The longer half-life aspect is shortened in younger animals and especially in animals in which the neuroepithelium is undergoing reconstitution following injury. Details of these earlier studies on the properties of the OMP, its distribution, and response to various experimental manipulations have been reviewed (Margolis, 1985, 1980*a*).

4.2. Amino Acid Sequence

The abundance, wide phylogentic distribution, developmental profile, response to lesion, and specificity of cellular occurrence outlined above all suggest that OMP plays a key role in the function of mature olfactory receptor neurons. However, although it has been studied for several years in many laboratories, its precise biological role is unknown.

Since functional portions of protein molecules tend to be conserved, it seemed possible that the existence of any homologies between the primary amino acid sequence of OMP with those of proteins of known function or structure might offer some insight as to the function of OMP. In addition, knowledge of the primary sequence would permit the synthesis of defined oligonucleotide that could be used for the cloning and characterization of the mRNA and the gene responsible for the synthesis of OMP.

Initial attempts at direct sequencing of OMP isolated from rat and mouse indicated the protein was blocked at the amino terminus. Therefore, chemical and enzymatic cleavages were used to generate peptide fragments amenable to amino acid sequencing, resulting in virtually the complete amino acid sequence of rat OMP. Furthermore, fast atom bombardment mass spectrometry demonstrated that the amino terminus was acetylated in common with other cytoplasmic proteins (Sydor *et al.*, 1986). The primary structures of the mouse and rat OMP are highly homologous based on similar amino acid composition, immunoreactivity, and high-pressure liquid chromatography (HPLC) elution profiles of peptides derived by protease action. As might be expected for a cytoplasmic protein, the amino acid sequence does not predict any transmembrane-spanning regions, and there are only two significant hydrophobic regions at residues 137–144 and at the carboxy terminal end of the protein. Of the 14 hydroxy amino acids, one half are clustered between residues 78–95. OMP has an α-helical content of 54% when analyzed by the Delphi secondary structure prediction program (Fig. 1). This may be related to the relative stability of OMP and to its ability to regain immunological reactivity on dialysis following denaturation in guanidinium hydrochloride.

A computer search of the sequence of rat OMP was performed in the N.B.R.F. Dayhoff Protein and Nucleic Acid Databases, using FASTP (Lipman and Pearson, 1985) at ktup = 1 and the mutation data matrix. The latter was performed using serial 15 amino acid segments of rat OMP with sequential five amino acid offsets. In neither case was any significant homology observed with any sequence in the database. These considerations, coupled with the absence of any evidence for OMP to be a protein kinase substrate (W.

```
      Pred: HHHHCHHHHHSSSSSHHHHHHHHHHHHHHHHHHHHHHHHHHHHHHHHHHH
      Safe:                HHHHHHHHHH HHHHHHHHHHHH  HHHHH HHHH
  Sequence: AEDGPQKQQLDMPLVLDQDLTKQMRLRVESLKQRGEKKQDGEKLLRPAES
            12345678901234567890123456789012345678901234567890
                     1         2         3         4         5

      Pred: HHHHHHHHHHHHCTTTTTHHHHTCTTSSSSSTCCCTCCCTCCCHHHHHHH
      Safe: H  HHHHHHH                SSSSS
  Sequence: VYRLDFIQQQKLQFDHWNVVLDKPGKVTITGTSQNWTPDLTNLMTRQLLD
            12345678901234567890123456789012345678901234567890
                     6         7         8         9         0

      Pred: HHHHHHHHHHTCHHHHHHHHHHHHHHHHHHHHHHHHHHHSSSSHCCCCHH
      Safe:  HHHHHHHH   HHHHHHHHHHHHHHHHHHHHHHHHH            H
  Sequence: PAAIFWRKEDSDAMDWNEADALEFGERLSDLAKIRKVMYFLITFGEGVEP
            12345678901234567890123456789012345678901234567890
                     1         2         3         4         5

      Pred: HHHHHHHHSHHH
      Safe: HHHHHHH
  Sequence: ANLKASVVFNQL
            123456789012
                     6
```

Figure 1. Secondary structure prediction for the rat olfactory marker protein (OMP). The 162-residue amino acid sequence is presented using the single-letter code. α-Helix is represented by H, β-sheet by S, turn by T, and coil by C. OMP has a predicted helical content of 54% and 13% β-sheet. The Delphi program used for this calculation derives from a program written by Morton Kjeldgaad (J. Jenson, personal communication), based on Garnier *et al.* (1978), Levitt (1978), and Lifson and Sanders (1979).

Wallace and P. Greengard, personal communication), to contain carbohydrate, to bind calcium, or to exhibit any evidence for the presence of a number of common enzymatic activities, suggest that it may play a unique role in olfactory neuron function.

4.3. Hypothetical Approaches to the Function of Olfactory Marker Protein

The cytosolic nature of OMP suggests that it does not participate in any membrane event associated with sensory transduction but might be a cytoplasmic enzyme. However, it is possible that some fraction of OMP redistributes among cellular compartments in response to specific cellular activity. Thus, it may be primarily or exclusively cytoplasmic under normal extraction conditions but transiently associated with membrane components when functional. The report by Pfister *et al.* (1985) suggests that a hypothetical analogy can be drawn between OMP and the 48,000-M_r retinal S antigen (arrestin). This latter protein is present in photoreceptor cells of the retina and in pinealocytes, where it behaves as a major cytoplasmic protein of dark-adapted retina based on extractability and immunocytochemical localization. However, it binds specifically to photoexcited phosphorylated rhodopsin in membranes and quenches the activity of the light-dependent cGMP–phosphodiesterase in rod outer segments. Thus, the 48,000-M_r protein plays a regulatory role in the light-induced cascade that controls cGMP hydrolysis during phototransduction by virtue of its redistribution among cellular compartments. The apparent subcellular distribution of transducin and the cGMP phosphodiesterase was also found to be dependent on extraction conditions. These comments are not intended to imply that OMP manifests one of these activities but are raised to suggest that OMP may well be involved at some step of the transduction sequence in olfactory neurons. The availability of isolated olfactory cilia (Rhein and Cagan, 1980; Pace *et al.*, 1985; Anholt *et al.*, 1986), patch-clamp olfactory neurons (Maue and Dionne, 1984; Trotier and MacLeod, 1986), and ciliary membrane fragments (Nakamura and Gold, 1987), as well as reconstituted lipid bilayers (Vodanoy and Murphy, 1983), suggests that these may be possible approaches for testing of either OMP is involved in the olfactory transduction cascade. One could evaluate the influence of either OMP or its antibody on odorant-induced changes in ion conductances, adenyl cyclase, or polyphosphoinositide hydrolytic activity (Huque and Bruch, 1986) to address its possible role in the early biochemical events associated with odorant transduction.

4.4. Characterization and Cloning of the mRNA and Gene

Clearly, additional methodology is required to address continuing questions of the function of OMP and the exquisite cellular and developmental regulation of its expression. In a broader sense, this relates to the general question of regulation of gene expression during neuronal development and differentiation. To pursue these questions, it would be useful to have cloned olfactory neuron cell lines, but this goal has yet to be realized (Schubert *et al.*, 1985; Goldstein and Quinn, 1981). Equally desirable would be the availability of strains of inbred organisms expressing quantitative or qualitative alterations in OMP, or the ability to manipulate the expression of OMP in a temporally or spatially anomalous manner. Thus, it became essential to clone and characterize the mRNA and gene for OMP (Rogers *et al.*, 1985, 1987).

Poly(A^+)mRNA was isolated from rat olfactory mucosa and translated *in vitro* in a rabbit reticulocyte system containing [^{35}S]methionine. Immunoprecipitation of the translated products with OMP antibody produced a single radioactive band that migrated to the 19,000-M_r position of isolated OMP on sodium dodecyl sulfate–polyacrylamide gel electrophoresis (SDS–PAGE). Approximately 0.5% of the total 35S-methionine that was incorporated was associated with immunoprecipitated OMP, consistent with the proportion of this protein in the tissue. Additional experiments confirmed that the primary *in vitro* translation product is indistinguishable in size from OMP isolated directly from tissue. Similar results were obtained following mRNA translation in a wheatgerm derived incorporation system. Thus, OMP is produced directly without the intermediate formation of a larger polypeptide precursor (Rogers *et al.*, 1985).

Olfactory marker protein consists of 162 amino acid residues (Fig. 1). The triplet code thus predicts a minimum mRNA coding length of 486 nucleotides. However, it was evident that the mRNA for OMP was significantly larger. Thus, size fractionated mRNA from rat olfactory mucosa was used to construct a cDNA plasmid library from which OMP clones were identified and characterized (Rogers *et al.*, 1987). Clones selected initially by hybridization to a synthetic oligonucleotide predicted from the primary amino acid sequence of rat OMP were further confirmed by hybrid selected translation and ultimately by nucleotide sequencing of the entire 2149 nucleotide cDNA insert (Fig. 2). The nucleotide sequence contained an open reading frame of 486 nucleotides that corresponded exactly to that expected for OMP, confirming the amino acid sequence obtained by protein chemistry. The open reading frame was followed by 1630 nucleotides of 3′-untranslated trailer sequence (of unknown function) including the 3′-polyadenylation signal 16 nucleotides upstream of the poly A tail. RNA blot analysis of olfactory mucosa RNA using this insert as a probe indicated that OMP mRNA was 2300 nucleotides in length. Thus, the insert in pOMP lacks less than 100 nucleotides from the 5′ leader sequence. The open reading frame for OMP is terminated by the pentanucleotide TGATG. This sequence contains a TGA termination signal frame shifted with an ATG initiator codon. Recent studies (Peabody and Berg, 1986) indicate that such termination starts to occur in the *trp* operon of *Escherichia coli*. The possible significance of this mechanism in eukaryotes is unknown, but one could search for the predicted, extended-OMP to test for the occurrence of this phenomenon.

Surgical or chemical lesion of the olfactory mucosa results in neuronal degeneration and loss of OMP. Axotomy or bulbectomy have similar effects on OMP (for review, see Margolis, 1985, 1980*a*). We have also shown that the mRNA for OMP virtually disappears within 3 days after bulbectomy and partially reappears after 1 month (Rogers *et al.*, 1987). This finding is consistent with recent reports that olfactory neurons can synthesize some OMP even in the absence of their normal target (Barber *et al.*, 1982; Chuah and Farbman, 1983; Monti-Graziadei, 1983). OMP cDNA does not manifest significant hybridization with RNA isolated from many nonolfactory tissues, consistent with our prior studies of immunological specificity and the lack of homology with any of the amino acid sequences in the NBRF data base. The inability of this nearly full-length cDNA to hybridize with RNA of nonolfactory tissues suggests that OMP mRNA is not derived from a primary gene transcript that is differentially processed in different tissues. An unexpected observation was the detection of a small quantity of OMP mRNA in the olfactory bulb, which is the site at which the olfactory neurons synapse. One explanation for this could be that, as in the squid giant axon (Giuditta *et al.*, 1986), some mRNAs are transported into axonal cytoplasm in the olfactory nerve. It is unknown whether this phenomenon is a property of OMP mRNA, of olfactory neuronal mRNAs, of sensory neuronal mRNAs, of mRNAs in neurons with long axons, or of abundant mRNAs, some small fraction of which enters the axoplasm along with other cytoplasmic components. Since ribosomes are thought not to be in the synaptic terminals, the function of this mRNA is unclear. An alternative explanation for the presence of OMP mRNA in the bulb is the existence of a small population of cells that contain OMP mRNA but have not been located by immunocytochemical staining. These alternative suggestions may ultimately be resolved by *in situ* hybridization studies. Cloning of additional olfactory neuron-specific genes would also help to address the generality of this observation.

Figure 2. Nucleotide sequence of the pOMP cDNA insert corresponding to rat OMP mRNA. The bases are numbered in the 5′–3′ direction. The translation of the open reading frame is illustrated and corresponds exactly to the amino acid sequence determined for the isolated protein. No other significant open reading frame is present. The canonical polyadenylation signal sequence is underlined. OMP, olfactory marker protein. (From Rogers *et al.*, 1987.)

```
                      1
CCCCCCCCCC CCG GCA GAG GAC GGG CCA CAG AAG CAG CAG CTG GAT ATG CCG CTG
               Ala Glu Asp Gly Pro Gln Lys Gln Gln Leu Asp MET Pro Leu
       50
GTT CTG GAC CAG GAC CTG ACT AAG CAG ATG CGG CTC CGA GTA GAG AGC CTG AAG
Val Leu Asp Gln Asp Leu Thr Lys Gln MET Arg Leu Arg Val Glu Ser Leu Lys
 100                                                                150
CAG CGC GGG GAG AAG AAG CAG GAT GGT GAG AAG CTG CTC CGG CCG GCT GAG TCT
Gln Arg Gly Glu Lys Lys Gln Asp Gly Glu Lys Leu Leu Arg Pro Ala Glu Ser
                                                            200
GTC TAC CGC CTT GAT TTC ATC CAG CAG CAG AAG CTG CAG TTC GAT CAC TGG AAC
Val Tyr Arg Leu Asp Phe Ile Gln Gln Gln Lys Leu Gln Phe Asp His Trp Asn
                                                        250
GTG GTT CTG GAC AAG CCG GGC AAG GTC ACC ATC ACG GGC ACC TCG CAG AAC TGG
Val Val Leu Asp Lys Pro Gly Lys Val Thr Ile Thr Gly Thr Ser Gln Asn Trp
                                                    300
ACG CCA GAC CTC ACC AAC CTC ATG ACA CGC CAG CTG CTG GAC CCT GCT GCC ATC
Thr Pro Asp Leu Thr Asn Leu MET Thr Arg Gln Leu Leu Asp Pro Ala Ala Ile
                                                350
TTC TGG CGC AAG GAA GAC TCC GAT GCC ATG GAT TGG AAT GAG GCA GAC GCC CTG
Phe Trp Arg Lys Glu Asp Ser Asp Ala MET Asp Trp Asn Glu Ala Asp Ala Leu
                                            400
GAG TTT GGG GAG CGC CTT TCT GAC CTG GCC AAG ATC CGC AAG GTC ATG TAT TTC
Glu Phe Gly Glu Arg Leu Ser Asp Leu Ala Lys Ile Arg Lys Val MET Tyr Phe
                                    450
CTC ATC ACC TTT GGC GAG GGT GTG GAG CCC GCC AAC CTA AAG GCC TCT GTG GTG
Leu Ile Thr Phe Gly Glu Gly Val Glu Pro Ala Asn Leu Lys Ala Ser Val Val
                                500
TTT AAC CAG CTC TGA TGGCAGCCGTG GCCTGCCTTCGGCCCCCACTCTCCCTTGGCTGGACCTCC
Phe Asn Gln Leu

       550                                                       600
TAGCTCATGTG TATTTTGGAAACATTCTTCTAGCTGTTCCTTCTGTGCTCATCTTGGCTAG AGGTCCCCT
                                               650
GAGTGCTACACCCGCTCTTTTTCCCTGGTGTCAGTGCCACGG CTCACAGGGATGTCCCATGGCTTCATAG
                  700
TCTAGAAGCTGGACGCTGCTA TCTCTAGACAGTAGAGGCCTTTTGGGTCCATGTGGCCAGAGGGATGAGC
                                                     800
C TCTTGGCCACCTGCCATCTCTGCTTTATTGTGGTGAAGAACAGGATTGAG AGAGAAAAGAGACTGACCA
                              850
AGAAATGCCAACGGCCATCATGATTCCTCCC TTTGGGGACAAGAGGCTGAGACTGGACAGGAACACCTTC
       900                                                       950
CAGGGATCCGG GGGAGAAGGCTTTTCCCTGCTGGCCAAAGCTGGAACCAGGAGGTGAATAC CCAGCAGCT
                                         1000
GCACATCGGCAGCAGGAAGGTGCTTCTTCCAGTGTTGGCAT CAGCCCGCGGTGACCTTAGGGCCTTCCAG
                1050
ACACTTGGCGGGATGACAGC AGGGCTTGATCTGACTGGTTTTCCAGGTCTGGCCCCGGTTTTTATGGAGT
                                                    1150
CGTGAGAGAACGCGTAGAAACGGAAACAGCCCTAAGCTACCTATACTCATA AGTATATTGAGAAATAGCC
                              1200
TGACTGTATCTGTATGGATGTGTGCCTGAGA GCTATTCCTAGTCAGGCGTAAGGCTAACTCTAGTTTAAT
       1250                                                     1300
TGTTGAGCTGG TACTGGTTTGTGGCCTTGGTGGAAGTGACCCTGGCTAAGCCTTCCTTGGT ACAGTGCTC
                                         1350
TTTGAACTGGGGGACTGAGGCTCAAATGGTGAAGCAGAGAA CTGCATTAGAGGGGTCCAGGACTTTGAGC
                 1400
TAGAAACACTTCCATTAGGAA GGCTGGCATTTGCTGATCACTGGTCATGGTGTTGTATCCTGCTGTCACC
                                                    1500
TCCCTGGGCATTTCCTAGCTGTTTTCCTGGAGTGAGGGCACATGGGTAAGG CTTGGGGGCAGTTATGATG
                              1550
CCTGACATCTGATGTGTGCTGGAGCTGTCCG GCTATGATGCCTAGTACTGGCCTCAGCTGTCCAGGACAG
       1600                                                     1650
CCACTCAGCAA ACTGACAGAAAAACTATGGCACAGTTATCAGCAGATTCAACCCTGCCCCA AGTCTCATT
                                         1700
GTGCCTCACCCTGCACATCCTGAGAGCCTCTCATTGAGGAG AAGCCTCACCTGTCACCTAGCATCAGCCA
                 1750
GGGGCACCCCAGCAAAGCCCT CGACTCTGTCTCAAGGCTGCCTCCATTGGGGATACCAAGATCTGAAGGT
                                                          1850
A GGAGTTGTCCCGGTGGGTGGGTTGTTAGAAGGGCAAGCCCTAGCTTAGATTCAGGAT TCTGGAGGCAGG
                              1900
AGATAGGCTGTGGTACCTACCGGCTCTTCTA CTGGTGCCTCTGCATCGGCCAGTGCTCTGCACTTGCTGA
       1950                                                     2000
CTCTAGGGAGC CATACCTAGACAGACCTACCTTTCTGCTTCTCTTTCTGCCTCTCCCTACA GCTTTAGAG
                                         2050
ACTCCTTTCACACTGCCAGACCCCCAATTCTGTCTCACTCC ATTTGCCCTATGGACAGTTGTGTCTCTG
                 2100
CTGTGCCTGTCACACAATAAA GACTGTATGCCCTCCCAAAAAAAAAAAAAAAAAAAA
```

Most recently, we have used the cDNA insert of pOMP to identify clones in human and rat genomic libraries constructed in λ-phage Charon 4a (K. Rogers, A. Tsai, and F. L. Margolis, unpublished observations). Nucleotide sequencing of a subcloned restriction fragment of the rat genomic clone that overlaps the 5′ end of pOMP has further confirmed the sequence of that portion of the cDNA clone. Additional preliminary results indicate that the rat OMP mRNA and the rat OMP gene are similar in size. If confirmed, this would suggest that the OMP gene consists primarily of exonic sequences.

4.5. Future Directions for Olfactory Marker Protein

We can now begin to characterize the gene for rat OMP and address the mechanisms that restrict its expression to a single class of neurons at a specific time in their development. Thus, we are in position to address certain critical questions relating to OMP but that have broader and more general implications with regard to both olfactory function and neuronal differentiation. Among the possibilities that can be considered are the following:

Is the OMP gene methylated or otherwise modified in olfactory neurons as compared to other cells?
Do olfactory neurons contain specific DNA binding proteins?
Are there DNA sequences that are preferentially associated with genes expressed in olfactory neurons?
Can we selectively induce tumors of olfactory neurons for conversion into cell lines?
What are the characteristics of nonolfactory cells that are induced to synthesize OMP?
Can we generate organisms that synthesize a selectively altered OMP in the olfactory neurons?
Can we block OMP synthesis by introducing a gene for antisense OMP mRNA?
Are there OMP-like genes in invertebrate olfactory systems?
Can we direct olfactory neurons to synthesize proteins that they do not normally express?

5. LESION-INDUCED CHANGES IN mRNAs

5.1. In Vitro Translation

A variety of approaches are available to permit identification of gene products that are differentially expressed in specific subsets of cells, in olfactory versus nonolfactory tissues, at particular times during cellular development or in response to selected perturbations in normal physiological homeostasis. Thus, a variety of molecules have been identified that are preferentially or exclusively expressed in olfactory tissue and that have been used subsequently as selective probes of function of this tissue (e.g., see Section 1, and Margolis *et al.*, 1985).

Since functional mRNAs are one of the initial products of gene activity, it seemed probable that significant differences in mRNA populations would occur in response to lesion or odorant stimulation. Therefore, RNA isolated from several tissues, including intact and lesioned olfactory mucosa, was used to direct [^{35}S]methionine incorporation in a rabbit reticulocyte lysate. The radiolabeled translation products were separated by one-

and two-dimensional acrylamide gel electrophoresis. Since the reticulocyte lysate system used for *in vitro* translation lacks membranes, it does not have the capability to glycosylate or process proteins as intact cells do, although it is capable of some modification reactions such as acetylation and phosphorylation. Thus, in the reticulocyte lysate, mRNAs direct amino acid incorporation into unmodified primary translation products. It was evident even in one-dimensional denaturing gels that the translation products derive from olfactory mucosa manifest a few easily identified novel bands of incorporation of radioactivity at 60,000, 50,000, and 20,000 M_r that are apparently absent from lung, liver, and brain (M. Grillo and F. L. Margolis, unpublished observations). Thus, some genes that are differentially expressed in this tissue are readily detected. The extent of incorporation also indicates that some of these mRNAs are quite abundant, implying that their translation products are of high abundance or rapid turnover or both. Curiously, when companion denaturing gel electrophoresis was performed on protein extracts of the tissues from which the mRNA was isolated, we could not identify a Coomassie blue-stained band that corresponded to the observed primary translation product at 60,000 M_r. This suggested either that the primary translation product has a short half-life and did not accumulate to any significant extent or that it was processed *in vivo* to a different molecular-weight species. This latter interpretation proved to be the case and is considered separately below.

More detailed evaluation of these translation products by two-dimensional gel electrophoresis (Fig. 3) indicated that not only are there tissue-specific differences but, as anticipated, there are changes in mRNA populations in response to olfactory nerve lesion and reconstitution as well (Khew-Goodall *et al.*, 1987; Wensley *et al.*, 1986). Rats were subjected to bilateral olfactory bulbectomy at 4 weeks of age. Olfactory bulbs and turbinates were obtained from intact rats at 4½ weeks of age and at 4 and 9 days after surgery. RNA isolated from these tissues was used to direct [^{35}S]methionine incorporation *in vitro* in a rabbit reticulocyte lysate. The radiolabeled translation products were subjected to two-dimensional gel electrophoresis followed by fluorography, essentially as described by O'Farrel (1975) and modified by Protein Databases, Inc. The first dimension was isoelectric focusing for separation by charge and the second dimension was in the presence of SDS for separation based on size.

Comparison of the autoradiograph (Fig. 3) indicated that there were several categories of translation products. First, are those translation products that are present in intact olfactory tissue and not in bulb and that manifest quantitative alterations in response to the lesion; second, are those that appear only in response to the lesion; and third, are those that are clearly present in both bulb and mucosa and whose level is altered or unchanged in response to the lesion.

This last group presumably includes and reflects the responses of a large number of housekeeping proteins present in many cell types. The autoradiographic spots clearly present in both bulb and mucosa from intact rats and that respond to the lesion are indicated by letters in Fig. 3a. Positions A–K, which include actin, all show an increase in intensity by 4 days after lesion, with little additional change by 9 days, except for the spot at Position h, which decreases in intensity at 4–9 days. At position L, which is probably calmodulin, the intensity is decreased at 4 days and then increased at 9 days to an intensity similar to the presurgical value. Not all translation products respond to the lesion, e.g., the group slightly larger in molecular weight and more basic than A–E seem unaltered.

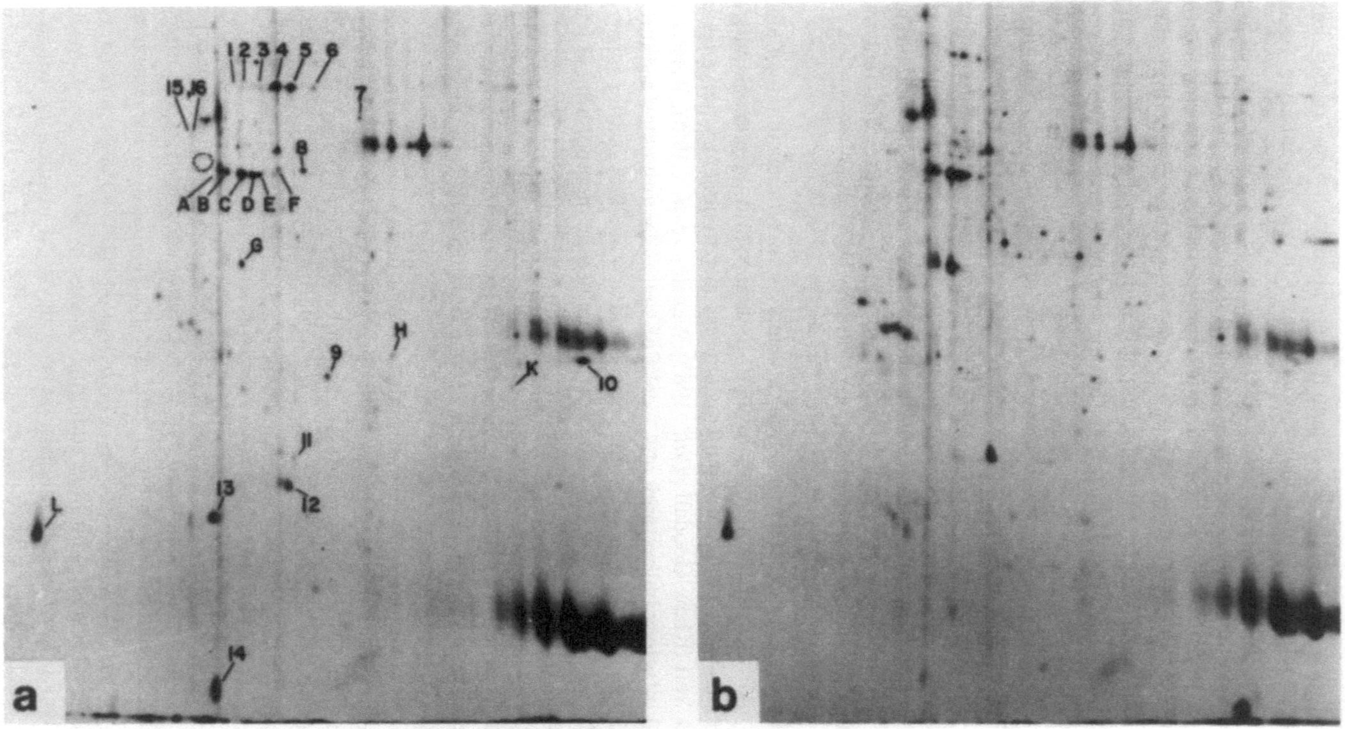

Figure 3. Fluorographs of two-dimensional gel electrophoresis of [^{35}S]methionine-radiolabeled products from *in vitro* translation of rat tissue mRNAs. Olfactory mucosa (a) and olfactory bulb (b) from intact rats. Olfactory mucosa, 4 days (c) and 9 days (d) after nerve lesion by bulbectomy. (a) The dotted circle indicates the location of the ovalbumin standard. One group of products (e.g., A–K) increase in intensity with the lesion but with different time courses and are seen in both mucosa and bulb. A second category (e.g., 1–6) are restricted to olfactory mucosa and increase after lesion, while components 15 and 16 seem to appear only after lesion. By contrast, products 11–14 all decline with time after lesion, especially product 13, which is due to OMP. Thus, the levels of specific expression of various mRNAs are individually regulated in this tissue in response to nerve lesion.

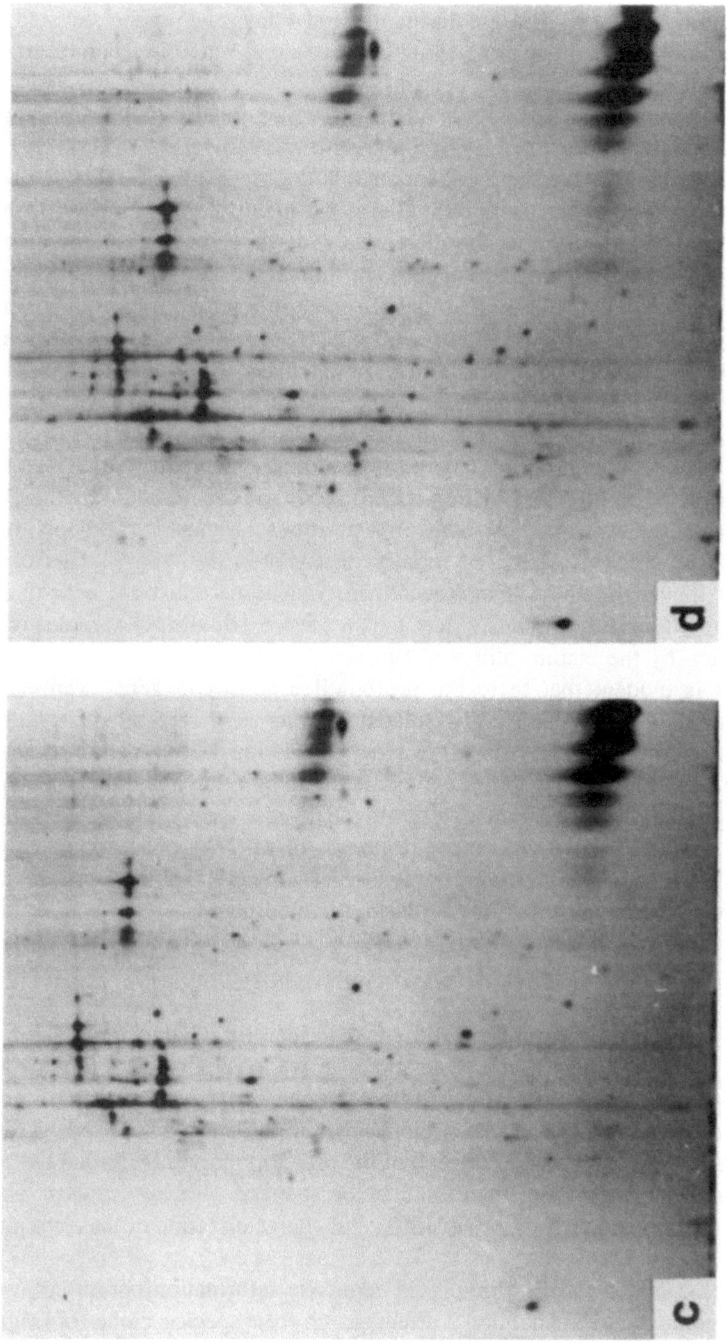
d
c

In the second category noted above, i.e., translation products that seem to appear only after lesion, there are two spots identified as 15 and 16. These primary translation products of about 50,000 M_r increase during the first 4 days and remain elevated at 9 days. These may possibly be cellular proteins that have been elevated in response to the lesion or proteins induced in the immature neurons, which have now been activated to undergo mitosis. The proteins associated with axon growth cones are also of this size range but are of more acidic pI (Zwiers *et al.*, 1985; Skene *et al.*, 1986).

In the third category are a group of spots that are present exclusively or primarily in olfactory mucosa as compared with bulb. These are identified by numbers. Inspection of the autoradiographs indicates that these are of a wide range of intensities reflecting their range of mRNA abundances. Many of these are increased at 4 days after lesion and are still at an elevated level at 9 days. This class is exemplified by the spots at positions 1–6, which represent a group of proteins of ~60,000 M_r. It is not clear whether they are a family of related proteins or are simply of similar mass. By contrast, at position 9 the elevated level of incorporation continues to increase further from 4 to 9 days, suggesting its involvement in olfactory neuron development. At position 12, there is little change in intensity by 4 days, and not until 9 days is there any obvious reduction. As expected, the primary example of an olfactory neuron-specific protein is represented by the OMP at 13 that has essentially disappeared by 4 days. At positions 11 and 14, two spots, of low and high intensity respectively, have essentially disappeared by 4 days. This time course suggests that they derived from mature olfactory neurons that have now degenerated. These last three translation products may thus represent additional examples of selective gene expression by the mature olfactory neurons.

Thus, it is evident that there are profound responses in gene expression by the olfactory mucosa when it is surgically isolated from its target, the olfactory bulb. mRNA activity as monitored by 2D gels of *in vitro* translation assays demonstrates that the transcription of tissue specific genes is modulated up and down and that the synthesis of novel gene products can be dramatically altered in this tissue. Thus, even though this approach is limited in sensitivity to the responses of mRNAs of reasonable abundance, significant differences in intertissue gene expression are apparent, as are intratissue changes in response to perturbations of normal homeostasis.

5.2. Cloning of Additional mRNAs from Nasal/Olfactory Tissue

We now have the opportunity and the task to identify, characterize, and understand these events in this complex tissue. One approach we have pursued is to generate cDNA libraries from intact and bulbectomized olfactory mucosa as well as from olfactory bulb and to use several strategies to identify clones that are olfactory tissue specific and whose levels are modulated in response to lesion of the olfactory nerve. Differential screening of cDNA libraries prepared from intact and bulbectomized olfactory mucosa should theoretically permit the identification, isolation, and characterization of these various classes of tissue-specific mRNAs.

In the case of the OMP, amino acid sequence information permitted synthesis of predicted oligonucleotides, enabling a direct search for a specific clone for OMP (Rogers *et al.*, 1985, 1987). This approach is also applicable if one is looking for a homologue of a known protein. Otherwise one could attempt to obtain amino acid sequence information from the radiolabeled translation products or use one of a variety of differential probing techniques to isolate corresponding cDNA clones from an olfactory tissue cDNA library. There are several options: (1) generate a cDNA library from the mRNA remaining after removal of nonolfactory messages by subtractive prereaction with cDNA from other tissues such as brain or liver (this should permit creation of a library containing primarily or exclusively olfactory derived clones); (2) create a complete olfactory cDNA library and probe it with a subtractive probe generated as above to identify only those clones that are primarily olfactory in origin; (3) differentially probe the complete library above with radiolabeled probes from olfactory and other tissues to identify clones of olfactory origin; or (4) use specific antibodies to probe an expression library to identify clones of interest. This latter approach presupposes knowledge of a specific protein or function that is being sought. Authentication of the cell type of origin of the isolated clone requires either the ability to generate antibodies, for immunocytochemical localization, directed against the predicted amino acid sequence or the ability to use *in situ* hybridization to identify the cellular origin of the mRNA from which the cDNA clone derived. Alternatively, if trying to identify components of the stimulus transduction or detection systems, then it might be appropriate to monitor functional activities following DNA transfection into cells or mRNA translation in frog oocytes.

Several of these approaches are being pursued in a number of laboratories that are studying the olfactory system (Khew-Goodall *et al.*, 1987; Dascal *et al.*, 1986; Lee *et al.*, 1987; Wensley *et al.*, 1986). A cDNA clone reported by Lee *et al.* (1987) predicts a protein sequence homologous to retinol binding proteins. *In situ* hybridization techniques have indicated that the mRNA corresponding to this clone is present in the cells of Bowman's gland. This cellular localization and the predicted binding properties suggest that it is related to the pyrazine binding proteins discussed earlier. A number of other clones have been identified, but their unambiguous derivation from the olfactory neuroepithelium, as opposed to the tissue of the olfactory/nasal mucosa, is not yet certain. Our experiences with p2/122 are detailed by way of illustration.

Encouraged by the changes in mRNA expression reflected in the polyacrylamide gel electrophoretic patterns of *in vitro* translation products (Fig. 3), we elected to pursue strategy 3 above. The rationale is diagrammed in Fig. 4. When an olfactory tissue cDNA library was probed with [^{32}P]-cDNA from several tissues, a subset of clones were obtained that initially fulfilled the criterion of being preferentially olfactory in origin. Plasmids from clones that did not hybridize with pOMP were isolated and their inserts obtained and compared by hybridization to eliminate replicate isolates. The cDNA inserts were then used to determine the size, sequence, protein product, and tissue distribution of the mRNAs from which they derive. Hybrid-select translation of a clone identified as p2/122 demonstrated its ability to select an abundant mRNA that directed the synthesis of a 60,000-M_r protein. When used to probe RNA blots, this clone reacted with a 2.1-kb band present only in nasal/olfactory mucosa and undetectable in more than a dozen nonolfactory tissues. These data suggested that it was the clone for the 60,000-M_r primary translation product initially observed on one- and two-dimensional gel electrophoresis (Fig. 3). Does this mRNA derive from olfactory neurons or from another of the cellular populations in the olfactory or nasal mucosa? The most direct approach to identify the cell type that contains the mRNA in question would be to apply the technique of *in situ* hybridization as used by Lee *et al.* (1987). This would be especially valuable if the total

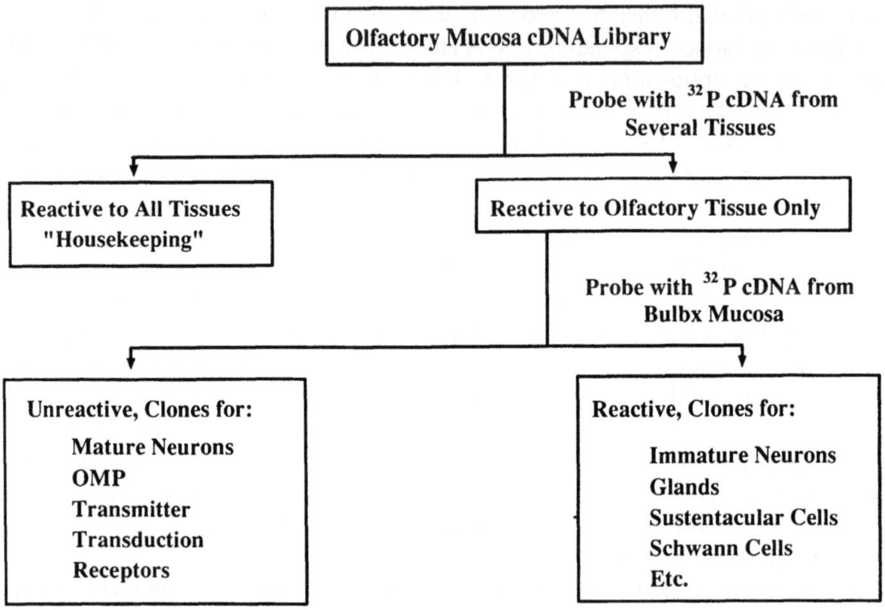

Figure 4. Strategy for selection of cDNA clones for specific cell types by differential probing of a library prepared from olfactory mucosal mRNA. Reciprocally, a cDNA library prepared from lesioned mucosal mRNA could be used to select clones representing functions expressed early in mitotically activated progenitor cells, immature neurons or non-neuronal cells. The selectivity of this approach would be greatly enhanced if the individual cell populations could be subjected to prior isolation and separation.

tissue level of the mRNA is low because it is restricted to a minor cell type where it is abundant (Mandel and Goodman, 1987; Schwartz and Costa, 1986). Our attempts at *in situ* hybridization of p2/122 to sections of olfactory mucosa were repeatedly frustrated by high nonspecific background when using either the cDNA insert or a cloned cRNA derived from it. This problem plagued us even for the known positive controls derived from pOMP. Therefore, we evaluated the ability of tissue RNA to hybridize to p2/122 after olfactory nerve lesion by bulbectomy. Decline of this mRNA following lesion would, in analogy to OMP, be *prima facie* evidence for the presence of 2/122 in mature olfactory neurons. Increase following such lesion might imply its presence in a population of activated stem cells or reflect a gene expressed in a non-neuronal cell type that responds to neuronal lesion; a treatment that is thought to have no effect on the non-neuronal cell populations in this tissue. If this mRNA is unchanged following lesion, we would be encouraged to conclude that it is expressed in a non-neuronal population of this tissue. The level of mRNA hybridizing to p2/122 changed in a manner that was largely independent of the interval after lesion or even of the presence of lesion. Subsequently, it became apparent that the critical consideration was the extent to which the dissected olfactory mucosa was contaminated with nonolfactory nasal mucosa. Thus, RNA hybridizing to p2/122 was also present in anterior nasal mucosa that was devoid of any olfactory tissue. The abundance of this mRNA coupled with its presence in nasal mucosa suggested that it might be present in one of the populations of gland cells contributing to the secretions of the nasal mucosa. If true, it might play a role in regulating access to, or removal of, stimuli from the olfactory neurons as has been suggested for the secretions of the sustentacular or Bowman gland cells (Getchell *et al.*, 1984*b*). To identify further the cellular product of this clone, a partial nucleotide sequence of the cDNA insert was determined. It contained a large incomplete open reading frame which predicted a 98-amino acid carboxy terminal translation product that was absent from the NBRF sequence data base. This was consistent with the RNA plot experiments indicating the tissue specificity of this material. Two peptides predicted from this open reading frame were synthesized by Dr. W. Danho, independently coupled to thyroglobulin and antipeptide rabbit antibodies produced in rabbits. Each of these antibodies was specific for the inducing peptide but, when tested against rat tissue extracts, both antisera gave the same pattern of immunoblot reactivity indicating that they were recognizing the same macromolecule. The results with one antiserum are presented in Fig. 5. Tissue extracts were electrophoresed in SDS–PAGE, transferred to nitrocellulose, and probed with either of the antipeptide antisera. For each antiserum, a single band was observed at about 71,000 M_r in soluble extracts of

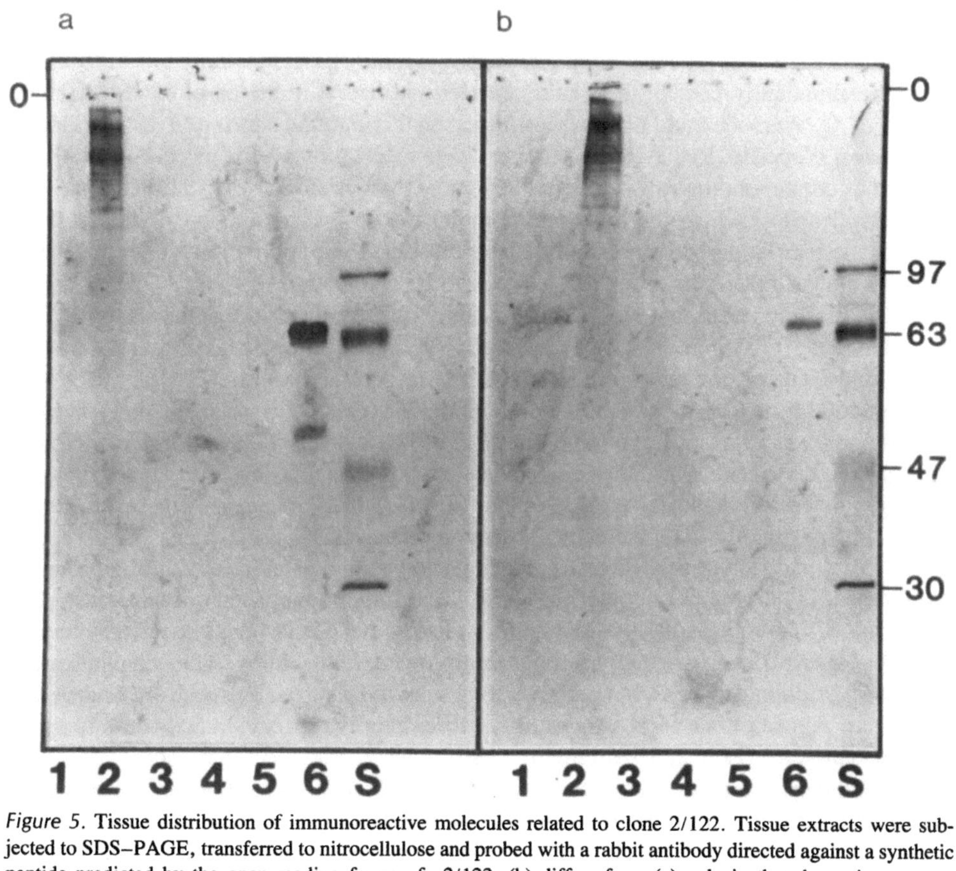

Figure 5. Tissue distribution of immunoreactive molecules related to clone 2/122. Tissue extracts were subjected to SDS–PAGE, transferred to nitrocellulose and probed with a rabbit antibody directed against a synthetic peptide predicted by the open reading frame of p2/122. (b) differs from (a) only in that the antiserum was pretreated with the immunizing peptide to demonstrate the specificity of the immunoreaction. Lane 1, esophagus; 2, trachea; 3, gut; 4, lung; 5, CNS; and 6, nasal mucosa. Note the high specificity of the antiserum for nasal mucosa in which the major immunoreactive tissue product is about 71,000 M_r.

nasal mucosa but not in lung, trachea, esophagus, gut, or CNS. In nasal mucosa the immunoreaction with either antiserum could be eliminated by pretreatment with the appropriate synthetic peptide. Incubation of the extracts with endoglycosidase-F prior to electrophoresis and immunoblotting resulted in a reduction of mobility consistent with a reduction in molecular mass to the size of the 60,000-M_r primary translation product, indicating that the material is an *N*-glycosylated glycoprotein (Freeze and Varki, 1986). This, coupled with its presence in soluble tissue extracts, and the abundance of its mRNA, suggested this cellular product is destined for secretion. Preliminary immunocytochemistry with these antisera indicates that they stain a population of submucosal, nongoblet secretory glands in the nonolfactory nasal respiratory mucosa of the rat (M. Grillo and F. L. Margolis, unpublished observations). This staining is also seen in surface patches and can be eliminated by pretreatment of the antisera with the appropriate synthetic peptide. A similar observation has been made in the nasal mucosa of the salamander using these antisera (M. Getchell, personal communication) indicating the phylogenetic conservation of this material. The relationship of this cellular product to those reacting with specific monoclonal antibodies identified by various investigators is currently unknown. Clone p2/122 is indistinguishable from one independently isolated and generously shared by R. Reed and E. Barbatos (personal communication) that they have shown by *in situ* localization to be abundantly present in an area anterior and lateral in the nasal cavity. A clone isolated by C. Wensley and D. Chikaraishi (personal communication) has very similar, if not identical properties to these clones. Its *in situ* hybridization profile has been tentatively assigned as corresponding to the zygomatic gland, which is similar in location to that seen by R. Reed and E. Barbatos (personal communication). The function of this cellular product, the regulation of its metabolism and presumed secretion and its significance to olfactory and/or respiratory function remains to be elucidated.

At least three additional clones have recently been isolated in this laboratory (Y.-S. Khew-Goodall, T. Goren, J. Miara, and F. L. Margolis, unpublished observations) by probing libraries constructed in λgt_{10}. Two of these, λ21 and λ55, react with mRNAs of 1.1 and 4400 M_r, respectively, and are preferentially expressed in nasal/olfactory mucosa and little or not at all in several other tissues (Fig. 6). However, their response to olfactory nerve lesion is variable and their cellular localization is as yet uncertain. A third clone, λ84, hybridizes to two mRNAs of 2.0 and 4.5 kb, which are present in several brain regions but are barely detectable in non-neural tissues. The proportions of these two mRNAs vary in different neural tissues, waxing and waning differentially during postnatal development (T. Goren, J. Verhaagen, and F. Margolis, unpublished observations). The 2.0-kb mRNA was particularly enriched in olfactory mucosa (which is nearly devoid of the 4500-M_r form) and it virtually disappears following bulbectomy. The behavior of the mRNAs hybridizing to this clone suggest they may be neuronally localized and derive from a gene whose primary RNA transcript is differentially processed in response to tissue site and ontogeny (Leff and Rosenfeld, 1986). The identity, cellular localization, ontogeny, and response to lesion and recovery of these two mRNAs are under investigation. The current status of our information about these several clones is summarized in Table 1. Thus, many genes are selectively expressed in this tissue and respond to olfactory nerve lesion. Their roles in stimulus detection and processing and in neural plasticity are under investigation. Equally important is the use of these clones in studying mechanisms regulating cell-specific gene expression in this neuroepithelium.

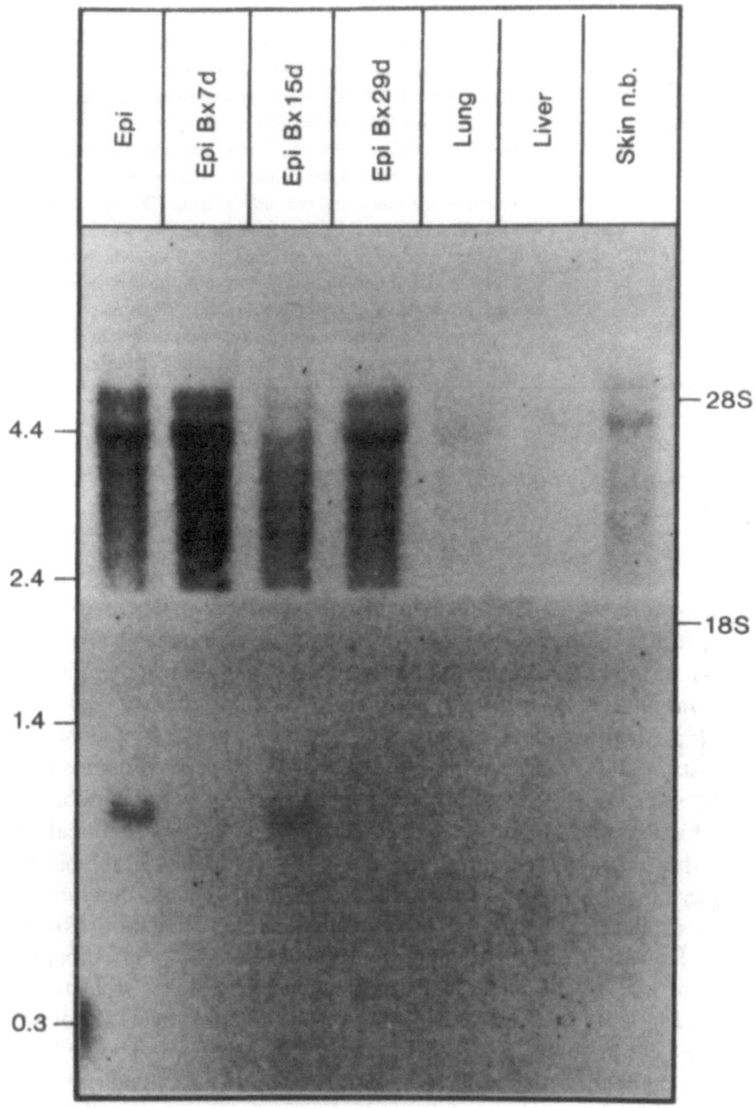

Figure 6. Tissue specificity and response to bulbectomy of two clones isolated from a λgt_{10} library screened by a strategy similar to that described in Fig. 4. RNA blots (2 μg/lane) were probed with cDNA inserts from clones λ21 and λ55. Both exhibit high tissue specificity and, in the olfactory epithelium, seem to respond reciprocally to the surgical removal of the olfactory bulb (Bx). These clones may represent non-neuronal mRNA in the nasal/olfactory mucosa. Under identical conditions, the mRNA for olfactory marker protein declines dramatically within three days and remains very low.

Table 1. Properties of cDNA Clones Isolated from Rat Nasal/Olfactory Mucosa

Clone	Insert	mRNA	Properties
pOMP	2.2 kb	2.3 kb	Codes for a 19,000-M_r, cytoplasmic protein, restricted to mature olfactory receptor neurons, declines on lesion, absent from 16 other tissues
p2/122	0.7	2.1	Codes for a 71,000-M_r, secretory glycoprotein restricted to nasal glands, absent from 18 other tissues
λ21	0.6	1.1	Present in olfactory mucosa, lesion response variable, absent from E17 CNS, liver, lung, spleen, and newborn skin
λ55	1.4	4.4	Present in olfactory mucosa, absent from CH, E17 CNS, spleen, liver, low in newborn skin, and trace in lung
λ84	0.4	2.0, 4.5	Olfactory mucosa highly enriched in 2.0 kb, which is developmentally regulated and declines on lesion; both present in several neural tissues but proportions of two mRNAs vary; levels of both mRNAs developmentally regulated in regions of CNS; in spleen, lung, liver, and respiratory mucosa, only the 2.0 kb is detectable at very low levels

6. CARNOSINE SYNTHETASE

6.1. Overview

Olfactory nerve axons terminate in the olfactory bulb, where they synapse with dendrites of mitral and tufted cells and with a population of biochemically diverse intrinsic neurons. None of the known neuropeptides or classic neurotransmitters has been demonstrated to be present in these olfactory terminals (Macrides and Davis, 1983; Halasz and Shepherd, 1983). Investigations to determine whether a novel compound was playing a transmitter role at this synapse demonstrated that the dipeptide carnosine (*b*-alanyl-L-histidine) was selectively present in the olfactory tissue of several vertebrates (Margolis, 1980*b*). Autoradiographic (Burd *et al.*, 1982) and immunocytochemical (Sakai *et al.*, 1987) studies have demonstrated the presence of this compound in the olfactory neurons and in their terminals in the olfactory bulb. In addition, synaptosomal carnosine can be released by a calcium dependent, potassium induced depolarization (Rochel and Margolis, 1982). However, direct attempts to determine whether this dipeptide has electrophysiological effects consistent with its postulated role as a transmitter have been equivocal (see Margolis *et al.*, 1987 and Margolis, 1980*b* for review). Possibly carnosine functions as a neuromodulator to regulate the efficacy of action of other neuroactive peptides in the complex synaptic glomerular complex of the olfactory bulb. The potent chelation properties of carnosine and the multiplicity of carnosinase activities (Margolis *et al.*, 1983, Margolis and Grillo, 1984*b*) are consistent with this possibility in analogy to the recent proposal by Hokfelt and Terenius (1987). It should be kept in mind that carnosine and its congeners anserine and homocarnosine, as well as their common biosynthetic enzyme, carnosine synthetase, are present in several excitable tissues such as brain, muscle, and retina (Crush, 1970; Margolis and Grillo, 1984*a;* Margolis, 1980).

Unlike many peptides of biological interest the dipeptide carnosine and its congeners are synthesized enzymatically directly from amino acid precursors (Horinishi *et al.*, 1978) and not by proteolytic processing of larger protein precursors. Thus, dipeptide metabolism must be regulated in part by the properties of the specific biosynthetic and degradative enzymes and by regulation of substrate access to them (Margolis *et al.*, 1985*a*). However, the short half-life of carnosine and the 10-fold higher level of the synthetase activity in olfactory as compared to muscle tissue (Horinishi *et al.*, 1978; Tamaki *et al.*, 1980) suggests that this cytoplasmic enzyme may play a significant role in the olfactory system. Thus, although carnosine synthetase is present in olfactory neurons, many questions with regard to the differential regulation of its expression in these cells and in the other tissues in which it occurs are unanswered.

6.2. Characterization

Characterization of partially purified carnosine synthetase indicates that the same enzyme is capable of synthesizing all of the related dipeptides, albeit at different efficiencies (Horinishi *et al.*, 1978; Bauer *et al.*, 1982*a,b*). How then do some areas of the nervous system, such as the olfactory neuron, contain exclusively carnosine while others contain homocarnosine or anserine (Margolis *et al.*, 1980*b*, 1984*a*)? It became of interest to evaluate whether the same or related enzyme proteins are involved in each tissue, whether the large tissue to tissue variations in synthetase activity are due to the differential modulation of the same gene by tissue specific regulators, or whether multiple genes exist for this enzyme that are selectively expressed in different cells resulting in the formation of a set of different but related carnosine synthetases.

One approach would be to isolate the homogeneous enzyme protein from several tissues and compare directly their physical and enzymological properties. Earlier studies from this laboratory (Horinishi *et al.*, 1978), in contrast to others (Skaper *et al.*, 1973) indicated that the partially purified carnosine synthetase activity of muscle and olfactory tissues exhibited similar enzymological and structural properties. Thus, the enzyme activities from these tissues of rat, mouse, and rabbit were all of similar size (300,000–400,000 M_r) on gel-filtration columns. Isolation from these sources for direct characterization and comparison was hampered by the low abundance of the enzyme protein and by its instability. One could try to identify cDNA clones for this enzyme by generating subtractive probes, subtractive libraries, by the use of differential tissue screens of cDNA libraries, or by transfection into cells that did not normally express this cytoplasmic enzymatic activity. However, the absence of specific antibodies or any amino acid sequence information and its existence as a cytoplasmic enzyme of very low abundance compelled us to conclude that a more direct approach was needed.

6.3. Monoclonal Antibodies

We elected to generate a panel of monoclonal antibodies directed against carnosine synthetase. These antibodies could be used as reagents that would permit comparison of structural aspects of the enzyme protein from different tissues and species even in the absence of homogeneous enzyme protein. They could also be used for immunoaffinity purification of the enzyme and possibly as probes of expression libraries, permitting isolation of cDNA clones for this low-abundance enzyme. In a collaborative study (Margolis *et al.*, 1987), mouse monoclonal antibodies were generated against rabbit muscle carnosine synthetase. Monoclonal supernatants were screened with a solid-phase second antibody protocol that selected for mouse monoclonal antibodies capable of binding enzymatically active carnosine synthetase. Selected clones were expanded *in vivo* in pristane primed mice and ascitic fluids collected.

When incubated with various tissue extracts, these antibodies formed immune complexes that exhibited carnosine synthetase activity. This ability of the antibodies to sequester active enzyme quantitatively from tissue extracts formed the basis for an assay to evaluate the amount and similarity of the synthetase protein in various tissues and species. The synthetase activity of olfactory, brain, and muscle tissue of rabbit was thus shown to be due to either the same protein or at least to a very similar set of proteins that had several epitopes in common (Fig. 7). This conclusion was true for the rat as well as the rabbit. Furthermore, these antibodies showed strong cross-species reactivity, demonstrating that some antigenic properties had been phylogenetically conserved among the synthetases from several mammalian species. Thus, antibody D effectively cross-reacts with the enzymatic activity in rabbit, cow, and monkey. If this cross-reactivity observed with primate extends to human tissue, we may be able to evaluate the possible involvement of carnosine synthetase in disorders of human muscle.

6.3.1. Immunoaffinity Purification

Despite their efficacy in sequestering carnosine synthetase activity from tissue extracts, these antibodies were less useful for immunocytochemical applications. Preliminary immunocytochemical localization of carnosine synthetase (A. Farbman, personal communication) in rabbit olfactory mucosa was consistent with our earlier autoradiographic localization of the dipeptide itself in the olfactory neurons, but the intensity of the reaction was quite weak. Attempts to use these antibodies for immunocytochemical localization in other tissues or species or on immunoblots were unsuccessful despite their ability to react with native enzyme from the same sources. This suggests that they would also be ineffective at probing expression libraries generated from mRNA of muscle or olfactory tissue constructed in λgt_{11}. Nevertheless, these antibodies have been used advantageously for isolation and purification of muscle carnosine synthetase from rabbit and should be applicable to other tissues and species.

The enzymatic activity exhibited in partially purified rabbit muscle extracts elutes from HPLC gel-filtration columns slightly behind the apoferritin standard; however, the activity associated with the enzyme–antibody complexes elutes much earlier, indicating a larger apparent molecular size. Initial attempts to use these antibodies for immunoaffinity purification of the synthetase were hampered by the need to use denaturing conditions to dissociate the complexes. The absence of pure enzyme to use as a standard precluded unambiguous identification of the now enzymatically inactive protein component(s) that had been responsible for synthetase activity. Thus, it was necessary to purify a small amount of synthetase to near homogeneity by traditional means while monitoring the enzyme activity prior to the utilization of immunoaffinity based isolation. This was facilitated by our observations that (1) the addition of EGTA was crucial in stabilizing the enzyme, and (2) the enzyme and the enzyme-antibody complex had very different elution profiles from ion exchange and sizing columns. Thus, with the assistance of the laboratory of Dr. C. Perley of Hoffmann LaRoche, enzyme was isolated from 10 kg of muscle and

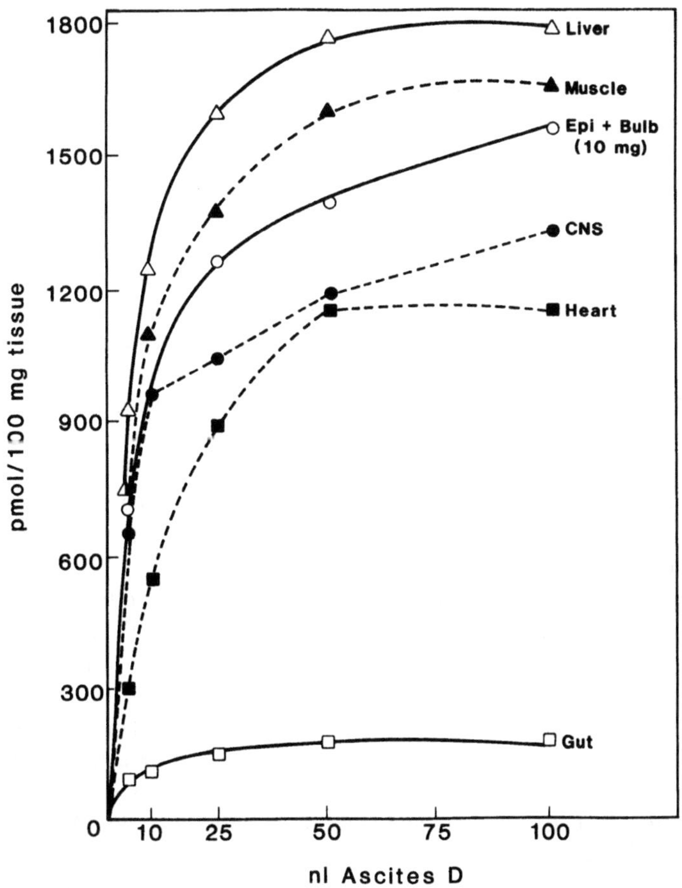

Figure 7. Extracts of several rabbit tissues immunotitrated with monoclonal antibody D to carnosine synthetase: brain (●), liver (△), muscle (▲), heart (■), olfactory mucosa and bulb (○), and gut (□). The enzyme activity represents that associated with the washed immunobead complex. Note that the activity in the olfactory tissue is derived from only 10 mg of tissue, in contrast to 100 mg for each of the others (Margolis *et al.*, 1987).

fractionation of the synthetase activity before and after incubation with the antibody permitted isolation of a very pure, enzymatically active, immune complex. Dissociation of this complex followed by analytical SDS–PAGE indicated that the synthetase was composed of 80,000–90,000-M_r subunits. Amino acid sequence of these subunits or peptide fragments derived from them will now permit the generation of appropriate oligonucleotide probes for identification of cDNA clones for carnosine synthetase. In addition, the immune complex itself may prove useful for enzymological and structural comparisons of the synthetase between tissues as well as for the generation of polyclonal antibodies. The study of tissue specific regulation of expression of the synthetase will then be feasible. Coupled with the availability of antipeptide antibodies for cellular localization by immunocytochemical techniques, it will be feasible to study this peptide system in the various tissues in which it occurs and probe its functional roles.

7. SUMMARY

The peripheral olfactory system presents us with a series of questions that are of general intrinsic biological interest. Although several of these have been considered in this chapter in very specific terms, virtually all of them ultimately relate to general cellular mechanisms of signal identification, information transfer, cellular differentiation, target selection, and cell specification. All these properties themselves derive from differential regulation of gene expression along temporal and spatial coordinates. Thus, this chapter has illustrated the application of contemporary techniques of molecular biology to identify genes differentially expressed in the peripheral olfactory tissue.

We have been able to identify several categories of genes and gene products in the olfactory mucosa that reflect steps along the continuum of differential regulation of gene expression. One category includes genes that are qualitatively regulated and whose expression is thus restricted to only a single-cell type. Other genes are quantitatively regulated and are expressed to a greater or lesser extent in different cell types. Finally, we have observed a pair of related mRNAs, possibly deriving from a single gene, that seem to be coordinately regulated during development but are differentially expressed in brain and olfactory tissue.

Thus, the expression of the olfactory marker protein is exquisitely regulated and is apparently present only in olfactory receptor neurons late in their differentiation from progenitor cells. A non-neuronal example of such qualitatively restricted gene expression is manifested by the secretory glycoprotein that derives from the mRNA cloned into p2/122. This glycoprotein is found in a subpopulation of nasal gland cells and is present in the mucous layer that odor stimuli must traverse in order to reach the OMP-containing receptor neurons. Thus, specific cell types in the nasal/olfactory tissue synthesize unique proteins that are not seen elsewhere and that presumably reflect highly specific functions of these individual cell types.

Quantitatively variable gene expression is certainly more common in general, and in this tissue is exemplified by the enzyme carnosine synthetase and also by the mRNAs which gave rise to clones λ55 and λ21. These are all present at much higher concentration in olfactory tissue than elsewhere. Thus, carnosine synthetase is present in several excitable tissues but is at least an order of magnitude more abundant in olfactory receptor neurons. In this tissue, the enzymatic product, β-alanyl-L-histidine, manifests a half-life very much shorter than that seen elsewhere attesting to its functional significance.

A more complex pattern is represented by the mRNAs identified by the clone λ84. This clone identifies two mRNAs, the smaller of which is the predominant form in olfactory tissue and that is decreased following olfactory nerve lesion. By contrast, in brain we have observed an age related switch from the smaller to the larger of this pair of mRNAs. These observations suggest that there may be cell specific regulation of the expression of two related genes or differential processing of the primary transcript of a single gene.

Thus, we now have in hand a first set of molecular tools derived from this intriguing tissue that represent different categories of gene expression. Our goals now need to be addressed to the roles these genes play during the various tissue and cellular events identified above. These include the process of neuronal reconstitution in this tissue and its role as a model of neural development and plasticity. Furthermore, their potential roles in sensory detection and processing and the possible implications for degenerative diseases of the nervous system need to be considered. The specific functions of the products of these genes in the overall activity of the olfactory system are largely unknown. In certain instances, this is also true for the specific cellular location. Thus, it is now essential to mount a coordinated effort that includes the application of both molecular and biological approaches to address these aspects. The future promises to be exciting as we unravel the mechanisms of gene regulation and the role they play in determining biological function of this primary sensory system.

ACKNOWLEDGMENTS. Some of the previously unpublished data presented here was generously provided by my co-workers W. Danho, T. Goren, M. Grillo, Y.-S. Khew-Goodall, J. Miara, and J. Verhaagen. Dedicated secretarial support was provided by Elizabeth Cahill.

REFERENCES

Allen, W. K., and Akeson, R., 1985, Identification of a cell surface glycoprotein family of olfactory receptor neurons with a monoclonal antibody, *J. Neurosci.* **5**:284–296.

Anholt, R. H., Murphy, K. M. M., Mack, G., and Snyder, S. H., 1984, Peripheral type benzodiazepine receptors in the central nervous system: Localization to olfactory nerves, *J. Neurosci.* **4**:593–603.

Anholt, R. H., Aebi, U., and Snyder, S. H., 1986, A partially purified preparation of isolated chemosensory cilia from the olfactory epithelium of the bullfrog, *Rana catesbeiana, J. Neurosci.* **6**:1962–1969.

Anholt, R. H., Mumby, S. M., Stoffers, D. A., Girard, P. R., Kuo, J. F., and Snyder, S. H., 1987, Transduction proteins of olfactory receptor cells: Identification of guanine nucleotide binding proteins and protein kinase C, *Biochemistry* **26**:788–795.

Baker, H., Kawano, T., Margolis, F. L., and Joh, T. H., 1983, Transneuronal regulation of tyrosine hydroxylase expression in olfactory bulb of mouse and rat, *J. Neurosci.* **3**:69–78.

Baker, H., Kawano, T., Albert, V., Joh, T. H., Reis, D. J., and Margolis, F. L., 1984, Olfactory bulb dopamine neurons survive deafferentation induced loss of tyrosine hydroxylase, *Neuroscience* **11**:605–615.

Barber, P. C., Jensen, S., and Zimmer, J., 1982, Differentiation of neurons containing olfactory marker protein in adult rat olfactory epithelium transplanted to the anterior chamber of the eye, *J. Neurosci.* **7**:2687–2695.

Bauer, K., Hallermayer, K., Salnikow, J., Kleinkauf, H., and Hamprecht, B., 1982*a*, Biosynthesis of carnosine and related peptides by glial cells in primary culture, *J. Biol. Chem.* **257**:3593–3597.

Bauer, K., Jungblut, P., and Kleinkauf, H., 1982*b*, Biosynthesis of carnosine and related peptides, in: *Peptide Antibiotics: Biosynthesis and Functions* (H. Kleinkauf and H. von Dohren, eds.), pp. 337–346, de Gruyter, Berlin.

Bignetti, E., Cavaggioni, A., Pelosi, P., Persaud, K. C., Sorbi, R. T., and Trindelli, R., 1985, Purification and characterization of an odorant binding protein from cow nasal tissue, *J. Biochem.* **149**:227–231.

Brunjes, P. C., Smith-Crafts, L. K., and McCarty, R., 1985, Unilateral odor deprivation: effects on development of olfactory bulb catecholamines and behavior, *Dev. Brain Res.* **22**:106.

Burd, G. D., Davis, B. J., Macrides, F., Grillo, M., and Margolis, F. L., 1982, Carnosine in primary afferents of the olfactory system: An autoradiographic and biochemical study, *J. Neurosci.* **2**:244–255.

Cavaggioni, A., Sorbi, R. T., Keen, J. N., Pappin, D. J. C., and Findlay, J. B. C., 1987, Homology between the pyrazine-binding protein from nasal mucosa and major urinary proteins, *FEBS Lett.* **212**:225–228.

Chen, Z., Ophir, D., and Lancet, D., 1986*a*, Monoclonal antibodies to ciliary glycoproteins of frog olfactory neurons, *Brain Res.* **368**:329–338.

Chen, Z., Pace, U., Ronen, D., and Lancet, D., 1986*b*, A unique glycoprotein of olfactory cilia with transmembrane receptor properties, *J. Biol. Chem.* **261**:1299–1305.

Chuah, M. I., and Farbman, A. I., 1983, Olfactory bulb increases marker protein in olfactory receptor cells, *J. Neurosci.* **3**:2197–2205.

Crush, K. G., 1970, Carnosine and related substances in animal tissues, *Comp. Biochem. Physiol.* **34**:3–30.

Dahl, A. R., Hadley, W. M., Hahn, F. F., Benson, J. M., and McClellan, R. O., 1982, Cytochrome P-450-dependent monooxygenases in olfactory epithelium of dogs: Possible role in tumorigenicity, *Science* **216**:57–59.

Dascal, N., Heldman, J., Gershon, E., and Lancet, D., 1986, Possible expression of odorant receptor proteins in *Xenopus* oocytes injected with rat olfactory epithelial mRNA, *Soc. Neurosci. Abst.* **12**:1354.

Farbman, A. I., 1986, Prenatal development of mammalian olfactory receptor cells, *Chem. Senses* **11**:3–18.

Farbman, A. I., and Margolis, F. L., 1980, Olfactory marker protein during ontogeny: immunohistochemical localization, *Dev. Biol.* **74**:205–215.

Freeze, H. H., and Varki, A., 1986, Endo-glycosidase and peptide N-glycosidase F release the great majority of total cellular N-linked oligosaccharides: Use in demonstrating that sulfated N-linked oligosaccharides are frequently found in cultured cells, *Biochem. Biophys. Res. Commun.* **140**:967–973.

Fujita, S. C., Mori, K., Imamura, K., and Obata, K., 1985, Subclasses of olfactory receptor cells and their central projections demonstrated by a monoclonal antibody, *Brain Res.* **326**:192–196.

Garnier, J., Osguthorpe, D. J., and Robson, B., 1978, Analysis of the accuracy and implications of simple methods for predicting the secondary structure of globular proteins, *J. Mol. Biol.* **120**:97–120.

Getchell, T. V., Heck, G. L., DeSimone, J. A., Price, S., 1980, The location of olfactory receptor sites. Inferences from latency measurements, *Biophys. J.* **29**:397–472.

Getchell, M. L., Rafols, J. A., and Getchell, T. V., 1984*a*, Histological and histochemical studies of the secretory components of the salamander olfactory mucosa: Effects of isoproterenol and olfactory nerve section, *Anat. Rec.* **208**:553–565.

Getchell, T. V., Margolis, F. L., and Getchell, M. L., 1984*b*, Perireceptor and receptor events in vertebrate olfaction, *Prog. Neurobiol.* **23**:317–345.

Goldstein, N. I., and Quinn, M. R., 1981, A novel cell line isolated from the murine olfactory mucosa, *In Vitro* **17**:593–598.

Graziadei, P. P. C., 1971, The olfactory mucosa of vertebrates, in: *Handbook of Sensory Physiology. Chemical Senses*, Vol. 4 (L. M. Beidler, ed.), pp. 27–58, Springer-Verlag, Berlin.

Graziadei, P. P. C., and Monti-Graziadei, G. A., 1978*a*, Continuous nerve cell renewal in the olfactory system, in: *Handbook of Sensory Physiology. Development of Sensory Systems*, Vol. 9 (M. Jacobson, ed.), pp. 55–83, Springer-Verlag, Berlin.

Graziadei, P. P. C., and Monti-Graziadei, G. A., 1978*b*, The olfactory system: A model for the study of neurogenesis and axon regeneration in mammals, in: *Neuronal Plasticity* (C. W. Cotman, ed.), pp. 131–153, Raven, New York.

Graziadei, P. P. C., and Monti-Graziadei, G. A., 1985, Neurogenesis and plasticity of the olfactory sensory neurons, *Ann. NY Acad. Sci.* **457**:127–142.

Guiditta, A., Hunt, T., and Santella, L., 1986, Evidence for the presence of mRNA in the axoplasm of the squid giant axon, *Neurochem. Int.* **8**:435–442.

Hadley, W. M., and Dahl, A. R., 1982, Cytochrome P-450 dependent monooxygenase activity in rat nasal epithelial membranes, *Toxicol. Lett.* **10**:417–422.

Halasz, N., and Shepherd, G. M., 1983, Neurochemistry of the vertebrate olfactory bulb, *Neuroscience* **10**:579–619.

Hempstead, J. L., and Morgan, J. I., 1983, Monoclonal antibodies to the rat olfactory sustentacular cell, *Brain Res.* **288**:289–295.

Hempstead, J. L., and Morgan, J. I., 1985*a*, A panel of monoclonal antibodies to the rat olfactory epithelium, *J. Neurosci.* **5**:438–449.

Hempstead, J. L., and Morgan, J. I., 1985*b*, Monoclonal antibodies reveal novel aspects of the biochemistry and organization of olfactory neurons following unilateral olfactory bulbectomy, *J. Neurosci.* **5**:2382–2387.

Hokfelt, T., and Terenius, L., 1987, More on receptor mismatch, *Trends Neurosci.* **10**:22–23.

Horinishi, H., Grillo, M., and Margolis, F. L., 1978, Purification and characterization of carnosine synthetase from mouse olfactory bulb, *J. Neurochem.* **31**:909–919.

Huque, T., and Bruch, R. C., 1986, Odorant- and guanine nucleotide-stimulated phosphoinositide turnover in olfactory cilia, *Biochem. Biophys. Res. Commun.* **137**:36–42.

Jones, D. T., and Reed, R. R., 1987, Molecular cloning of five GTP-binding protein cDNA species from rat olfactory neuroepithelium, *J. Biol. Chem.* **262**:14241–14249.

Kawano, T., and Margolis, F. L., 1982, Transsynaptic regulation of olfactory bulb catecholamines in mice and rats, *J. Neurochem.* **39:**342–348.

Kessel, R. G., and Kardon, R. H. (eds.), 1979, *Tissues and Organs. A Text–Atlas of Scanning Electron Microscopy,* W. H. Freeman, San Francisco.

Khew-Goodall, Y. S., Goren, T., Grillo, M., and Margolis, F. L., 1987, Identification and cloning of olfactory specific mRNAs (Abstr.) *J. Neurochem.* **48**(Suppl.):S144.

Kream, R. M., Davis, B. J., Kawano, T., Margolis, F. L., and Macrides, F., 1984, Substance P and catecholaminergic expression in neurons of the hamster main olfactory bulb, *J. Comp. Neurol.* **222:**140–154.

Kream, R. M., and Margolis, F. L., 1983, Olfactory marker protein: Turnover and transport in normal and regenerating neurons, *J. Neurosci.* **4:**868–879.

Lancet, D., 1986, Vertebrate olfactory reception, *Annu. Rev. Neurosci.* **9:**329–355.

Lee, K-H., Wells, R. G., and Reed, R. R., 1987, Isolation of an olfactory cDNA: Similarity to retinol-binding protein suggests a role in olfaction, *Science* **235:**1053–1056.

Leff, S. E., and Rosenfeld, M. G., 1986, Complex transcriptional units: Diversity in gene expression by alternative RNA processing, *Annu. Rev. Biochem.* **55:**1091–1117.

Levitt, M., 1978, Conformational preferences of amino acids in globular proteins, *Biochemistry* **17:**4277–4285.

Lifson, S., and Sanders, C., 1979, Antiparallel and parallel β-strands differ in amino acid residue preferences, *Nature (Lond.)* **282:**109–111.

Lipman, D. J., and Pearson, W. R., 1985, Rapid and sensitive protein similarity searches, *Science* **227:**1435–1441.

Macrides, F., Davis, B. J., 1983, The olfactory bulb, in *Chemical Neuroanatomy* (P. C. Emson, ed.), pp. 391–426, Raven, New York.

Mandel, G., and Goodman, R. H., 1987, Using the brain to screen cloned genes, *Trends Neurosci.* **10:**101–104.

Margolis, F. L., 1972, A brain protein unique to the olfactory bulb *Proc. Natl. Acad. Sci. USA* **69:**1221–1224.

Margolis, F. L., 1980*a,* A marker protein for the olfactory chemoreceptor neuron, in: *Proteins of the Nervous System* (R. A. Bradshaw and D. Schneider, eds.), pp. 59–84, Raven, New York.

Margolis, F. L., 1980*b,* Carnosine: An olfactory neuropeptide, in: *Role of Peptides in Neuronal Function* (J. L. Barker and T. Smith, eds.), pp. 545–572, Dekker, New York.

Margolis, F. L., 1974, Carnosine in the primary olfactory pathway, *Science* **184:**909–911.

Margolis, F. L., 1981, Neurotransmitter biochemistry of the mammalian olfactory bulb, in: *Biochemistry of Taste and Olfaction* (R. H. Cagan and M. R. Kare, eds.), pp. 369–394, Academic, New York.

Margolis, F. L., 1985, Olfactory marker protein: from PAGE band to cDNA clone, *Trends Neurosci.* **8:**542–546.

Margolis, F. L., and Grillo, M., 1984*a,* Carnosine, homocarnosine and anserine in vertebrate retinas, *Neurochem. Int.* **6:**207–209.

Margolis, F. L., and Grillo, M., 1984*b,* Inherited differences in mouse kidney carnosinase activity, *Biochem. Genet.* **22:**444–451.

Margolis, F. L., Grillo, M., Grannot-Reisfeld, N., and Farbman, A. I., 1983, Purification, characterization and immunocytochemical localization of mouse kidney carnosinase, *Biochim. Biophys. Acta* **744:**237–248.

Margolis, F. L., Grillo, M., Kawano, T., and Farbman, A. I., 1985*a,* Carnosine synthesis in olfactory tissue during ontogeny: effect of exogenous β-alanine, *J. Neurochem.* **44:**1459–1464.

Margolis, F. L., Grillo, M., Hempstead, J., and Morgan, J. I., 1987, Monoclonal antibodies to mammalian carnosine synthetase, *J. Neurochem.* **48:**593–600.

Margolis, F. L., Sydor, W., Teitelbaum, Z., Blacher, R., Grillo, M., Rogers, K., Sun, R., and Gubler, U., 1985*b,* Molecular biological approaches to the olfactory system: Olfactory marker protein as a model, *Chem. Senses* **10:**163–174.

Maue, R. A., and Dionne, V. E., 1984, Ion channel activity in isolated murine olfactory receptor neurons, *Soc. Neurosci. Abst.* **10:**655.

Menco, B. P. M., 1980, Quanlitative and quantitative freeze-fracture studies on olfactory and nasal respiratory epithelial surfaces of frog, ox, rat, and dog, *Cell Tissue Res.* **211:**5–29.

Miragall, F., and Monti-Graziadei, G. A., 1982, Experimental studies on the olfactory marker protein. II. Appearance of the olfactory marker protein during differentiation of the olfactory sensory neurons of mouse: an immunohistochemical and autoradiographic study, *Brain Res.* **239:**245–250.

Monti-Graziadei, G. A., 1983, Experimental studies on the olfactory marker protein. III. The olfactory marker protein in the olfactory neuroepithelium lacking connections with the forebrain, *Brain Res.* **262:**303–308.

Monti-Graziadei, G. A., Stanley, R. S., and Graziadei, P. P. C., 1980, The olfactory marker protein in the olfactory system of mouse during development, *Neuroscience* **5:**1239–1252.

Mori, K., 1987, Monoclonal antibodies (2C5 and 4C9) against lactoseries carbohydrates identify subsets of olfactory and vomeronasal receptor cells and their axons in the rabbit, *Brain Res.* **408:**215–221.

Mori, K., Fujita, S. C., Imamura, K., and Obata, K., 1985, Immunohistochemical study of subclasses of olfactory nerve fibers and their projections to the olfactory bulb in the rabbit, *J. Comp. Neurol.* **242:**214–229.

Nadi, N. S., Head, R., Grillo, M., Hempstead, J., Grannot-Reisfeld, N., and Margolis, F. L., 1981, Chemical deafferentation of the olfactory bulb; plasticity of the levels of tyrosine hydroxylase, dopamine and norepinephrine, *Brain Res.* **213:**365–371.

Nakamura, T., and Gold, G. H., 1987, A cyclic nucleotide-gated conductance in olfactory receptor cilia, *Nature (Lond.)* **325:**442–444.

Nakashima, T., Kimmelman, C. P., and Snow, J. P., 1984, Structure of human fetal and adult olfactory neuroepithelium, *Arch. Otolaryngol. (Stockh.)* **110:**641–646.

O'Farrell, P., 1975, High-resolution two-dimensional electrophoresis of proteins, *J. Biol. Chem.* **250:**4007–4021.

Pace, U., Hanski, U. E., Salomon, Y., and Lancet, D., 1985, Odorant-sensitive adenylate cyclase may mediate olfactory reception, *Nature (Lond.)* **316:**255–258.

Peabody, D. S., and Berg, P., 1986, Termination-reinitiation occurs in the translation of mammalian cell mRNAs, *Mol. Cell Biol.* **6:**2695–2703.

Pevsner, J., Trifiletti, R. R., Strittmatter, S. M., and Synder, S. H., 1985, Isolation and characterization of an olfactory receptor protein for odorant pyrazines, *Proc. Natl. Acad. Sci. USA* **82:**3050–3054.

Pfister, C., Chabre, M., Plouet, J., Tuyen, V. V., De Kozak, Y., Faure, J. P., and Kuhn, H., 1985, Retinal S antigen identified as the 48k protein regulating light-dependent phosphodiesterase in rods, *Science* **228:**891–893.

Reed, C. J., Lock, E. A., and DeMatteis, F., 1986, NADPH: cytochrome P-450 reductase in olfactory epithelium, *Biochem. J.* **240:**585–592.

Rhein, L. D., and Cagan, R. H., 1980, Biochemical studies of olfaction: Isolation characterization and odorant binding activity of cilia from rainbow trout olfactory rosettes. *Proc. Natl. Acad. Sci. USA* **77:**4412–4416.

Rochel, S., and Margolis, F. L., 1980, The response of ornithine decarboxylase during neuronal degeneration and regeneration in olfactory epithelium, *J. Neurochem.* **35:**850–860.

Rochel, S., and Margolis, F. L., 1982, Carnosine release from olfactory bulb synaptosomes is calcium-dependent and depolarization-stimulated, *J. Neurochem.* **38:**1505–1514.

Rogers, K., Grillo, M., Poonian, M., and Margolis, F. L., 1985, Olfactory neuron-specific protein is translated from a large poly(A)$^+$mRNA, *Proc. Natl. Acad. Sci. USA* **82:**5218–5222.

Rogers, K., Dasgupta, P., Gubler, U., Grillo, M., Khew-Goodall, Y. S., and Margolis, F. L., 1987, Molecular cloning and sequencing of a cDNA for rat OMP, *Proc. Natl. Acad. Sci. USA* **84:**1704–1708.

Sakai, M., Yoshida, M., Karasawa, N., Teramura, M., Ueda, H., and Nagatsu, I., 1987, Carnosine-like immunoreactivity in the primary olfactory neuron of the rat, *Experientia* **298:**300.

Schneiderman, A. M., Matsumoto, S. G., and Hildebrand, J. G., 1982, Trans-sexually grafted antennae influence development of sexually dimorphic neurons in moth brain, *Nature (Lond.)* **298:**844–846.

Schubert, D., Stallcup, W., LaCorbiere, M., Kidokoro, Y., and Orgel, L., 1985, The ontogeny of electrically excitable cells in cultured olfactory epithelium, *Proc. Natl. Acad. Sci. USA* **82:**7782–7786.

Schwartz, J. P., and Costa, E., 1986, Hybridization approaches to the study of neuropeptides, *Annu. Rev. Neurosci.* **9:**277–304.

Shepherd, G. M., 1972, Synaptic Organization of the mammalian olfactory bulb, *Physiol. Rev.* **52:**864–917.

Shepherd, G. M., 1977, The olfactory bulb: a simple system in the mammalian brain, in: *Handbook of Physiology,* Vol. 1 (J. M. Brookhart, V. B. Mountcastle, E. R. Kandel, and S. R. Geiger, eds.), pp. 945–968, American Physiological Society, Bethesda, Maryland.

Simmons, P. A., and Getchell, T. V., 1981, Neurogenesis in olfactory epithelium: loss and recovery of transepithelial voltage transients following olfactory nerve section, *J. Neurophysiol.* **45:**516–528.

Skaper, S. D., Das, S., and Marshall, F. D., 1973, Some properties of a homocarnosine-carnosine synthetase isolated from rat brain, *J. Neurochem.* **21:**1429–1445.

Skene, J. H. P., Jacobson, R. D., Snipes, G. J., McGuire, C. B., Nordn, J. J., and Freeman, A., 1986, A protein induced during nerve growth (GAP-43) is a major component of growth-cone membranes, *Science* **233:**783–785.

Sklar, P. B., Anholt, R. H., and Snyder, S. H., 1986, The odorant-sensitive adenylate cyclase of olfactory receptor cells. Differential stimulation by distinct classes of odorants, *J. Biol. Chem.* **261:**15538–15543.

Sydor, W., Teitelbaum, Blacher, Z., Sun, R., Benz, W., and Margolis, F. L., 1986, Amino acid sequence of a unique neuronal protein: Rat olfactory marker protein, Arch. Biochem. Biophys. **249:**351–362.

Tamaki, N., Morioka, S., Ikeda, T., Harada, M., and Hama, T., 1980, Biosynthesis and degradation of carnosine and turn-over rate of its constituent amino acids in rats, *J. Nutr. Sci. Vitaminol. (Tokyo)* **26:**127–139.

Trotier, D., and MacLead, P., 1986, cAMP and cGMP open channels and depolarize olfactory receptor cells, *Chem. Senses* **11:**674.

Vinnikov, Y. A., 1982, Evolution of receptor cells, *Mol. Biol. Biochem. Biophys. Ser.* **34:**1–141.

Vodyanoy, V., and Murphy, R. B., 1983, Single-channel fluctuations in bimolecular lipid membranes induced by rat olfactory epithelial homogenates, *Science* **220:**717–719.

Voigt, J. M., Guengerich, F. P., and Baron, J., 1985, Localization of a cytochrome P-450 isozyme (cytochrome P-450 PB-B) and NADPH-cytochrome P-450 reductase in rat nasal mucosa, *Cancer Lett.* **27:**241–247.

Wang, R. T., and Halpern, M., 1982, Neurogenesis in the vomeronasal epithelium of adult garter snakes. 1. Degeneration of bipolar neurons and proliferation of undifferentiated cells following experimental vomeronasal axotomy, *Brain Res.* **237:**23–39.

Wensley, C. H., Fung, B. P., and Chikaraishi, D. M., 1986, Cloning of neural specific genes from the olfactory epithelium and developing rat brain, *Sco. Neurosci. Abst.* **12:**212.

Zwiers, H., Verhaagen, J., van Dongen, C. J., de Graan, P. N. E., and Gispen, W. H., 1985, Resolution of rat brain synaptic phosphoprotein B-50 into multiple forms by two-dimensional electrophoresis: Evidence for multisite phosphorylation, *J. Neurochem.* **44:**1083–1090.

IV

Development and Differentiation

12

Monoclonal Antibody Mapping of the Rat Olfactory Tract

James I. Morgan

1. INTRODUCTION

1.1. Scope

While odor perception *per se* most probably occurs at the level of the olfactory receptor neuron, the overall process of olfaction almost certainly involves other non-neuronal elements (e.g., Bowman's glands and sustentacular cells) of the olfactory epithelium (Getchell *et al.*, 1984). Thus, in attempting to investigate the complex signal recognition–transduction mechanism that is olfaction, it would be anticipated that approaches that could simultaneously provide evidence relating to molecular structure, cellular localization, and biological function would provide key information relating to this sense. Monoclonal antibodies (MAbs) represent one such a technique.

The intentions of this chapter are twofold. First, a number of strategies are discussed for the production of MAbs to either cellular elements of the rat olfactory epithelium or to defined proteins of this structure. Second, several examples are drawn from our own studies to highlight specific applications of hybridoma technology to the olfactory system. A final section gives some prospects for the further application of MAb technology to the study of olfaction.

2. STRATEGIES FOR MONOCLONAL ANTIBODY PRODUCTION

2.1. General Considerations

To detail the precise protocols used to generate antibody-secreting hybridomas would clearly be beyond the scope of this article, and they have been published in detail elsewhere (Fazekas De St. Groth and Scheidegger, 1980; Kohler and Milstein, 1975; Morgan, 1984). Rather, this section deals with immunization and screening strategies

James I. Morgan • Department of Neuroscience, Roche Institute of Molecular Biology, Roche Research Center, Nutley, New Jersey 07110.

relevant to detecting MAbs directed against olfactory tissues and proteins. Fundamentally, two strategies are considered. The first describes raising MAbs to whole-tissue preparations and screening these by immunocytochemical means. The second strategy involves generating MAbs to defined enzymes and screening for enzymatic activity.

2.2. Monoclonal Antibodies to Adult and Embryonic Epithelium

2.2.1. Immunization

CD Sprague-Dawley-derived rats were killed by decapitation and their olfactory epithelia removed, freed of cartilage, and immediately frozen in liquid nitrogen. Embryonic epithelia (E14 and E20) were obtained by dissection under a microscope and freed of as much adjacent tissue as possible. The samples were also frozen as above until required. For immunization the epithelia were homogenized using either a Polytron or Sonicator in phosphate buffered saline at 4°C. It was our experience that when immunizing BALB/c mice with rat tissue that large immunogen doses were required. We attempted to give 50–100 mg wet weight of tissue per mouse. For the primary injection, the homogenate was emulsified 50 : 50 in Freund's complete adjuvant; 4–5 weeks later, the animals were boosted with an equivalent amount of tissue in Freund's incomplete adjuvant IP using all-glass syringes. The mice were sacrificed 4 days later and their spleens removed and fused with PAI-O myeloma as described (Morgan, 1984; Hempstead and Morgan, 1985*a*).

Because lesser amounts of tissue were obtained from fetuses, the mice received an additional primary injection seven days after the first, also in Freund's complete adjuvant. Approximately 16 fetal epithelia were used per immunization.

2.2.2. Screening

When screening MAbs raised against adult olfactory epithelia, rats were anesthetized with Nembutal and perfused via the aorta with 4% buffered paraformaldehyde. The epithelium was removed and, after a 2-hr postfix, was cryoprotected in 30% sucrose at 4°C. When the tissue sank, it was removed from the sucrose and sectioned on a cryostat (5- to 10-μm sections) after blocking and rapid freezing in freon-dry ice. Fetal tissues were removed and fixed directly in paraformaldehyde due to their small size. Further processing was as for adult epithelium.

Sections were kept at 4°C until required for screening; in our hands, approximately 1000 sections were required per fusion. For immunocytochemistry, sections were immersed in PBS for 30 min to remove sucrose. We found it convenient to ring sections with rubber cement prior to use to minimize the volumes of reagents required. After PBS, the sections were treated with PBS containing 0.2% NP40 for 10 min and then blocked with 10% normal goat serum in PBS (v/v) for a further 30 min. Hybridoma supernatants, buffered to pH 7.2 with 0.5 M sodium phosphate, were next added, usually neat, and allowed to incubate for 2–3 hr at room temperature. Occasionally, we also employed overnight incubations at 4–8°C. However, these longer times of incubation involving temperature shifts sometimes caused the sections to begin detaching from the subbed slide. The slides were next washed three times in PBS before incubation for 1 hr with fluorescein-conjugated goat antimouse immunoglobulin (Cappel), diluted 1 : 20 with

PBS containing 0.2% NP40 and 10% normal goat serum. After washing three times in PBS and mounting in glycerol or other appropriate medium, the slides were viewed under fluorescence optics. Obviously higher numerical aperture objectives (>0.6 NA) are preferred as is a tight bandwidth fluorescein filter. Trimethylrhodamine reporters also work well; however, at low signal-to-noise ratios a red autofluorescence may become a problem. This may be improved by including lysine in the fixative, or employing agents that suppress the autofluorescence (e.g., Evans blue).

When examining the sections, a few precautions of interpretation should be observed. First, most secondary antibodies give some form of background staining, the precise nature of which varies from batch to batch and supplier to supplier. Usually this is seen as a nonspecific staining of the connective tissue of the lamina propria mucosae and immune cells of the sinuses in this region. Often the walls of blood vessels will also appear labeled. With one batch of FITC-labeled rabbit antimouse serum, we have also seen weak staining of nerve bundles. Clearly, it is worth screening several second antibody batches to find one with tolerable background. It should be emphasized, however, that such background is only of significance when dealing with MAbs that yield a weak signal. A second point of interpretation is the distinction between olfactory and respiratory epithelia under fluorescence optics. While those familiar with the histology can readily distinguish these two adjacent tissues, the less experienced can easily be mislead. To identify olfactory epithelium unambiguously, we routinely include an antiserum to OMP in the primary antibody step. Since this antibody was raised in a different species, a cocktail of two second antibodies may be used, the OMP being revealed by a TRITC antibody and the MAb of interest by a FITC antibody (Hempstead and Morgan, 1983*a*, 1985*a*). This permits the localization of both antibodies in the same section by merely flipping between FITC and TRITC emission filters.

2.3. Monoclonal Antibodies Raised against Membranes from Olfactory Epithelium

2.3.1. Methods

In a recent collaborative study with Dr. A. I. Farbman of Northwestern University, MAbs were raised to membrane preparations from neonatal and adult olfactory epithelium. The ultimate goal of this investigation was to obtain MAbs useful in probing the development and function of olfactory receptor cilia in the epithelium. Crude membranes were prepared, as follows:

1. Neonatal or adult rats were sacrificed and their olfactory epithelia removed and placed into 20–30 vol 50 mM sodium phosphate buffer, pH 7.4.
2. The tissue was next homogenized on a Brinkmann Polytron at maximum setting for 10 sec on ice.
3. Following low-speed centrifugation to remove debris (120 × g_{av}/10 min), membranes were pelleted at 30,000 × g_{av}/30 minutes at 4°C.
4. For immunization, the pellet was resuspended in PBS and injected IP without adjuvant. One mouse received the membranes derived from approximately three adult or eight neonatal rats. The mice were boostered 4 weeks later with the same

amount of immunogen and the fusion performed as described (Morgan, 1984; Hempstead and Morgan, 1985*a;* Kohler and Milstein, 1975; Fazekas De St. Groth and Scheidegger, 1980).

Most of the positive clones obtained in these experiments reacted variously with the luminal boundaries of the olfactory and respiratory epithelia. Only a small number of clones were obtained to the receptor neuron perikarya or nerve bundles when starting with membranes; this is in contrast to the large number seen when total homogenate was used (Hempstead and Morgan, 1985*a*). The reason for this may be that a major 210,000-M_r membrane antigen of receptor neurons is extracted even under the mild conditions used to prepare membranes and is discarded in the supernatant (J. L. Hempstead and J. I. Morgan, unpublished observations). Thus, this procedure tends to favor the generation of MAbs to intrinsic membrane antigens. It is likely that this strategy will be useful in obtaining MAbs for use in studies of receptor cilia structure and function.

2.4. Antibodies to Carnosine Synthetase

2.4.1. General Considerations

When developing MAbs to defined proteins, there are a number of considerations to be taken into account. The two major points are (1) the intended application of the antibody, and (2) the options open for screening. A detailed account of these considerations has been given previously (Morgan, 1984). An outline is given here of the strategies employed by our laboratory in a collaborative effort with the group of Dr. Frank Margolis to produce MAbs to carnosine synthetase (Margolis *et al.*, 1987). This example has been chosen because of the known relevance of this enzyme to the olfactory tract (Margolis, 1980; Rochel and Margolis, 1982). However, the same basic approach has been used in producing MAbs to the lysosomal protease, cathepsin M (Erickson-Viitanen *et al.*, 1985), and human placental alkaline phosphatase. For our application, we required MAbs that would react with the mature enzyme, without necessarily inhibiting it, and having sufficient affinity to perform subsequent enzyme purification. Furthermore, since this enzyme had not been purified to homogeneity by conventional means, we were restricted in our options for both immunization schedules and screening procedures. Thus, the following protocols were employed.

2.4.2. Immunization

Immunization was performed with a 40% ammonium sulfate precipitate of 100,000 $\times$ g supernatant of rabbit skeletal muscle homogenate. The precipitate was resuspended in PBS and then absorbed onto Maalox, a stabilized aluminum hydroxide suspension used commercially as an antacid. Briefly, equal volumes of washed, packed Maalox and protein solution (at least 1 mg/ml) were mixed and left at room temperature for 30 min. They were then injected IP (200 μl) as a primary innoculation. At the same time, *B. pertussis* antigen, 10^9 IU, was given intramuscularly as a adjuvant. Mice were boosted several times over a few months with crude homogenate IP and fusion performed 4 days after the final immunization. It can be calculated that the content of carnosine synthetase in the original homogenate could only be 0.1% or less. This highlights one power of this technique, i.e., that one does not necessarily require pure antigen as starting material.

2.4.3. Screening

The screen used the native enzymatic activity of carnosine synthetase, thus circumventing the difficulty of having no pure antigen, which precluded enzyme-linked immunosorbent assay (ELISA)-type tests or radioimmunoassay (RIA). Furthermore, it ensured that only noninactivating MAbs were cloned and also avoided the ambiguities of screens that measure indirect functions of the antigen. A further point is that the assay involved affinity precipitation of the active antibody–enzyme complex. This ensured that the MAb had a sufficiently high affinity to permit future purification of the enzyme. Details of the assay may be found in Margolis *et al.* (1987).

3. APPLICATIONS OF MONOCLONAL ANTIBODIES

3.1. Studies of the Structure of the Olfactory Epithelium

3.1.1. General Anatomy

The olfactory epithelium lines the turbinate bones high in the nasal vault. In mammals, the epithelium has two components: a neuroepithelium composed of receptor neurons, basal cells, and sustentacular cells that is separated by a basement membrane from the lamina propria mucosae consisting of Bowman's glands, axonal bundles, and connective tissue (Graziadei, 1971).

3.1.2. Results

Our initial studies described the production of MAbs that reacted with all the major cell types of the rat olfactory epithelium (Hempstead and Morgan, 1985*a*). Time will be taken at this juncture to catalogue the various known markers of the olfactory mucosa. These data are also presented in Table 1.

3.1.2.a. Olfactory Neurons. Various biochemical and histochemical studies have shown the receptor neurons to contain the following molecules that may be used as marker substances: carnosine (Burd *et al.*, 1980; Margolis, 1980), carnosine synthetase (Margolis, 1980), carbonic anhydrase (Brown *et al.*, 1984), olfactory marker protein (OMP) (Margolis, 1972; Hartman and Margolis, 1975), H and B blood group antigens (Mollicone *et al.*, 1985), and vimentin (Schwob *et al.*, 1986). In addition, several MAbs have been described that react variously with all, subsets, or regions of receptor neurons. For example, Allen and Akeson (1985*a,b;* see also Chapter 13, this volume) have described a MAb, termed 2B8, that reacts with a subset of receptor neurons. Fujita and colleagues have also reported a MAb that delineates receptor subsets (Fujita *et al.*, 1985). Two MAbs from our laboratory also reveal subclasses of receptor neurons (NEU-4, NEU-9) (Hempstead and Morgan, 1985*a,b*), while others (NEU-5) (Hempstead and Morgan, 1985*a,b*) stain all

Table 1. Markers of the Olfactory Mucosa

Cell type	Marker
Receptor neurons	Proteins
	Carnosine synthase[a]
	Carbonic anhydrase[b]
	Blood group antigens[c]
	Olfactory marker protein[d,e]
	Vimentin[f]
	Peptides
	Carnosine[a,g]
	Monoclonal antibodies
	2B8[h,i]
	Unspecified[j]
	NEU-3[k]
	NEU-4[k,l]
	NEU-9[k,l]
	LUM-3[k]
Sustentacular cells	Proteins
	Cytochrome P_{450}[t]
	Monoclonal antibodies
	SUS-1[k,m]
	SUS-2[k]
	SUS-3[k]
Bowman's glands	Proteins
	Pyrazine-binding protein[n]
	Carnosinase[o,p]
	Cytochrome P_{450}[t]
	Monoclonal antibodies
	GLA-1 thru GLA-8[k]
	SUS-1[k,m]
	Miscellaneous
	Peanut lectin-binding substance[p]
Perineuronal sheath cells (Schwann cells)	Proteins
	GFAP[q]
	S100[r,s]
	Monoclonal antibodies
	NEU-6[k]
	NEU-8[k]

[a]Burd *et al.* (1980); [b]Brown *et al.* (1984); [c]Mollicone *et al.* (1985); [d]Margolis (1972); [e]Hartman and Margolis (1975); [f]Schwob *et al.* (1986); [g]Margolis (1980); [h]Allen and Akeson (1985*a*); [i]Allen and Akeson (1985*b*); [j]Fujita *et al.* (1985); [k]Hempstead and Morgan (1985a); [l]Hempstead and Morgan (1985b); [m]Hempstead and Morgan (1983a); [n]Pevsner *et al.* (1985); [o]Farbman and Margolis (1982); [p]Hempstead and Morgan (1983b); [q]Barber and Lindsay (1982); [r]Takahashi *et al.* (1984); [s]Fujita *et al.* (1985); [t]P. Thomas, J. I. Morgan, and A. I. Farbman (unpublished observations).

receptors. The NEU-4 and NEU-9 Mabs react with both soma and axon of the neuron, however, NEU-5 binds predominantly to the axon (Hempstead and Morgan, 1985*a,b*). More striking is the localization of a number of MAbs to olfactory receptor cilia, such as LUM-3 (Hempstead and Morgan, 1985*a*), since they stain no other region of the neuron.

3.1.2.b. Supporting (Sustentacular) Cells. These cells are reactive with SUS-1 through SUS-3 MAbs (Hempstead and Morgan, 1983*b*, 1985*a*), however, the antigen(s) are unknown. Sustentacular cells contain the enzyme cytochrome P450b and its reductase (Farbman, Morgan and Thomas, unpublished). In the mouse, weak immunoreactivity to carnosinase was observed in sustentacular cells (Farbman and Margolis, 1982); however, this was not seen in the rat (Hempstead and Morgan, 1983*b*).

3.1.2.c. Bowman's Glands: The olfactory mucosa contains a large number of glands that have been partially characterized by immunocytochemistry. The secretory products of the Bowman's glands include carnosinase (Farbman and Margolis, 1982; Hempstead and Morgan, 1983*b*), a peanut lectin-binding substance (Hempstead and Morgan, 1983*b*), and a pyrazine-binding protein (Pevsner *et al.*, 1985). In addition, we have obtained eight independent MAbs to the glands, termed GLA-1 through GLA-8 (Hempstead and Morgan, 1985*a*); again, the antigens are undefined. The acinar cells of the glands are positive for a number of antigens. Thus they react with the SUS-1 MAb (Hempstead and Morgan, 1983*a*, 1985*a*) as well as antibodies to cytochrome P450 (A. I. Farbman, J. I. Morgan, and P. Thomas, unpublished observations).

Beyond the documentation of the MAbs, it was further shown that sustentacular cells and the cells comprising Bowman's glands have some antigenic homologies (Hempstead and Morgan, 1983*a*, 1985*a*). This contention was based on the fact that both cell types reacted with a series of MAbs (SUS-1, SUS-3) that did not stain other cell types of the epithelium (Hempstead and Morgan, 1983*a*, 1985*a*). More recently we have shown that the enzymes cytochrome P450 and cytochrome P450 reductase are also exclusively localized in sustentacular cells and Bowman's gland cells (A. I. Farbman, J. I. Morgan, and P. Thomas, unpublished observations). This strongly suggests at least a functional homology between these two cell populations. One common function of sustentacular cells and Bowman's glands could be secretion. However, MAbs to the secretory product of Bowman's glands (GLA-1 to GLA-8) do not seem to react with sustentacular cells (Hempstead and Morgan, 1985*a*), indicating that the production of a common secretory substance is not the link between the two cell populations. The SUS antigen is apparently cytosolic and nonsecreted since it is absent in the lumen of the gland (Hempstead and Morgan, 1983*a*, 1985*a*). By contrast, a secretory product of Bowman's glands is apparently also present in secretory cells of the respiratory epithelium (Hempstead and Morgan, 1985*a*).

A notion was presented several years ago that sustentacular cells were derived from cells lining Bowman's glands (Mulvaney and Heist, 1971) but this has not been validated by [^{3}H]thymidine-labeling studies. In addition, we have reported the presence of cells penetrating the basement membrane of the olfactory epithelium that are reactive with the SUS-1 MAb (Hempstead and Morgan, 1983*a*). It is our contention that these cells may be moving to the surface of the epithelium where they morphologically transform into sustentacular cells. To the best of this author's knowledge, there are no rigorous studies in the primary literature clearly defining the origin of sustentacular cells in adult animals. Thus, the antibodies described to these cells may prove useful in determining their mode of generation.

MAbs to olfactory neurons have also revealed aspects of axonal organization and neuronal heterogeneity. The studies of Allen and Akeson (1985*a,b*) using the 2B8 MAb have clearly shown two subsets of olfactory receptor neurons. One set of neurons is only positive for the olfactory marker protein (OMP) while the other, presumably immature, subset is positive for OMP and 2B8. In studies from this laboratory, the NEU-9 MAb has also revealed the presence of a subset of neurons (Hempstead and Morgan, 1985*b*). These NEU-9-positive neurons have the morphology of the carbonic anhydrase-positive neurons recently described by Brown *et al.* (1984). Yet other workers have recently reported MAbs that reveal a spatial organization of olfactory receptor axonal projections to the olfactory bulb (Fujita *et al.*, 1985). This finding is significant, as it presents the first possible biochemical basis for the known spatial and functional organization of olfactory projections (Jasterboff *et al.*, 1984; Lancet *et al.*, 1982). Our own MAbs have also revealed antigenic differences in the axonal projections of the primary olfactory and vomeronasal pathways (Hempstead and Morgan, 1985*a*).

From these various findings it is clear that MAbs are providing valuable insights into the more subtle aspects of the structure and organization of the olfactory epithelium. This is perhaps exemplified by a series of MAbs (NIS-1 to NIS-4) that reveal a previously undescribed structure in the olfactory epithelium (Hempstead and Morgan, 1985*a*). These elements appear to consist of globular clusters of cells, perhaps of a secretory nature. While they are observed throughout the epithelium, they appear to be present in some areas in much higher density (Hempstead and Morgan, 1985*a*). These structures also seem to have a single basal process that projects toward the basal lamina; their function is quite unknown.

3.2. Monoclonal Antibody Studies of Ontogeny in the Olfactory Epithelium

3.2.1. Background

The olfactory epithelium and vomeronasal organ are believed to be derived from the olfactory placode, a paired thickening of the anterior portion of the cranial ectoderm (Robecchi, 1972; Breipohl and Mendoza, 1977; Cuschieri and Bannister, 1975). At least in amphibians, the olfactory placode contains both a population of neural precursors and non-neuronal cells that give rise to Bowman's glands and sustentacular cells (Klein and Graziadei, 1983). During ontogeny, the precursor neuroblasts proliferate and differentiate and begin to send axons toward the neural tube, where they appear to be able to organize the forebrain (Stout and Graziadei, 1980). Evidence in rodents (Bossy, 1980; Farbman and Squinto, 1985) and chickens (Mendoza *et al.*, 1982) also shows that at the time of axogenesis a group of cells migrates ahead of the axons and may fulfill a pathfinding role (Farbman and Squinto, 1985).

A critical time for neuronal maturation appears to occur around embryonic day 18. At this time, a marked increase in cilia length occurs and the apical knobs and receptor neurons assume their mature adult morphology (Farbman, 1986). By this time many neurons express the OMP (Allen and Akeson, 1985b; Farbman and Margolis, 1980; Monti-Graziadei *et al.*, 1980), which correlates with synapse formation within the developing olfactory bulb (Monti-Graziadei *et al.*, 1980). At this age, the neurons also have the capacity to synthesize the dipeptide carnosine (Margolis *et al.*, 1985), a candidate for the neurotransmitter of the olfactory neuron (Rochel and Margolis, 1982). It was therefore of some interest to establish whether MAbs could be generated that detected this important transition period for the olfactory receptor neurons. To this end, we have attempted to raise MAbs to embryonic day E14 olfactory epithelium, which might reveal antigens expressed early in development and prove useful in pathfinding, as well as to compare these with MAbs to older tissue, which might reveal antigens switched on at later critical times such as E18.

3.2.2. Results

Table 2 summarizes the expression of known markers of olfactory cells. This is a compendium of published data and results obtained in collaboration with Dr. A. Farbman. Immunocytochemical analysis has shown the OMP to first be present at approximately day E15 (Allen and Akeson, 1985b) in rat olfactory neurons (E14 in mouse). The appearance of OMP is preceded by the expression of the 2B8 antigen, which is first detectable on day E13 (Allen and Akeson, 1985*b*). The 2B8 antigen is retained into adulthood (see Section 3.1.2), where it is always coexpressed with OMP (Allen and Akeson, 1985*b*) In a collaborative study with Dr. A. I. Farbman, we have obtained several MAbs directed against E13/E14 olfactory tissue. One interesting MAb, 1A6,

Table 2. The Developmental Appearance of Marker Substances in the Rodent Olfactory Epithelium[a]

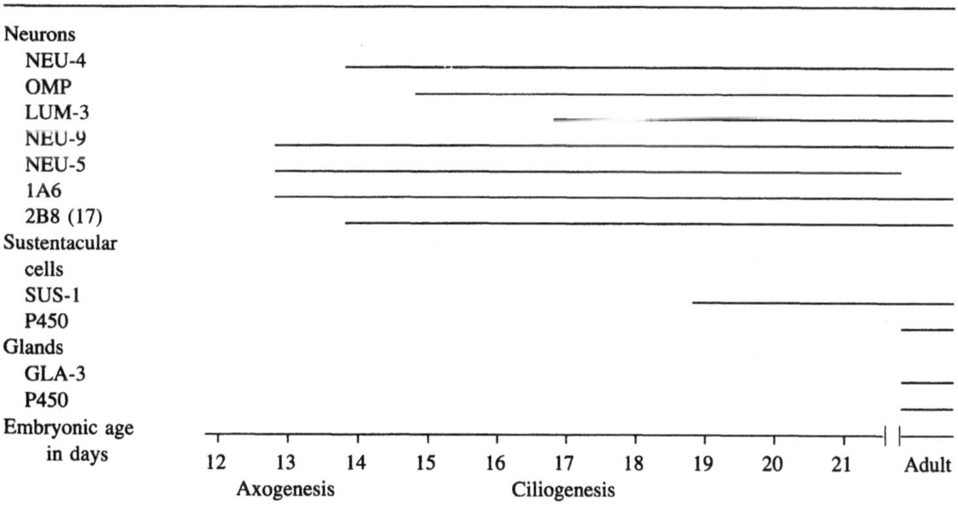

[a]This table represents the developmental appearance of antigens in the rat olfactory epithelium as determined by immunocytochemistry. The figure is expanded and modified from that presented in Farbman *et al.* (1987).

reacts with presumptive olfactory neuron precursors in sections of E13 rat head. The same antibody reacts with neural tube and the migrating cell population that appear to pathfind for axons of olfactory neurons (unpublished observations). Interestingly, in contrast to the 2B8 antigen, the 1A6 antigen is lost during fetal development and is essentially absent at birth.

Two other MAbs, NEU-5 and NEU-9 (Hempstead and Morgan, 1985*a*), which were raised to adult olfactory epithelium, also react with fetal neurons. Both MAbs stain E13/E14 neurons and early axonal bundles; thus, their expression also preceeds that of OMP. Preliminary studies suggest that the immunocytochemical localization of NEU-5 and NEU-9 MAbs are not completely coincident with reactivity for 1A6 (A. I. Farbman, J. L. Hempstead, and J. I. Morgan, unpublished observations), and of course staining with the former MAbs does persist throughout adulthood (Hempstead and Morgan, 1985*a*). Thus. neither NEU-5 nor NEU-9 recognizes the same epitopes as either 1A6 or 2B8. Evidence presented in the next section also demonstrates that NEU-5 and NEU-9 have distinct antigenic determinants.

Also present in our library of MAbs is a clone, NEU-4, that has a staining distribution rather like that of OMP (Hempstead and Morgan, 1985*a*). This neuronal antigenic determinant appears to be expressed later than the previously cited clonal antibodies, first appearing around day E16 (A. I. Farbman, J. L. Hempstead, and J. I. Morgan; see Table 2). Thus, like OMP, the NEU-4 antigen seems to be expressed at a time when olfactory axons begin to synapse in the olfactory bulb (Monti-Graziadei *et al.*, 1980).

Preliminary results from a collaborative project with Dr. A. Farbman have established the sequence of expression of a number of non-neuronal antigenic determinants for which we have MAbs. Antibodies to Bowman's glands (e.g., GLA-1) do not react up to day E20 in sections of rat olfactory epithelium; thus, their expression is essentially post partum. The SUS-1 MAb does react with occasional E20 sustentacular cells but is absent on day E18. We conclude, therefore, that this antigen is first expressed between days E18 and E20. Based upon intensity of staining, only low levels of the SUS-1 antigen are present *in utero*. No evidence was seen of staining of glandular cells by the SUS-1 MAb in fetal olfactory epithelium (unpublished observations).

Using MAbs to the enzyme cytochrome P450 provided by Dr. Paul Thomas of the Roche Institute, it can be shown that this enzyme does not appear until near the time of birth in sustentacular cells and the acinar cells of Bowman's glands (unpublished). Taking these data together, several major antigens of sustentacular cells and Bowman's glands are only expressed late in development, indeed, essentially after birth. Two of these antigens are known, one being a secretory product of Bowman's glands (Hempstead and Morgan, 1985*a*) and the other an important enzyme involved in the oxidative modification of molecules (cytochrome P450) (see Chapter 3). Therefore, it is suggested that certain major functional roles of sustentacular cells and Bowman's glands are only required postpartum.

We have reported one other major group of MAbs that react with the luminal surface of the olfactory epithelium (Hempstead and Morgan, 1985*a*). These have been augmented by several new clones to the luminal surface obtained by immunization with crude membranes derived from neonatal and adult olfactory epithelium (see Section 2.3). The MAbs can be divided into three categories; those that only react with olfactory epithelium; those that react with olfactory and respiratory epithelium; and those that only react with respiratory epithelium. In our original paper (Hempstead and Morgan, 1985*a*), we designated the latter group of MAbs RES, while the former two groups were referred to by the prefix LUM. Only LUM-type MAbs are discussed here. Fundamentally, most LUM MAbs first stain at about day E18. The staining appears suddenly and is intense, essentially equivalent to adult tissue. While we had initially thought that LUM-type MAbs might be binding to the apical surface of sustentacular cells (Hempstead and Morgan, 1985*a*), these recent findings suggest that they may be reactive with olfactory cilia. This would account for the appearance of immunoreactivity at the time of ciliogenesis and the reaction of some clones with respiratory cilia (Hempstead and Morgan, 1985*a*). A more indirect argument against sustentacular cells being the site of all LUM MAb binding is the fact that all other known markers of these cells appear subsequent to day E20. Thus, we believe that the LUM MAbs should be viewed as late markers of olfactory receptor neuron differentiation rather than early developmental markers of sustentacular cells. It will be interesting to perform electron microscopic studies with these LUM MAbs to define their localization to cilia and substructures thereof more fully.

3.3. Monoclonal Antibody Studies of Regeneration in the Olfactory Epithelium

3.3.1. Background

In the adult olfactory epithelium of the rat and mouse, there is a population of precursor neuroblasts, termed basal cells, that appear to proliferate throughout the life of the animal (Thornhill, 1970; Moulton, 1974; Graziadei, 1973; Graziadei and Monti-Graziadei, 1979). It is a matter of some controversy as to the fate of these cells. Some workers believe that mature olfactory receptor neurons have a finite lifespan and undergo programmed death and are replaced by newly formed neurons (Graziadei, 1973; Graziadei and Monti-Graziadei, 1979). Thus, there would be a cellular equilibrium between, on the one hand, neuron production by the division and differentiation of neuroblasts and, on the other hand, mature neuron loss by programmed death. By contrast, other workers believe that receptor neurons are essentially long lived and that they are only replaced upon death provoked by exogenous causes, such as viral or bacterial infection (Hinds *et al.*, 1984). Since in the mouse the olfactory epithelium does not expand after the second month of life (Gross, 1982), this would have to mean that the progeny of the proliferating basal cells must die, presumably without making synapses in the olfactory bulb.

Mature olfactory neurons also undergo retrograde degeneration when their axons are severed, either by axotomy or olfactory bulbectomy (Graziadei and Monti-Graziadei, 1979; Graziadei *et al.*, 1979; Harding *et al.*, 1977; Monti-Graziadei, 1983). Subsequently, new neurons are formed by the proliferation and differentiation of neuroblasts (Graziadei *et al.*, 1979; Harding *et al.*, 1977; Monti-Graziadei, 1983; Samanen and Forbes, 1984; Costanzo, 1984; Oley *et al.*, 1975). Furthermore, olfactory function appears to be normal in the regenerated olfactory epithelium (Oley *et al.*, 1975; Simmons and Getchell, 1981*b*). By contrast, where the target for the olfactory receptor neurons (the olfactory bulb) has been removed, the regenerated epithelium shows a number of marked changes from control. Thus, there are fewer neurons expressing OMP in the decentralized epithelium and an increased number of immature neurons (Monti-Graziadei, 1983; Samanen and Forbes, 1984). We have studied this phenomenon in unilaterally olfactory bulbectomized (OBX) rats. Following regeneration (≈3 months in our studies), staining with a number of MAbs was assessed in the decentralized epithelium and compared to staining in the contralateral mucosa in the same section.

3.3.2. Results

Following unilateral OBX and regeneration, there is a marked attenuation in the levels of staining of some MAbs, notably NEU-4 and NEU-9. In contrast, others MAbs (e.g., NEU-5 and NEU-3) show no difference in staining intensity when one compares contralateral, control, and ipsilateral, decentralized, olfactory mucosae in the same animal (Hempstead and Morgan, 1985*b*). Thus, there appear to be antigenic determinants whose level of expression is controlled, directly or indirectly, by the olfactory bulb. It will be recalled that OMP expression is also modified by the olfactory bulb (Monti-Graziadei, 1983; Samanen and Forbes, 1984; Chuah *et al.*, 1985). However, in the case of OMP, it appears that the modulation of OMP levels is an indirect consequence of the olfactory bulb in that the olfactory bulb controls the rate of neuron differentiation rather than the direct expression of OMP itself (Monti-Graziadei, 1983; Samanen and Forbes, 1984; Chuah *et al.*, 1985). Thus, the amount of OMP per neuron is not significantly altered by bulbectomy only the number of neurons synthesizing the protein is reduced (Monti-Graziadei, 1983; Samanen and Forbes, 1984; Chuah and Farbman, 1983). The issue of a direct control of OMP expression by the olfactory bulb is clouded by studies employing transplants of olfactory epithelium to the anterior chamber of the eye (Barber *et al.*, 1982; Heckroth *et al.*, 1983), the parietal cortex (Morrison and Graziadei, 1983) and cerebral ventricle (Morrison and Graziedei, 1983). For in these studies OMP was only expressed in a relatively small number of neurons. Thus, while it is clear that some receptor neurons can unequivocally express OMP in the absence of the bulb, there may be a second-order modulation by this structure. It is clearly of importance to assess the levels of NEU-4, NEU-5, and NEU-9 expression in primary organotypic cultures of olfactory epithelium and transplants of this tissue.

In the decentralized olfactory epithelium, the NEU-9 MAb detects a subset of presumptive neurons that are only infrequently observed in the control mucosa (Hempstead and Morgan, 1985*b*). These cells bear a resemblance to a neuronal phenotype, characterized by the presence of carbonic anhydrase, recently described by Brown *et al.* (1984). It will also be recalled that the NEU-9 MAb reacts with early fetal olfactory neurons, thus, the presence of these cells may be a reflection of a more primitive neuronal phenotype being produced upon basal cell differentiation. This possibility would also imply that the olfactory bulb could direct the development of neuron differentiation into several pathways. Another possibility is that these cells are one state in the normal life history of an olfactory neuron. It could then be argued that this phenotype is transient under normal conditions but becomes prolonged in the absence of the olfactory bulb, thus increasing its abundance. These cells could also represent a terminal phenotype, being essentially dying cells, their abundance then reflecting an overall increase in the equilibrium turnover of neurons and their precursors. This implies that neurons that fail to make a synapse in the olfactory bulb (as in unilaterally OBX rats) die prematurely. This appears to be, and would be indistinguishable from, programmed neuronal death.

A recent finding in the decentralized olfactory epithelium is the presence of misplaced axonal bundles (Hempstead and Morgan, 1985*b*). The axons are reactive with a range of MAbs to neurons as well as to OMP antibodies but are negative for MAbs to non-neuronal elements of the olfactory mucosa (Hempstead and Morgan, 1985*b*). Interestingly, these axonal bundles or neuromas reside in the receptor neuron cell layer and rarely invade the substentacular cell layer. This has led us to suggest that olfactory neurons and their axons have some form of selective adhesion mechanism, which perhaps becomes overproduced during regeneration in the absence of a target (Hempstead and Morgan, 1985*b*). Such a mechanism may be the Ng-CAM system known to be present in the olfactory epithelium (Chuong and Edelman, 1984). Indeed, the NEU-5 MAb recognizes a protein that is indistinguishable from N-CAM on gel electrophoresis (see Section 3.6.2.).

While unilateral OBX produces pronounced effects on neuron organization, dynamics, and biochemistry, we have observed no obvious effects upon non-neuronal cells. Thus, following OBX, glands appear normal with respect to staining with a series of MAbs to their secretory products (unpublished observations). Furthermore, sustentacular cells are unaffected either in number (confirming previous studies, Simmons and Getchell, 1981b) or in expression of the SUS-1 antigen (Hempstead and Morgan, 1983*a*). Apparently, sustentacular cells are rather insensitive to OBX both in their biochemistry and gross cell kinetics.

3.4. Monoclonal Antibodies to Specific Proteins

3.4.1. MAbs to Carnosine Synthetase

In collaboration with Dr. Frank Margolis of the Roche Institute, we have succeeded in producing 6 distinct monoclonal antibody clones to carnosine synthetase (Margolis *et al.*, 1987), using the procedures outlined above (see Section 2.4). While the antibodies were generated to carnosine synthetase from rabbit muscle, various clones showed a wide species cross-reactivity, permitting their use in routine laboratory species, including rat. Chapter 11 (this volume) deals with the enzymological aspects of carnosine synthetase revealed by these MAbs. However, it could be shown that the enzymes from muscle and olfactory epithelium were indistinguishable and indicated the potential for the use of these MAbs in the olfactory system.

Immunocytochemical analysis has shown only one clonal antibody, designated CS-E, to be of significant use in sections of rat olfactory epithelium. This MAb revealed carnosine synthetase reactivity in the perikarya and axonal bundles of olfactory receptor neurons (unpublished observations). This finding corroborates earlier biochemical studies of carnosine synthetase levels in the olfactory epithelium (Margolis, 1980; Burd *et al.*, 1980), and supports the notion that carnosine is indeed synthesized in olfactory neurons (Burd *et al.*, 1980). One other MAb, CS-D, has shown a staining of occasional cells in the sustentacular cell layer (unpublished observations). Since this MAb did not stain olfactory neurons, the significance or specificity of this reactivity is unclear.

3.4.2. MABs to Proteins of the Cytoskeleton

We have examined the distribution of several proteins of the cytoskeleton and contractile machinery in the olfactory pathway. A MAb to tubulin showed the protein to be highly concentrated in the cilia and axonal tract of olfactory receptor neurons. It was also present in the perikarya of olfactory neurons. Other cell types were positive for tubulin but, because of the intense staining of neurons, they tended to be overlooked as background staining.

Antibodies to actin showed a diffuse distribution with no particular cell type being preferentially stained. It is possible that this distribution could be a function of the antibody rather than the actual presence of the antigen. That is, further actin antibodies must be examined before precise conclusions regarding distribution are drawn.

We were unable to see any specific localization of antibody binding with MAbs to cytokeratin and prekeratin. By contrast, some staining was observed with MAbs to the other intermediate filament proteins, vimentin and neurofilaments. Staining was weak for the anti-neurofilament MAb in the axonal bundles while staining with the antivimentin MAb was strongest in axonal bundles. However, this latter MAb also reacted with cells in the lamina propria mucosae (Schwob *et al.*, 1986).

3.4.3. MAbs to the Cytochrome P450 System

Dr. Paul Thomas has made available MAbs to various cytochrome P450 enzymes and to cytochrome P450 reductase. These antibodies have shown cytochrome P450b to be present in sustentacular cells and acinar cells of Bowman's glands. The reductase enzyme also localizes in these same cell populations and in addition appears to react with olfactory neurons. The presence of the cytochrome P450 enzymes might be interpreted as indicating a role for sustenacular cells and Bowman's glands in odor molecule modification. Such chemical modification of odorants could be important not only for odor elimination but also potentially as a mechanism of presentation of odorants. If the distribution of P450 reductase is correct, the data would also suggest that a P450 exists in neurons that does not react with MAbs to the known P450 enzymes. This assumption is based upon the observation that P450 and P450 reductases are usually present in the same cell. It is interesting to speculate that olfactory neurons possess a novel cytochrome P450 enzyme involved in odorant modification, perhaps as an elimination process.

3.4.4. MAbs to Glial Fibrillary Acidic Protein

A commercial MAb to glial fibrillary acidic protein (GFAP) stains the cells ensheathing the axonal bundles, often referred to as perineuronal sheath cells (Fig. 1). This finding supports the previous report of the presence of GFAP reactivity in perineuronal sheath cells (Barber and Lindsay, 1982). Furthermore, it demonstrates a difference between perineuronal sheath cells and Schwann cells of the peripheral nervous system that do not express GFAP.

3.5. Monoclonal Antibody Studies of Cultures of the Olfactory Epithelium

This section deals with our preliminary production and characterization of dissociated cultures of the rat olfactory epithelium. The goal of this study is to produce clonal lines of olfactory basal cells that retain the capacity for induction to yield neurons expressing more differentiated functions of receptor cells. These cells would then be used to define the molecules and mechanisms controlling basal cell differentiation, to follow the

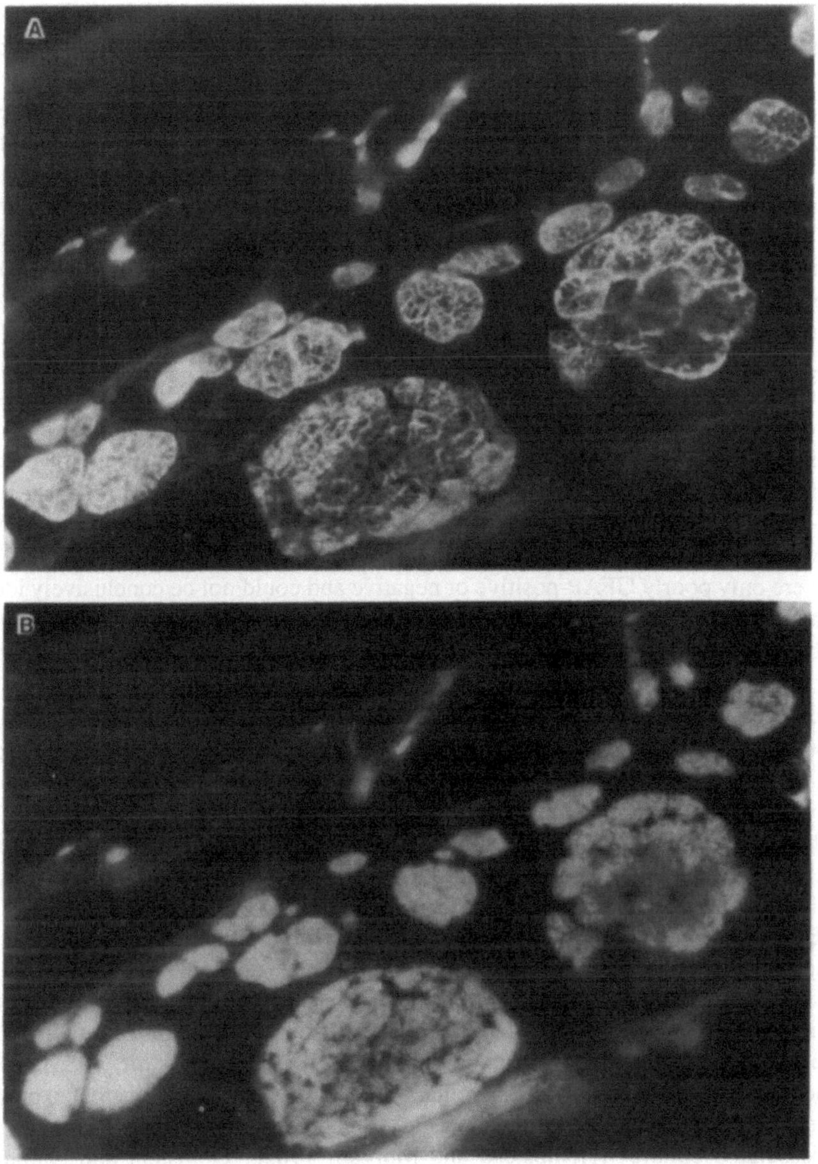

Figure 1. Double immunostaining of adult rat olfactory epithelium by (A) antibodies to GFAP and (B) the NEU-5 MAb. The anti-GFAP is seen to react with the olfactory perineuronal sheath cells that fasciculate axonal bundles that are revealed by their reactivity with NEU-5. It can also be seen that the GFAP staining does not penetrate the basal lamina into the neuroepithelium proper. NEU-5 staining is also predisposed to axonal bundles after fasiculation in the lamina propria mucosae. Magnification is × 250.

molecular and biochemical sequelae of development, and perhaps to isolate odorant receptors by molecular genetic techniques.

To produce dissociated cells from the olfactory epithelium, we have started with young (4-week-old) CD rats reared in our own facility to minimize respiratory infections. Epithelia were removed and disaggregated using collagenase/hyaluronidase as described by Hirsch and Margolis (1979). Following the DNAase step described by the above investigators, we subjected the cells to centrifugation on a 10–80% SIP Percoll gradient. This reproducibly yielded several bands of cells and removed debris and dead cells which were arrested at the top of the gradient. Erythrocytes were pelleted under the conditions used (30 min at 350 $\times g_{av}$) and lymphocytes were recovered as a dense banding population and discarded. Examination of the other two major bands showed them to be mixtures of single and small aggregates of cells. These were taken for culture (details of media and substrata will be presented elsehwere).

Upon culture, two major cell types were observed, the first a fibroblastlike attached cell and the second a round weakly or nonadherent cell. Many of the flat cells in the culture were reactive with antibodies to GFAP (see Section 3.4.4) and were classified as filamentous astrocytes in initial studies (Hempstead and Morgan, 1983*c*). However, these cells may have to be subsequently designated as GFAP-positive perineuronal sheath cells. Over longer periods of culture the round, nonadherent cells attached to the substratum whereupon they assumed a stellate morphology (Hempstead & Morgan, 1983c). These cells were only poorly GFAP-positive or negative and could not be conclusively identified with any of the MAbs present in our library. However, based on their morphology, we feel that they may be the equivalent of stellate astrocytes, or nonmyelinating Schwann cells.

Besides these two major cell types, we have observed several minor cellular phenotypes. The most interesting of these is a small, round, adherent cell that often appears to be OMP positive. Attempts at cloning this cell have failed, perhaps because it is a terminal phenotype, or its progenitor may have a quite different morphology. Another infrequent morphology is a small process-bearing cell that has the appearance of a cultured neuron. These cells react with fluorescent peanut lectin, which may be ultimately useful in isolating this phenotype. They do not express OMP.

Besides our own preliminary study of rat olfactory epithelium in dissociated culture (Hempstead and Morgan, 1983*c*), two other reports have appeared in the literature (Goldstein and Quinn, 1981; Schubert *et al.*, 1985). This is in addition to the work of Farbman and his colleagues who have developed primary organotypic cultures of fetal rat olfactory epithelium (Farbman, 1977; Chuah and Farbman, 1983*b;* Chuah *et al.*, 1985). While the organotypic cultures unambiguously contain olfactory receptor neurons that express OMP (Farbman, 1977; Chuah and Farbman, 1983; Chuah *et al.*, 1985), this cannot be said for the dissociated cultures (Hempstead and Morgan, 1983*c;* Goldstein and Quinn, 1981; Schubert *et al.*, 1985). Thus, while these latter studies report cells with a remarkably similar morphology and phenotype none conclusively identify the cell types present. Both the original study of Goldstein and Quinn (1981) and the recent report of Schubert *et al.* (1985) identified cultures that could synthesize carnosine, the dipeptide present in high concentrations in olfactory epithelium (Margolis, 1980; Burd *et al.*, 1980). This was taken as direct biochemical evidence for the cells being olfactory neurons or their immediate precursors (Goldstein and Quinn, 1981; Schubert *et al.*, 1985). However, several studies have shown that carnosine is synthesized in many tissues and cell types quite unrelated to olfactory neurons (Margolis, 1980; Burd *et al.*, 1980; Bauer *et al.*, 1979, 1982; Margolis and Grillo, 1984; Margolis *et al.*, 1987). Of particular importance is the fact that astrocytes are also a source of carnosine synthetase activity (Bauer *et al.*, 1979, 1982) and, as we have suggested (Hempstead and Morgan, 1983*c*), the cultured cells may be a form of astrocyte or nonmyelinating Schwann cell. Thus, it is interesting that two of the studies (Hempstead and Morgan, 1983*c;* Schubert *et al.*, 1985) present evidence for GFAP-positive cells in cultures of rat olfactory epithelium. Whereas our interpretation of this is that the cells are related to perineuronal sheath cells (Hempstead and Morgan, 1983*c*); Schubert *et al.* (1985) consider this an indicator of a pluripotent neuronal–glial precursor. It should be recalled, however, that the only resident GFAP-positive cells in the olfactory epithelium are the perineuronal sheath cells (Barber and Lindsay, 1982) (Fig. 1). Since we have found the resident sheath cells to contain S100 protein, this would appear to be a relevant marker to assess in these cultures.

Both our study (Hempstead and Morgan, 1983*c*) and that of Schubert *et al.*, (1985) are in agreement on the fact that the cultures generate round cells that are GFAP negative. Interestingly the latter workers also showed these cells to be positive for neuron-specific enolase and the T200, T-cell antigen (Schubert *et al.*, 1985). In our hands, we cannot show reactivity of resident cells of the rat olfactory epithelium with the neuron-specific enolase antiserum in our possession (J. L. Hempstead and J. I. Morgan, unpublished observations). Clearly, it awaits further characterization of these cells to define their true phenotype.

Recently, we have obtained tissue samples from a human olfactory neuroesthesioblastoma and subjected this to an immunocytochemical analysis (L. Zbar, J. L. Hempstead and J. I. Morgan, unpublished observations). Two major findings resulted from this study that are relevant both to the ontogeny of the olfactory epithelium and its culture. First, the tumors were found to be bounded by process-bearing nonmyelinating Schwann cells that were positive for S100b protein (Fig. 2). These Schwann cells also permeated the tumor and formed quasifasciculations reminiscent of the pattern seen with S100 antibodies in the olfactory mucosa *in vivo* (Fig. 3). Indeed it has been pointed out that olfactory neurons organize their Schwann cells into this somewhat atypical configuration (Barber & Lindsay, 1982; Barber, 1982). Thus, it may be that the cells within the tumor have retained this organizational ability. A second finding may also bear upon this contention. It has been found that the NEU-5 MAb (Hempstead & Morgan, 1985a) reacts not with the S100 positive Schwann cells but rather with the tumor cells themselves (Fig. 2). It will be recalled that NEU-5 reacts primarily with the axons of olfactory receptor neurons (Hempstead and Morgan, 1985*a*) (Figs. 1 and 3). This might indicate, since NEU-5 is a membrane antigen, that this glycoprotein is involved in the control of Schwann cell organization.

A further point of speculation is also worthy of mention. We have shown that NEU-5 reacts not only with adult and developing axons (Hempstead and Morgan, 1985*a;* Farbman *et al.*, 1986), but also with the migratory cells that accompany axogenesis in the fetal olfactory epithelium (A. I. Farbman, V. McM. Carr, J. I. Morgan, and J. L. Hempstead, unpublished observations). It has been suggested that these migratory cells might be Schwann cell mothers that ultimately give rise to the perineuronal sheath cell population (Farbman and Squinto, 1985). Thus it may be that this olfactory neuroesthesio-

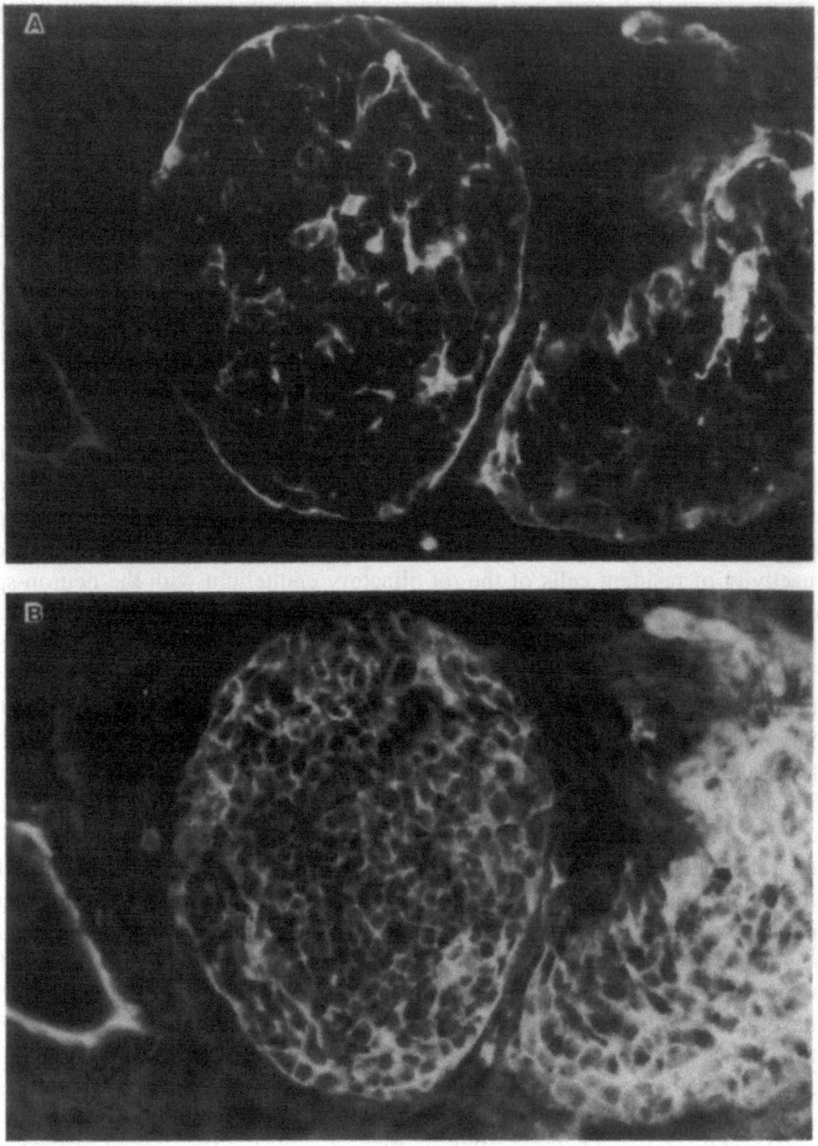

Figure 2. Double immunostaining of a human neuroepithelioma with (A) antibodies to S100 protein and (B) the NEU-5 MAb. Anti-S100 antibodies stain process-bearing cells that encapsulate and permeate the tumor. NEU-5 appears to react with all cells in the tumor mass. Magnification is × 360.

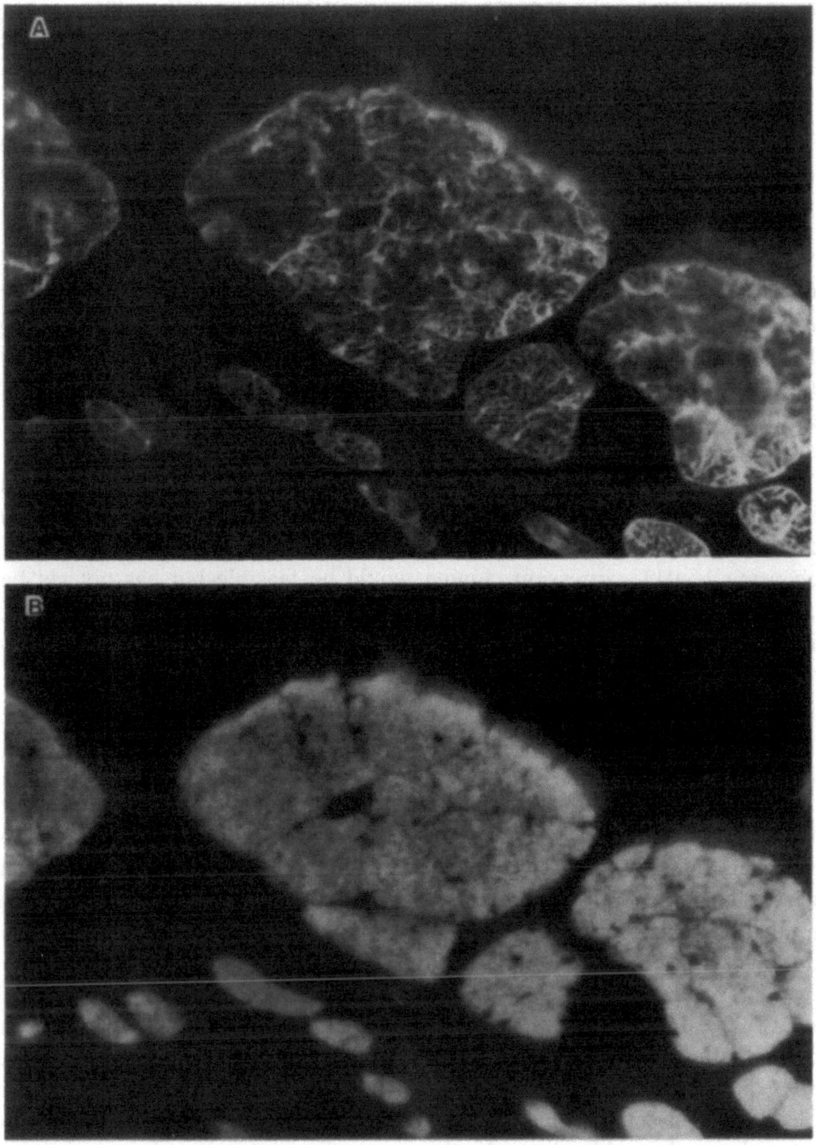

Figure 3. Double immunostaining of adult rat olfactory epithelium by (A) antibodies to the S100 protein and (B) NEU-5 MAb. The staining distribution for S100 is indistinguishable from that of GFAP, and is also confined to ensheathing cells of the fasciculated axons in the lamina propria mucosae. Magnification is × 360.

blastoma is in reality a Schwannoma or a tumor of Schwann cell mothers. The poorly differentiated, round, NEU-5-positive cells would then be the precursors of the more mature process-bearing S100-positive Schwann cells.

3.6. The Use of Monoclonal Antibodies in the Identification and Isolation of Novel Olfactory Tract Antigens

3.6.1. General Considerations

An important extension of the foregoing studies would be the biochemical identification and isolation of the antigens stained by the individual MAbs. This would then permit more extensive functional and biochemical analyses of the roles of these substances in the process of olfaction. Conceptually there are many approaches to establishing the identity of these antigens, ranging from classical protein chemistry to molecular genetic methods. This section deals with three methods applied in this laboratory to the analysis of the NEU-5 and NEU-9 antigens. Two are biochemical (Western blotting and immunoprecipitation of labeled proteins) and a third procedure involves gene transfer and has been performed in collaboration with the laboratory of Dr. A. Blume. A final molecular approach using λ gt11 expression cloning would be feasible but has not yet been attempted in this laboratory. The main reason for this being the reasonable possibility that the various MAbs may recognize carbohydrates on proteins which will not be present on the bacterially synthesized protein.

3.6.2. Molecular and Biochemical Studies of the NEU-5 and NEU-9 Antigens

In an initial study, high-speed supernatants from homogenates of rat olfactory epithelium were subjected to SDS–PAGE and then Western blot analysis using the NEU-5 MAb and a peroxidase reporter system. This revealed that the NEU-5 antigen was a high-molecular-weight (210,000 M_r) protein (T. Kwiatkowski, J. L. Hempstead, and J. I. Morgan, unpublished observations (Figs. 4–6). While non-neuronal tissues had no immunoreactive material, other brain regions did have positive bands, although at lower molecular weights (160,000 and 120,000 M_r) (Figs. 4 and 5). The olfactory bulb had all three bands at 210,000, 160,000, and 120,000 M_r. Thus, it was concluded that in the adult, the olfactory epithelium and olfactory nerve layer of the olfactory bulb possessed a unique 210,000-M_r protein. Other neural regions must possess antigenically related smaller proteins. Clearly, all three bands may represent the same gene product with varying degrees of post-translational modification. Furthermore, the molecular characteristics of the NEU-5 antigen and its distribution are similar to those of rat Ng-CAM, which is known to be present in olfactory epithelium (Chuong and Edelman, 1984). In a further study, we have shown the 210,000-M_r NEU-5 band to be present in fetal and neonatal cortex (Fig. 5). Indeed, fetal and neonatal neural tissues show a progressive shift to the lower-molecular-weight forms, the olfactory epithelium and bulb being the only tissues to retain the high-molecular-weight species into adulthood.

During the study, it was determined that several rat neuroblastoma lines (B103 and B104) and an astrocytoma (C6) expressed the 160,000- and 120,000-M_r forms of the NEU-5 antigen. In addition, the human neuroblastoma, SY5Y, expressed the same anti-

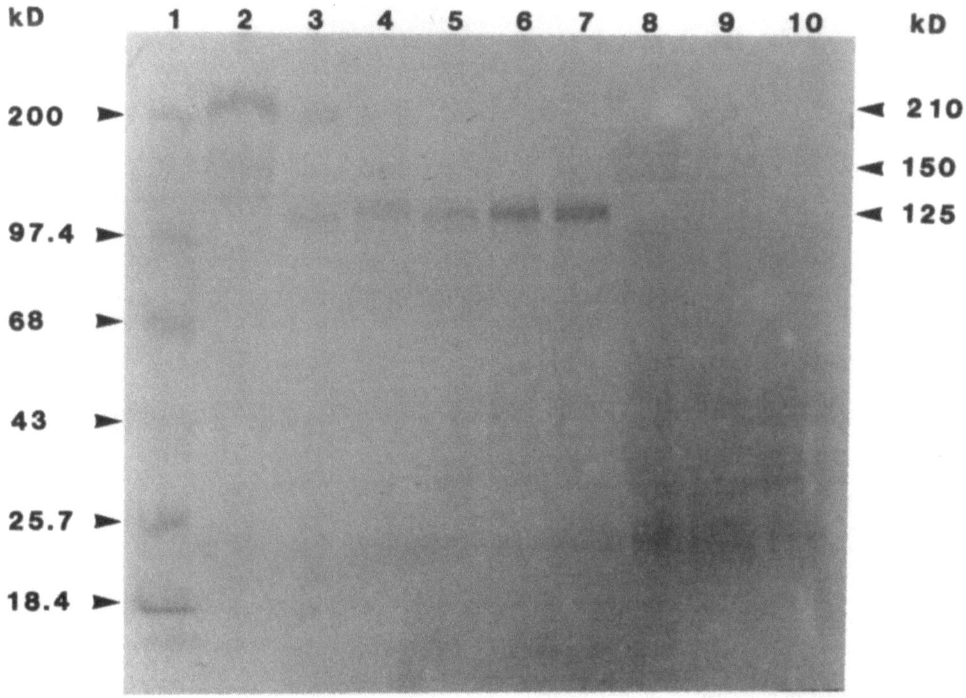

Figure 4. Western blot analysis of the NEU-5 antigen in rat tissues. Various rat brain regions and extraneural tissues were dissolved in Laemmli dissociation buffer and equivalent amounts of sample subjected to sodium dodecylsulfate polyacrylamide gel electrophoresis. Subsequently the proteins were transferred to nitrocellulose, blocked with gelatin, and exposed to the NEU-5 MAb. Following washing, bound antibody was revealed by incubation of the filter with peroxidase-coupled anti-mouse immunoglobulin, washing and exposure to 2-chloronaphthol, a peroxidase substrate. Coomassie blue prestained molecular weight standard proteins were run on the same gel and transferred with the test samples (Lane 1). The olfactory epithelium (Lane 2) and olfactory bulb (Lane 3) contain prominent NEU-5-positive bands at approximately 210 KDa. However, the bulb as well as all other brain regions, contain a second and principal band of immunoreactivity at about 125 KDa (Lane 4 is cerebral cortex; Lane 5 is cerebellum; Lane 6 is brain stem; Lane 7 is spinal cord). A minor band of around 150 KDa is also seen in some brain regions. There is no overt reaction to NEU-5 with liver (Lane 8), adrenals (Lane 9), or kidney (Lane 10).

gens. Furthermore, these cells expressed sufficient antigen on their surfaces to permit detection by immunofluorescence. With this information, it proved possible to immunoprecipitate the NEU-5 antigen from both biosynthetically labeled neuroblastoma and iodinated high speed supernatants from homogenates of rat olfactory bulb and epithelium, cerebellum, cerebral cortex and brain stem (Fig. 6). The basic protocol involved preclearing the labeled supernatant with rabbit antimouse IgG bound to protein A Sepharose, before precipitating specific proteins with the MAb bound to the antibody–protein A–Sepharose complex. The pellets were washed and run on SDS–PAGE and the gels dried and autoradiographed. Only the 120,000- and 160,000-M_r bands were observed in these studies, indicating that (1) these cells never generated the 210,000-M_r species seen in fetal neurons and adult olfactory epithelium and bulb and (2) there are no detectable lower- (or indeed higher-) molecular-weight species present that were missed on Western blots.

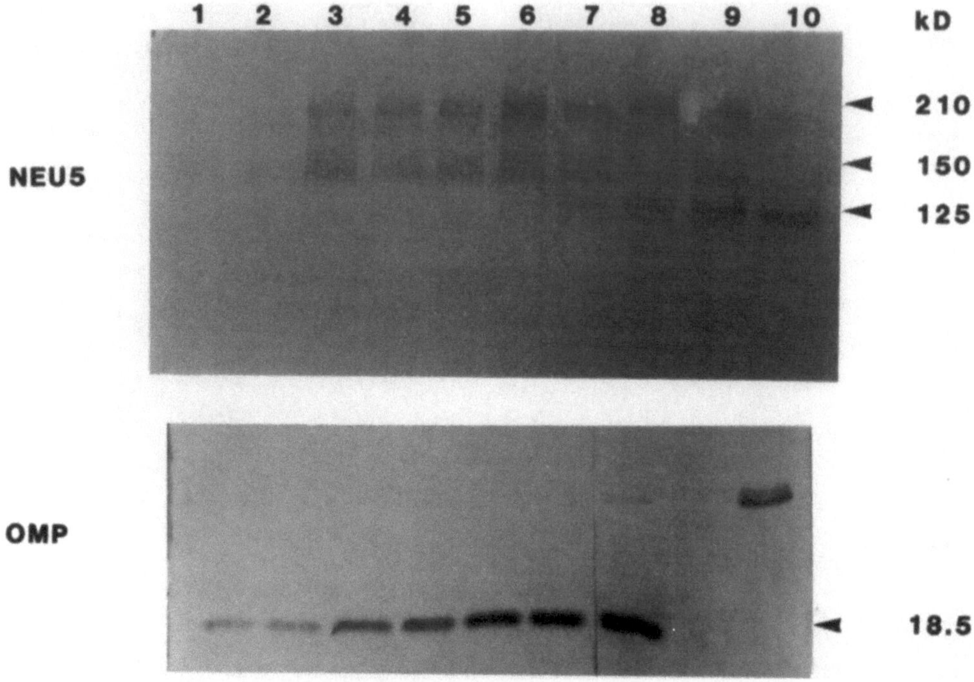

Figure 5. Western blot analysis of the developmental expression of immunoreactivity to the NEU-5 MAb and anti-OMP antiserum in brain (upper panel) and olfactory epithelium (lower panel). Rats of the indicated fetal or adult ages were processed for Western blot analysis as detailed in the legend to Figure 4. The ages were: for the upper panel (NEU-5 reactivity), Lane 1, E13; Lane 2, E15; Lane 3, E17; Lane 4, E20; Lane 5, P1; Lane 6, P3; Lane 7, P11; Lane 8, P17, Lane 9, P25; Lane 20, P52. (Note that in the fetus and newborn, the predominant form of the NEU-5 antigen is at 210 and 150 KDa. However, during postnatal life the 210 KDa band disappears, to be replaced by another band at 125 KDa which is the major adult reactive species. As shown in Figure 4, the 210 KDa form of the antigen is only retained into adulthood in significant quantities in the olfactory epithelium and olfactory bulb. For comparison, the development of expression of a known olfactory receptor neuron-specific protein (OMP) is shown. Lane 1, P4; Lane 2, P7; Lane 3, P11; Lane 4, P14; Lane 5, P18; Lane 6, P26; Lane 7, P80; Lane 8, prestained molecular weight standard.)

These findings permitted us to attempt a relatively novel approach to obtaining the primary structure and function of the NEU-5 and NEU-9 antigens. The strategy was as follows and has been performed in collaboration with the group of Dr. A. Blume of this Institute. Since the NEU-5 MAb reacts with human neuroblastoma (SY5Y), it is inferred that the MAb recognizes the human counterpart of the rat NEU-5 antigen. The attempt has then been made to transfer the human gene encoding the NEU-5 antigen into a mouse fibroblast (L cell) that does not express NEU-5. Technically, sheared human DNA is transfected by the calcium phosphate method along with a viral thymidine kinase gene. The small number of L cells that take up a piece of human DNA also assimilate the thymidine kinase gene. This is important, since the normal L cell lacks the enzyme thymidine kinase and therefore dies in HAT selection medium. Thus, cells transfected with human DNA and the viral kinase gene survive in HAT medium, while the large majority of the L cells die. The surviving cells have of course picked up diverse fragments of human DNA, and only a small number, perhaps 1 in 2000 or less, will have stably incorporated the human NEU-5 gene into its DNA. Thus, the transfectants must be screened for the specific presence of the NEU-5 antigen. This has been done by a rosetting assay. Basically, the cells are incubated with the NEU-5 MAb and then washed. Subsequently the cells are exposed to erythrocytes coated with an anti-mouse immunoglobulin. Cells that express the NEU-5 antigen, having the MAb on their surface, bind to the erythrocytes forming rosettes. These cells can be grown up using cloning cylinders and represent the primary transfectant.

Unfortunately, the lengths of human DNA inserts usually are too long for gene analysis. Thus, a second round of transfection is performed as above into naive L cells except that the source of sheared DNA is the primary transfectant. After HAT selection and screening, a secondary transfectant is obtained that has human DNA inserts about one

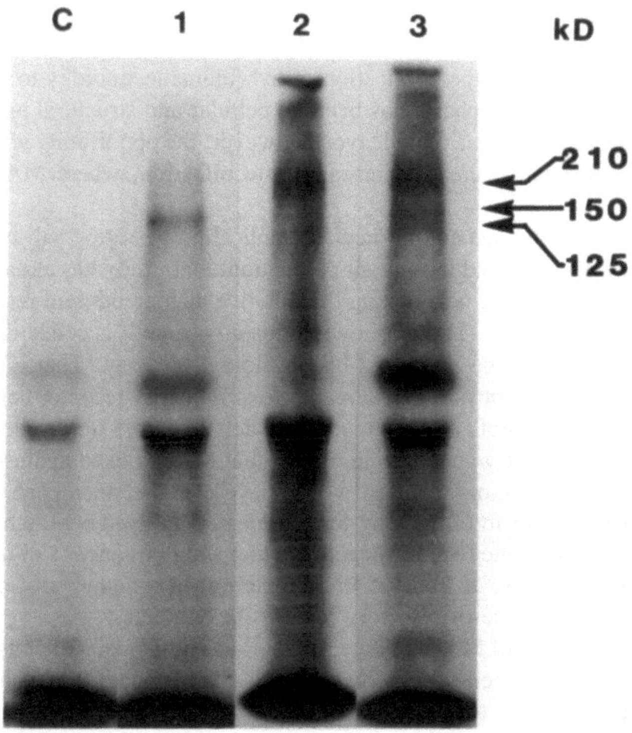

Figure 6. Immunoprecipitation of iodinated proteins from olfactory epithelium and brain regions by the NEU-5 MAb. 30,000 ×g supernatant of homogenates of olfactory epithelium (Lanes C and 2), brain stem (Lane 1) and olfactory bulb (Lane 3) were labelled with ^{125}I onto tyrosine residues using iodogen. Samples in Lanes 1 thru 3 were exposed to the NEU-5 MAb and precipitated with protein A-sepharose, while the control lane received no MAb but all other reagents. The pellets were washed and dissolved in Laemmli buffer, run on polyacrylamide gel electrophoresis (as for Figure 4), dried and autoradiographed. In olfactory epithelium a specifically immunoprecipitated band was observed at 210 KDa (Lane 2) which was not evident when the NEU-5 MAb was omitted (Lane C). A 210 KDa band was also seen in olfactory bulb (Lane 3) but in addition bands were observed at 150 and 125 KDa (Lane 3). Brain stem only showed a single band at 125 KDa. These results confirmed the western blot analysis in Figure 4.

tenth the size of the primary transfectant. To find the human DNA encoding the NEU-5 gene, DNA from the secondary transfectant is cloned into cosmid vectors. The cosmid library is then screened for human DNA sequences by their hybridization to radioactive probes (e.g., the blur probe) to the highly characteristic repetitive DNA sequences of humans, the so-called alu repeats. Having identified clones containing human DNA, these are screened against mRNA from a cell known to be expressing the NEU-5 antigen by Northern analysis. Positive clones must then be sequenced to determine the structure of the NEU-5 gene. En passant, one obtains a fibroblast expressing a neuronal surface protein. Thus, the primary and secondary transformants represent potentially exciting cells with which to investigate the possible function of these antigens in the olfactory tract. In the future, the use of this form of molecular genetic technology could provide a method with which to isolate individual odorant receptors in a cultured fibroblast.

4. DISCUSSION

This chapter documents the application of hybridoma technology to the study of the olfactory epithelium. The emphasis has been on cellular and structural analysis, since to date this has been the major area of endeavor. However, the opportunity will now be taken to outline some prospective areas of research into olfaction, where MAbs could be of some value.

Perhaps the most obvious and immediate application of MAbs will be in identifying the molecules directly involved in odorant perception. One study has already reported the production of a MAb capable of blocking the putative alanine odorant receptor in catfish (Goldstein and Cagan, 1982). Thus, following this precedent it is conceivable that MAbs to odorant receptors might be identified by physiological assay. That is, can MAbs be found that block the electrophysiological response of olfactory receptor neurons to specific odorants or groups thereof? If, as suggested, such receptors lie on the olfactory cilia (Gesteland *et al.*, 1982), it would be anticipated that MAbs raised against cilia or membrane preparations of olfactory mucosa would have a higher incidence of antireceptor clones. This was in fact a part of the long-term strategy employed here, where membranes were used as immunogen, the fusions indeed yielding a high frequency of luminal surface-reactive (LUM) antibodies. It has yet to be determined whether any of these MAbs actually modify the response of olfactory neurons to odorants.

A second application of MAbs to outer membrane components of cells of the olfactory epithelium would be as cell-specific cytotoxins. For instance, by covalently linking a toxic ligand such as Ricin to a particular MAb, it should be possible to specifically eliminate one cell type *in vivo*. This might then make it possible to elucidate the role of non-neuronal cells in olfaction. A particularly attractive candidate would be the sustentacular cell since it is exposed to the external milieu. Thus it is conceivable that one could destroy sustentacular cells, or at least inhibit their functioning, by the intranasal administration of cytotoxic MAbs. In this instance, one could isolate the role played by sustentacular cells in the response to odorants. A further application of cytotoxic MAbs could also be in cultures derived from the olfactory epithelium. Here the MAbs could be added directly to the culture vessel to obtain pure cultures of individual cell types.

A largely untapped use of MAbs is in the purification of cells from the olfactory epithelium. One immediate use of MAbs to membrane antigens of olfactory neurons will be to coat culture dishes with the antibody and pan neurons from dissociated cell preparations. Candidates for attaching and panning olfactory neurons would be the NEU-4, NEU-5, and NEU-9 MAbs, all of which bind to membrane antigens of these neurons. The possibility of obtaining isolated and functional olfactory neurons in cultures would open new vistas in the study of odorant perception and processing and would greatly facilitate electrophysiological recording from these cells. Clearly, such cultures would also provide a powerful model system with which to monitor the biochemistry of olfactory neurons and to study the binding and fate of odorant molecules. This approach could also be carried out by way of a fluorescence-activated cell sorter (FACS) to enhance cell yields for biochemistry or genetic purposes.

Our own preliminary studies have shown the feasibility of transferring genetic material normally only expressed in olfactory neurons into fibroblasts of another species. This procedure is made possible by the availability of MAbs to surface antigens. Thus, it might be possible to transfer odorant receptors to tumor cell lines, particularly if a suitable MAb is forthcoming. Even in the absence of such an antibody, other procedures for assessing the transfer of specific genes are already at hand and, with little modification, could be used to screen for odorant receptor expression.

To conclude, this chapter has described a number of strategies for deriving MAbs to the rat olfactory epithelium. A number of applications of these reagents have been discussed, ranging from the descriptive characterization of the cellular elements of the epithelium to molecular genetic approaches for cloning genes expressed in this tissue. Furthermore, the results reported define a number of novel antigens of the epithelium whose expression is apparently regulated by the olfactory bulb and that may therefore be important in the process of olfaction. This type of information has been further complemented by studies using MAbs to known enzymes and proteins, which have defined possible biochemical functions for several of the non-neuronal elements of the olfactory epithelium. Given the burgeoning interest in the olfactory system, it is likely that the future use of MAbs in this field of research will far outstrip the initial steps described to date.

ACKNOWLEDGMENTS. I wish to express my gratitude to Dr. V. Carr, Dr. A. Farbman, Dr. F. Margolis, Dr. P. Thomas, and Dr. L. Zbar, for permission to cite results of collaborative studies, and to Jim Hempstead, who generated much of the MAb data contained herein.

REFERENCES

Allen, W. K., and Akeson, R., 1985*a*, Identification of a cell surface glycoprotein family on olfactory receptor neurons with a monoclonal antibody, *J. Neurosci.* **5:** 284–296.

Allen, W. K., and Akeson, R., 1985*b*, Identification of an olfactory receptor neuron subclass: Cellular and molecular analysis during development, *Dev. Biol.* **109:** 393–401.

Barber, P. C., 1982, Regeneration of olfactory sensory axons into transplanted segments of peripheral nerve, *Neuroscience* **7:** 2677–2685.

Barber, P. C., Jensen, S., and Zimmer, J., 1982, Differentiation of neurons containing olfactory marker protein in adult rat olfactory epithelium transplanted to the anterior chamber of the eye, *Neuroscience* **7:** 2687–2695.

Barber, P. C., and Lindsay, R. M., 1982, Schwann cells of the olfactory nerves contain glial fibrillary acidic protein and resemble astrocytes, *Neuroscience* **7:** 3077–3090.

Bauer, K., Hallermayer, K., Salnikow, J., Kleinkauf, H., and Hamprecht, B., 1982, Biosynthesis of carnosine and related peptides by glial cells in primary culture, *J. Biol. Chem.* **257:** 3593–3597.

Bayer, K. J., Salnikow, J., DeVitry, F., Tixier-Vidal, A., and Kleinkauf, H., 1979, Characterization and biosynthesis of omega-aminoacyl amino acids from rat brain and the C6 glioma cell line, *J. Biol. Chem.* **254:** 6402–6407.

Bossy, Y., 1980, Development of olfactory and related structures in staged human embryos, *Anat. Embryol.* **161:** 225–236.

Breipohl, W., and Mendoza, A., 1977, Die Entwicklung der olfaktorischen Plakode beim Haushuhn, *Beitr. Electonenmikroskop. Direktabb. Oberft.* **10:** 551–556.

Brown, D., Garcia-Segura, L-M., and Orci, L., 1984, Carbonic anhydrase is present in olfactory receptor cells, *Histochemistry* **80:** 307–309.

Burd, G. D., Davis, B., Macrides, F., and Margolis, F. L., 1980, Distribution of label in the hamster main and accessory olfactory bulb after administration of β-alanine *Soc. Neurosci. Abst.* **6:** 243.

Chuah, M., and Farbman, A. I., 1983, Olfactory bulb increases marker protein in olfactory receptor cells, *J. Neurosci* **3:** 2197–2205.

Chuah, M. I., Farbman, A. I., and Menco. B. Ph. M., 1985, Influence of olfactory bulb on dendritic knob density of rat olfactory receptor neurons in vitro, *Brain Res.* **338:** 259–266.

Chuong, C-M., and Edelman, G. M., 1984. Alterations in neural cell adhesion molecules during development of different regions of the nervous system, *J. Neurosci.* **4:** 2354–2368.

Costanzo, R. M., 1984, Comparison of neurogenesis and cell replacement in the hamster olfactory system with and without a target (olfactory bulb), *Brain Res.* **307:** 295–301.

Cuschieri, A., and Bannister, L. H., 1975, The development of the olfactory mucosa in the mouse: Electron microscopy, *J. Anat.* **119:** 471–498.

Erickson-Viitanen, S., McDermott, M.. Hempstead, J., Balestreri, E., Morgan, J., and Horecker, B. L., 1985, Monoclonal antibodies against cathepsin M from rabbit liver lysosomes, *Fed. Proc.* **44:** 2724.

Farbman, A. I., 1977, Differentiation of olfactory receptor cells in organ culture, *Anat. Rec.* **189:** 187–200.

Farbman, A. I., 1986, Prenatal development of mammalian olfactory receptor cells, *Chem. Senses* **11:** 3–18.

Farbman, A. I., Carr, V. McM., Morgan, J. I., and Hempstead, J. L., 1986, Immunofluorescent studies of the development of rat olfactory epithelium, *Proc. NY Acad. Sci.*

Farbman, A. I., and Margolis, F. L., 1982, Immunohistochemical localization of carnosinase in olfactory and other tissues of the mouse, *Anat. Rec.* **202:** 53A.

Farbman, A. I., and Margolis, F. L., 1980, Olfactory marker protein during ontogeny: Immunohistochemical localization, *Dev. Biol.* **74:** 205–215.

Farbman, A. I., and Squinto, L. M., 1985, Early development of olfactory receptor axons, *Dev. Brain Res.* **19:** 205–213.

Fazekas De St. Groth, S., and Scheidegger, D., 1980, Production of monoclonal antibodies: Strategy and tactics, *J. Immunol. Methods* **35:** 1–21.

Fujita, S. C., Mori, K., Imamura, K., and Obata, K., 1985, Subclasses of olfactory receptor cells and their segregated central projections demonstrated by a monoclonal antibody, *Brain Res.* **326:** 192–196.

Goldstein, N. I., and Cagan, R. H., 1982, Biochemical studies of taste sensation: Monoclonal antibody against L-alanine binding activity of catfish taste epithelium, *Proc. Natl. Acad. Sci. USA* **79:** 7595–7597.

Getchell, T. V., Margolis, F. L., and Getchell, M. L., 1984, Perireceptor and receptor events in vertebrate olfaction, *Progress in Neurobiology* **23:** 317–345.

Gesteland, R. C., Yancey, R. A., and Farbman, A. I., 1982, Development of olfactory receptor neuron selectivity in the rat fetus, *Neuroscience* **7:** 3127–3136.

Goldstein, N. I., and Quinn, M. R., 1981, A novel cell line isolated from the murine olfactory mucosa, *In Vitro* **17:** 593–598.

Graziadei, P. P. C., 1973, Cell dynamics in the olfactory mucosa, *Tissue Cell* **5:** 113–131.

Graziadei, P. P. C., 1971, The olfactory mucosa of vertebrates, in: Handbook of sensory physiology. Volume IV: Chemical Senses. 1. Olfaction (L. M. Beidler ed.), pp 27–58, Springer-Verlag, Berlin.

Graziadei, P. P. C., Levine, R. R., and Monti Graziadei, G. A., 1979, Plasticity of connections of the olfactory sensory neuron: Regeneration into the forebrain following bulbectomy in the neonatal mouse, *Neuroscience* **4:** 713–717.

Graziadei, P. P. C., and Monti Graziadei, G. A., 1979, Neurogenesis and neuron regeneration in the olfactory system of mammals: 1. Morphological aspects of differentiation and structural organization of the olfactory sensory neurons, *J. Neurocytol.* **8:** 1–18.

Gross, E. A., 1982, Comparative morphometry of the nasal cavity in rats and mice, *J. Anat.* **135:** 83–88.

Harding, J., Graziadei, P. P. C., Monti Graziadei, G. A., and Margolis, F. L., 1977, Denervation in the primary olfactory pathway of mice: IV. Biochemical and morphological evidence for neuronal replacement following nerve section, *Brain Res.* **132:** 11–28.

Hartman, B. K., and Margolis, F. L., 1975, Immunofluorescence localization of the olfactory marker protein, *Brain Res.* **96:** 176–180.

Heckroth, J. A., Monti Graziadei, G. A., and Graziadei, P. P. C., 1983, Intraocular transplants of olfactory neuroepithelium in rat, *Int. J. Devl. Neuroscience* **1:** 273–287.

Hempstead, J. L., and Morgan, J. I., 1983*a*, Monoclonal antibodies to the rat olfactory sustentacular cell, *Brain Res.* **288:** 289–295.

Hempstead, J. L., and Morgan, J. I., 1983*b*, Fluorescent lectins as cell-specific markers for the rat olfactory epithelium, *Chem. Senses* **8:** 107–120.

Hempstead, J. L., and Morgan, J. I., 1983*c*, Culture and immunocytochemical characterization of the rat olfactory epithelium, *Soc. Neurosci Abst.* **9:** 464.

Hempstead, J. L., and Morgan, J. I., 1985*a*, A panel of monoclonal antibodies to the rat olfactory epithelium, *J. Neurosci.* **5:** 438–449.

Hempstead, J. L., and Morgan, J. I., 1985*b*, Monoclonal antibodies reveal novel aspects of the biochemistry and organization of olfactory neurons following unilateral olfactory bulbectomy, *J. Neurosci.* **5:** 2382–2387.

Hinds, J. W., Hinds, P. L., and McNelly, N. A., 1984, An autoradiographic study of the mouse olfactory epithelium: evidence for long-lived receptors, *Anat. Rec.* **210:** 375–383.

Hirsch, J. D., and Margolis, F. L., 1979, Cell suspensions from rat olfactory neuroepithelium: Biochemical and histochemical characterization, *Brain Res.* **161:** 277–292.

Jastreboff, P. J., Pedersen, P. E., Greer, C. A., Stewart, W. B., Kauer, J. S., Benson, T. E., and Shepherd, G. M., 1984, Specific olfactory receptor populations projecting to identified glomeruli in the rat olfactory bulb, *Proc. Natl. Acad. Sci. USA* **81:** 5250–5254.

Klein, S. L., and Graziadei, P. P. C., 1983, The differentiation of the olfactory placode in *Xenopus laevis:* A light and electron microscope study, *J. Comp. Neurol.* **217:** 17–30.

Kohler. G., and Milstein, C., 1975. Continuous culture of fused cells secreting antibody of defined specificity, *Nature* **256:** 495–497.

Lancet, D., Greer, C., Kauer, J. S., and Shepherd, G. M., 1982. Mapping of odor-related neuronal activity in the olfactory bulb by high resolution 2-deoxyglucose autoradiography, *Proc. Natl. Acad. Sci. USA* **79:** 670–674.

Margolis, F. L., 1972, A Brain protein unique to the olfactory bulb, *Proc. Natl. Acad. Sci., USA* **69:** 1221–1224.

Margolis, F. L., 1980, Carnosine: An Olfactory Neuropeptide, in Role of peptides in neuronal function (J. L. Barker and T. Smith, eds.), pp. 545–572. Marcel Dekker, New York.

Margolis, F. L., and Grillo, M., 1984, Carnosine, homocarnosine and anserine in vertebrate retinas, *Neurochem. Intl.* **6:** 207–207.

Margolis, F. L., Grillo, M., Kawano, T., and Farbman, A. I., 1985. Carnosine synthesis in olfactory tissue during ontogeny: Effect of exogenous β alanine, *J. Neurochem.* **44:** 1459–1464.

Margolis, F. L., Grillo, M., Hempstead, J., and Morgan, J. I., 1987, Monoclonal antibodies to mammalian carnosine synthetase, *J. Neurochem.*, **48:** 593–600.

Mendoza, A. S., Breipohl., W., and Miragall, F., 1982, Cell migration from the chick olfactory placode: A light and electron microscopic study, *J. Embryol. Exp. Morphol.* **69:** 47–59.

Mollicone, R., Trojan, J., and Oriol, R., 1985, Appearance of H and B antigens in primary sensory cells of the rat olfactory apparatus and inner ear, *Dev. Brain Res.* **17:** 275–279.

Monti Graziadei, G. A., 1983, Experimental studies on the olfactory marker protein: III. The olfactory marker protein in the olfactory neuroepithelium lacking connections with the forebrain, *Brain. Res.* **262:** 303–308.

Monti Graziadei, G. A., Stanley, R. S., and Graziadei, P. P. C., 1980, The olfactory marker protein in the olfactory system of the mouse during development *Neuroscience* **5:** 1239–1252.

Morgan, J., 1984, Monoclonal antibody production, in, Modern methods in pharmacology (S. Spector and N. Bach, eds.), Vol. 2, pp. 29–67, Alan R. Liss, New York.

Morrison, E. E., and Graziadei, P. P. C., 1983, Transplants of the olfactory mucosa in the rat brain: 1. A light microscopic study of transplant organization, *Brain Res.* **279:** 241–245.

Moulton, D. G., 1974, Cell renewal in the olfactory epithelium of the mouse, *Ann. N. Y. Acad. Sci.* **237:** 52–61.

Mulvaney, B. D., and Heist, H. E., 1971. Regeneration of rabbit olfactory epithelium, *Amer. J. Anat.* **131:** 241–252.

Oley, N., Dettan, R. S., Tucker, D., Smith, J. C., and Graziadei, P. P. C., 1975, Recovery of structure and function following transection of the primary olfactory nerves in pigeons, *J. Comp. Physiol. Psychol.* **88:** 477–495.

Pevsner, J., Trifiletti, R. R., Strittmatter, S. M., and Snyder, S. H., 1985, Isolation and characterization of an olfactory receptor protein for odorant pyrazines, *Proc. Natl. Acad. Sci. USA* **82:** 3050–3054.

Robecchi, M. G., 1972, Ultrastructure of differentiating sensory neurons in the olfactory placode of the chick embryo, *Minerva Otorinolaringologica* **22:** 195–204.

Rochel, S., and Margolis, F. L., 1982, Carnosine release from olfactory bulb synaptosomes is calcium-dependent and depolarization stimulated, *J. Neurochem.* **38:** 1505–1514.

Samanen, D. W., and Forbes, W. B., 1984. Replication and differentiation of olfactory receptor neurons following axotomy in the adult hamster: A morphometric analysis of postnatal neurogenesis, *J. Comp. Neurol.* **255:** 201–211.

Schubert, D., Stallcup, W., LaCorbiere, M., Kidokoro, Y., and Orgel, L., 1985, Ontogeny of electrically excitable cells in cultured olfactory epithelium, *Proc. Natl. Acad. Sci. USA* **82:** 7782–7786.

Schwob, J. E., Farber, N. B., and Gottlieb, D. I., 1986, Neurons of the olfactory epithelium in adult rats contain vimentin, *J. Neurosci.* **6:** 208–217.

Simmons, P. A., and Getchell, T. V., 1981*a*, Neurogenesis in olfactory epithelium: Loss and recovery of transepithelial voltage transients following olfactory nerve section, *J. Neurophysiol.* **45:** 516–528.

Simmons, P. A., and Getchell, T. V., 1981*b*, Physiological activity of newly differentiated olfactory receptor neurons correlated with morphological recovery from olfactory nerve section in the salamander, *J. Neurophysiol.* **45:** 529–549.

Stout, R. P., and Graziadei, P. P. C., 1980, Influence of the olfactory placode on the development of the brain in *Xenopus laevis, Neuroscience* **5:** 2175–2186.

Takahashi, S., Iwanaga, T., Takahashi, Y., Nakano, Y., and Fujita, T., 1984, Neuron-specific enolase, neurofilament protein and S100 protein in the olfactory mucosa of human fetuses: An immunohistochemical study, *Cell. Tiss. Res.* **238:** 231–234.

Thornhill, R. A., 1970, Cell division in the olfactory epithelium of the lamprey: *Lampetra fluviatilis*. *Z. Zellforsch. Mikrosk. Anat.* **109:** 147–157.

13

Primary Olfactory Neuron Subclasses

Richard A. Akeson

1. INTRODUCTION

Vertebrate nervous systems contain hundreds of groups of neurons that can be identified as distinct categories by their anatomical location and cellular morphology. In recent years, the increasing number of physiological and molecular parameters that can be determined has permitted many of these groups to be divided into subsets with discrete characteristics. On the basis of gross anatomical and cell morphological classifications, the peripheral olfactory system contributes only a single recognized cell type to neuronal diversity: the primary olfactory neuron or olfactory receptor neuron. In contrast to this single anatomical cell type, most theories of olfactory function suggest that multiple physiological groups of primary olfactory neurons are present. Individual theories imply that the actual number of functionally distinct categories of olfactory neurons is either 1 or 4 or 20 or 10^6. Using more recently developed morphological and biochemical criteria, about 10^1 different olfactory neuron groups can be defined. These groups are identified by dividing the olfactory neuron population into subsets with or without a particular characteristic. As these characteristics have been defined in several laboratories, the relationships among these olfactory neuron subsets are presently not known. Of more importance, the significance of these subsets for olfactory function is also unknown. This chapter reviews current information on olfactory neuron subclasses and then suggests additional approaches to determining whether the class of olfactory neurons is a unitary group of purebred and interchangeable units or whether like many other neuronal systems it is a mongrel pack of individuals with diverse capabilities.

Neurons are generally characterized by their highly asymmetrical shapes and ability to generate action potentials. Although biochemical characteristics, such as those related to neurotransmitters, are frequently used to subtype neurons, some biochemical markers are shared by most neurons. Before discussion of markers that subdivide olfactory recep-

Abbreviations used in the text: CNS, central nervous system; DRG, dorsal root ganglia; GFAP, glial fibrillary acidic protein; HLA, histocompatibility antigens; HNK, a carbohydrate group covalently attached to some polypeptides from neural tissue and human natural killer cells; H-2, the murine major histocompatibility locus; Mab, monoclonal antibody; NCAM, neural cell adhesion molecule; OMP, olfactory marker protein; PNS, peripheral nervous system; SSEA, stage-specific embryonic antigen.

Richard A. Akeson • Division of Basic Science Research, Children's Hospital Research Foundation, Cincinnati, Ohio 45229.

tor neurons, it is useful to discuss those markers these neurons have in common with other neurons. Neuron-specific enolase is an isozyme of a glycolytic enzyme found primarily in neurons, including those of the human olfactory mucosa (Takahashi *et al.*, 1984). A second cytoplasmic marker for neurons are the unique intermediate filament polypeptides of neurons, the neurofilaments. Although Takahashi *et al.* (1984) demonstrated neurofilaments histologically in second-trimester (but not first-trimester) human fetal olfactory epithelium, Schwob *et al.* (1986) found neurofilaments essentially absent from the adult rat epithelium. Rather, these workers found vimentin, an intermediate filament protein characteristic of immature neurons and other dividing cells. Thus, neurofilament expression by olfactory receptor neurons may be species dependent.

At the cell surface, olfactory receptor neurons and other neurons all express very low levels of histocompatibility antigens compared with other tissues (Whelan *et al.*, 1986). Olfactory receptor neurons also lack the cell-surface glycoprotein Thyl, which is found in most other neurons (Moriss and Barber, 1983). NILE is a larger neural cell surface glycoprotein that appears to be found in high levels on the axons of embryonic rat olfactory receptor neurons but in much lower levels on the cell bodies (Stallcup *et al.*, 1985). A third neural cell-surface glycoprotein, neural cell-adhesion molecule (N-CAM), is expressed by olfactory receptor neurons. In summary, olfactory receptor neurons share some but not all the biochemical characteristics typical of most other neurons.

2. OLFACTORY NEURON SUBCLASSES

2.1. Morphological Subclasses

Most olfactory neurons are morphologically identical. This consensus statement is based on the recognition of varying states of differentiation of olfactory receptor neurons—from basal cell to mature neuron—within the same area of epithelium (Graziadei and Metcalf, 1971). However, two morphological categories of mature olfactory neurons have been recognized, those that have cilia and those that have microvilli. Most vertebrate sensory structures, such as the retinal rod cell outer segment, are formed from modified cilia. Most olfactory neurons extend cilia into the mucosal space. However, Moran *et al.* (1982*a,b*) have identified a population of cells in human olfactory epithelium that have elongated bodies and axonlike cytoplasmic processes extending through the epithelium toward the basal lamina but have microvilli rather than cilia. Thus, these microvillous cells have morphological characteristics of olfactory receptors. Definitive proof of this identity is difficult but, in view of the microvillous olfactory neurons in other species, it seems a reasonable working assumption. In the human, microvillar cells are about 10% of the total olfactory neuron population, and small numbers of cells with similar appearance have been observed in other mammals. It should be noted that all the sensory neurons of the vomeronasal organ have microvilli rather than cilia. No specific function for the microvillar cells in the main human olfactory epithelium has been proposed.

In fishes, ciliated and microvillous neuron populations are present in similar numbers and are intermixed in the olfactory epithelium. However, Thommesen (1983) demonstrated that the density of ciliated receptor cells is greater nearer the peripheral margin of each olfactory lamella; he has inferred a potential physiological specificity from this distribution. Two major categories of olfactants in fishes are bile-saltlike compounds and amino acids. Electro-olfactograms recorded from the peripheral portions of olfactory lamella show higher responses to bile salts, while central portions show greater responses to methionine. These results suggest that ciliated receptor cells are selectively responsive to bile salts, whereas microvillous receptors respond to amino acids. Although single-cell recordings for rigorous testing of this hypothesis have not been reported, the two morphological olfactory neuron types represent a clear test system for examining olfactory neuron subclass specificity.

The bile acids are structurally related to steroids. The vomeronasal organ of tetrapods is believed to mediate some sexual and maternal behaviors based on olfactory cues that may be steroidal. These observations might predict that the bile salt-sensitive subclass of fishes would be microvillous. Unfortunately, this prediction is not fulfilled in Thommesen's study; in this instance there appears to be no clear cross-species correlation between morphological subclass and potential olfactant specificity.

2.2. Carbohydrate Expression Subclasses

Many monoclonal antibodies react with carbohydrate groups (Feizi, 1985). Experiments with these reagents combined with advances in carbohydrate analysis techniques have emphasized that discrete cell types have individual patterns of expression of carbohydrate differentiation antigens on cell surface and intracellular glycolipids and glycoproteins. Some, but not all, of the glycoprotein antigens commonly found on central nervous system (CNS) neurons are also found on olfactory neurons. Olfactory cells in general also express soybean agglutinin-binding components not found in CNS structures (Key and Giorgi, 1986). Examples of the differential expression of carbohydrate antigens in olfactory neuron subsets are described here and compared with similar studies of dorsal root ganglion (DRG) and CNS neurons. For the latter groups of neurons. correlations between carbohydrate antigen expression and physiological cell subclass and also inferences of the functions of the carbohydrate antigens are also discussed. For olfactory neurons, the biological significance of the observed variations in carbohydrate antigen expression is unknown.

2.2.1. Subclasses Expressing Blood Groups Antigens

The best current example of the expression of carbohydrate differentiation antigens by olfactory neurons is the demonstration that the human blood group A, B, O antigens are found on olfactory subpopulations. Blood group antigen carbohydrate groups have been classically demonstrated on glycoproteins and glycolipids of erythrocytes and epithelial cells and in the glycoprotein secretions of humans with the Se gene (Watkins, 1980). The human blood group A and B substances are formed by the glycosyltransferase-mediated addition of a single sugar group to the precursor H (or O) substance. Thus, the presence of these groups in a cell type is dependent on both the expression of the gene for the specific glycosyltransferase and also all the genes necessary for glycosyltransferases and other polypeptides that cooperate to synthesize the precursor substance. Furthermore, expression of blood group antigens is not only restricted to some cell types but within a cell type is developmentally regulated as well. For example, human HeLa epithelial cells in culture express the H blood group substance concomitant with what appears to be a terminal differentiation event (Kuhns and Pann, 1973; Pann and Kuhns, 1976).

More recently, blood group or blood grouplike substances* have been identified on two specific rat neuronal populations—olfactory neurons and dorsal root ganglion neurons. Blood group antigens are present in detectable levels on very few other CNS or peripheral nervous system (PNS) neuronal populations (Mollicone *et al.*, 1986). Their presence on olfactory neurons could be a result of the embryologic derivation of the cells from the epithelial cells of the olfactory placode. Consistent with this interpretation is the observation that the sensory cells of the inner ear that are derived from the otic placode also express blood group antigens (Mollicone *et al.*, 1985). This reasoning is clearly not applicable to DRG neurons of neural crest origin, but perhaps the recent observations that somewhat different structures are actually expressed by DRG neurons are particularly relevant to this generalization.

In the embryonic rat, olfactory neurons expressing blood group H were detectable at days 15–17 of gestation but were not observed by postnatal days 18–20 (Mollicone *et al.*, 1985). Neurons expressing blood group B were detected on day 16 of gestation. Some of these B-positive cells arise from cells previously expressing H substance, while others appear to arise from H-negative cells. Blood group A-positive cells were not detected despite the fact that the strain of rats used expresses A substance in their intestinal epithelium. Olfactory axons leaving the epithelium and in glomeruli in the bulb express H substance for a longer period than do the cell bodies; in neonatal animals, B substance-positive glomeruli are also observed. Both markers are found on only a portion of glomeruli during this period. In our own experiments, a small number of adult olfactory neurons were found to express B or H substance (see Fig. 2). These results are in partial contrast to those of Mollicone *et al.* (1985), but in both studies the proportion of olfactory epithelial neuronal cell bodies that express blood group substances in adult rats is very low.

In humans, blood group antigens are also expressed in some cells of sympathetic ganglia and also the mesencephalic nucleus of the trigeminal nerve. No other cranial nerve nuclei are reported positive, to the disappointment of those hoping for similarities between olfactory and gustatory pathways. Is there any evidence indicating the subsets of olfactory neurons or other neurons which express blood group antigens are functionally significant? Within the DRG, blood group substance expression is limited primarily to small- and intermediate-size neurons that project to lamina II of the spinal cord posterior horn. These data and complementary data on the expression of other carbohydrate differentiation antigens by rat DRG neurons clearly suggest a relationship between surface carbohydrate

*In the discussion of these results, we have assumed that the structure of the reactive antigen within olfactory tissue is identical to that used as the immunizing antigen or that which shows optimal reactivity with the antibody in other assays. This oversimplification is not necessarily always true. The antibodies may be reacting with olfactory molecules with similar but not identical structures. One related observation that illustrates the point is the reactivity of monoclonal but not polyclonal antiblood group B antibodies with rat dorsal root ganglion neurons. In an exchange of reagents between the two laboratories, the polyclonal reagents were found specific for the regular α (1–2)-fucosylated blood group B determinant while the monoclonal reagent reacted with this structure and also the linear α Gal (1–3) β Gal (1–3) β GLcNAc-R blood group B structure (Mollicone *et al.*, 1986). Rat DRG neurons express only the latter. Olfactory neurons express the former. Thus, in the absence of actual isolation and structural analysis of the olfactory antigens, the identification of a particular structure on olfactory tissue must be provisional. Nevertheless, in this presentation, for simplicity of form, antibodies that react with a molecular species are termed as identifying that species in olfactory tissue.

expression and functional DRG neuron subclass (see Section 2.2.3.). However the available data do not permit any attribution of functional significance to olfactory neurons expressing blood group substances. This inability is due in part to a lack of detailed mapping of blood group substance distribution in the olfactory epithelium and bulb and in part to the limited physiological data that could potentially be correlated with such mapping. There are also no observations in the clinical literature suggesting differences in either olfactory or DRG-mediated sensory function among individuals of different ABO blood type or those rare individuals entirely lacking these blood group substances. Such differences are thus either subtle or nonexistent. Therefore, the strongest possible contribution to the analysis of olfactory function that can be hoped for from the study of blood group antigens is a correlation. That is, the subset of olfactory neurons expressing a given blood group antigen may correlate with a functional subset even though the antigen itself is not involved in the function. This type of correlation does exist in dorsal root ganglion neurons.

2.2.2. *Subclasses Expressing the 2B8 Antigen*

Our research interests in cell–cell and cell–substrate interaction in the nervous system led us to produce a number of monoclonal antibodies (Mabs) to neural cell-surface antigens. We have primarily used purified rat brain membrane fractions or intact cultured neural cells as immunogens. The rat pheochromocytoma cell line PC12 is of particular interest because of its ability to form synapses in culture (Schubert *et al.*, 1977). After encouragement by Dr. Michael Shipley, a number of Mabs previously developed in our laboratory to neural cell-surface components were tested by immunofluorescence with olfactory bulb cryostat sections. One Mab to PC12 cells, designated 2B8, bound only to glomeruli and the olfactory nerve layer in these sections (Fig. 1). Several Mabs, including 2B8, were also reactive with olfactory epithelium. 2B8 was selected for more complete characterization based on the clear initial impression that it reacted with a subpopulation of olfactory epithelial neurons.

Further analysis of Mab 2B8 proceeded in three areas: (1) rigorous analysis of the proportion of olfactory neurons that bind Mab 2B8, (2) determination of the anatomical distribution and molecular nature of the 2B8 antigen, and (3) analysis of the developmental appearance of 2B8 positive olfactory neurons (Allen and Akeson 1985*a*,*b*).

The determination of the percentage of olfactory neurons that express 2B8 used double-label immunofluorescence techniques with Mab 2B8 and goat anti-OMP (donated by Dr. Frank Margolis). In adult rats. about 25% of OMP-positive epithelial cells were found to express 2B8, confirming the initial impression that 2B8 binds to a subclass of olfactory neurons. By contrast, all vomeronasal neurons express 2B8 antigenic activity. In the olfactory bulb, individual glomeruli tend to be either completely positive or completely negative (Fig. 1), although occasional mixed glomeruli are seen. In the epithelium, in addition to the strongly reactive olfactory neurons, cells in the basal layer also appear to bind lower levels of 2B8 Mab. These may be basal cell precursors of 2B8-positive neurons, but definitive markers for basal cells have not been identified, making this hypothesis difficult to test.

2B8 antigenic activity is not unique to olfactory tissue and PC12 cells, as several non-neural tissues bind 2B8 Mab. However, when membrane proteins from these tissues

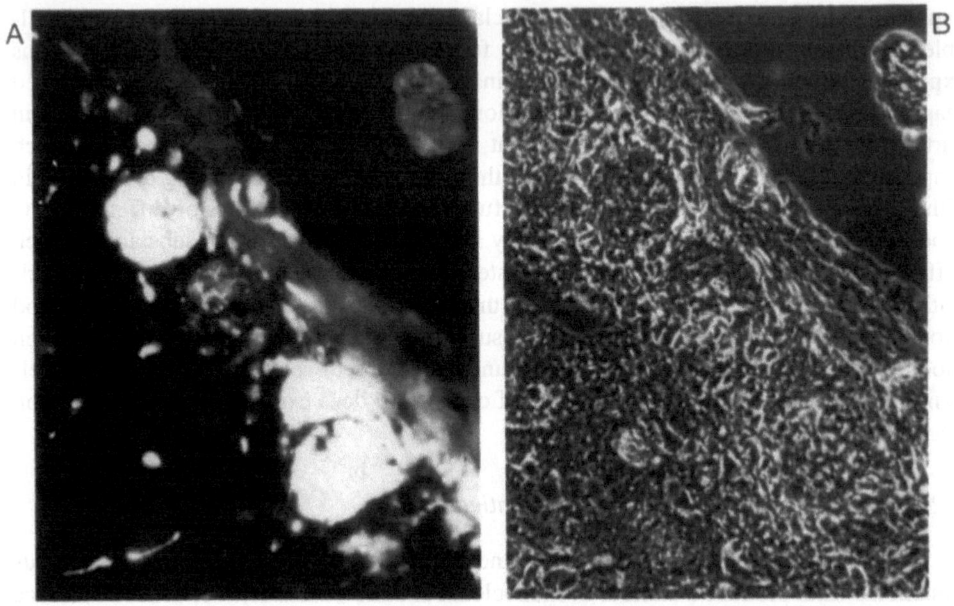

Figure 1. **Immunofluorescence (A) and phase (B) micrographs of a cryostat section of olfactory bulb reacted with Mab 2B8. Note the varying immunofluorescence intensity of individual glomeruli for 2B8.**

are electrophoresed on sodium dodecyl sulfate (SDS) gels and reacted with Mab 2B8 (immunoblotting), the apparent molecular weights of the 2B8-reactive antigens vary widely among tissues. In olfactory epithelium, two major bands of apparent molecular weights of 200,000 and 160,000 are seen. No other tissue expresses these molecular-weight forms of the 2B8 antigen. Thus, Mab 2B8 identifies a molecule apparently unique to olfactory tissue.

Cells in the olfactory epithelium that express 2B8 antigen can be detected by embryonic day 13 in the rat. At day 15, cells expressing OMP are observed. Both populations increase rapidly and at birth 90% of the $2B8^+$ cells also express OMP. From embryonic day 15 to postnatal day 2, the percentage of OMP^+ cells that also expressed 2B8 was relatively constant ranging from 23% to 33%. Similar percentages are found in adult epithelium, where $2B8^+$ olfactory neurons are more abundant than blood group B or H-positive neurons (Fig. 2, cf. A–C). These observations suggest, but do not prove, that the $2B8^+$ population represents a discreet olfactory neuron group rather than a developmental phase through which all olfactory neurons pass. During this same period, the observed molecular weight of the 2B8 antigens in olfactory epithelium varied somewhat from those of adult epithelium. 2B8 antigen was first detectable in the olfactory bulb by immunoblot on day E19.5. At this time, bands of 215,000 and 160,000 were observed and in contrast to the epithelium the apparent molecular weight of these antigens did not change during development. The specific activity of these bands increased three- to four-fold and 10-fold, respectively, from birth to adulthood.

The diverse molecular weights of 2B8 antigens among tissues suggest that the 2B8 antigenic site is either a protein modification such as a carbohydrate group or a linear amino acid sequence shared by several different polypeptides. To test the first alternative, 2B8 immunoprecipitates were electrophoresed and treated with either neuraminidase, sodium *m*-periodate, endoglycosidase H (which cleaves complex N-linked carbohydrate groups), endoglycosidase F (which cleaves both high mannose and complex N-linked carbohydrate groups), or 0.1 N NaOH (which cleaves O-linked carbohydrate groups). The precipitates were then tested with fresh 2B8 antibody for retention of antigenic activity. Antigenic activity was lost after both periodate and endoglycosidase F treatment, suggesting that the 2B8 antigenic site is composed at least in part of a high mannose carbohydrate group (Allen, 1984).

In summary, the 2B8$^+$ and 2B8$^-$ olfactory neuronal sets are another example of olfactory neuronal groups distinguished by the expression of carbohydrate differentiation

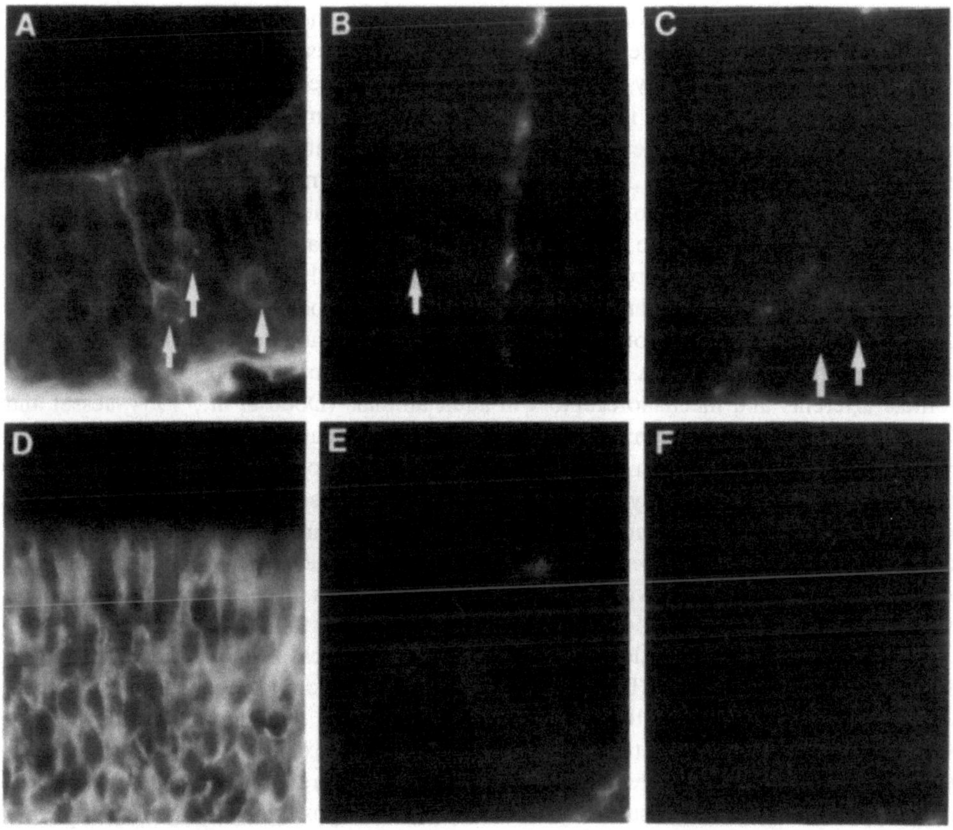

Figure 2. Immunofluorescence analysis of cell-surface antigens on olfactory epithelial neurons. Cryostat sections of adult rat olfactory epithelium were incubated with a monoclonal antibody and then rhodamine conjugated rabbit-antimouse-IgG. Photographic and printing conditions were identical for all antibodies, but the observed intensities may not be reflective of actual antigen densities as antibody excess conditions were not verified for all antibodies. The antibodies used were (A) 2B8; (B) anti-B blood group substance; (C) anti-H blood group substance; (D) 3F4.18 anti-N-CAM; (E) HNK-1; and (F) media from Ag8.653 nonsecreting myeloma cells as a negative control. 2B8 and 3F4.18 were developed in this laboratory, anti-B and H were obtained from Accurate Scientific and HNK-1 from the American Type Culture Collection.

antigens. The subsets defined by 2B8 appear to be relatively stable subpopulations in development. Similar data are not yet available for the other subsets defined by carbohydrate antigens. The physiological significance of these subsets is unknown. However, as these carbohydrate differentiation antigens are expressed on the cell surface, immunoaffinity purification techniques could be used to isolate these subpopulations specifically. In this regard, the 2B8 antigen may be particularly useful, as it is expressed both by mature olfactory neurons and by a population of cells in the basal layer of the epithelium, which may be the basal cell precursors of the neurons.

2.2.3. Carbohydrate Differentiation Antigens of Dorsal Root Ganglion Neurons

The significance of blood group antigen expression to the function of olfactory neurons is certainly unclear; indeed, the general role of complex cell-surface carbohydrates in cellular function is not understood. Nevertheless, the identification of subclasses of olfactory neurons which express individual complex carbohydrate groups may be useful in elucidating olfactory function. This possibility is exemplified by recent studies of DRG neurons. Efforts to identify stage-specific cell-surface antigens on very early mouse embryos led to the development of monoclonal antibodies, i.e., Stage-specific embroyonic antigen (SSEA 1, 3, and 4), which were shown to react with glycolipids of the lacto- and globoseries (Soltor and Knowles, 1978; Kannagi *et al.*, 1983). Although within the early embryo these glycolipids are expressed in a stage-specific fashion, they are also expressed on a variety of cell types in older animals. Jessel and Dodd and their collaborators recently showed that these and additional monoclonal antibodies to carbohydrate differentiation antigens of the globoside series bind to functional subsets of rat dorsal root ganglion neurons. SSEA-3 and SSEA-4 epitopes are present on low-threshold hair follicle afferents and mechanoreceptors in adult animals (Dodd *et al.*, 1984; Jessel and Dodd, 1985). This conclusion is based on the strong correlation between the distribution of DRG neuronal terminals in the spinal cord and the previous physiological mapping of the distribution of these two types of sensory afferents. These neurons are not topographically organized within the ganglion itself, thus leaving unanswered the question of how the selective distribution of their termini arises. The attractive hypothesis that the selective expression of surface carbohydrates by these cells underlies a selective fasciculation and consequent selective termination of their axons has not been thoroughly tested. However, endogenous lactose-binding lectins have recently been demonstrated in DRG neuron subsets (Regan *et al.*, 1986). These glycolipids are expressed by DRG neurons on embryonic days 15–16 soon after the final mitosis has occurred lending strength to the possibility of selective fasciculation. These and other complex carbohydrates are also expressed on subsets of cultured DRG neurons. Their expression correlates with other phenotypic markers of DRG subsets. For example, SSEA positive neurons are completely distinct from those that contain substance P, somatostatin, or the fluoride-resistant acid phosphatase enzyme. By contrast, a set of monoclonal antibodies, including LD2, reacts with only 25% of small DRG neurons, but 100% of this subpopulation expresses somatostatin.

Rat nociceptive neurons also express unique cell-surface carbohydrates of the lactoside series (Dodd and Jessel, 1985). Like the globoside expressing DRG neurons, the lactoside-positive DRG subset has a distinct profile of expression of other phenotypic markers. Thus two subsets of DRG neurons express distinct carbohydrate differentiation antigens and more subsets defined in this fashion can be anticipated. This expression correlates with other phenotypical markers of DRG neurons and most importantly with physiologically defined subsets of DRG neurons. Regardless of the validity of the hypothesis that these carbohydrate molecules may mediate the selective association of the axons of these cells, the practical use of these cells surface markers in the isolation and analysis of individual subpopulations of DRG neurons is clear. The extension of these results to the olfactory system is equally clear and initial experiments indicate olfactory neurons express the lactoside series of carbohydrates (J. Dodd, unpublished data). Subsets of olfactory neurons expressing these and other carbohydrate differentiation antigens have been identified. Are these subsets physiologically significant?

2.2.4. The HNK Carbohydrate Group

Another interesting recent example of carbohydrate differentiation antigens on neuronal cells has been developed from studies of the N-CAM and L1 molecules. Both N-CAM and L1 are large cell-surface glycoproteins that exist in more than one molecular-weight-form (Chuong and Edelman, 1984; Rathjen and Schachner, 1984). Antibodies to both molecules can block cell–cell interaction in model culture systems (Keilhauer *et al.*, 1985), and antibodies to N-CAM can inhibit the formation (Thanos *et al.*, 1984) or regeneration (Fraser *et al.*, 1984) of the retinotectal tract *in vivo*. Thus, N-CAM and also L1 are strong candidates for mediating cell–cell interaction during neural development. N-CAM on one cell can bind directly to N-CAM on another cell to cause adhesion via a homotypic interaction (Cunningham *et al.*, 1983; see also Cole *et al.*, 1986). The molecular mechanisms of L1-mediated adhesion are less well understood.

On the submolecular level, the contributions of specific N-CAM amino acid and carbohydrate sequences in N-CAM to N-CAM adhesion have not been entirely established. One complex carbohydrate group recognized by monoclonal antibody L2 is found on some but not all N-CAM polypeptides and also some but not all L1 polypeptides (Kruse *et al.*, 1984). This specific glycosylation heterogeneity is not related to the previously identified immunologic heterogeneity of N-CAM molecules (Williams *et al.*, 1985; R. Allison and R. Williams, unpublished data). Surprisingly the same or a very closely related carbohydrate group reacts with monoclonal antibody HNK-1 originally developed against human natural killer (HNK) cells.

Mab L2 will partially inhibit cell–cell adhesion in *in vitro* assays (Keilhaur *et al.*, 1985). Neuron–astrocyte and astrocyte–astrocyte adhesion is inhibited by L2, but no effect on neuron–neuron adhesion is seen. Polyclonal antibodies to N-CAM partially inhibit neuron–neuron adhesion. Combinations of monoclonal anti-L2 and polyclonal anti-N-CAM show synergistic inhibition. Thus, the L2/HNK-1 carbohydrate group is expressed by some but not all N-CAM and L1 molecules on several neural cell types and represents a clear example of the potential role of carbohydrate groups in neural cell–cell interaction. We have used the HNK Mab to test adult rat olfactory epithelium for this carbohydrate group. (HNK is a carbohydrate group covalently attached to some polypeptides from neural tissue and human natural killer cells.) Surprisingly, even though olfactory bulb and many other CNS areas were strongly HNK$^+$, no activity was detectable in the olfactory epithelium (Fig. 2, cf. E, F). Many cells in the olfactory epithelium do express the N-CAM polypeptide as indicated by reaction with another monoclonal antibody to rat N-CAM (Fig. 2D). Therefore, it appears that rat olfactory neurons specifically lack the L2/HNK carbohydrate group. Data are not available on whether N-CAM molecules that lack the L2/HNK carbohydrate group contain an alternative carbohydrate group. Likewise the homotypic interaction capability of HNK^- N-CAM is unknown. The potential of HNK^- N-CAM and carbohydrate differentiation antigens to mediate selective axon fasciculation, glomerulus formation, or other aspects of olfactory organization remains to be evaluated.

2.3. Other Defined Subclasses

2.3.1. Subclasses Expressing Carbonic Anhydrase

Carbonic anhydrase is a zinc-requiring metalloenzyme that catalyzes the hxdration of CO_2 to produce HCO_3^- and H^+ ions. It is believed to play a critical role in the control of tissue pH. In neural tissue carbonic anhydrase has been extensively studied in glial cells (Linser, 1985) but is generally absent in neurons. Brown *et al.* (1984*b*) stained olfactory epithelium for carbonic anhydrase anticipating reactivity of the mitochondria rich sustentacular cells as mitochondria rich cells in other tissues have high carbonic anhydrase levels. To their surprise, these support cells were negative, but a subpopulation of olfactory neurons and their axons stained strongly. Carbonic anhydrase-containing cells were more common adjacent to the cribriform plate and less common near the transition to respiratory epithelium. The significance of carbonic anhydrase expression by this olfactory neuron subpopulation is unknown. It might be informative to test the effects of metabolic stimulation on carbonic anhydrase expression. Although its significance is unknown, this observation is intriguing in part because of the recent demonstration of carbonic anhydrase staining in the chemosensory cells of the taste pore in rat circumvallate taste buds (Brown *et al.*, 1984*b*). Furthermore, reminiscent of the complex carbohydrate antigens, subpopulations of large- and medium-sized neurons in the rat spinal, nodose, and trigeminal ganglia also express carbonic anhydrase (Wong *et al.*. 1983).

2.3.2. Additional Subclasses Defined by Monoclonal Antibodies

Monoclonal antibody approaches have recently been used to identify subpopulations of olfactory and vomeronasal neurons in the rabbit. Using olfactory bulb homogenates as immunogens, Mori and collaborators developed a number of monoclonal antibodies and have presented the characterization of three of them. Mab R2D5 is of limited interest here, as it reacts with a low-molecular-weight cytoplasmic protein found in the cell bodies and axons of all rabbit olfactory neurons. Mab R4B12, however, reacts with a subset of olfactory neuron axons found in thorough mapping studies to be in the ventrolateral but not the dorsomedial areas of the adult olfactory epithelium (Fujita *et al.*, 1985). Surprisingly, although this antibody reacts strongly with olfactory neuron axons, the bodies of these cells do not react. The axonal reactivity extends to the olfactory bulb, where positive glomeruli are present in the ventrolateral and caudal but not dorsomedial regions. Using light microscopy most individual glomeruli tend to be either completely positive or completely negative for Mab R4B12: 63% of the area of the glomerular region expressed R4B12, 31% was negative, and only 6% of the glomerular region was comprised of mixed fibers (Mori *et al.*, 1986). Thus, Mab R4B12 delineates a discreet subset of olfactory neurons and termini which has been anatomically mapped with more precision than any other olfactory subset.

This antibody also reacts with a subset of glomeruli and axons in the accessory olfactory bulb and the vomeronasal organ. Again, about two thirds of the area of the accessory olfactory bulb reacts with R4B12. In this case, the authors have identified another Mab, R5A10, which has the very interesting property of reacting with an exactly reciprocal distribution of termini within the accessory olfactory bulb. Thus the rabbit accessory olfactory system is comprised of two nonoverlapping subsets: R4B12 reactive R5A10 nonreactive cells whose axons terminate in the rostralateral two thirds of the accessory olfactory bulb and R5A10 reactive R4B12 nonreactive cells whose axons terminate in the caudomedial one third of the accessory olfactory bulb.

Data on R5A10 reactivity in the main olfactory bulb have not been reported, so it is not known whether the reciprocal distribution between this antigen and that identified by R4B12 extends to this system. Mab R4B12 does not react with formaldehyde-fixed olfactory bulbs of several species, including rat and human, so it seems unlikely that it reacts with a blood group antigen. The molecular nature of the olfactory antigens that react with either R4B12 or R5A10 have not been reported. As Mab R4B12 does not react with structures in the olfactory epithelium itself, it seems unlikely that the antigen is directly involved in chemoreception. However, the detailed mapping of R4B12 terminals in the bulb will permit future test of correlations between these terminals and functionally significant subgroups.

Complementary data have been obtained with Mab RB-8 in the rat (Schwob and Gottlieb, 1986). This Mab reacts with a 125,000-M_r polypeptide identified in several CNS and PNS neuronal cells and olfactory epithelium, but not other tissues. RB-8 immunohistochemical staining is more intense in olfactory bulb glomeruli and the olfactory nerve than on the olfactory receptor cell bodies in the epithelium. Within the epithelium, staining is distinctly nonuniform: a contiguous region composed of the ventral and lateral epithelium is RB-8 positive while a complementary nonoverlapping contiguous layer composed of the dorsal and medial epithelium is RB-8 negative. Thus, distinct Mabs identify anatomically discrete regions of the olfactory epithelium in both the rat and rabbit. This result is in contrast to the receptor cell heterogeneity observed with Mab 2B8 or antibodies to blood group substances; in these cases, positive and negative cells are intermixed within the epithelium.

In summary, with the potential exception of the morphological subclasses in fishes, none of the olfactory neuron subclasses discussed here is known to be functionally significant. One approach to testing such significance directly is isolation of cells (or membranes from cells) expressing a particular marker using immunoaffinity techniques. Electrophysiological methods could then be used to determine the olfactant sensitivity of these subsets. For this and other reasons, monoclonal antibodies will clearly be used to define olfactory neuron cell-surface antigens and additional olfactory neuron subsets. (See also Chapters 2 and 12, this volume.) The characteristics used to date to define olfactory neuron subsets are compared in Table 1 with those of OMP and of these additional monoclonal antibody approaches.

Table 1. Molecular Markers for Olfactory Neurons[a]

Marker	Proportion of olfactory neurons positive	Subcellular distribution	Olfactory glomeruli distribution	VNO	Species	Comments
OMP	All	Cytoplasmic	All	Weakly positive	All vertebrates	Best-characterized olfactory marker
Microvilli	Fish 1/3; human 1/5 (?)	—	—	All cells have microvilli	Vertebrates	Functionally significant group?
Blood group antigens	Major populations, regulated	Cell surface and cytoplasm	Some	NR	Rat, human, baboon	
2B8 antigen	One third, developmentally regulated	Cell surface	Dorsomedial more prominent	All neurons positive	Rat	Antigen, probably carbohydrate
R2D5 antigen	All	Cytoplasmic	All	All	Rabbit	Antigen is small polypeptide
R4B12 antigen	2/3	Axons react but not cell bodies	Ventrolateral	Two thirds positive	Rabbit	Rostrolateral AOB
R5A10 antigen	NR	NR	NR	One third positive	Rabbit	Reciprocal to R4B12
Mab RB-8	2/3	Axons strong, membrane associated	Ventrolateral	NR	Rat	Neural, but not olfactory specific
Carbonic anhydrase	Major population	Cytoplasmic, nuclear	NR	NR	Rat	
Mab 18.1	All?	95,000-M_r surface glycoprotein	Absent	NR	Frog	See Chapter 2
Mab	Several specificities	Cell surface and cytoplasmic	—	Some positive	Rat	See Chapter 13

[a]NR, not reported.

3. PERSPECTIVES ON RESEARCH DIRECTIONS

3.1. Clonal Olfactory Neurons

Analogies have often been made between the proposed molecular recognition mechanisms of the olfactory system and those of the immune system. Regardless of the ultimate validity of these analogies, experimental approaches to the study of olfaction that are analogous to those used to analyze the immune system may be useful. During the past three decades, decisive evidence has shown that many aspects of immune system function are based on clonal cell populations. Individual clonal populations of B cells produce individual immunoglobulin species. Analyses of the polypeptide products of clonal cells—Bence Jones proteins—were key aspects of the elucidation of the structure of immunoglobulin molecules. Individual clonal populations of T cells also have antigen specific receptors. These T-cell receptors are cell surface polypeptides. One approach to the recent molecular identification of the T-cell receptor used monoclonal antibodies specific for clonal T-cell populations (Haskins *et al.*, 1983). These antibodies inhibited antigen dependent activation of the T cells and were subsequently used to immunoprecipitate the solubilized T-cell receptor. Using amino acid sequence data from the purified receptor and also with independent recombinant DNA techniques (Yanagi *et al.*, 1985, Hedrick *et al.*, 1985), a family of T-cell receptor genes has been identified and the extent of molecular diversity of this family is being determined.

Functionally competent clonal cell lines of olfactory neurons could be useful reagents in establishing the molecular basis of olfactant–receptor neuron interaction, the diversity of categories of olfactory neurons, and other aspects of olfactory function. One continuous cell line that arose spontaneously in a culture of mouse olfactory epithelium has been reported (Goldstein and Quinn, 1981). Based on the presence of high levels of carnosine and carnosine synthetase, peripheral-type benzodiazepine receptors, and also action potential generation (Goldstein *et al.*, 1986), this line has been proposed to be of neuronal phenotype. It does not express OMP, but this may be due to a relative lack of maturity of the cells. Unfortunately, this line has not been studied in several laboratories; thus, its value in the analysis of olfaction has not been fully determined.

An underutilized resource is the spontaneous neuroblastomas of the nasal mucosa. These are relatively rare and represent a small proportion of all nasal tumors (Reznick and Stinson, 1983). Since their initial description as a clinical entity in 1924 approximately 200 case reports have appeared in the clinical literature. However, with increasing sophistication of diagnosis the reported incidence of these tumors appears to be increasing (Tamada *et al.*, 1984). Many of these tumors have differentiated histological properties common to other neuronal tumors (retinoblastoma, neuroblastoma, and pheochromocytoma), particularly the circular grouping of cells around a central lumen termed a rosette. Others have a less differentiated appearance, including the nasal teratoma (Heffner and Hyams, 1983). The potential usefulness of these tumors is best illustrated by the observations that some of them undergo olfactory differentiation. These cells can form rosettes composed of tall columnar cells, some of which extend extensive microvilli into the central lumen and others that have terminal swellings (olfactory vesicles?) at the lumen (Silva *et al.*, 1982). If these tumors can be maintained and will differentiate in culture, they could be a very useful resource.

A large number of CNS clonal cell lines of murine origin have been isolated Schubert *et al.* (1974) by applying the transplacental carcinogen ethylnitrosourea to pregnant rats and isolating cell lines from cultures of the tumors that arose in the progeny. Individual cells lines were isolated that expressed phenotypical markers of neurons, of glia, and of both cell types. The latter have been proposed to be primitive stem cells (Wilson *et al.*, 1981). More recently, the same laboratory has applied a similar approach to the isolation cultured lines of neuroepithelial tumors. Nasal tumors were induced by the inclusion of *N*-nitroso-*N'*-methylpiperazine in the drinking water of young Fischer rats, and a number of cell lines were isolated from the tumors. Analysis of the phenotypical characteristics of these cell lines has not yet permitted clear phenotypical categorization of these cells (D. Schubert, unpublished data). Thus the usefulness of this direct approach using carcinogens to obtain olfactory neuronal cell lines requires further testing.

One source of difficulty in obtaining clonal cell lines is precise identification of the properties of basal precursor cells (or clonal lines thereof) in culture. No unique phenotypical properties of basal cells have been identified. In a promising approach, two Mabs that react with elements of the basal layer have been isolated (Hempstead and Morgan, 1985). Confirmation that these Mab react with basal cells themselves will require additional analysis. Methods to dissociate frog (Kleene and Gesteland, 1981) and to dissociate and culture mammalian (Hirsch and Margolis, 1979; Noble *et al.*, 1984; Schubert *et al.*, 1985; Gonzales *et al.*, 1985) epithelial cells have been established. However none of these methods seems to yield large numbers of cells reliably with the properties of olfactory neurons. Olfactory epithelia prepared in the author's laboratory contain some cells that express either the 2B8 antigen or N-CAM and have morphology of olfactory neurons (Fig. 3 A–E). However, many cells with a non-neuronal morphology that express these markers as well as blood group B antigen are also obtained in culture (Fig. 3 F–H). These cells could be the basal cell neuronal precursors, but this relationship has not been established. Thus, these markers cannot presently be used to identify cultured olfactory neurons uniquely. Since properties such as OMP expression and electrical excitability seem to be found only in more mature neurons, it is difficult to distinguish between cultured cells that may be neuronal precursors and those derived from the glial and supporting cells of the epithelium. Although it may be possible to obtain cell lines that constitutively express olfactory neuronal properties, it is also possible that the cell lines obtained may be derived from the primitive basal precursor cells that maintain the ability to divide.

Progress toward basal cell identification has been made in a recent study of olfactory primary cultures (Schubert *et al.*, 1985). In these cultures, neuronlike cells were observed after the initial dissociation and culture of epithelium from 2-month-old rats. The initial surviving cells were flat and ciliated and expressed the astrocyte-specific intermediate filament protein glial fibrillary acidic protein (GFAP). After 3–5 days, small round cells appeared that were electrically excitable and contained neuron-specific enolase but lacked GFAP. These cells were hypothesized to be derived from the flat GFAP-positive precursor cells by processes involving both cell division and cell differentiation. If this hypothesis can be confirmed, this observation represents a significant step forward in identifying at least one phenotypical marker of basal precursor cells—GFAP. Unfortunately, GFAP is not a cell-surface marker. Thus, GFAP assays require sacrifice of the culture, and antibodies to GFAP cannot be used to sort epithelial cell populations. Furthermore, the

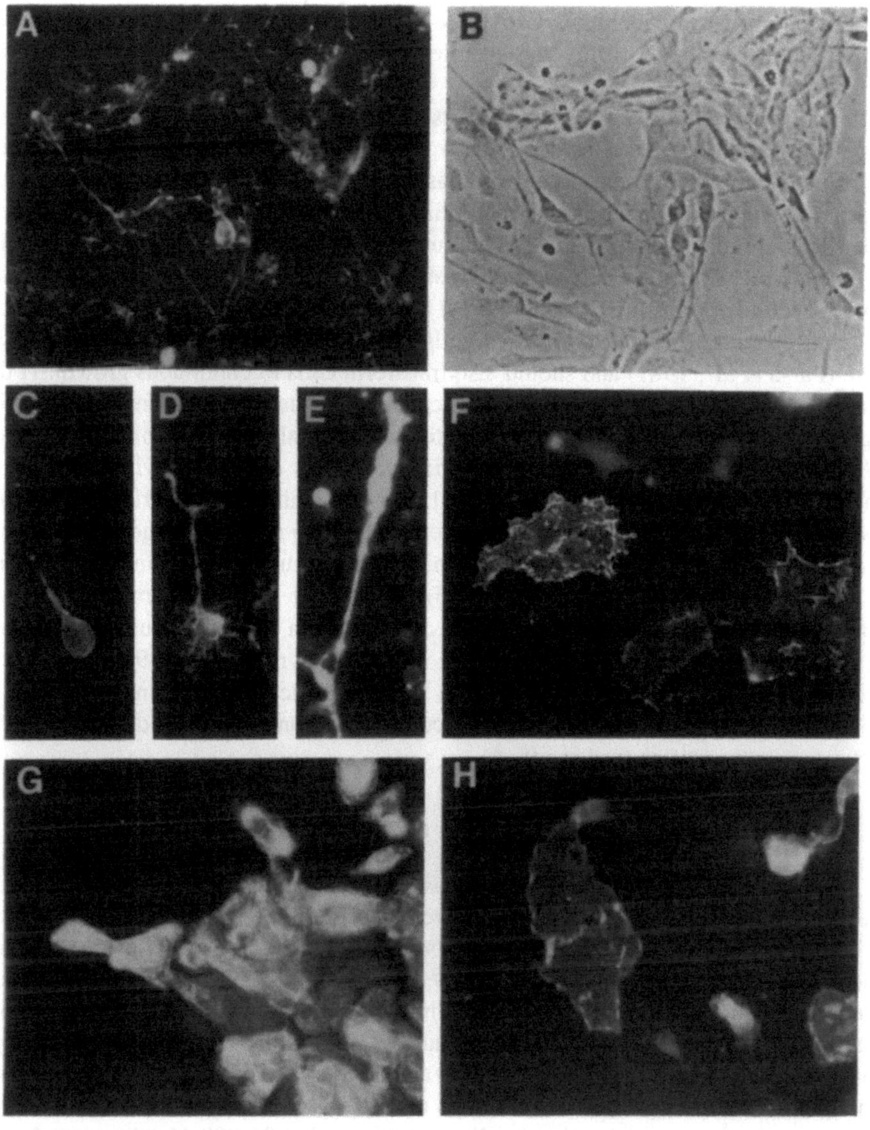

Figure 3. N-CAM, 2B8, and blood group B antigens are found on cultured olfactory epithelial cells with varying morphologies. Cultures were prepared from embryonic day 21 rats and analyzed for cell surface antigens by indirect immunofluorescence after 6 days in culture. (A, B) Fluorescence and phase photographs of a large clump of cells with neuronal morphology reacted with 3F4 monoclonal anti-rat-N-CAM antibody. (C, D, F) Fluorescence micrographs of cultures reacted with Mab 3F4. (E, G) Mab 2B8. (H) Monoclonal anti-blood group B.

criterion of GFAP expression cannot be used alone to identify basal cells as the Schwann cells of the olfactory nerve also express GFAP (Barber and Lindsay, 1982; Doucette, 1986). Such cells could easily contaminate olfactory epithelial cultures. A number of phenotypical properties, including GFAP, N-CAM, 2B8, OMP, and others, will need to be analyzed to establish firmly the identity of cultured olfactory neurons and their precursors.

As culture methods are refined, efforts to immortalize cultured olfactory neurons directly may be productive. One approach is to fuse the normal neuron with a previously transformed neuronal line, such as a neuroblastoma to combine, hopefully, the phenotypic properties of the normal neuron with the unlimited capacity of the tumorigenic partner to divide. This approach has previously been used to develop murine PNS and CNS lines with phenotypically differentiated properties (Greene *et al.*, 1975; Hammond *et al.*, 1986). A second approach may be the direct application of transforming chemicals, such as ethylnitrosourea, directly to cultured cells. The nitrosoureas are potent mutagens in culture (Parsa *et al.*, 1981) and should also transform cells to continuous growth. A third approach may be the use of DNA sequences capable of causing transformation to continuous growth (oncogenes). Transfection of cultures of normal bronchial epithelial (Yoakum *et al.*, 1985) and other cells (Bishop, 1985; Bechade *et al.*, 1985) with oncogene sequences has resulted in continuous clonal cell lines that maintain appropriate phenotypical properties. In similar experiments, it has been possible to couple oncogene sequences to tissue-specific promoter sequences to obtain directed tissue-specific tumor formation *in vivo* (Hanahan, 1985; Adams *et al.*, 1985). A number of rat and mouse lines with astrocytic characteristics obtained by transfection of brain secondary cultures with polyoma large T- or adenovirus E1A-containing plasmids (Evrard *et al.*, 1986). Despite the uncertainties of the best technical approach and the difficulties of identification of cell phenotype in culture, the potential of clonal cell lines to contribute to understanding of the olfactory system justifies further efforts.

3.2. Direct Approaches to Molecular Olfactory Receptors

Discussion of olfactory neuron subclasses often emphasized the possibility of a correlation between a morphologically or molecularly defined subclass and a functional subclass. Can we hope to extend beyond correlations to rigorous tests of the functional capacity of olfactory neuronal subclasses? Clearly, one fundamental approach is to attempt to identify the olfactory receptor molecules directly and then to examine the distribution of these molecules among olfactory neurons. Lancet (1986) suggested criteria for identifying such molecules and the laboratories of Lancet and Snyder, and others (see Chapters 1 and 2, this volume) have initiated such efforts. Two recently discovered candidates for molecular olfactory receptors are a pyrazine binding protein and a 95,000-M_r glycoprotein. The pyrazine-binding protein has been localized to olfactory glandular tissue and secretions (Pevsner *et al.*, 1985, 1986). The possibility that the 95,000-M_r glycoprotein is on an olfactory receptor cell subclass has not been examined in detail, but it appears to be on a substantial proportion of olfactory neurons (Chen *et al.*, 1986*a,b*). Earlier studies also identified candidate olfactory receptor polypeptides. A monoclonal antibody that inhibited alanine binding to homogenates of catfish taste epithelium has been reported (Goldstein and Cagan, 1982; Rhein and Cagan, 1983). Amino acids are potent olfactants in fish. Similiarly polyclonal antibodies to an anisole-binding protein that, when applied to the mucosal surface of a rat, decreased the amplitude of the electrofactogram have been reported (Goldberg *et al.*, 1979; Price, 1982). Control IgG preparations at the same concentrations had significantly lesser effects on the observed response, suggesting the feasibility of this antibody-intervention approach *in vivo*. However, the effects of this immune IgG were not limited to phenolic odorants, indicating that interpretation of the effects of antibody nasal sprays on olfaction may be complex. In both cases, no histological data demonstrating the distribution of antibody-binding sites among the appropriate olfactory neurons have been reported. Lacking additional characterization of these latter components, the strength of their candidacy as olfactory receptor molecules diminishes with time. Nevertheless, efforts to identify the putative olfactory receptor molecules are sure to increase. Antibodies are attractive tools in these efforts due to their potential value in manipulating and isolating individual olfactory classes *in vitro* as well as potentially modifying cellular function in the living animal.

3.3. Genetic Approaches

Studies of olfactory specificity have not generally used the power of classic and current genetic techniques. One of the simplest predictions made by the hypothesis that individual olfactory cell types have unique olfactant specificities is that specific anosmias should exist in some individuals due to the loss of that class of olfactory cells. Analogous specific anosmias would also exist if specific olfactant receptors are widely distributed among olfactory neurons rather than confined to a single subgroup. In either case, such anosmias should show classic mendelian recessive inheritance patterns in the simplest situations. Studies of anosmia in humans have yielded varying results. Amoore (1971) analyzed reports of specific anosmias to 62 different olfactants and concluded that they represent 20–30 primary odors, each of which might be expected to correspond to a unique molecular and/or cellular receptor class. Inheritance patterns of the inability to detect *n*-butyl mercaptan, hydrogen cyanide, pentadecalactone, and Freesia scent have been reported and appear to be mendelian recessive in most but not all studies (Amoore, 1971). The tentative nature of such human pedigree data could be used to question whether indeed there is any genetic component to olfactory sensitivity. A recent study (Wysocki and Beauchamp, 1985) of the ability of identical and fraternal twins to smell the C_{19} androgen androstenone addressed this question directly. Androstenone was chosen because its sensory threshold in the population is extremely variable. This threshold had a much greater concordance for identical twin pairs than for fraternal twin pairs, clearly suggesting a genetic component in sensory detection of this compound. No such correlation was detected for pyridine, suggesting that general olfactory sensitivity differences were not responsible for the observed concordance. Although the action of this genetic component in androstenone detection could potentially be at any level from initial reception through CNS processing, clearly variations in molecular or cellular receptor populations are an attractive site of gene action.

Studies of anosmias or olfactory sensitivity in more genetically defined organisms have not contributed as much as might be anticipated when one considers the substantial role genetic approaches have played in the molecular analysis of chemosensory systems in procaryotes (Kleene, 1986; Hazelbauer and Harayama, 1983). Initial efforts to analyze genetics of olfactory reception in rodents have not been further pursued (Price, 1977; Wysocki *et al.*, 1977). However, an interesting set of experiments has demonstrated that the production of scents has a genetic component. Inbred mice can discriminate between the natural body odors or urine of other strains of mice. Using congenic strains, this difference in scent has been shown to be detectable between $H\text{-}2^b$ and $H\text{-}2^{b_{m1}}$ mice (Yamazaki *et al.*, 1983). That is, mice that differ genetically only at the portion of chromosome 17, which specifies one of the genes for a cell-surface histocompatibility antigen, have mutually detectable different scents. As the vapor pressure of histocompatibility antigens is rather low it seems likely that this scent difference is due to an unknown influence of H-2 histocompatibility antigens (the murine major histocompatibility locus) on metabolic pathways. It is interesting to note that in the study of human ability to smell androstenone, no linkage to HLA histocompatibility antigen haplotype was observed (Pollack *et al.*, 1982). No analyses of specific anosmias or olfactory discrimination in congenic mouse strains have been reported.

In conclusion, it seems likely that analyses of human olfactory capabilities will continue to define categories of olfactants and provide confirmatory evidence for genetic control of olfactory sensitivity. Amphibia are favorite organisms for analysis of olfaction *in situ,* but amphibian genetics is relatively primitive. Studies of mammalian systems have the potential to identify genes that control the expression of olfactory neuron subclasses and the range of olfactory sensitivities. However, given the complexities involved in such experiments in a mammalian system, it is possible that the first insights will be derived from the analysis of insect systems. Insects have clearly specialized categories of olfactory neurons, particularly those responding to pheromones. Methods of genetic manipulation are highly sophisticated, particularly in *Drosophila,* allowing relatively easy identification and isolation of single genes or groups of genes such as might code for specific cell types. Such an approach would then be followed by searches for homologous genes in mammals. It is interesting to note that in an analogous cross-species search, about one half of a set of 146 monoclonal antibodies raised to *Drosophila melanogaster* brain cross-reacted with human brain (Miller and Benzer, 1983).

4. SUMMARY

A number of molecular markers have been identified that divide mammalian olfactory neurons into subclasses of unknown physiological significance. In fish olfactory neurons, the expression of either microvilli or cilia does appear to correlate with cellular olfactant specificity. Clearly, the potentially most powerful approach to defining physiologically significant olfactory neuron subclasses is direct identification and characterization of the long postulated molecular olfactory receptors. A key prediction of many theories of olfaction, and thus a key test of molecular receptor candidates, is that only subclasses of olfactory neurons will express individual molecular receptors. More refined cell culture methods would significantly aid the analysis of many olfactory functions, including the identification of the molecular olfactory receptor. These studies would be facilitated by the development of continuous lines of functionally differentiated olfactory neurons. Direct approaches to the olfactory receptor will be supplemented by studies of carbohydrate differentiation antigens on olfactory neurons. These antigens have been previously shown to mediate selective cell–cell interactions and their expression correlates with functional subclasses in other neural systems. Combined with genetic approaches, these methods of identifying and analyzing olfactory neuron subclasses will contribute significantly to understanding olfactory function.

ACKNOWLEDGMENTS. Work in my laboratory has been largely carried out by Dr. W. Allen and Ms. S. Warren and is supported by grant PO1-NS23348 from the National Institutes of Health. I would like to thank many colleagues, particularly Dr. J. Dodd, Dr. K. Mori, Dr. J. Morgan, Dr. M. Noble, Dr. R. Oriel, Dr. D. Schubert, and Dr. G. Thommesen, for contributions of preprints and unpublished data; Dr. R. Gesteland, Dr. S. Kleene, and Dr. M. Shipley and Ms. S. Warren for manuscript review; and Ms. D. Bernhard for manuscript preparation.

REFERENCES

Adams, J. M., Harris, A. W., Pinkert, C. A., Corcoran, L. M., Alexander, W. S., Cory, S., Palmiter, R. D., and Brinster, R. L., 1985, The c-myc oncogene driven by immunoglobulin enhancers induces lymphoid malignancy in transgenic mice, *Nature (Lond.)* **318:** 533–538.

Allen, W. K., 1984, Identification of an olfactory receptor neuron subclass: Cellular and molecular analysis during development, Ph.D. thesis, University of Cincinnati.

Allen, W. K., and Akeson, R. A., 1985*a*, Identification of a cell surface glycoprotein family of olfactory receptor neurons with a monoclonal antibody, *J. Neurosci.* **5:** 284–296.

Allen, W. K., and Akeson, R., 1985*b*, Identification of an olfactory receptor neuron subclass: Cellular and moleular analysis during development, *Dev. Biol.* **109:** 393–401.

Amoore, S. E., 1971, Olfactory genetics and anosmia, in: *Handbook of Sensory Physiology. Chemical Senses.* Vol. 1: Olfaction (L. M. Beidler, ed.), pp. 245–256, Springer-Verlag, New York.

Amoore, S. E., 1977, Specific anosmia and the concept of primary odors, *Chem. Senses Flavour* **2:** 267–281.

Barber, P. C., and Lindsay, R. M., 1982, Schwann cells of the olfactory nerves contain glial fibrillary acidic protein and resemble astrocytes, *Neuroscience* **7:** 3077–3090.

Bechade, C., Calothy, G., Pessac, B., Martin, P., Coll, J., Denhez, F., Saule, S., Ghysdael, J., and Stehelin, D., 1985, Inducton of proliferation or transformation of neuroretina cells by the *mil* and *myc* viral oncogenes, *Nature (Lond.)* **316:** 559–562.

Bishop, J. M., 1985, Viral oncogenes, *Cell* **42:** 23–38.

Brown, D., Garcia-Segura, L. M., and Orci, L., 1984*a*, Carbonic anhydrase is associated with taste buds in rat tongue, *Brain Res.* **324:** 346–348.

Brown, D., Garcia-Segura, L.-M., and Orci, L., 1984*b*, Carbonic anhydrase is present in olfactory receptor cells, *Histochemistry* **80:** 307–309.

Chen, Z., Ophir, D., and Lancet, D., 1986*a*, Monoclonal antibodies to ciliary glycoproteins of frog olfactory neurons, *Brain Res.* **368:** 329–338.

Chen, Z., Pace, U., Ronen, D., and Lancet, D., 1986*b*, Polypeptide gp95, *J. Biol. Che.* **261:** 1299–1305.

Chuong, C-M., and Edelman, G. M., 1984, Alteractions in neural cell adhesion molecules during development of different regions of the nervous system, *J. Neurosci.* **4:** 2354–2368.

Cole, G. J., Loewy, A., Cross, N. V., Akeson, R., and Glaser, L., 1986, Topographic localization of the heparin-binding domain of the neural cell adhesion molecule N-CAM, *J. Cell Biol.* **103:** 1739–1744.

Cunningham, B. A., Hoffman, S., Rutishauser, U., Hemperly, J. J., and Edelman, G. M., 1983, Molecular topography of the neural cell adhesion molecule N-CAM: Surface orientation and location of sialic acid-rich and binding regions, *Proc. Natl. Acad. Sci. USA* **80:** 3116–3120.

Dodd, J., Solter, D., and Jessell, T. M., 1984, Monoclonal antibodies against carbohydrate differentiation antigens identify subsets of primary sensory neurons, *Nature (Lond.)* **311:** 469–472.

Doucette, J. R., 1986, Astrocytes in the olfactory bulb, *Astrocytes* (S. Fedoroff and A. Vernadakis, eds.), Vol. 1, pp. 293–310, Academic,

Evrard, C., Galiana, E., and Rouget, P., 1986, Establishment of "normal" nervous cell lines after transfer of polyoma virus and adenovirus early genes into murine brain cells, *EMBO J.* **5:** 3157–3162.

Feizi, T., 1985, Demonstration by monoclonal antibodies that carbohydrate structures of glycoproteins and glycolipids are onco-developmental antigens, *Nature (Lond.)* **314:** 53–57.

Fraser, S. E., Murray, B. A., Chuong, C-M., and Edelman, G. M., 1984, Alteration of the reinotectal map in *Xenopus* by antibodies to neural cell adhesion molecules, *Proc. Natl. Acad. Sci. USA* **81:** 4222–4226.

Fujita, S. C., Mori, K., Imamura, K., and Obata, K., 1985, Subclasses of olfactory receptor cells and their segregated central projections demonstrated by a monoclonal antibody, *Brain Res.* **326:** 192–196.

Goldberg, S. J., Turpin, J., and Price, S., 1979, Anisole binding protein from olfactory epithelium: Evidence for a role in transduction, *Chem. Senses Flavour* **4:** 207–214.

Goldstein, N. I., and Cagan, R. H., 1982, Biochemical studies of taste sensation: Monoclonal antibody against L-alanine binding activity of catfish taste epithelium, *Proc. Natl. Acad. Sci. USA* **79:** 7595–7597.

Goldstein, N. I., and Quinn, M. R., 1981, A novel cell line isolated from the murine olfactory mucosa, *In Vitro* **17:** 593–598.

Goldstein, N. I., Quinn, M. R., and Teeter, J. H., 1986, Differentiation of a cell line from olfactory epithelium is induced by dibutyryladenosine 3′,5′-cyclic monophosphate and 12-0-tetradecanoylphorbol-13-acetate, *Brain Res.* **370:** 205–214.

Gonzales, F., Farbman, A. I., and Gesteland, R. C., 1985, Cell and explant culture of olfactory chemoreceptor cells, *J. Neurosci. Methods* **14:** 77–90.

Graziadei, P. P., and Metcalf, J. F., 1971, Autoradiographic and ultrastructural observatons on the frog's olfactory musoca, *Z. Zellforsch,* **116:** 305–318.

Greene, L. A., Shain, W., Chalazonitis, A., Breakfield, X., Minna, J., Coon, H. G., and Nirenberg, M., 1975, Neuronal properties of hybrid neuroblastoma X sympathetic ganglion cells, *Proc. Natl. Acad. Sci. USA* **72:** 4923–4927.

Hammond, D. N., Wainer, B. H., Tonsgard, J. H., and Heller, A., 1986, Neuronal properties of clonal hybrid cell lines derived from central cholinergic neurons, *Science* **234:** 1237–1240.

Hanahan, D., 1985, Heritable formation of pancreatic β-cell tumours in transgenic mice expressing recombinant insulin/simian virus 40 oncogenes, *Nature (Lond.)* **315:** 115–122.

Haskins, K., Kubo, R., White, J.. Pigeon, M., Kappler, J., and Marrack, P. (1983), The major histocompatibility complex restricted antigen receptor on T cells. I. Isolation with a monoclonal antibody, *J. Exp. Med.* **57:** 1149–1169.

Hazelbauer, G. L., and Harayama, S., 1983, Sensory transduction in bacterial chemotaxis, *Int. Rev. Cytol.* **81:** 33–70.

Hedrick, S. M., Cohen, D. I., Nielsen, E. A., and Davis, M. M., 1984, Isolaton of cDNA clones encoding T cell-specific membrane-associated proteins, *Nature (Lond.)* **308:** 149–158.

Heffner, D. K., and Hyams, V. J., 1984, Teratocarcinosarcoma (malignant teratoma?) of the nasal cavity and paranasal sinuses, *Cancer* **53:** 2140–2154.

Hempstead, J. L., and Morgan, J. I., 1985, A panel of monoclonal antibodies to the rat olfactory epithelium, *J. Neurosci.* **5:** 438–449.

Hirsch, J. D., and Margolis, F. L., 1979, Cell suspensions from rat olfactory neuroepithelium: biochemical and histochemical characterization, *Brain Res.* **161:** 277–291.

Imamura, K., Mori, K., Fujita, S. C., and Obata, K, 1985, Immunochemical identification of subgroups of vomeronasal nerve fibers and their segregated terminations in the accessory olfactory bulb, *Brain Res.* **328:** 362–366.

Jessell, T. M., and Dodd, J., 1985, Structure and expression of differentiation antigens on functional subclasses of primary sensory neurons, *Philos. Trans. R. Soc. Lond. [Biol.]* **308:** 271–281.

Kannagi, R., Cochran, N. A., Ishigami, F., Hakomori, S. I., Andrews, P. W., Kowles, B. B., and Solter, D., 1983, Stage-specific embryonic antigens (SSEA-3 and -4) are epitopes of a unique globoseries ganglioside isolated from human teratocarcinoma cells, *EMBO J.* **2:** 2355–2361.

Keilhauer, G., Faissner, A., and Schachner, M., 1985, Differential inhibition of neurone–neurone, neurone–astrocyte and astrocyte–astrocyte adhesion by L1, L2 and N-CAM antibodies, *Nature (Lond.)* **316:** 728–730.

Key, B., and Giorgi, P. P., 1986, Selective binding of soybean agglutinin to the olfactory system of *Xenopus, Neuroscience* **18:** 507–515.

Kleene, S. J., 1986, Bacterial chemotaxis and vertebrate olfaction, *Experientia* **42:** 241–250.

Kleene, S. J., and Gesteland, R. C., 1981, Dissociation of frog olfactory epithelium with N-ethylmaleimide, *Brain Res.* **229:** 536–540.

Kruse, J., Mailhammer, R., Wernecke, H., Faissner, A., Sommer, I., Goridis, C., and Schachner, M., 1984, Neural cell adhesion molecules and myelin-associated glycoprotein share a common carbohydrate moiety recognized by monoclonal antibodies L2 and HNK-1, *Nature (London.)* **311:** 153–155.

Kuhns, W. J., and Pann, C., 1973, Growth kinetics in cultured epithelioid cells and phenotypic expression of blood group H, *Nature New Biol.* **245:** 217–219.

Lancet, D., 1986, Vertebrate olfactory reception, *Annu. Rev. Neurosci.* **9:** 329–356.

Linser, P., 1985, Multiple marker analysis in the avian optic tectum reveals three classes of neuroglia and carbonic anhydrase-containing neurons, *J. Neurosci.* **5:** 2388–2396.

Miller, C. A., and Benzer, S., 1983, Monoclonal antibody cross-reactions between *Drosophila* and human brain, *Proc. Natl. Acad. Sci. USA* **80:** 7641–7645.

Mollicone, R., Davies, D. R., Evans, B., Dalix, A. M., and Oriol, R., 1985, Cellular expression and genetic control of ABH antigens in primary sensory neurons of marmoset, baboon and man, *J. Neuroimmunol.* **10:** 255–269.

Mollicone, R.. Trojan, J., and Oriol, R., 1985, Appearance of H and B antigens in primary sensory cells of the rat olfactory apparatus and inner ear, *Dev. Brain Res.* **17:** 275–279.

Moran, D. T., Rowley, J. C. III, and Jafek, B. W., 1982a, Electron microscopy of human olfactory epithelium reveals a new cell type: The microvillar cell, *Brain Res.* **253:** 39–46.

Moran, D. T., Rowley, J. R. III, and Jafek, B. W., 1982b, The fine structure of the olfactory mucosa in man, *J. Neurocytol.* **11:** 721–746.

Mori, K., Fujita, S. C., Imamura,, K., and Obata, K., 1985, Immunohistochemical study of subclasses of olfactory nerve fibers and their projections to the olfactory bulb in the rabbit, *J. Comp. Neurol.* **242:** 214–229.

Morris, R. J., and Barber, P. C., 1983, Fixation of Thy-1 in nervous tissue for immunohistochemistry: A quantitative assessment of the effect of different fixation conditions upon retention of antigenicity and the cross-linking of Thy-1, *J. Histochem. Cytochem.* **32:** 263–274.

Noble, M., Mallaburn, P. S., and Klein, N. 1984, The growth of olfactory neurons in short-term cultues of rat olfactory epithelium, *Neurosci. Lett.* **45:** 193–198.

Pann, C., and Kuhns, W. J., 1976, Regulation of glycoconjugate expression on cloned mammalian cells, *Exp. Cell Res.* **98:** 73–78.

Parsa, I., Marsh, W. H., and Sutton, A. L., 1981, An *in vitro* model of human pancreas carcinogenesis: Effects of nitrosourea compounds, *Cancer* **47:** 1543–1551.

Pevsner, J., Trifiletti, R. R.. Strittmatter, S. M., and Snyder, S. H., 1985, Isolation and characterization of an olfactory receptor protein for odorant pyrazines, *Proc. Natl. Acad. Sci. USA* **82:** 3050–3054.

Pollack, M. S., Whsocki, C. J., Beauchamp, G. K., Braun, D., Jr., Callaway, C., and Dupont, B, 1982, Absence of HLA association or linkage for variations in sensitivity to the odor of androstenone, *Immunogenetics* **15:** 579–589.

Price, S., 1977, Specific anosmia to geraniol in mice, *Neurosci. Lett.* **4:** 49–50.

Price, S., 1981, Receptor proteins in vertebrate olfaction, in: *Biochemistry of Taste and Olfaction* (R. H. Cagan and M. R. Kare, eds.), Vol. 4, pp. 69–84, Academic, New York.

Rathjen, F. G., and Schachner, M., 1984, Immunocytological and biochemical characterization of a new neuronal cell surface component (L1 antigen) which is involved in cell adhesion, *EMBO J.* **3:** 1–10.

Regan, L. J., Dodd, J., Barondes, S. H., and Jessell, T. M., 1986, Selective expression of endogenous lactose-binding lectins and lactoseries glycoconjugates in subsets of rat sensory neurons, *Proc. Natl. Acad. Sci. USA* **83:** 2248–2252.

Reznick, G., and Stinson, S. F., 1983, *Nasal Tumors in Animals and Man,* **Vols. I–III,** CRC Press, Boca Raton, Florida.

Rhein, L. D., and Cagan, R. H., 1983, Biochemical studies of olfaction:binding specificity of odorants to a cilia preparation from rainbow trout olfactory rosettes, *J. Neurochem.* **41:** 569–577.

Schubert, D., Heinemann, S., Carlisle, W., Tarikas, H., Kimes, B., Patrick, J., Steinbach, J. H., Culp, W., and Brandt, B. L., 1974, Clonal cell lines from the rat central nervous system, *Nature (Lond.)* **249:** 224–227.

Schubert, D., Stallcup, W., LaCorbiere, M., Kidokoro, Y., and Orgel, L., 1985, The ontogeny of electrically excitable cells in cultured olfactory epithelium, *Proc. Natl. Acad. Sci. USA* **82:** 7782–7786.

Schwob, J. E., and Gottlieb, D. I., 1986, The primary olfactory projection has two chemically distinct zones, *J. Neurosci.* **6:** 3393–3404.

Schwob, J. E., Farber, N. B., and Gottlieb, D. I., 1986, Neurons of the olfactory epithelium in adult rats contain vimentin, *J. Neurosci.* **6:** 208–217.

Silva, E. G., Butler, J. J., Mackay, B.. and Goepfert, H., 1982, Neuroblastomas and neuroendocrine carcinomas of the nasal cavity, *Cancer* **50:** 2388–2405.

Solter, D., and Knowles, B. B., 1978, Monoclonak antibody defining a stage-specific mouse embryonic antigen (SSEA-1), *Proc. Natl. Acad. Sci. USA* **75:** 5565–5569.

Stallcup, W. B., Beasley, L. L., and Levine, J. M., 1985, Antibody against nerve growth factor-inducible large external (NILE) glycoprotein labels nerve fiber tracts in the developing rat nervous system, *J. Neurosci.* **5:** 1090–1101.

Takahashi, S., Iwanaga, T., Takahashi, Y., Nakano, Y., and Fujita, T., 1984, Neuron-specific enolase, neurofilament protein and S-100 protein in the olfactory mucosa of human fetuses. An immunohistochemical study, *Cell Tissue Res.* **238:** 231–234.

Tamada, A., Makimoto, K., Okawa, M., Hirono, Y., and Yamabe, H., 1984, Olfactory neuroblastoma: Presentation of a case and review of the Japanese literature, *Laryngoscope* **94:** 252–256.

Thanos, S., Bonhoeffer, F., and Rutishauser, R., 1984, Fiber-like interaction and tectal cues influence the development of the chicken retinotectal projection, *Proc. Natl. Acad. Sci. USA* **81:** 1906–1910.

Thommesen, G., 1983, Morphology, distribution, and specificity of olfactory receptor cells in salmonid fishes, *Acta Physiol. Scand.* **117:** 241–249.

Watkins, W. M., 1980, Biochemistry and genetics of the ABO, Lewis and P blood group systems, *Adv. Hum. Genet.* **10:** 1–136.

Whelan, J. P., Wysocki, C. J., and Lampson, L. A., 1986, Distribution of β_2-microglobulin in olfactory epithelium: A proliferating neuroepithelium not protected by a blood–tissue barrier, *J. Immunol.* **137:** 2567–2571.

Williams, R. K., Goridis, C., and Akeson, R., 1985, Individual neural cell types express immunologically distinct N-CAM forms, *J. Cell Biol.* **101:** 36–42.

Wilson, S-S., Baetge, E. E., and Stallcup, W. B., 1981, Antisera specific for cell lines with mixed neuronal and glial properties, *Dev. Biol.* **83:** 146–153.

Wong, V., Barrett, C. P., Donati, E. J., Eng, L. F., and Guth, L., 1983, Carbonic anhydrase activity in first-order sensory neurons of the rat, *J. Histochem. Cytochem.* **31:** 293–300.

Wysocki, C. J., and Beauchamp, G. K., 1984, Ability to smell androstenone is genetically determined, *Proc. Natl. Acad. Sci. USA* **81:** 4899–4902.

Wysocki, C. J., Whitney, G., and Tucker, D., 1977, Specific anosmia in the laboratory mouse, *Behav. Genet.* **7:** 171–188.

Yamamoto, M., Boyer, A. M., and Schwarting, G. A., 1985, Fucose-containing glycolipids are stage- and region-specific antigens in developing embryonic brain of rodents, *Proc. Natl. Acad. Sci. USA* **82:** 3045–3049.

Yamazaki, K., Beauchamp, G. K., Egorov, I. K., Bard, J., Thomas, L., and Boyse, E. A., 1983, Sensory distinction between H-2^{b} and H-2^{bml} mutant mice, *Proc. Natl. Acad. Sci USA* **80:** 5685–5688.

Yanagi, Y., Yoshikai, Y., Leggett, K., Clark, S. P., Aleksander, I., and Mak, T. W., 1984, A human T cell-specific cDNA clone encodes a protein having extensive homology to immunoglobulin chains, *Nature (Lond.)* **308:** 145–149.

Yoakum, G. H., Lechner, J. F., Gabrielson, E. W., Korba, B. E., Malan-Shibley, L., Willey, J. C., Valerio, M. G., Shamsuddin, A. M., Trump, B. F., and Garris, C. C., 1985, Transformation of human bronchial epithelial cells transfected by Harvey *ras* oncogene, *Science* **227:** 1174–1175.

14

Cellular Interactions in the Development of the Vertebrate Olfactory System

Albert I. Farbman

1. INTRODUCTION

Developmental biologists agree that cellular or tissue interactions play important roles in development of the organism and its many parts. The first direct proof of this was reviewed by Spemann (1938), when he discussed experiments on the two-cell stage of amphibian embryos. In the normal course of events, the two-celled amphibian embryo develops into a whole tadpole, as each of the cells, in effect, gives rise to half of the tadpole. If the two cells are separated from one another in this early stage of development, each gives rise to a whole tadpole of half-normal size. In other words, separation results in change in the developmental potential of each cell and they behave as if each were a fertilized ovum, whereas if they remain in contact, each contributes to only one half the embryo. This was the first evidence showing the effect of cell–cell interaction on development. At later stages of development, cellular or tissue interactions play a role in organogenesis; e.g., interactions between epithelium and mesenchyme are important in development of teeth, glands, kidney, skin and skin derivatives, etc. (Fleischmajer and Billingham, 1968).

The purpose of this chapter is to review some of the experimental evidence for the existence of tissue interactions in the formation of the primary olfactory pathway in vertebrates. The kinds of questions asked by experimental embryologists include the following. What tissues participate in the induction of the olfactory placode? Does the forming nasal sac, including olfactory epithelium, have an effect on development of the olfactory bulb? At the cellular level, do olfactory receptor cells have the capacity to differentiate by themselves, i.e., as an epithelium in isolation from the bulb, or as individual receptor cells isolated from others? After development is completed, are tissue interactions required for maintenance? For example, what happens to mature cells in the bulb when they are isolated from olfactory receptor cells, i.e., when interactions between the two types of tissue are prevented? (see Chapter 9, this volume).

Albert I. Farbman • Department of Neurobiology and Physiology, Northwestern University, Evanston, Illinois 60208.

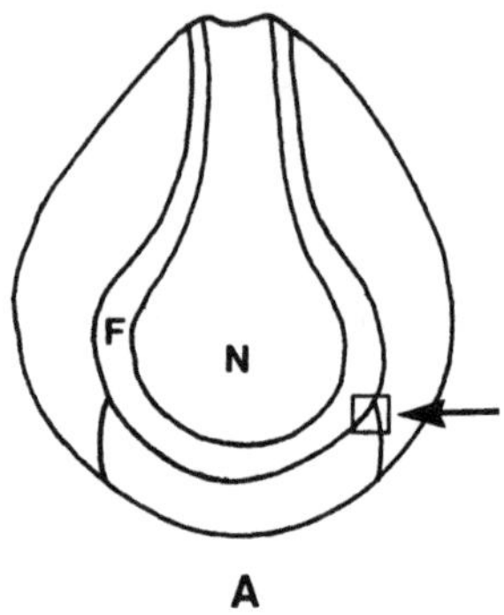

Figure 1. Diagram of the open neural plate stage of development in a frog (*Rana*) embryo. A indicates the anterior end of the embryo. The arrow indicates the approximate position of an olfactory placode, overlying the edge of the neural plate. N, neural plate; F, neural fold.

2. EARLY DEVELOPMENT AND FORMATION OF THE OLFACTORY PLACODE

The experimental analysis of tissue interactions involved in formation of the olfactory placode in early development has been an active area of investigation for several years. Most of these studies were done on amphibian embryos by using the classic experimental embryological techniques of ablation, transplantation or explantation. In some studies, transplantation was done using one amphibian species as donor and another as host. The rationale for this was based on the fact that some species had unique cell markers, such as pigment granules, so that, after a period of growth, cells derived in the donor animal could be distinguished from those in the host. The major question asked was: What tissue(s) were responsible for the induction of the olfactory placode?

The developing olfactory placode in amphibians is detectable at the early neural plate stage, when bilateral ectodermal thickenings are seen overlying the lateral portion of the transverse medullary ridge (of the neural plate) in the head region (Zwilling, 1940). At this stage, the placode is in very close proximity to the neural ectoderm and the mesodermal tissue associated with the roof of the archenteron (primitive gut) (Fig. 1). Because of this proximity, both tissues, the telencephalic end of the neural plate and the roof of the archenteron, have been implicated as inducers of the placode.

2.1. Experimental Analysis of Induction of the Olfactory Placode

The early experiments conducted by Zwilling (1934) exemplify the kind of work that has led many investigators to conclude that the telencephalic portion of the neural plate induces the olfactory organ. Zwilling transplanted pieces of the anterior neural plate of donor *Rana pipiens* embryos to various parts of the body of recipient embryos and showed that they could induce formation of nasal sacs from ectoderm other than head ectoderm. These transplants were homotypic; i.e., the donor and host belonged to the same species. The results of later studies, however, made Zwilling change his mind (Zwilling, 1940). When he made heterotypic transplants, in which the donor and host tissue could be clearly distinguished by different pigmentation patterns, he found that the donor neural plate tissue often included fragments of "determined" olfactory epithelium; i.e., the genome in these cells had already been primed to express olfactory epithelium. This became evident because the nasal sac growing at the transplantation site was composed largely of epithelial cells with the donor pigmentation pattern, indicating that the developing sac was made up mostly of donor cells. When care was taken to remove all determined olfactory tissue from the neural plate transplants, no nasal sacs appeared at the transplant site. In other words, the donor neural plate did not induce formation of olfactory epithelium in the host.

Zwilling (1940) also showed that the ectodermal cell region destined to become olfactory placode in *Rana pipiens* differentiates into the placode before the neural plate ectoderm is distinguishable in the embryo. He suggested that the induction of placode probably occurs in the late gastrula stage and concluded that both the olfactory organ and the neural plate are probably induced by the mesoderm associated with the roof of the archenteron.

Zwilling's results were substantiated by Emerson (1944), who showed that nasal sacs were formed by transplants of the presumptive neural ectoderm region of mid-gastrula *Rana pipiens* embryos to regenerating tadpole tails. These transplants were taken from embryos before the neural plate developed. Emerson showed further that when fragments of presumptive neural plate ectoderm from late gastrula embryos were explanted and grown on agar in saline solution, nasal sacs would develop *in vitro* as well (Emerson, 1945). The cells that were to give rise to the olfactory placode and nasal sac were already determined in *Rana pipiens* embryos as stage 13 (the mid-gastrula stage) before the neural plate was evident. These studies supported Zwilling's conclusions that (1) the presence of the neural plate was not necessary for the early determination of the olfactory organ, and (2) the neural plate and the olfactory placodes probably developed simultaneously under the influence of the archenteron roof.

However, Haggis (1956) favored the notion that neural tissue does indeed play a role in placode formation. He argued that in experiments of earlier workers no situation was found in which nasal tissue appeared in explants or transplants without the concomitant appearance of neural tissue. His experiments were carried out on a different species, *Amblystoma punctatum.* Although a previous study on this species by Carpenter (1937) had shown the prospective nasal ectoderm can self-differentiate when transplanted to other head regions as early as stage 13 (in agreement with the experiments of Zwilling and Emerson described above), Haggis's results showed that when nasal primordia were extirpated and grown *in vitro* nasal organs did not develop unless the explants were taken from embryos of stage 21 or later. Moreover, in all cases, the neural tube was implicated as the inducer tissue. This series of experiments involving transplantation, ablation, explantation, etc. supported "the view that inductive stimuli emanating from the prospective forebrain are solely responsible for the emergence of the olfactory organs" (Haggis, 1956).

Experiments with chick embryos showed that the olfactory placode is not visible until the 24-somite stage, well after neurulation has occurred (Street, 1937). When grafts of ectoderm from the heads of 9–12 somite embryos were grown on the chorioallantoic membrane, olfactory epithelium was clearly expressed (Street, 1937). In other words, the olfactory epithelial cells had been "determined" by this stage, well after neurulation had begun, but well before the placode was demonstrable. This is quite different from the situation in amphibian embryos.

Because the literature is clouded by the different sets of results from experiments done in different species, generalizations concerning the origin of the inductive stimulus cannot readily be made. It is quite possible that the stimulus is present in both the primitive forebrain and in the mesenchymal tissue associated with the roof of the archenteron or with the head mesenchyme and that some or all of these tissues participate in induction of the nasal organ (Yntema, 1955; Jacobson, 1963). This issue will probably not be resolved satisfactorily until the factor or factors responsible for inducing the olfactory placode and nasal cavity are identified and their sites of synthesis localized.

3. EFFECT OF OLFACTORY RECEPTOR CELLS ON EARLY FORMATION OF THE OLFACTORY BULB

Removal of the olfactory placode in amphibian (Burr, 1916; Piatt, 1951; Stout and Graziadei, 1980) or chick (Venneman *et al.*, 1982) embryos results in a reduction in size of the anterior portion of the telencephalon. When transplants of additional placodes are made to the head region of amphibian embryos, the axons growing from these transplants to the brain can promote cell proliferation in the diencephalon and telencephalon (Burr, 1924; Stout and Graziadei, 1980).

Under normal circumstances, the growing olfactory axons invade the bulb and have the capacity to organize glomeruli, where they synapse with the second order neurons in the bulb. Several recent works show that olfactory axons can organize glomeruli and form synapses with neurons in other parts of the forebrain as well. This ability has been demonstrated in both amphibians (Stout and Graziadei, 1980) and in mammals (Graziadei *et al.*, 1978, 1979; Graziadei and Kaplan, 1980; Graziadei and Samanen, 1980; Morrison and Graziadei, 1983). One can conclude that formation of the bulb in early development of the central nervous system is dependent on influences exerted by the growing axons of olfactory receptor cells. Moreover, even after the telencephalon is formed, the olfactory receptor cell axons have the capacity to induce formation of glomeruli in the olfactory bulb and other regions of the forebrain.

4. EFFECT OF THE OLFACTORY BULB ON DIFFERENTIATION OF RECEPTOR CELLS IN A REGENERATING SYSTEM

Studies in several systems have shown that target organs may have major effects on the survival, differentiation, and maturation of neuronal cells. For example, embryonic motoneurons are unable to survive if they fail to make connections with muscle cells (Hamburger, 1934, 1958; Barron, 1943, 1946, 1948; Hamburger and Keefe, 1944; Cowan and Wenger, 1967; Oppenheim *et al.*, 1978). In sensory systems, the dorsal root ganglion neurons of chick embryos fail to survive if their target tissue in the limbs is removed (Hamburger and Levi-Montalcini, 1949). Similarly, the neurons of the ciliary ganglion are dependent on their target tissues for survival (Landmesser and Pilar, 1974; Pilar and Landmesser, 1976). *In vitro* studies have shown evidence of trophic interactions between the cochleovestibular ganglion of chick embryos and their sensory epithelia (Ard *et al.*, 1985).

Target tissues may also have a modulating influence on neurons. For example, superior cervical ganglion grown with target salivary glands *in vitro* showed greater elaboration and directionality of nerve fiber outgrowth than did control explants (Coughlin *et al.*, 1978). Clearly, the totality of evidence suggests that target organs do indeed supply factors necessary for the maintenance or survival of neurons.

The olfactory bulb is the sole target of olfactory receptor neurons. The axons of all these neurons synapse within the glomerular layer of the bulb, primarily on dendrites of mitral or tufted cells. In light of the studies from other neural systems, it might well be asked, therefore, whether the olfactory bulb influences the survival or differentiation of the receptor cells.

The evidence for a bulbar influence on differentiation of olfactory receptor cells is derived primarily from the results of both *in vivo* studies on degeneration and regeneration and *in vitro* experiments on development. The regeneration experiments took advantage of the fact that olfactory receptor cells can be selectively damaged by surgery, either by transecting the olfactory nerve(s) (Harding *et al.*, 1977; Harding and Wright, 1979; Monti Graziadei and Graziadei, 1979; Graziadei and Monti Graziadei, 1980; Graziadei *et al.*, 1980; Monti Graziadei et al., 1980; Booth *et al.*, 1981; Kawano and Margolis, 1982; Baker *et al.*, 1984; Camara and Harding, 1984; Costanzo, 1984; Gozzo and Fülöp, 1984; Samanen and Forbes. 1984), or by removing the olfactory bulb (Ferriero and Margolis, 1975; Graziadei *et al.*, 1978, 1979; Booth *et al.*, 1981; Costanzo and Graziadei, 1983; Butler *et al.*, 1984; Hempstead and Morgan, 1985). Following each of these surgical interventions, olfactory receptor cells degenerate and are replaced, after a period of time, by new receptor cells. The critical difference between the two types of experiments is the presence or absence of the bulbar synaptic target for the new olfactory axons growing from the replacement receptor neurons.

Most of the experiments done to examine directly the effect of bulb on recovery of the olfactory receptors have been carried out in mammals. Before discussing the results in detail, it is worth noting that there are major methodological difficulties in doing an olfactory axotomy in mammals. The mammalian olfactory nerve is organized into hundreds or possibly thousands of tiny nerve bundles (fila olfactoria), which pass through openings in the cribriform plate of the ethmoid bone to reach the bulb. Because of the large number of fila olfactoria in mammals, and because it is particularly difficult to gain access to those nerve bundles innervating the ventral aspect of the bulb, it is almost impossible to ensure that all olfactory axons have been severed (Costanzo, 1984). In most experiments, the axons to the ventral bulb may have been left intact and any conclusions from the results must be guarded. Costanzo (1984) made quantitative comparisons of epithelial cell number and thickness in hamsters that had been either axotomized or bulbectomized. He used a specially designed microdissection instrument to perform olfactory axotomy and tried to verify the extent of success of the surgical procedure by placing a Teflon marker in the wound site. He was able to confirm that, on the average, transections were complete to 86% of the total bulb depth, and in a few cases, the transections involved 100% of the fila olfactoria. The significance of this study is that it presents a critical analysis of the methods used in studying the recovery of olfactory receptor cells in mammals.

Given the difficulty of ensuring a complete axotomy, it is not surprising that the results of experiments in which this technique was used have produced conflicting data. In some experiments, this procedure permitted recovery of chemical markers (Harding *et al.*, 1977) and cell numbers (Samanen and Forbes, 1984) to near-control values. In others (Costanzo, 1984) recovery was somewhat less complete. Further complications of axotomy experiments include the possibility that other factors, such as postsurgical scar formation or mechanical blockage of the openings in the cribriform plate, possibly brought about by ossification (Graziadei and Monti-Graziadei, 1979) could prevent the developing axons of newly formed receptor cells from reaching their target (Costanzo, 1984).

Bulbectomy was the second type of surgical procedure used to promote degeneration and regeneration of the olfactory epithelium. Because of the anatomy of the main olfactory bulb in mammals, removal of this structure is difficult to accomplish without damaging the accessory olfactory bulb and adjacent parts of the forebrain. Nevertheless, the results of bulbectomy studies have been fairly consistent. When the bulb is removed, the observed recovery of the epithelium is incomplete. The number of replacement receptor cells reaches only 60–70% of control values (Costanzo and Graziadei, 1983). Other studies have shown that biochemical markers, including the amount of olfactory marker protein (OMP) (Harding and Margolis, 1976), also are considerably lower than those of control tissue. The results of immunohistochemical studies are consistent with the biochemical and anatomical data (Monti Graziadei, 1983; Hempstead and Morgan, 1985). On the basis of these studies, genesis and differentiation of receptor neurons seem to occur in the absence of the bulb, up to a point, but the healing is not complete. It has been suggested that recovery to control values is somehow linked to the presence of an intact connection with the olfactory bulb which may influence later stages of differentiation (Doucette *et al.*, 1983; Monti-Graziadei, 1983; Hempstead and Morgan, 1985).

Because of the methodological caveats (Costanzo, 1984), one must be cautious in making conclusions on the effect of bulb on regenerating epithelium of mammals. However, most of the evidence to date suggests that such an effect does exist. It should be emphasized that neither removal of the bulb nor axotomy compromises neurogenesis in olfactory epithelium. Moreover, neither procedure totally prevents some differentiation of receptor cells from taking place.

Olfactory axotomy is much easier to perform in amphibians or birds because the axons are bundled into a single easily accessible nerve of variable length. Although, to my knowledge, no experiments specifically addressing the question of cellular interaction in development have been done in these animal species, some information bearing on the problem is available from studies concerned with other issues. For example, salamanders have been used to examine morphological and physiological recovery of olfactory epithelium after axotomy (Simmons and Getchell 1981*a,b;* Simmons *et al.*, 1981). In these studies, the bulb was not removed. Although no epithelial cell counts were done in these studies, the general conclusions indicate that full functional and anatomical recovery occurred when new axons grew into the distal stump of the severed olfactory nerve and reinnervated the bulb. Moreover, it was "concluded that receptor neurons pass through two phases of functional maturity: the first independent of bulbar contact and the second dependent on presumed synaptic contact with bulbar neurons" (Simmons and Getchell, 1981*b*).

In some studies on regeneration of olfactory epithelium of birds after axotomy, it was noted that after regeneration had occurred, the reconstituted epithelium was indistinguishable from controls (Oley *et al.*, 1975; Bedini *et al.*, 1976; Graziadei and Okano.

1979). Functional tests indicated that the reconstituted receptor cells had made appropriate central connections (Oley *et al.*, 1975; Bedini *et al.*, 1976; Kiyohara and Tucker, 1978). However, quantitative estimates of the number of axons in the reconstituted olfactory nerve indicated that it contained fewer than one half the axons of the nerve on the unoperated side (Bedini *et al.*, 1976). These apparently conflicting data make sense if reconstitution of the epithelium is patchy; i.e., it may be almost fully recovered in some regions, and incompletely reconstituted in others. This would account for the apparently normal histological appearance of epithelium reported by some authors (Graziadei and Okano, 1978) and the low count of the total number of olfactory axons reported by others (Bedini *et al.*, 1976). The data on birds are not consistent with the notion that the presence of the bulb promotes full epithelial recovery after axotomy.

A recurrent theme in the experiments on regeneration is the ability of the olfactory epithelium and nerve to become reconstituted after either axotomy or bulbectomy. In neither case does the recovery restore the primary olfactory pathway to preexperimental conditions, but the axotomized animal can recover olfactory function to a large extent. As far as the receptor cells themselves are concerned, they probably do not become structurally or chemically mature until or unless they have synapsed with other neurons (cf. Doucette *et al.*, 1983). After all is said and done, however, the experiments on degeneration and regeneration are probably not ideally suited to answer the question of whether the bulb influences differentiation of receptor cells. There are too many difficulties with technique, both in ensuring completeness of the axotomy of bulbectomy and in problems such as scar formation in the wound healing process.

5. EFFECT OF OLFACTORY BULB ON RECEPTOR CELL DEVELOPMENT

5.1. Organ Culture Method

The influence of the bulb on development and maturation of receptor cells has been studied *in vitro*. We have devised a method by which explanted olfactory mucosa can be grown in organ culture, either in isolation or in combination with olfactory bulb. When mucosa is grown alone, the receptor neurons undergo differentiation, but full maturation of all receptors does not occur *in vitro*. Axons grow out, a few cilia are expressed at the dendrite. olfactory marker protein (OMP) is synthesized by some cells, and electro-olfactograms can be recorded (Farbman and Gesteland, 1975; Farbman, 1977; Chuah and Farbman, 1983, 1986; Chuah *et al.*, 1985; Gonzales *et al.*, 1985). Similar findings have been reported in experiments in which olfactory mucosa is transplanted to the anterior chamber of the eye (Barber *et al.*, 1982; Morrison and Graziadei, 1983; Novoselov *et al.*, 1984). It seems clear from *in vivo* and *in vitro* experiments that receptor neurogenesis and differentiation can occur in the absence of the bulb. However, the evidence that full maturation is achieved under these circumstances is lacking.

5.2. Bulbar Effects on OMP Synthesis

In vitro experiments have been done to examine the effect of the bulb on differentiation and maturation of receptor cells. Two criteria were used as signs of maturation, i.e., synthesis of OMP and ciliogenesis. Fragments of olfactory mucosa were dissected from

Table 1. Olfactory Marker Protein Ratio of Pooled Explants (Cultured for 7 Days)[a]

Groups of explants	OMP ratio[b]
Olfactory mucosa + olfactory bulb	1.00
Olfactory mucosa alone	0.51
Olfactory mucosa + cerebrum	0.44
Olfactory mucosa + cerebellum	0.45
Olfactory mucosa + spinal cord	0.58
Olfactory mucosa + heart	0.33
Olfactory mucosa + olfactory bulb, separated and recombined	0.97
Olfactory mucosa + olfactory bulb, separted by	
Millipore filter, 0.45-μm pore size	0.53
Nucleopore filter, 3.0-μm pore size	0.46

[a]After Chuah and Farbman (1983).
[b]The average amount of OMP in a combined explant of E15 fetal olfactory mucosa and olfactory bulb was arbitrarily set at 1.0; all values of OMP in other groups were expressed as a ratio of this value.

E15 rat embryos, either alone or *en bloc* with presumptive olfactory bulb attached, and explanted into organ culture dishes. After four to ten days in culture, radioimmunoassays were done on pooled explants in both groups to determine how much OMP was present. In the cultures of mucosa alone, the amount of OMP increased slightly from 4 to 10 days, but in the en bloc cultures of mucosa and bulb, the amount of OMP increased on the sixth day to twice that of the mucosa alone (Table 1). This ratio was maintained for the remainder of the experimental period (Chuah and Farbman, 1983). Using an immunocytochemical technique to detect OMP in tissue sections of the explants, we determined that this twofold increase was attributable to a doubling of the number of receptor cells expressing OMP (Chuah and Farbman, 1983). Furthermore, we showed that the effect on differentiation of receptor cells was specifically exerted by the bulb; i.e., when other nerve tissue, including cerebrum, cerebellum, and spinal cord, was substituted for the bulb in the cultures, the amount of OMP was essentially the same as that in cultures of mucosa alone (Table 1).

It was possible that in the organ cultures of mucosa alone, separation of the mucosa from the bulb during the dissection procedures resulted in damage to the axons or epithelium and this could have accounted for the reduced amount of OMP in the explants of mucosa alone, in which this separation was done before explantation. In order to investigate this, we separated olfactory epithelium from the bulb of E15 rat embryos and then recombined the two tissues in the culture. The amount of OMP after 7 days was twice that found in explants of epithelium alone (Table 1) (Chuah and Farbman, 1983). This experiment showed that damage by dissection cannot account for the low OMP values in cultures of mucosa alone. The bulbar target, then, must have a specific enhancing effect on OMP synthesis in the cultures.

These results from *in vitro* experiments indicated that olfactory neurons can express OMP "on their own," i.e., without any influence from the bulb. However, when the bulb is present, the expression of OMP is significantly enhanced. It should be noted that whether or not the bulb was present, the degree of maturation of olfactory epithelium was considerably less than that *in vivo* at an equivalent age (Chuah and Farbman, 1983; Chuah *et al.*, 1985). The possibility was investigated that a diffusible factor from the target organ might be responsible for this enhancing effect. Filter sandwich cultures of olfactory mucosa and bulb were devised in which the two tissue types were separated by a thin porous filter. The filter was thin enough to maintain a short diffusion distance and had pores large enough to permit macromolecules to pass. We tried several different pore sizes, from 0.45 to 12.0 μm and two different filter materials: Millipore (a cellulose material), 25 μm thick, and Nucleopore (a polycarbonate material), about 12 μm thick. The results showed that the amount of OMP was not increased in the transfilter cultures (see Table 1), suggesting that the effect of bulb on receptor cell differentiation was not attributable to a diffusible factor. Electron microscopic examination of these sandwich cultures revealed no nerve axons growing through even the largest pores of the filters (Chuah and Farbman, 1983). Whatever is responsible for the specific bulbar enhancement of epithelial OMP production is probably mediated by direct contact between receptor cells and their target tissues (Chuah and Farbman, 1983).

5.3. Bulbar Effects on Ciliogenesis

In a morphological study on development of mouse olfactory epithelium, it was observed that ciliogenesis begins at about the time when olfactory axons reach the bulb, thus suggesting there may be some retrograde effect of bulb on ciliary growth (Cuschieri and Bannister, 1975). In the rat, as well, the first signs of ciliogenesis are correlated with the time when axons reach the bulb (Menco and Farbman, 1985). Using our organ culture method, we investigated the effect of olfactory bulb on ciliogenesis. Ciliated olfactory dendritic knobs developed in cultures of olfactory mucosa alone, as in earlier studies (Farbman and Gesteland, 1975; Farbman, 1977) but there was a twofold increase in their number in those cultures containing olfactory bulb, again an *enhancement* of maturation (Chuah *et al.*, 1985). It is apparent, therefore, that although the target tissue is not absolutely essential for ciliogenesis, it can exert a strong influence.

5.4. Does the Bulb Exert a Tropic Effect?

Another way in which target organs influence growing nerves is to attract the growing fibers toward them. In other words, the target tissue can exert a tropic effect on growing axons. Whether this occurs in the primary olfactory pathway was investigated in a culture system of olfactory mucosa/olfactory bulb. We explanted fragments of olfactory bulb and olfactory mucosa at a distance 0.5–0.7 mm apart. Neurites grew out from both explants in a more or less radial direction. The direction of growth from each explant seemed unaffected by the presence of the other (Gonzales *et al.*, 1985).

6. CELL-CULTURE STUDIES

It seems clear that the olfactory receptor neurons can develop up to a point independently of the bulb. It is interesting to consider what other factors might be operating to affect development of these cells, particularly those operating in early development.

Although, to my knowledge, there are no published data on experiments designed to determine what these other factors might be, some information can be extracted from recent efforts to grow dissociated olfactory epithelium in cell culture. The early experiments on amphibian development discussed above were designed to determine what influences were operative on the epithelium as a whole. The aim of cell-culture experiments is to develop a cell line of olfactory receptor cells that will continue to divide and differentiate in culture in isolation from other cells and tissues. The yield from these experiments is low, but certain common threads emerge. For example, Noble *et al.* (1984) claimed that differentiation of receptor cells could occur in cultures of dissociated rat olfactory epithelium plated on a monolayer of purified astrocytes. After one day in culture these cells displayed some of the characteristic ultrastructural features of olfactory receptor cells *in situ,* i.e., a bipolar cell with a dendrite (15–30 μm long) containing many microtubules and slender mitochondria and centrioles at or near the club shaped terminal. A single axon grew from the other end of the cell to lengths as great as 250 μm. Other workers grew dissociated olfactory mucosal cells on noncellular substrates. They were unable to show morphological differentiation but were able to show that one of the biochemical markers, carnosine, did appear in the cultures and that the cells displayed electric activity (Goldstein and Quinn, 1981; Schubert *et al.*, 1985).

If criteria for olfactory receptor cell differentiation and maturation are rigorously applied, none of the cell-culture experiments has been successful. These criteria include, e.g., morphological differentiation of dendritic knobs with several ciliary appendages, growth of axons of reasonable length from the opposite pole of the cell, and synthesis of a marker for mature cells, i.e., olfactory marker protein (Gonzales *et al.*, 1985). We have suggested that in the cell-culture experiments conducted by Noble *et al.* (1984), some degree of olfactory receptor maturation was obtained, probably because the substrate consisted of a monolayer of astrocytes. It is possible that astrocytes have some characteristics in common with supporting or other non-neuronal cells of olfactory mucosa (cf. Getchell, 1977; Barber and Lindsay, 1982; Miragall, 1983; Rafols and Getchell, 1983). In other words, in those culture systems that promoted contact between presumptive receptor cells and neighbors, it was possible to achieve a degree of morphological differentiation and maturation. We propose, as a reasonable working hypothesis, that contact with non-neuronal cells plays a major role in differentiation of olfactory receptor neurons. We are currently testing this hypothesis. In preliminary experiments, we have made a few observations that lend it some support. When the olfactory epithelium of a chick embryo is separated from its underlying mesenchyme, and the epithelial cells are dissociated, plated on a noncellular substrate in a culture dish, and encouraged to reaggregate, morphological differentiation of receptor cells can be achieved. We have not yet determined whether the differentiation is dependent only on contact with non-neuronal epithelial cells.

7. SUMMARY

Cellular interactions play an important role in development of many organs and organ systems in the vertebrate. The olfactory system is no exception. In early development, the formation of the olfactory placode is influenced by signals from surrounding tissues, as has been shown clearly by experimental embryologists working primarily on amphibian embryos. Later, the growing olfactory axons play a major role in development and organization of the olfactory bulb. In experimental situations, moreover, the olfactory axons can organize glomeruli in brain tissue other than its usual target organ. It is therefore fairly clear that receptor cells affect development of the target bulb. At later stages of cell differentiation, it has been shown convincingly that the bulb affects olfactory receptor cell maturation. In organ-culture experiments, the presence of bulb enhances the expression of olfactory marker protein and the growth of dendritic cilia. Other kinds of *in vitro* experiments have shown that, at the cellular level, olfactory receptor epithelial cells may require contact with non-neuronal cells before they express some of their qualities, including axon formation, etc. It is highly probable that the receptor cell does not reach a fully mature structural and biochemical state unless it makes synaptic contact with the appropriate cells in its target organ, the olfactory bulb.

ACKNOWLEDGMENTS Some of the work described in this chapter was supported by grants NS-06181, NS-18490, and NS-233348 from the National Institutes of Health. I am grateful to my colleagues, Virginia Carr, Federico Gonzales, and Bert Menco for their helpful suggestions in preparing this chapter, and to Ms. Lynn Colletti for the drawing.

REFERENCES

Ard, M. D., Morest, D. K., and Hauger, S. H., 1985, Trophic interactions between the cochleovestibular ganglion of the chick embryo and its synaptic targets in culture, *Neuroscience* **16:** 151–170.

Baker, H., Kawano, T., Albert, V., Joh, T. J., Reis, D. J., and Margolis, F. L., 1984, Olfactory bulb dopamine neurons survive deafferentation-induced loss of tyrosine hydroxylase, *Neuroscience* **11:** 605–615.

Barber, P. C., Jensen, S., and Zimmer, J., 1982, Differentiation of neurons containing olfactory marker protein in adult rat olfactory epithelium transplanted to the anterior chamber of the eye, *Neuroscience* **7:** 2687–2695.

Barber, P. C., and Lindsay, R. M., 1982, Schwann cells of the olfactory nerves contain glial fibrillary acidic protein and resemble astrocytes, *Neuroscience* **7:** 3077–3090.

Barron, D. H., 1943, The early development of the motor cells and columns in the spinal cord of the sheep, *J. Comp. Neurol.* **78:** 1–26.

Barron, D. H., 1946, Observations on the early differentiation of the motor neuroblasts in the spinal cord of the chick, *J. Comp. Neurol.* **85:** 149–169.

Barron, D. H., 1948, Some effects of amputation of the chick wing bud on the early differentiation of the motor neuroblasts in the associated segments of the spinal cord, *J. Comp. Neurol.* **88:** 93–127.

Bedini, C., Fiaschi, V., and Lanfranchi, A., 1976, Olfactory nerve reconstitution in the homing pigeon after resection: ultrastructural and electrophysiological data, *Arch. Ital. Biol.* **114:** 1–22.

Booth, W. S., Baldwin, B. A., Poynder, T. M., Bannister, L. H., and Gower, D. B., 1981, Degeneration and regeneration of the olfactory epithelium after olfactory bulb ablation in the pig: A morphological and electrophysiological study, *Q. J. Exp. Physiol.* **66:** 533–540.

Burr, H. S., 1916, The effects of the removal of the nasal pits in Amblystoma embryos, *J. Exp. Zool.* **20:** 27–57.

Burr, H. S., 1924, Some experiments on the transplantation of the olfactory placode in Amblystoma. I. An experimentally produced aberrant cranial nerve, *J. Comp. Neurol.* **37:** 455–479.

Butler, A., Graziadei, P. P. C., Monti-Graziadei, G. A., and Slotnick, B. M., 1984, Neonatally bulbectomized rats with new olfactory-neocortical connections are anosmic, *Neurosci. Lett.* **48:** 247–254.

Camara, C. G., and Harding, J. W., 1984, Thymidine incorporation in the olfactory epithelium of mice: Early exponential response induced by olfactory neurectomy, *Brain Res.* **308:** 63–68.

Carpenter, E., 1937, The head pattern in Amblystoma studied by the vital staining and transplantation methods, *J. Exp. Zool.* **75:** 103–130.

Chuah, M. I., and Farbman, A. I., 1983, Olfactory bulb increases marker protein in olfactory receptor cells, *J. Neurosci.* **3:** 2197–2205.

Chuah, M. I., and Farbman, A. I., 1986. Mitral cell differentiation and synaptogenesis in organ culture of rat presumptive olfactory bulb, *Cell Tissue Res.* **24:** 359–365.

Chuah, M. I., Farbman, A. I., and Menco, B. P. M., 1985, Influence of olfactory bulb on dendritic knob density of rat olfactory receptor neurons *in vitro, Brain Res.* **338:** 259–266.

Costanzo, R. M., 1984, Comparison of neurogenesis and cell replacement in the hamster olfactory system with and without a target (olfactory bulb), *Brain Res.* **307:** 295–301.

Costanzo, R. M., and Graziadei, P. P. C., 1983, A quantitative analysis of changes in the olfactory epithelium following bulbectomy in hamster, *J. Comp. Neurol.* **215:** 370–381.

Coughlin, M. D., Dibner, M. D., Boyer, D. M., and Black, I. B., 1978, Factors regulating development of an embryonic mouse sympathetic ganglion, *Dev. Biol.* **66:** 513–528.

Cowan, W. M., and Wenger, E., 1967, Cell loss in the trochlear nucleus of the chick during normal development and after radical extirpation of the optic vesicle, *J. Exp. Zool.* **164:** 267–280.

Cuschieri, A., and Bannister, L. H., 1975, The development of the olfactory mucosa in the mouse: Electron microscopy, *J. Anat.* **119:** 471–498.

Doucette, J. R., Kiernan, J. A., and Flumerfelt, B. A., 1983. Two different patterns of retrograde degeneration in the olfactory epithelium following transection of primary olfactory axons, *J. Anat.* **136:** 673–689.

Emerson, H. S., 1944, Embryonic grafts in regenerating tissue. III. The development of dorsal and ventral ectoderm of *Rana pipiens* larvae, *J. Exp. Zool.* **97:** 1–19.

Emerson, H. S., 1945, The development of late gastrula explants of *Rana pipiens* in salt solution. *J. Exp. Zool.* **100:** 497–521.

Farbman, A. I., 1977, Differentiation of olfactory receptor cells in organ culture, *Anat. Rec.* **189:** 187–200.

Farbman, A. I., and Gesteland, R. C., 1975, Developmental and electrophysiological studies of olfactory mucosa in organ culture, in: *International Symposium on Olfaction and Taste,* Vol. V (D. A. Denton and J. P. Coghlan, eds.), pp. 107–110, Academic, New York.

Ferriero, D., and Margolis, F. L., 1975, Denervation in the primary olfactory pathway of mice. II. Effects on carnosine and other amine compounds, *Brain Res.* **94:** 75–86.

Fleischmajer, R., and Billingham, R. E., 1968, *Epithelial–Mesenchymal Interactions,* Williams & Wilkins, Baltimore.

Getchell, T. V., 1977, Analysis of intracellular recordings from salamander olfactory epithelium, *Brain Res.* **123:** 275–286.

Goldstein, N. I., and Quinn, M. R., 1981, A novel cell line isolated from the murine olfactory mucosa, *In Vitro* **17:** 593–598.

Gonzales, F., Farbman, A. I., and Gesteland, R. C., 1985, Cell and explant culture of olfactory chemoreceptor cells, *J. Neurosci. Methods* **14:** 77–90.

Gozzo, S., and Fülöp, 1984, Transneuronal degeneration in different inbred strains of mice—A preliminary study of olfactory bulb events after olfactory nerve lesion, *Int. J. Neurosci.* **23:** 187–194.

Graziadei, P. P. C., and Kaplan, M. S., 1980, Regrowth of olfactory sensory axons into transplanted neural tissue. 1. Development of connections with the occipital cortex, *Brain Res.* **201:** 39–44.

Graziadei, P. P. C., and Monti-Graziadei, G. A., 1979, Neurogenesis and neuron regeneration in the olfactory system of mammals. I. Morphological aspects of differentiation and structural organization of the olfactory sensory neurons, *J. Neurocytol.* **8:** 1–18.

Graziadei, P. P. C., and Monti Graziadei, G. A., 1980, Neurogenesis and neuron regeneration in the olfactory system of mammals. III. Deafferentation and reinnervation of the olfactory bulb following section of the fila olfactoria in rat, *J. Neurocytol.* **9:** 145–162.

Graziadei, P. P. C., and Okano, M., 1979, Neuronal degeneration and regeneration in the olfactory epithelium of pigeon following transection of the first cranial nerve, *Acta Anat.* **104:** 220–236.

Graziadei, P. P. C., and Samanen, D. W., 1980, Ectopic glomerular structures in the olfactory bulb of neonatal and adult mice, *Brain Res.* **187:** 467–472.

Graziadei, P. P. C., Levine, R. R., and Monti Graziadei, G. A., 1978, Regeneration of olfactory axons and synapse formation in the forebrain after bulbectomy in neonatal mice, *Proc. Natl. Acad. Sci. USA* **75:** 5320–5324.

Graziadei, P. P. C., Levine, R. R., and Monti Graziadei, G. A., 1979, Regeneration into the forebrain following bulbectomy in the neonatal mouse, *Neuroscience* **4:** 713–727.

Graziadei, P. P. C., Karlan, M. S., Monti Graziadei, G. A., and Bernstein, J. J., 1980, Neurogenesis of sensory neurons in the primate olfactory system after section of the fila olfactoria, *Brain Res.* **186:** 289–300.

Haggis, A. J., 1956, Analysis of the determination of the olfactory placode in *Amblystoma punctatum, J. Embryol. Exp. Morphol.* **4:** 120–139.

Hamburger, V., 1934, The effects of wing bud extirpation on the development of the central nervous system in chick embryos, *J. Exp. Zool.* **68:** 449–494.

Hamburger, V., 1958, Regression versus peripheral control of differentiation in motor hypoplasia, *Am. J. Anat.* **102:** 365–410.

Hamburger, V., and Keefe, E. L., 1944, The effects of peripheral factors on the proliferation and differentiation in the spinal cord of the chick embryo, *J. Exp. Zool.* **96:** 223–242.

Hamburger, V., and Levi-Montalcini, R., 1949, Proliferation, differentiation and degeneration in the spinal ganglia of the chick embryo under normal and experimental conditions. *J. Exp. Zool.* **111:** 457–501.

Harding, J., and Margolis, F. L., 1976, Denervation in the primary olfactory pathway of mice. III. Effect on enzymes of carnosine metabolism, *Brain Res.* **110:** 351–360.

Harding, J. W., and Wright, J. W., 1979, Reversible effects of olfactory nerve section on behavior and biochemistry in mice, *Brain Res. Bull.* **4:** 17–22.

Harding, J., Graziadei, P. P. C., Monti Graziadei, G. A., and Margolis, F. L., 1977, Denervation in the primary olfactory pathway of mice. IV. Biochemical and morphological evidence for neuronal replacement following nerve section, *Brain Res.* **132:** 11–28.

Hempstead, J. L., and Morgan, J. I., 1985, Monoclonal antibodies reveal novel aspects of the biochemistry and organization of olfactory neurons following unilateral olfactory bulbectomy, *J. Neurosci.* **5:** 2382–2387.

Jacobson, A. G., 1963, The determination and positioning of the nose, lens and ear. I. Interactions within the ectoderm, and between the ectoderm and underlying tissues. *J. Exp. Zool.* **154:** 273–283.

Kawano, T., and Margolis, F. L., 1982, Transsynaptic regulation of olfactory bulb catecholamines in mice and rats, *J. Neurochem.* **39:** 342–348.

Kiyohara, S., and Tucker, D., 1978, Activity of new receptors after transection of the primary olfactory nerve in pigeons, *Physiol. Behav.* **21:** 987–994.

Landmesser, L., and Pilar, G., 1974, Synaptic transmission and cell death during normal ganglionic development, *J. Physiol. (Lond.)* **241:** 738–750.

Menco, B. P. M., and Farbman, A. I., 1985, Genesis of cilia and microvilli of rat nasal epithelia during prenatal development. I. Olfactory epithelium, qualitative studies, *J. Cell Sci.* **78:** 283–310.

Miragall, F., 1983, Evidence for orthogonal arrays of particles in the plasma membranes of olfactory and vomeronasal sensory neurons of vertebrates, *J. Neurocytol.* **12:** 567–576.

Monti Graziadei, G. A., 1983, Experimental studies on the olfactory marker protein. III. The olfactory marker protein in the olfactory neuroepithelium lacking connections with the forebrain. *Brain Res.* **262:** 303–308.

Monti Graziadei, G. A., and Graziadei, P. P. C., 1979, Neurogenesis and neuron regeneration in the olfactory system of mammals. II. Degeneration and reconstitution of the olfactory sensory neurons after axotomy, *J. Neurocytol.* **8:** 197–213.

Monti-Graziadei, G. A., Karlan, M. S., Bernstein, J. J., and Graziadei, P. P. C., 1980, Reinnervation of the olfactory bulb after section of the olfactory nerve in monkey (*Saimiri sciureus*), *Brain Res.* **189:** 343–354.

Morrison, E. E., and Graziadei, P. P. C., 1983, Transplants of olfactory mucosa in the rat brain. I. A light microscopic study of transplant organization, *Brain Res.* **279:** 241–245.

Noble, M., Mallaburn, P. S., and Klein, N., 1984, The growth of olfactory neurons in short-term cultures of rat olfactory epithelium, *Neurosci. Lett.* **45:** 193–198.

Novoselov, V. I., Bragin, A. G., Novikov, J. V., Nesterov, V. I., and Fesenko, E. E., 1984, Transplants of olfactory mucosa in the anterior chamber of the eye: Morphology, electrophysiology and biochemistry, *Dev. Neurosci.* **6:** 317–324.

Oley, N., DeHan, R. S., Tucker, D., Smith, J. C., and Graziadei, P. P. C., 1975, Recovery of structure and function following transection of the primary olfactory nerves in pigeons, *J. Comp. Physiol. Psychol.* **88:** 477–495.

Oppenheim, R. W., Chuwang, I.-W., and Maderdrut, J. L., 1978, Cell death of motoneurons in chick embryo spinal cord. II. The differentiation of motoneurons prior to their induced degeneration following limb bud removal, *J. Comp. Neurol.* **177:** 87–112.

Piatt, J., 1951, An experimental approach to the problem of pallial differentiation, *J. Comp. Neurol.* **94:** 105–121.

Pilar, G., and Landmesser, L., 1976, Ultrastructural differences during embryonic cell death in normal and peripherally deprived ciliary ganglia, *J. Cell. Biol.* **68:** 339–356.

Rafols, J. A., and Getchell, T. V., 1983 Morphological relations between the receptor neurons, sustentacular cells and Schwann cells in the olfactory mucosa of the salamander, *Anat. Rec.* **206:** 87–101.

Samanen, D. W., and Forbes, W. B., 1984. Replication and differentiation of olfactory receptor neurons following axotomy in the adult hamster: A morphometric analysis of postnatal neurogenesis, *J. Comp. Neurol.* **225:** 201–211.

Schubert, D., Stallcup, W., LaCorbiere, M., Kidokoro, Y., and Orgel, L., 1985, Ontogeny of electrically excitable cells in cultured olfactory epithelium, *Proc. Natl. Acad. Sci. USA* **82:** 7782–7786.

Simmons, P. A., and Getchell, T. V., 1981*a*, Neurogenesis in olfactory epithelium: Loss and recovery of transepithelial voltage transients following olfactory nerve section, *J. Neurophysiol.* **45:** 516–528.

Simmons, P. A., and Getchell, T. V., 1981*b*, Physiological activity of newly differentiated olfactory receptor neurons correlated with morphological recovery from olfactory nerve section in the salamander, *J. Neurophysiol.* **45:** 529–549.

Simmons, P. A., Rafols, J. A., and Getchell, T. V., 1981, Ultrastructural changes in olfactory receptor neurons following olfactory nerve section, *J. Comp. Neurol.* **197:** 237–257.

Spemann, H., 1938, *Embryonic Development and Induction,* Hafner, New York, repr. in 1962.

Stout, R. P., and Graziadei, P. P. C., 1980, Influence of the olfactory placode on the development of the brain in Xenopus laevis (Daudin). I. Axonal growth and connections of the transplanted placode, *Neuroscience* **5:** 2175–2186.

Street, S. F., 1937, The differentiation of the nasal area of the chick embryo in grafts, *J. Exp. Zool.* **77:** 49–85.

Venneman, W., Van Nie, C. J., and Tibboel, D., 1982, Developmental abnormalities of the olfactory bulb: A comparative study of the pig and chick embryo, *Teratology* **26:** 65–70.

Yntema, C. L., 1955, Ear and nose, in: *Analysis of Development* (B. H. Willier, P. A. Weiss, and V. Hamburger, eds.), pp. 415–428, W. B. Saunders, Philadelphia.

Zwilling, E., 1934, Induction of the olfactory placode by the forebrain in *Rana pipiens, Proc. Soc. Exp. Biol. Med.* 31: 933–935.

Zwilling, E., 1940, An experimental analysis of the development of the Anuran olfactory organ, *J. Exp. Zool.* **84:** 291–318.

15

Olfactory Tissue Interactions Studied by Intraocular Transplantation

Peter C. Barber and Steen Jensen

1. INTRODUCTION

In normal adult animals, the population of olfactory sensory neurons is in a state of continuous turnover, with mature elements constantly replaced by division and differentiation of stem cells present in the basal layers of the epithelium. If a catastrophic loss of mature neurons occurs, the stem cells are able to increase their rate of division and repopulate the epithelium within a relatively short period. In both cases, axons are produced by the newly formed sensory neurons, and these are capable of growing to reach the central nervous system (CNS) and of reestablishing synaptic contact with central neurons (Barber, 1982; Monti-Graziadei and Graziadei, 1979, 1984, and references cited therein).

Such unusual features raise many questions related not only to the mechanisms controlling the size, turnover rate, and functional stability of the olfactory neuron population, but to the means by which the CNS deals with the unusual problem of a constantly changing population of afferent synapses. In addition, the developmental problems of this neuronal system must be rather different from those of other highly organized but (in terms of cell population) ultimately static central nervous regions. Investigation of the factors that permit continuous axon growth and synaptogenesis in the primary olfactory pathway might shed light on the relative lack of regenerative ability elsewhere in the CNS.

The mechanisms that control the establishment and maintenance of the olfactory neuron population must include some parameters that are determined in the neuroblast at a very early stage, and thus appear later as innate characteristics of the developing neuron, and other parameters (probably many and varied) that result from interactions of the neurons with their environment at particular stages. In addition, the neurons will themselves affect surrounding tissues, producing potential feedback loops that may be important in overall control. Studies of anatomical dynamics in the primary olfactory pathways have thus included many experiments of the "confrontation" type, in which olfactory

---

Peter C. Barber • Department of Morbid Anatomy and Histology, Addenbrooke's Hospital, Cambridge CB2 2QQ, England. *Steen Jensen* • Institute of Anatomy B, University of Aarhus, Aarhus 8000-C, Denmark.

neurons have been isolated wholly or partly from their normal situation and observed when confronted by an unusual, more or less experimentally defined, neural environment. Such studies range from *in vitro* dissociated (Noble *et al.*, 1984; Goldstein and Quinn, 1981) or organotypic (Chuah and Farbman, 1983; Farbman, 1977) cultures to more direct transplantation experiments *in vivo* (Morrison and Graziadei, 1983; Monti-Graziadei and Graziadei, 1984). *In vitro* studies have the advantage of a fairly well-defined medium for growth of the tissues under investigation; thus, any tissue interactions are not masked by unknown effects of the host environment, which may complicate *in vivo* transplantation studies. However, the culture dish is an abnormal environment, and maturation or growth processes may be so distorted as to be useless for experimental manipulation. On the other hand, direct tissue transplantation leaves many more factors uncontrolled. The mechanical effects of differences in surgical technique and tissue trauma may, for instance, introduce a great deal of variability into experimental results.

Intraocular transplantation offers an interesting alternative method of setting up tissue interaction or confrontation experiments. This technique has recently been successfully adapted and developed by Olson and Sieger and colleagues for transplantation of a variety of neural tissues. It is their method that is now generally used (Olson *et al.*, 1981) and that we have followed ourselves. Olfactory mucosa can be transplanted into the anterior chamber of the eye either alone or in combination with other co-transplanted tissue. Interactions between the olfactory tissue and the host eye, as well as with the co-transplant, may be of interest.

Intraocular transplantation is convenient for observation and access to the isolated tissue. The tissue is bathed in the aqueous humor and has space in which to grow; it also receives rapid vascularization and protection from infection. The success rate for producing viable transplants is thus high, and graft survival can be of long duration, if the eye itself is not damaged surgically. When carefully performed, the operation is among the least harmful types of experimental surgery as far as the host is concerned. However, more aspects of the graft environment are uncontrolled, when compared with an *in vitro* culture (including immunological effects on allo- and xenografted tissue), and the range of novel environments to which the transplant can be exposed is less than with direct transplantation. Results obtained by intraocular transplantation must therefore be carefully compared with those obtained by other techniques.

2. ORGANIZATION OF OLFACTORY SENSORY EPITHELIAL GRAFTS IN OCULO

Several studies have contributed to establishing the essential sequence of events when olfactory sensory epithelium taken from immature or adult donors is transplanted to the anterior eye chamber (Barber *et al.*, 1982; Heckroth *et al.*, 1983; Monti-Graziadei and Graziadei, 1984; Novoselov *et al.*, 1984). The graft rapidly attaches to either the iris or the cornea, or both, and becomes highly vascularized. During the same period, the tissue undergoes disorganization and degeneration, with loss of the mature sensory neurons. Subsequently, the graft becomes reorganized and by 8 weeks postoperative consists of a series of vesicles of a range of sizes, partially interconnected, with a varied epithelial lining. In some regions, the epithelium is of simple cuboidal or flattened structure. In

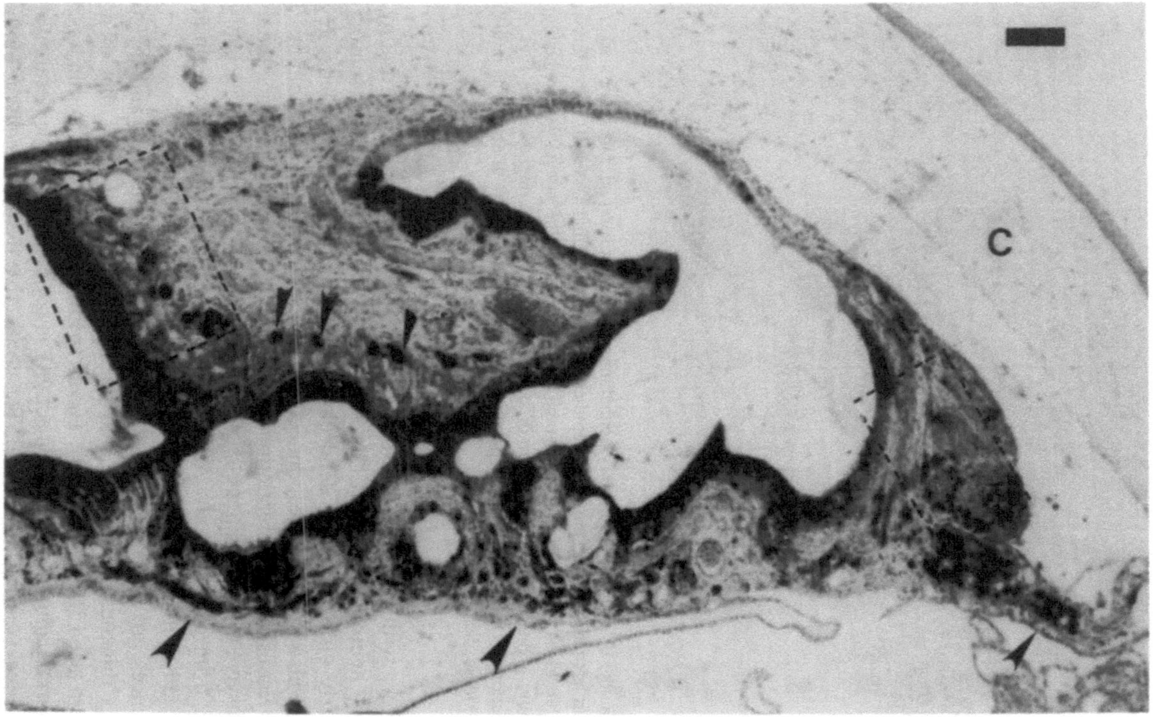

Figure 1. Intraocular autograft of adult rat olfactory epithelium, at 8 weeks survival. The graft is arranged as a series of vesicles of various sizes. The section is stained for olfactory marker protein (OMP) (black). OMP-containing neurons are present in many regions of the epithelium and OMP-positive nerve fascicles are seen in the subepithelial connective tissue and running into the host iris (arrowheads). The epithelium attached directly to the cornea (c) is of simple nonsensory type. Boxes indicate areas shown at higher power in Figs. 5 and 6. Immunoperoxidase. Scale bar: 100 μm.

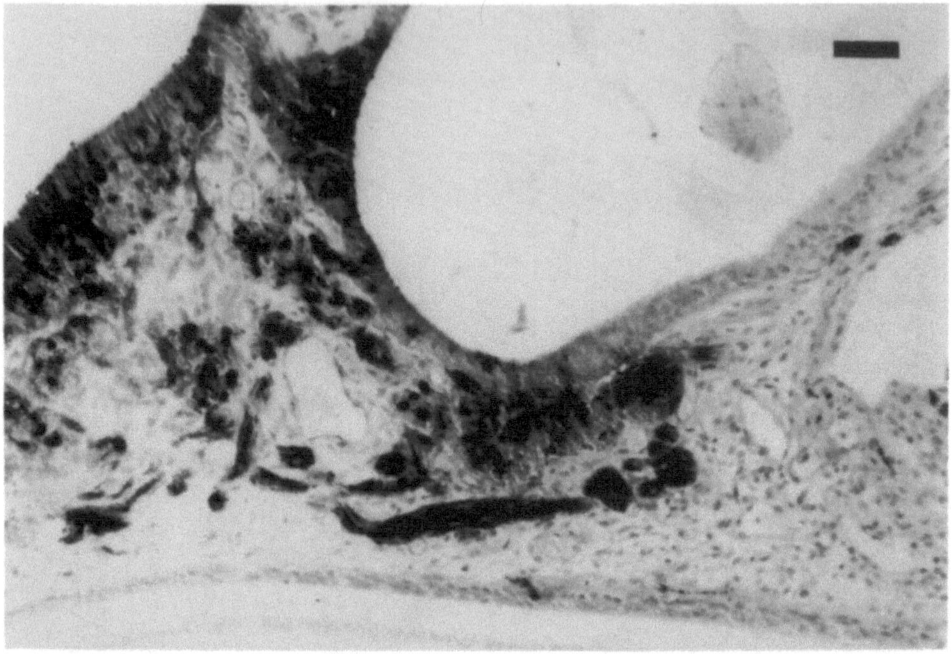

Figure 2. Intraocular autograft of adult rat olfactory epithelium, at 8 weeks survival. Olfactory marker protein (OMP)-containing neurons (dark-staining) are seen in epithelium of variable thickness at 10 weeks survival. The iris attachment is at the lower border of the photograph. In some places the convoluted epithelium is cut obliquely or tangentially, giving the false impression of subepithelial neuroblast migration (cf. Figs. 2 and 3, which are adjacent sections). OMP-positive nerve bundles of various sizes run in all directions in the *lamina propria*. Immunoperoxidase. Scale bar: 50 μm.

others, it is thicker, with multiple layers of nuclei visible by light microscopy, and resembles olfactory neuroepithelium. Cells with the appearance of mature sensory neurons, bearing clearly visible sensory dendrites, can be seen within these latter regions; axon fascicles are also seen in the connective tissue deep to the epithelium. These appearances must represent redifferentiation of neurons from stem cells, since the mature elements do not survive transplantation. The proportion of epithelium with a sensory type of structure varies from graft to graft. All surviving transplants seem to contain at least some areas of it, and in many grafts it is abundant (Figs. 1, 2, and 3).

Some differences are apparent between the graft neuroepithelium and that of the normal animal. Its thickness is variable and does not in general reach that of normal olfactory epithelium (see Figs. 1, 2, and 3). The arrangement of the cells within it is not as regular as in the normal animal, and mitotic activity is not confined to the basal layers but can be seen at any level (see Figs. 4 and 10). The connective tissue beneath the epithelium is also of variable thickness and structure, and the glandular elements in it are scattered and disorganized. Olfactory nerve fascicles of the transplant have a much wider range of sizes than normal; many are small, running in all directions within the connective tissues (see Fig. 2). In addition, the subepithelial connective tissue contains groups of immature neuroblastic elements (Figs. 4 and 5) resembling basal cells that have apparently migrated

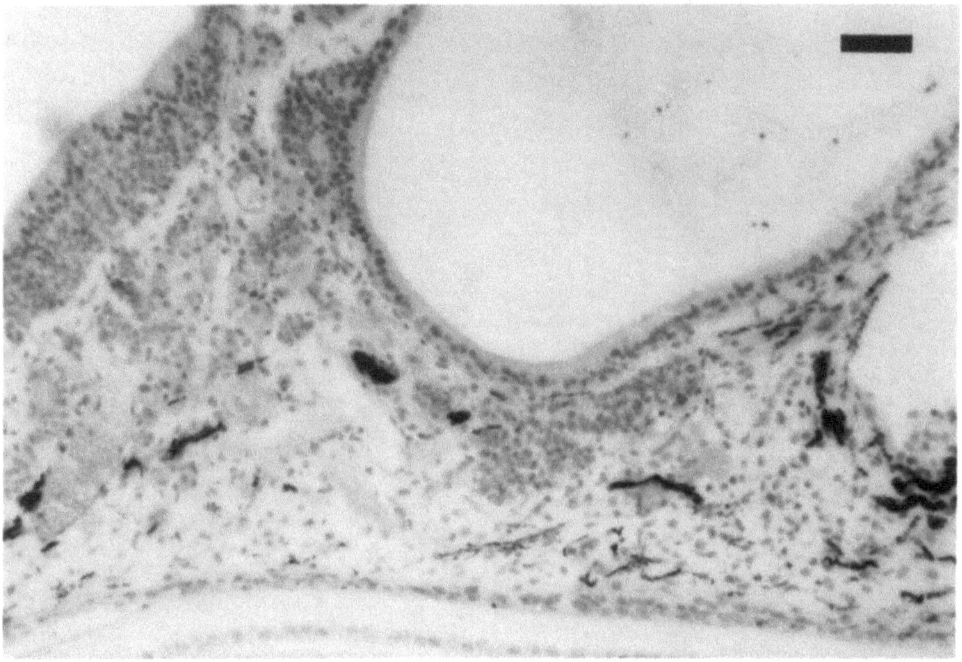

Figure 3. Adjacent section to Fig. 2, stained for neurofilament protein. Nerve fibers of host origin are seen growing into the transplant from the iris (dark staining). Small nerve twigs either approach close to, or grow into, epithelial of both nonsensory (right) and sensory (center) types. On the left, the sensory epithelium has no detectable innervation. Immunoperoxidase. Scale bar: 50 μm.

deep to the epithelial basement membrane (never seen in normal animals), although this is sometimes difficult to distinguish from a tangential section through the convoluted vesicular epithelium by light microscopy (see Figs. 2 and 3). Even at long survival times, the graft gives the impression of a dynamic structure, with vesicles in all stages of development and mitoses visible within the epithelia.

3. INDICATORS OF NEURONAL MATURATION IN OCULO

Newly formed olfactory sensory neurons in the grafted epithelium exhibit other features of maturation as well as the growth of axons and sensory dendrites. For example, large numbers of neurons express the olfactory marker protein (OMP) in most grafts (Barber *et al.*, 1982) (see Figs. 1 and 2). This protein, unique to olfactory sensory neurons, appears only at a relatively advanced stage in their development (Farbman and Margolis, 1980; Miragall and Monti Graziadei, 1982; Monti Graziadei *et al.*, 1977)—although it is still not quite clear exactly *what* stage. In addition, a recent report (Novoselov *et al.*, 1984) suggests that by 8 weeks the new sensory neurons have developed carnosine synthetase activity (Harding and Margolis, 1976; Harding *et al.*, 1977) and can produce receptor potentials in response to odors.

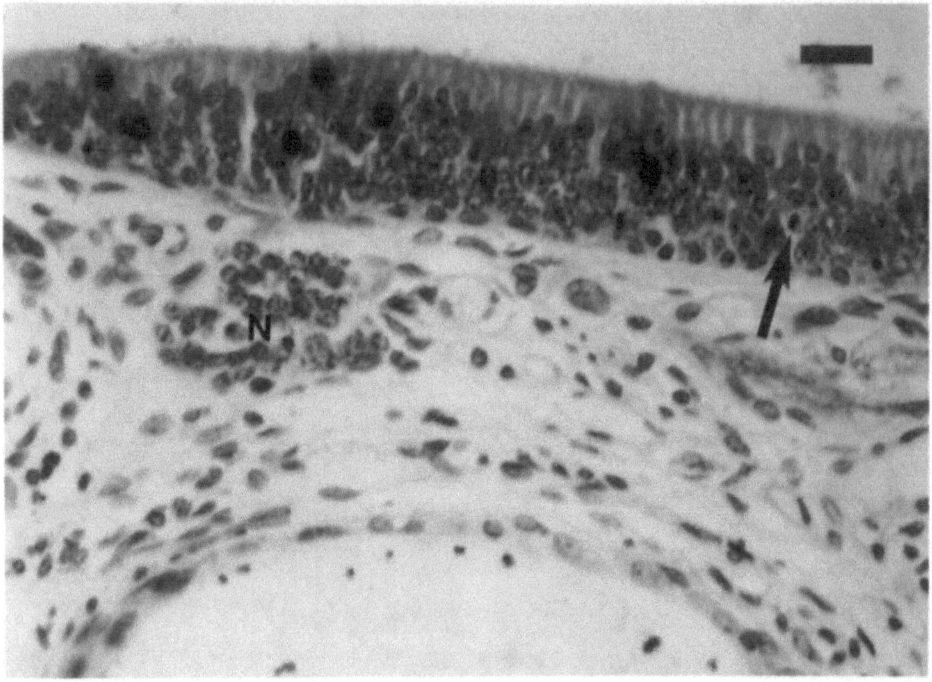

Figure 4. Neuroepithelium from a region of the transplant shown in Fig. 1. The epithelium is slightly thinner than normal and contains only a few olfactory marker protein (OMP)-positive neurons (dark). A mitosis is present in the middle layer of the epithelium (arrow). A group of neuroblasts (N) is present in the sub-epithelial connective tissue. Immunoperoxidase. Scale bar: 20 μm.

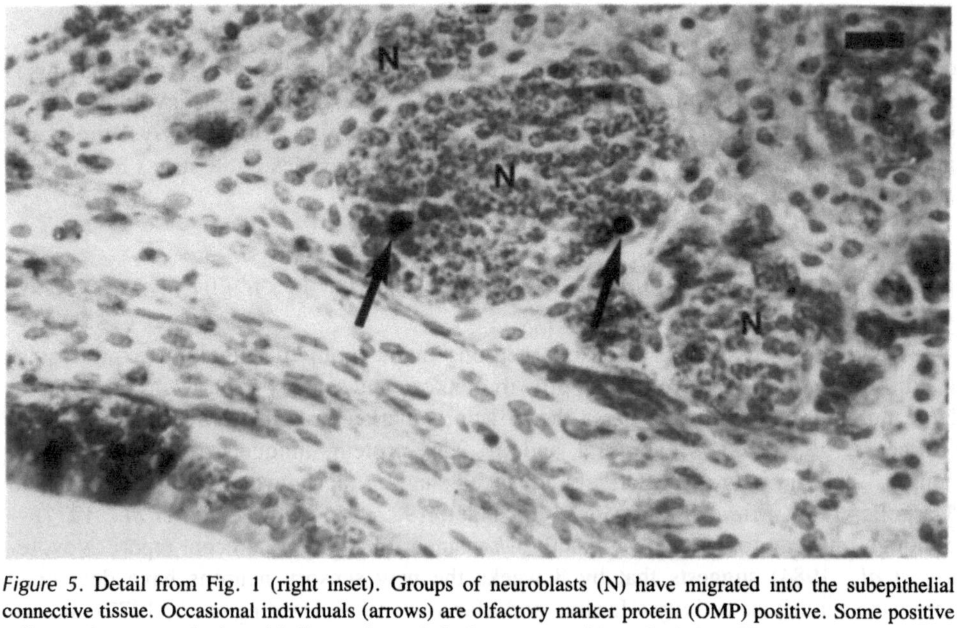

Figure 5. Detail from Fig. 1 (right inset). Groups of neuroblasts (N) have migrated into the subepithelial connective tissue. Occasional individuals (arrows) are olfactory marker protein (OMP) positive. Some positive cells are also present in the epithelium at lower left. Immunoperoxidase. Scale bar: 20 μm.

4. FACTORS AFFECTING NEURONAL MATURATION IN OCULO

Factors influencing neuronal maturation are potentially numerous and not immediately obvious. Taking OMP expression as an indicator of maturation, it can be seen at once by light microscopy-enhanced immunohistochemistry that the development of OMP-positive cells is quite variable, not to say completely haphazard, both from graft to graft and within a single graft. In some regions, a well-organized epithelium of neurosensory appearance contains relatively few cells that stain for OMP (Fig. 4). In other regions, a single dendrite-bearing OMP-positive cell may appear in the middle of an area of cuboidal ciliated epithelium containing no other neural elements (Fig. 7). Where OMP-positive cells are seen in large numbers, they tend to occur not only in the middle layers of the epithelium but also in the basal and superficial layers (see Fig. 6). These features contrast sharply with the situation in the normal animal, where OMP expression is confined to mature neurons in the middle layers of the neuroepithelium (Monti-Graziadei *et al.*, 1977). Much of the variability in number of newly formed neurons from region to region of a graft can probably be accounted for in terms of such general factors as differential speed of vascularization, rate of removal of degenerate debris, local inflammatory reac-

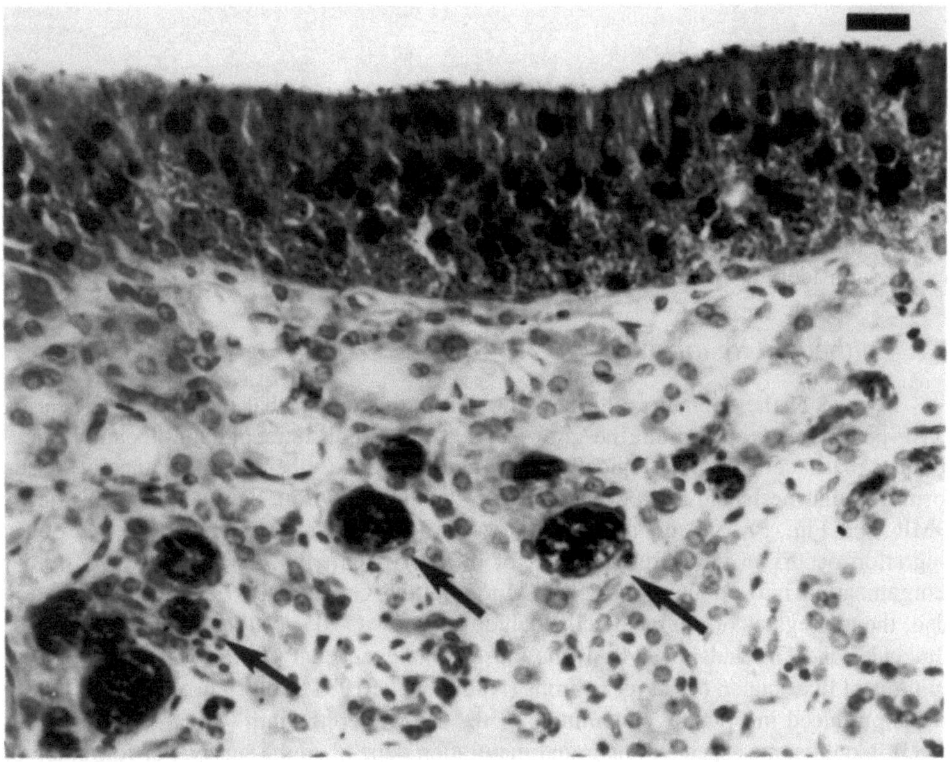

Figure 6. Detail from Fig. 1 (left inset). Neurosensory-type epithelium in the graft contains olfactory marker protein (OMP)-positive neurons (dark) in the basal and superficial regions as well as in the middle layers. OMP-positive nerve fascicles are seen in the lamina propria (arrows). Immunoperoxidase. Scale bar: 20 μm.

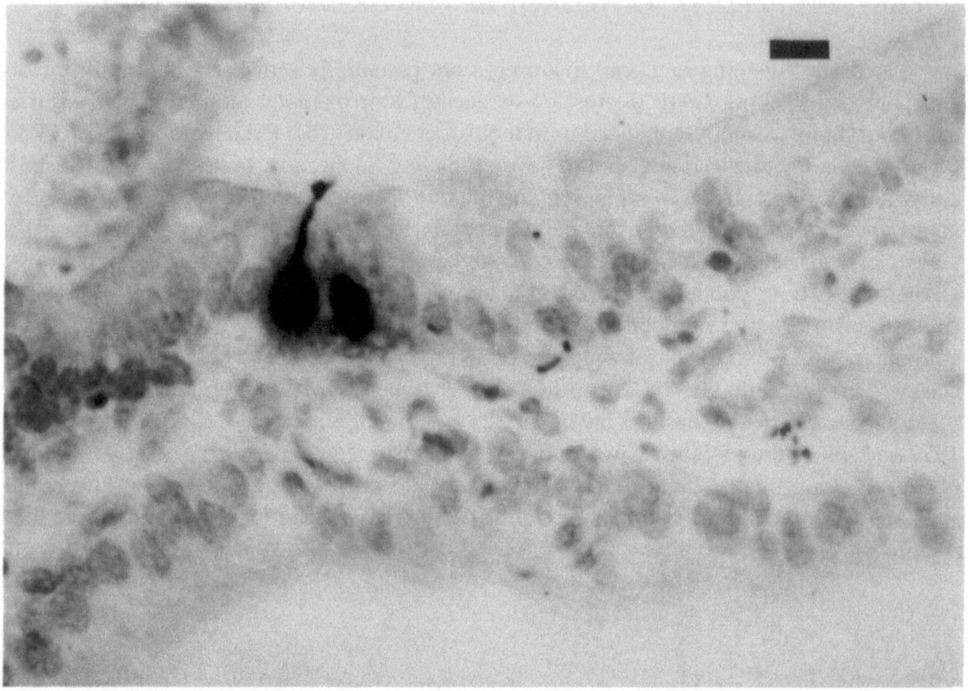

Figure 7. Two isolated olfactory marker protein (OMP)-positive neurons appear in an otherwise simple nonsensory epithelium. Immunoperoxidase. Scale bar: 10 μm.

tions to infecting organisms, etc. However, the possibility of more specific influences may be considered.

4.1. Epithelial Organization

In spite of the anomalies in maturation, it is fair to say that most dendrite-bearing OMP-positive neurons within the graft develop only in an intraepithelial location. Neuroblasts (basal cells, morphologically) that migrate into the subepithelial connective tissue form irregular nests and aggregates. Only scattered cells within these aggregates express OMP (see Fig. 5), and no dendrites are seen. A similar phenomenon of neuroblast migration occurs in intracerebral grafts of olfactory epithelium, which undergo similar reorganization into epithelial vesicles (Morrison and Graziadei, 1983); in this situation also, the aggregates of migrated cells include only few OMP-positive individuals (Monti-Graziadei and Graziadei, 1984). It may also be relevant that OMP does not develop in cultures of dissociated olfactory neurons (Noble *et al.*, 1984; Goldstein and Quinn, 1981) but is produced in organotypic cultures of the sensory epithelium (Chuah and Farbman, 1983). Could some aspects of neuronal maturation depend on the survival or formation of an epithelial architecture, or perhaps on the presence of a specific cell type, such as epithelial supporting cells near or in contact with the developing neurons?

It is apparent that development of neurosensory-type epithelium is consistently associated with the presence of loose subepithelial connective tissue reminiscent of that seen beneath the normal neuroepithelium. Where this connective tissue is present, the graft epithelium usually contains at least some neurons. Where it is absent, a simple cuboidal or flattened epithelium is found with no neurosensory component (see Figs. 1, 8, and 9). Developing neuroblasts in the transplant might be expected to make contact with other structures (e.g., cornea, iris, co-transplants) fairly frequently and have the opportunity to form epithelial vesicles directly on such substrates. However, this does not seem to occur. We may speculate that subepithelial connective tissue exerts some effect (nutritive? inductive?) that permits epithelial organization and development of mature sensory neurons and that other substrates do not. However, there is no good evidence either way, and the observation is so far unexplained.

4.2. Neurotrophic Influence of the Iris

Sensory neuroepithelium often develops better on that side of the graft vesicle adjacent to the iris than on its free borders or on those attached to the cornea (see Fig. 1). It is probable that this phenomenon is associated with good prompt vascularization from iris capillaries, but other factors may be operative as well. The iris is known to produce nerve growth factor (NGF) and a factor promoting survival and neurite outgrowth of cholinergic neurons in culture (Ebendal *et al.*, 1980; Adler *et al.*, 1979). Neurotrophic activity is enhanced by denervation of the iris (Ebendal *et al.*, 1980; Olson *et al.*, 1978), expecially sensory (trigeminal) denervation, and it would perhaps be instructive to examine growth of grafted olfactory epithelium in an eye which has been similarly denervated. However, attempts to co-transplant olfactory epithelium with iris tissue to other sites in the brain and to co-culture olfactory epithelium with iris have so far been unsuccessful in revealing any enhancement of growth or maturation of olfactory neurons (P. C. Barber, S. Jensen, and J. Zimmer, unpublished results).

4.3. Innervation of the Graft from the Iris

Successful attachment and survival of the intraocular graft are followed by an ingrowth of autonomic and sensory nerve fibres from the host iris. Any or all of these might exert some effect on the graft, but ingrowth of sensory fibers of trigeminal origin (Huhtala, 1976) might be of particular interest, since the trigeminal innervation has been suggested many years ago to have some sort of trophic effect on the nasal mucosa (Takata, 1929), and peripheral nerve tissue (both sensory and motor) has been shown to be an abundant source of at least one type of neurotrophic activity (Williams *et al.*, 1984).

Ingrowth of nerve fibers from the iris may be conveniently visualized at the light microscopic level by immunohistochemistry for neurofilament protein (NFP) which preferentially stains fibers of trigeminal origin (Seiger *et al.*, 1984) and is absent from olfactory sensory axons (Schwob *et al.*, 1986). The distribution of NFP-positive fibers in the graft can then be compared with the regional development of OMP-positive neurons in the sensory epithelium (Barber and Jensen, in preparation) in adjacent sections.

It is apparent that NFP-positive fibers, while never very dense, do grow into the graft from the iris and in some regions approach the base of the sensory epithelium closely, even growing into the epithelium (Fig. 3). A loose correspondence is observed between ingrowing nerve fibers and OMP-positive areas of graft epithelium, but in some areas with well-developed OMP-positive neurons there are no visible NFP-positive fibers nearby, and NFP-positive fibers also innervate areas of nonsensory epithelium (see Figs. 2 and 3). It would appear that close proximity of ingrowing fibers from the iris is not necessary for OMP development in the graft. A similar pattern is seen in olfactory mucosa transplanted into the sciatic nerve (Barber, in preparation). Here, also, close proximity of large numbers of host nerve fibers to the base of the transplanted epithelium does not appear to enhance differentiation of olfactory sensory neurons, and overall the development of such transplants is inferior to intraocular grafts.

4.4. Presence of Co-transplanted Central Nervous Tissue

A co-transplant of homografted embryonic CNS (hippocampus, cerebellum, cerebral cortex, olfactory bulb) attaches firmly to an intraocular graft of adult or immature olfactory epithelium, if placed within close proximity to it. If both components survive, a single complex tissue mass is formed. According to one report (Monti-Graziadei and Graziadei, 1984), the vesicular organization of the regenerating olfactory epithelium in such a case is more complex than in the absence of CNS tissue. This is also our own impression (Barber

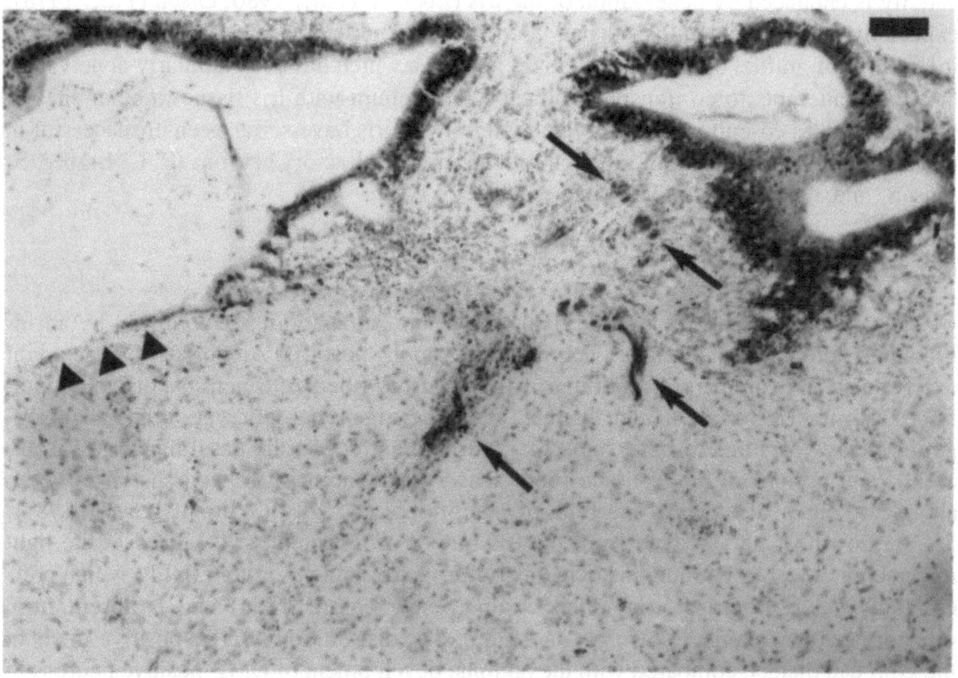

Figure 8. Combined intraocular autograft of adult olfactory epithelium and co-transplanted embryonic CNS at 10 weeks survival. Epithelial vesicles of the olfactory tissue contain olfactory marker protein (OMP)-positive sensory neurons (dark staining). OMP-positive nerve bundles run through the subjacent connective tissue and some penetrate the co-transplant (arrows). No glomerular formations are seen. Epithelium attached directly to the CNS tissue at the left of the figure (arrowheads) is of simple, nonsensory type. Immunoperoxidase. Scale bar: 100 μm.

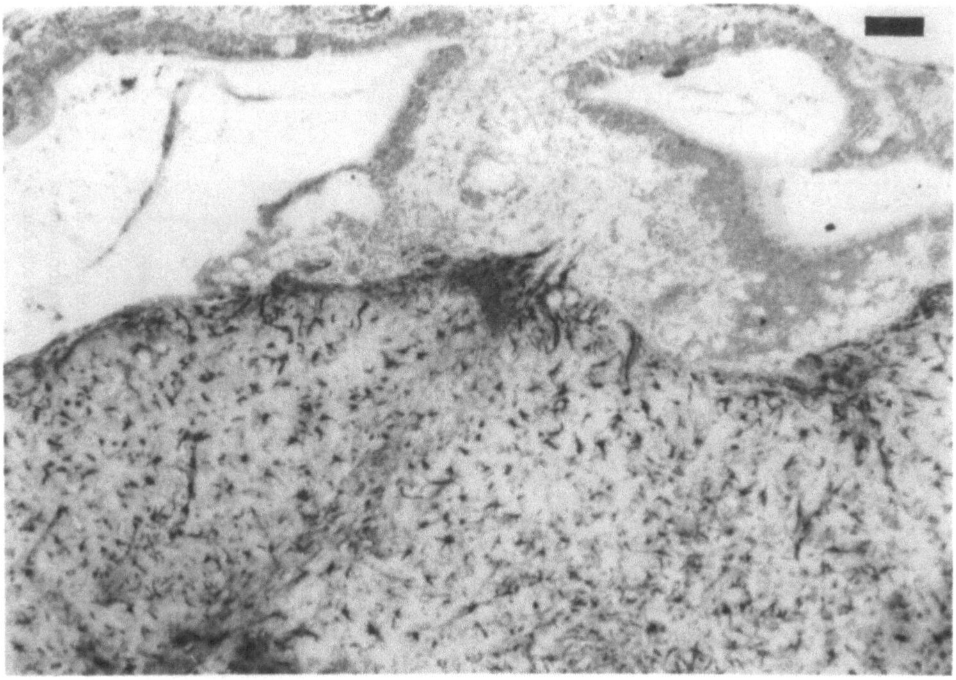

Figure 9. Adjacent section to Fig. 8, stained for glial fibrillary acidic protein (GFAP). Hypertrophied astrocytes are present throughout the CNS co-transplant and are concentrated at the interface with olfactory tissue. No GFAP-containing cells are seen in the olfactory tissue itself. Immunoperoxidase. Scale bar: 100 μm.

and Jensen, in preparation), although no systematic study has been performed. The proportion of epithelium of sensory type and the number of OMP-positive cells that develop also appears to be greater in the presence of co-transplanted CNS, supporting previous observations in organotypic cultures (Chuah and Farbman, 1983). However, we are uncertain as to whether olfactory bulb tissue, the normal target of the sensory neurons, has any greater effect than other CNS regions, as has been shown in cultures (Chuah and Farbman, 1983).

Direct contact between the CNS and olfactory epithelium does not seem to induce development of mature OMP-positive sensory neurons preferentially. Rather, the reverse is the case. As with epithelium transplanted alone, sensory neurons are seen to develop in areas in which there is also a lamina propria of vascular connective tissue beneath the epithelium. Where epithelium contacts other structures directly, including the co-transplanted CNS, it is of simple cuboidal or flattened type (Figs. 8, 9) (see also Section 4.1). However, migration of astrocytes out of CNS co-transplants into the subepithelial connective tissue of olfactory grafts does occur and does not appear to suppress development of sensory epithelium, even when these astrocytes reach high densities (Figs. 10 and 11). It seems to be the presence or absence of the lamina propria that is important rather than the presence or absence of adjacent or even intimately related CNS tissue. Olfactory neuroblasts that migrate into the co-transplanted CNS tissue (Monti-Graziadei and Graziadei, 1984) away from their lamina propria also do not appear to undergo differentiation.

Similar phenomena occur in intracerebrally grafted olfactory epithelium. Olfactory

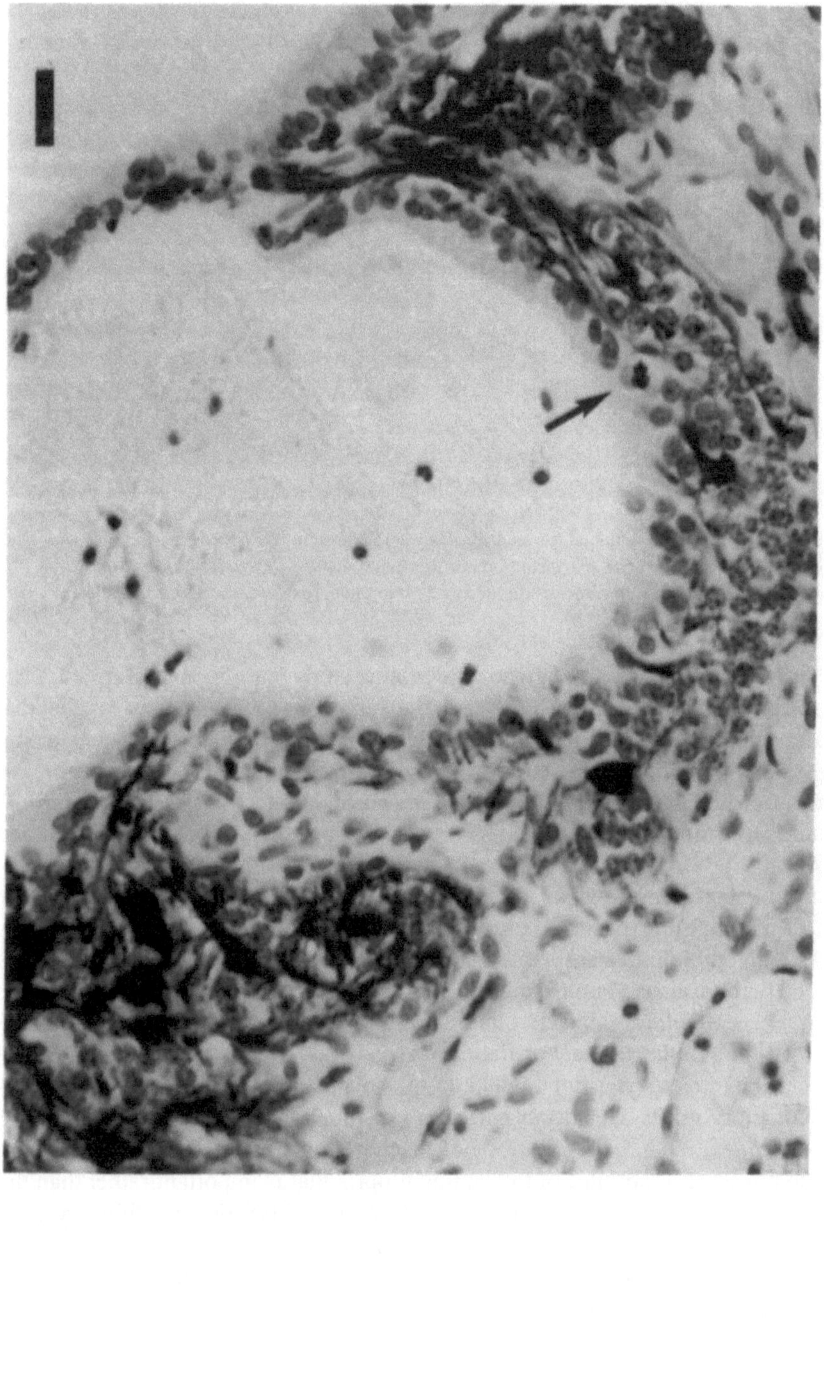

Figure 10. Olfactory neuroepithelium in a graft vesicle invaded by astrocytes (dark-staining) from adjacent co-transplanted CNS, which is farther behind the plane of this section. The epithelium is cut tangentially at the left and right of the photograph, due to formation of small subsidiary outpouchings of the main vesicle. A cell in mitosis is indicated by the arrow. GFAP immunoperoxidase. Scale bar: 20 μm.

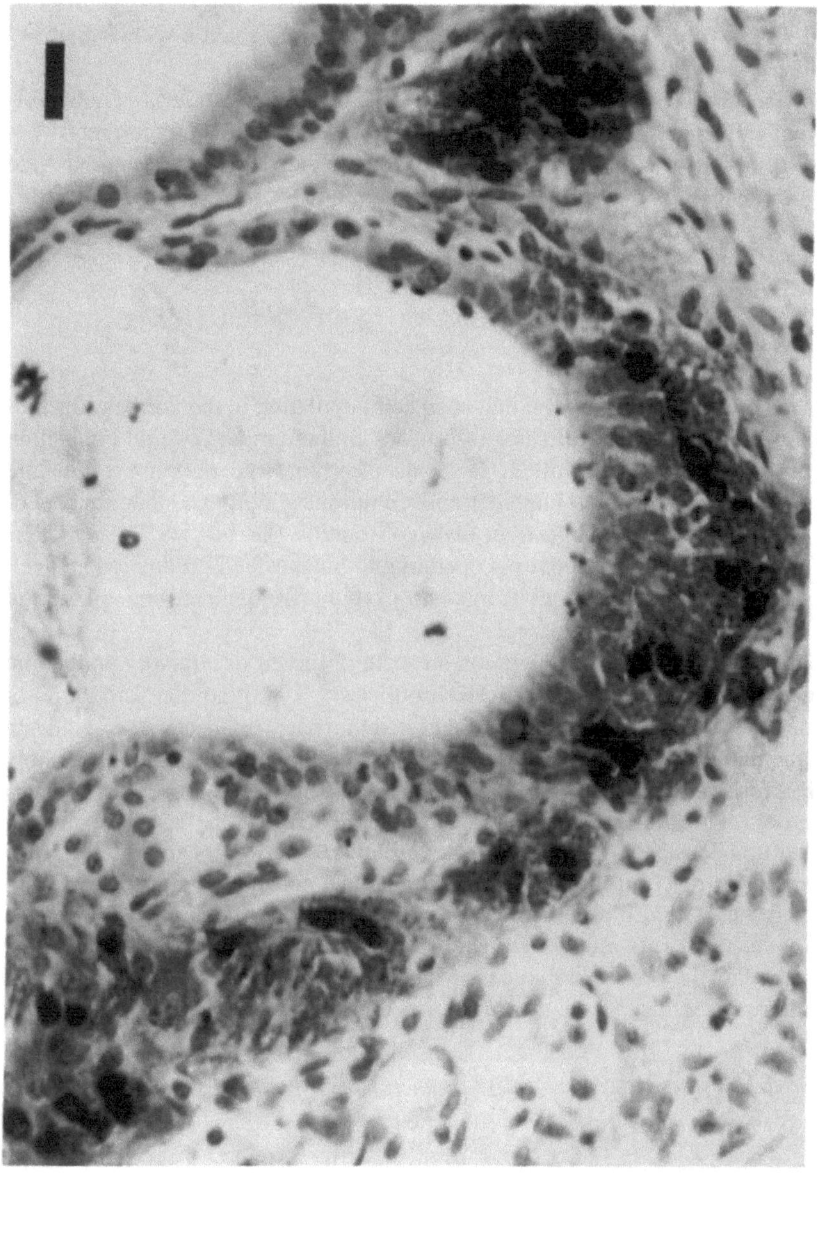

Figure 11. Adjacent section to Fig. 10, stained for olfactory marker protein (OMP). OMP-positive neurons (dark) are scattered throughout the epithelium in both astrocyte-containing and astrocyte-free areas. Tangential cuts gave the false impression of isolated aggregates of neurons in some regions (cf. Fig. 10). Immunoperoxidase. Scale bar: 20 μm.

epithelium transplanted intracerebrally or into the olfactory bulb grows rather poorly compared to grafts *in oculo*. The epithelial vesicles that form are smaller, less complex and contain fewer OMP-positive cells compared with intraocular grafts. However, as with intraocular grafts, absence of a lamina propria between olfactory and CNS tissue is associated with a simple nonsensory epithelium. We have seen this in our own material, and it is well shown in published reports (e.g., Fig. 1 of Morrison and Graziadei, 1983). Migrating neuroblasts from the olfactory tissue also seem not to mature within the CNS (Monti-Graziadei and Graziadei, 1984).

It thus appears that while co-transplanted or co-cultured CNS exerts a rather ill-defined inductive influence on the regenerating olfactory epithelium, direct contact between olfactory neuroblasts and CNS in the absence of connective tissue does not provide a suitable environment for formation of neuroepithelium nor for olfactory neuron maturation.

5. FACTORS AFFECTING OVERALL GROWTH OF OLFACTORY EPITHELIUM IN OCULO

The presence of a continuously dividing stem cell population in the normal olfactory epithelium raises questions of how the area of sensory epithelium and overall population size of olfactory neurons are controlled. It seems clear that the presence of mature elements in the epithelium somehow inhibits stem cell mitotic activity and that removal of those mature elements releases active stem cell proliferation (Barber, 1982, and references cited therein). The details of controls operating to balance cell production and cell death, preventing olfactory epithelium from dying out or from increasing in area to overgrow the nasal cavity, are less clear.

Mitotic activity and cellular degeneration are both observed in olfactory epithelium *in oculo* at long survival times (6 months) (Heckroth *et al.*, 1983), so that some form of turnover is apparently occurring. It has been suggested that early or accelerated cell death occurs among the mature neuronal elements compared with the normal situation, possibly as a result of lack of suitable target tissues. However, it is observed that some transplants show expansion of their epithelial vesicles to a very large size, filling the entire anterior chamber (Barber and Jensen, in preparation; Novoselov *et al.*, 1984), and the final size such a structure might reach in the absence of mechanical constraints is unclear. Quantitative studies are necessary to establish the growth of the neural population *in oculo* as well as the turnover rate of its individual members. However, it is entirely possible that the grafting procedure has removed some constraints on epithelial growth, so that it behaves in some respects like a benign neoplasm within the eye.

6. EFFECTS OF OLFACTORY EPITHELIUM IN OCULO ON CO-TRANSPLANTED CNS

6.1. Maturation of CNS

Immature CNS tissue transplanted to the eye or to another CNS region retains the capacity to mature normally to a surprising degree (Das *et al.*, 1980; Hallas *et al.*, 1980).

However, neural connectivity will obviously be abnormal (Oblinger *et al.*, 1980), as may those aspects of neural morphology that are dependent on extrinsic connections (Olson *et al.*, 1979). Glial development may also be abnormal in isolated pieces of immature CNS, and severe astrocytic gliosis may occur. This is well described in intraocular CNS transplants (Hedlund *et al.*, 1984; Bjorklund and Dahl, 1982), and it has been reported that the abnormal gliosis in these transplants is reduced by co-transplantation of a second adjacent piece of CNS tissue (Bjorklund and Dahl, 1984). However, in our own experiments, we find that astrocytic gliosis in CNS tissue co-transplanted *in oculo* with olfactory epithelium is not obviously affected by the presence of the epithelium. Many reactive intensely GFAP-positive astrocytes are seen (Fig. 9), which tend to invade the iris (as described in Hedlund *et al.*, 1984) and the lamina propria of the adjacent olfactory epithelium (Fig. 10). Neuronal maturation in the CNS tissue also appears unaffected by the adjacent sensory epithelium, at least by light microscopic examination.

6.2. Innervation of Co-transplant by Olfactory Axons

Newly formed olfactory sensory neurons in an intraocular graft form plentiful axons that run for comparatively long distances in the host iris without apparently contacting the central nervous structures of the retina (Barber *et al.*, 1982). We observe that, in the presence of a CNS co-transplant, large OMP-positive fiber bundles often grow over the surface of the co-transplant (Figs. 12 and 13), and sometimes into it—especially along lines of discontinuity produced by scarring (see Fig. 8). The OMP-positive fiber bundles may be irregular in shape and may form aggregates that indent the CNS tissue and appear reminiscent of glomeruli, although these are often not well defined (Fig. 13).

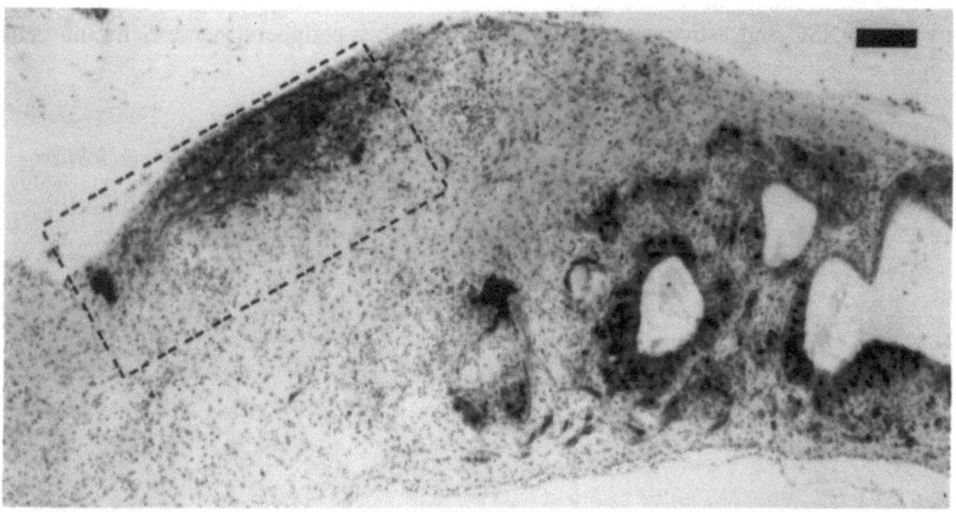

Figure 12. Vesicles of an olfactory epithelium autograft (right) fused to co-transplanted CNS (left). Olfactory marker protein (OMP)-containing sensory neurons are present in the olfactory tissue, with OMP-positive nerve bundles (dark staining). In addition a mass of OMP-positive fibers runs through the CNS tissue in its superficial layers (area within box). See also Figs. 13 and 14. Immunoperoxidase. Scale bar: 100 μm.

Do these OMP-positive aggregates represent olfactory innervation of the co-transplant? Published reports have shown that axons from olfactory neurons that have regenerated *in situ* after bulbectomy will grow into, and form synapses with, fragments of CNS tissue transplanted into the bulbectomy cavity next to the cribriform plate plate (Graziadei and Kaplan, 1980). In such cases, recognizable glomeruli are formed. By contrast, it is also reported that olfactory epithelium transplanted to an abnormal intracerebral or intraventricular site does not produce recognizable glomeruli, although OMP-positive axons may penetrate the host brain (Morrison and Graziadei, 1983). A similar result has been briefly mentioned for intraocular grafts of olfactory epithelium and co-transplanted CNS (Monti-Graziadei and Graziadei, 1984).

Our own observations at the light microscopic level suggest that aggregates of OMP-positive fibers associated with co-transplanted CNS *in oculo* are largely neuromatous. Immunohistochemical analysis of neurofilament protein demonstrates no significant growth of neuronal processes from the CNS tissue into these OMP-positive structures (Figs. 13 and 14). However, it is still possible that synaptic contacts are made along the CNS–neuroma interface, and these would only be seen by electron microscopy. We have also examined olfactory epithelium grafted intracerebrally into adult and immature hosts and find limited outgrowth of OMP-positive fibers into the host brain with no glomerular formations, in broad agreement with Morrison and Graziadei (1983).

The discrepancy between the obvious glomerular formation by axons of olfactory neurons formed *in situ* that innervate fragments of CNS tissue and the lack of such a process in neurons grafted to an abnormal site is puzzling. However, the phenomenon may be related to poor survival of olfactory nerve Schwann cells (ONSC) after transplantation. These cells are GFAP positive (Barber and Lindsay, 1982) and of a distinctive morphological type. However, they are difficult to find in olfactory epithelium transplanted intraocularly or intracerebrally, using antibodies to GFAP (Fig. 9) that are selective for ONSC and astrocytes but do not detect other peripheral nerve Schwann cells

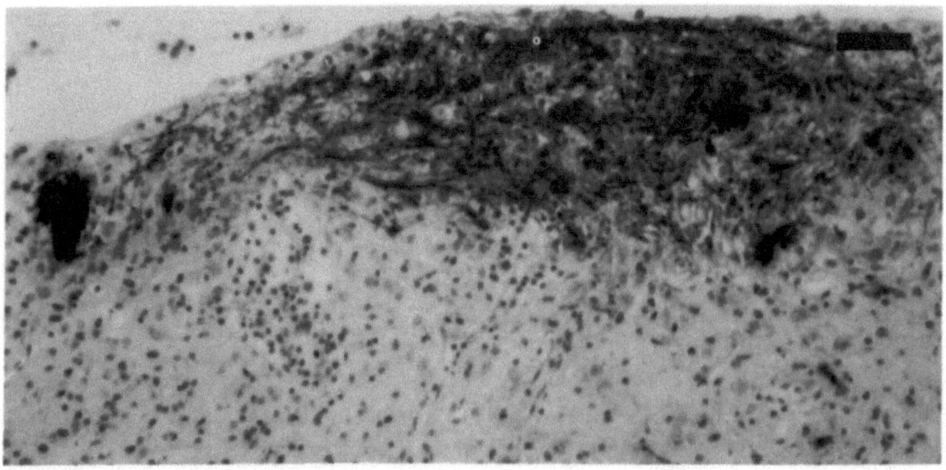

Figure 13. Detail from Fig. 12, area within box. The mass of olfactory marker protein (OMP)-positive fibres forms aggregates reminiscent of glomeruli. Immunoperoxidase. Scale bar: 50 μm.

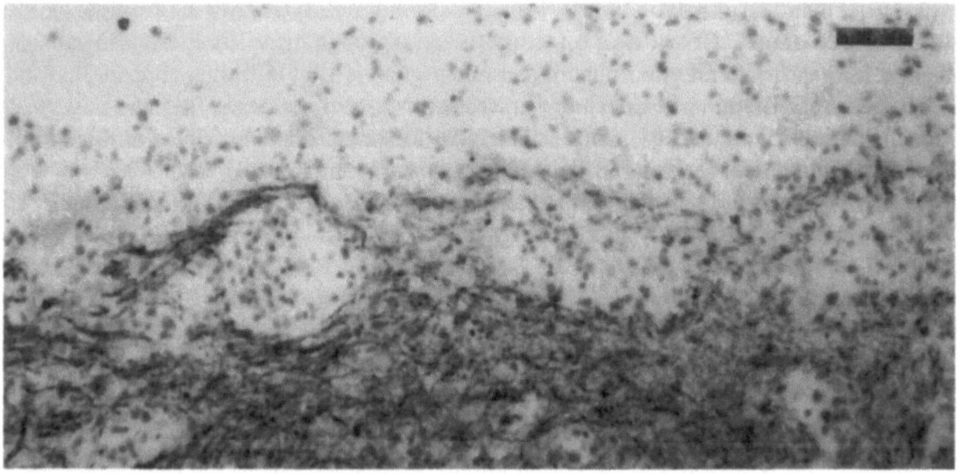

Figure 14. Adjacent section to Fig. 13, stained for neurofilament protein (NFP). A plexus of NFP-positive processes is seen in the CNS tissue (dark staining). These do not extend into the mass of olfactory marker protein (OMP)-positive structures in the superficial layers, i.e., in contrast to the normal olfactory bulb, where NFP-positive dendrites are seen extending into glomeruli. Immunoperoxidase. Scale bar: 50 μm.

(Barber and Dahl, 1987). It may well be that successful innervation of CNS tissue by olfactory axons is dependent on the presence of ONSC, which would be abundant around a CNS graft placed in a bulbectomy cavity, but may not survive transplantation in sufficient numbers. Current studies are directed toward investigation of this possibility and the enhancement of ONSC survival in grafted tissue.

Outgrowth of olfactory fibers occurs in all directions and does not seem to be preferentially directed toward a piece of co-transplanted CNS, indicating a lack of distant attraction of deafferented CNS tissue for the growing axons. A similar lack of attractant activity is seen when olfactory axons are transected *in situ;* for example, denervation of the accessory olfactory bulb by section of the vomeronasal nerves does not stimulate ingrowth of sensory fibers from the adjacent main bulb, where a continuous supply of growing axons is presumably available (Barber, 1981). In addition, those regenerating vomeronasal axons that do succeed in growing back into the olfactory bulb seem to do so at random; many of them take aberrant pathways into adjacent connective tissue and are not preferentially directed toward the bulb. The results of intraocular transplantation are thus consistent with findings *in situ* in failing to demonstrate a neurotropic effect of CNS tissue on olfactory axons.

6.3. Innervation of Olfactory Tissue from the CNS

Some axons originating from neurons in the co-transplanted CNS might be expected to grow out from it into the iris (Olson *et al.*, 1978) and possibly into the adjacent olfactory tissue. As far as can be seen from immunohistochemical staining for neurofilament protein, such growth is limited, although it has been shown that CNS axons, including those of olfactory bulb neurons, are capable of growing into peripheral nerve tissue (Friedman and Aguayo, 1985). We have also not observed growth of retinal axons toward or into intraocular grafts of olfactory tissue, although these fibers are also capable of growth in a peripheral environment (So and Aguayo, 1985). It thus appears that the grafted olfactory tissue itself does not attract ingrowth of other nerve fibers to any great extent, although our staining method is selective for larger-diameter fibers and may not detect fine unmyelinated axons (e.g., adrenergic type).

7. CONCLUSIONS AND DIRECTIONS

Explanted olfactory epithelium *in oculo* is a convenient preparation for anatomical, biochemical, and physiological studies. Undoubtedly the use of this technique will expand. The main conclusions to emerge from work performed to date imply that olfactory neurons can develop in these explants to a state of functional and structural maturity but that control of turnover and overall population size may be abnormal. The transplanted neurons also seem reluctant to innervate co-transplanted CNS tissue when compared to their counterparts *in situ,* although development of the olfactory neurons seems to be enhanced by the presence of a co-transplant. Areas that we believe deserve exploration include all of the above, in particular those properties of the olfactory neuron that permit it to reinnervate CNS tissue.

ACKNOWLEDGMENTS. This work was partly supported by grants from the Wellcome Foundation, East Anglian Regional Health Authority, and Danish Medical Research Council. The authors wish to thank Dr. J. Zimmer for help and encouragement at all stages of this work. Skilled technical assistance was provided by the staff of Anatomy Institute, B, Aarhus, and by Mrs. S. Gower, in Cambridge. Mrs. M. Wright contributed expert secretarial help.

REFERENCES

Adler, R., Landa, K. B., Manthorpe, M., and Varon, S., 1979, Cholinergic neuronotrophic factors: Intraocular distribution of trophic activity for ciliary neurons, *Science* **204:** 1434–1436.

Barber, P. C., 1981, Regeneration of vomeronasal nerves into the main olfactory bulb in the mouse, *Brain Res.* **216:** 239–251.

Barber, P. C., 1982, Neurogenesis and regeneration in the primary olfactory pathway of mammals, *Bibl. Anat.* **23:** 12–25.

Barber, P. C., and Dahl, D., 1987, Glial fibrillary acidic protein (GFAP)-like immunoreactivity in normal and transected rat olfactory nerve, *Experimental Brain Research,* **65:** 681–685.

Barber, P. C., and Lindsay, R. M., 1982, Schwann cells of the olfactory nerves contain glial fibrillary acidic protein and resemble astrocytes, *Neuroscience* **7:** 3077–3090.

Barber, P. C., Jensen, S., and Zimmer, J., 1982, Differentiation of neurons containing olfactory marker protein in adult rat olfactory epithelium transplanted to the anterior chamber of the eye, *Neuroscience* **7:** 2687–2695.

Björklund, H., and Dahl, D., 1982, Glial disturbances in isolated neocortex: Evidence from immunohistochemistry of intraocular grafts, *Dev. Neurosci.* **5:** 424–435.

Chuah, M. I., and Farbman, A. I., 1983, Olfactory bulb increases marker protein in olfactory receptor cells, *J. Neurosci.* **3:** 2197–2205.

Das, G. D., Hallas, B. H., and Das, K. G., 1980, Transplantation of brain tissue in the brain of rat. I. Growth characteristics of neocortical transplants from embryos of different ages, *Am. J. Anat.* **158:** 135–145.

Ebendal, T., Olson, L., Seiger, Å., and Hedlund, K. O., 1980, Nerve growth factors in the rat, iris, *Nature (London)* **286:** 25–28.

Farbman, A. I., 1977, Differentiation of olfactory receptor cells in organ culture, *Anat. Rec.* **189:** 187–200.

Farbman, A. I., and Margolis, F. L., 1980, Olfactory marker protein during ontogeny: Immunohistochemical localisation, *Dev. Biol.* **74:** 205–215.

Friedman, B., and Aguayo, A. J., 1985, Injured neurons in the olfactory bulb of the adult rat grow axons along grafts of peripheral nerve, *J. Neurosci.* **5:** 1616–1625.

Goldstein, N. I., and Quinn, M. R., 1981, A Novel cell line isolated from the murine olfactory mucosa, *In vitro* **17:** 593–598.

Graziadei, P. P. C., and Kaplan, M. S., 1980, Regrowth of olfactory sensory axons into transplanted neural tissue. 1. Development of connections with the occipital cortex, *Brain Res.* **201:** 39–44.

Hallas, B. H., Das, G. D., and Das, K. G., 1980, Transplantation of brain tissue in the brain of rat. II. Growth characteristics of neocortical transplants in hosts of different ages, *Am. J. Anat.* **158:** 147–159.

Harding, J., and Margolis, F. L., 1976, Denervation in the primary olfactory pathway of mice. III. Effect on enzymes of carnosine metabolism, *Brain Res.* **110:** 351–360.

Harding, J., Graziadei, P. P. C., Monti-Graziadei, G. A., and Margolis, F. L., 1977, Denervation in the primary olfactory pathway of mice. IV. Biochemical and morphological evidence for neuronal replacement following nerve section. *Brain Res.* **132:** 11–28.

Heckroth, J. A., Monti-Graziadei, G. A., and Graziadei, P. P. C., 1983, Intraocular transplants of olfactory neuroepithelium in rat, *Int. J. Dev. Neurosci.* **1:** 273–287.

Hedlund, K. O., Dahl, D., Björklund, H., and Seiger, Å., 1984, Ultrastructural and histochemical evidence for differentiation of intraocular locus coeruleus grafts and invasion of the host iris by central neurites and glia, *J. Neurocytol.* **13:** 989–1011.

Huhtala, A., 1976, Origin of myelinated nerves in the rat iris, *Exp. Eye Res.* **22:** 259–265.

Miragall, F., and Monti Graziadei, G. A., 1982, Experimental studies on the olfactory marker protein. II. Appearance of the olfactory marker protein during differentiation of the olfactory sensory neurons of mouse: An immunohistochemical and autoradiographic study, *Brain Res.* **239:** 245–250.

Monti-Graziadei, G. A., and Graziadei, P. P. C., 1979, Studies on neuronal plasticity and regeneration in the olfactory system: Morphologic and functional characteristics of the olfactory sensory neuron, in: *Neural Growth and Differentiation* (E. Meisami and M. A. B. Brazier, eds.), pp. 373–396, Raven, New York.

Monti-Graziadei, G. A., and Graziadei, P. P. C., 1984, The olfactory organ: Neural transplantation, in: *Neural Transplants* (J. R. Sladek, Jr., and D. M. Gash, eds.), pp. 167–186, Plenum, New York.

Monti-Graziadei, G. A., Margolis, F. L., Harding, J. W., and Graziadei, P. P. C., 1977, Immunocytochemistry of the olfactory marker protein, *J. Histochem. Cytochem.* **25:** 1311–1316.

Morrison, E. E., and Graziadei, P. P. C., 1983, Transplants of olfactory mucosa in the rat brain. 1. A light microscopic study of transplant organisation, *Brain Res.* **279:** 241–245.

Noble, M., Mallaburn, P. S., and Klein, N., 1984, The growth of olfactory neurons in short-term cultures of rat olfactory epithelium, *Neurosci. Lett.* **45:** 193–198.

Novoselov, V. I., Bragin, A. G., Novikov, J. V., Nesterov, V. I., and Fesenko, E. E., 1984, Transplants of olfactory mucosa in the anterior chamber of the eye: Morphology, electrophysiology and biochemistry, *Dev. Neurosci.* **6:** 317–324.

Oblinger, M. M., Hallas, B. H., and Das, G. D., 1980, Neocortical transplants in the cerebellum of the rat: Their afferents and efferents, *Brain Res.* **189:** 228–232.

Olson, L., Seiger, Å., and Ålund, M., 1978, Locus coeruleus fibre growth in oculo induced by trigeminotomy, *Med. Biol.* **56:** 23–27.

Olson, L., Seiger, Å., Ålund, M., Freedman, R., Hoffer, B., Taylor, D., and Woodward, D., 1979, Intraocular brain grafts: A method for differentiating between intrinsic and extrinsic determinants of structural and functional development in the central nervous system. in: *Neural Growth and Differentiation* (E. Meisami and M. A. B. Brazier, eds.), pp. 223–235, Raven, New York.

Olson, L., Seiger, Å., and Stromberg, I., 1981, Intraocular transplantation in rodents: A detailed account of the procedure and examples of its use in neurobiology with special reference to brain tissue grafting, in: *Neural Transplantation and Explanation: Techniques and Applications, Workshop of the Fifth European Neuroscience Meeting, Liège, Belgium.* pp. 51–89.

Schwob, J. E., Farber, N. B., Gottlieb, D. I., 1986, Neurons of the olfactory epithelium in adult rats contain vimentin, *Journal of Neuroscience* **6:** 208–217.
Seiger, Å., Dahl, D., Ayer-LeLievre, C., and Björklund, H., 1984, Appearance and distribution of neurofilament immunoreactivity in iris nerves, *J. Comp. Neurol.* **223:** 457–470.
So, K. F., and Aguayo, A. J., 1985, Lengthy regrowth of cut axons from ganglion cells after peripheral nerve transplantation into the retina of adult rats, *Brain Res.* **328:** 349–354.
Takata, N., 1929, Riechnerv und Geruchsorgan, *Arch. Ohren Nasen Kehlkopf.* **121:** 31–78.
Williams, L. R., Manthorpe, M., Barbin, G., Nieto-Sampedro, M., Cotman, C. W., and Varon, S., 1984, High ciliary neuronotrophic specific activity in peripheral nerve, *Int. J. Dev. Neurosci.* **2:** 177–180.

V

Biological Relevance of Olfactory Function

16

Age-Related Alterations in Olfactory Structure and Function

Richard L. Doty and James B. Snow, Jr.

1. INTRODUCTION

The sense of smell, in addition to contributing to aesthetics and to the overall quality of life, largely determines the flavor of foods and protects the organism from such hazards as leaking gas, spoiled food, and smoke. It is therefore of both practical and medical importance to understand the nature of age-related changes in human olfactory function and, when possible, to correct underlying dysfunctions. Age-related alterations in this primary sensory system are of particular significance when one considers that the number of persons in the American population over the age of 65 years has increased from 3.1 million in 1900 to 24.1 million in 1978 and will likely exceed 32 million by the year 2000 (Brody and Brock, 1985).

Until recently, little was known about age related changes in the ability to smell. Thus, Engen wrote in 1982:

> Most assume that aging is correlated with decreased ability in all sense modalities. Actually, the effect of aging on odor perception is not clear and has been debated since the first observations were reported around the turn of the century.

This statement no longer applies, since studies based on recent advances in human olfactory testing procedures—coupled with histological studies of aging olfactory systems of humans and other mammals—clearly demonstrate that olfactory function and structure are markedly altered in later life. However, causal relations have yet to be demonstrated between the perceptual and structural alterations, and the types of olfactory dysfunction occurring in the later years are only beginning to be fully appreciated.

This chapter reviews studies of age-related changes in olfactory structure and function, with particular emphasis on the human subject. In addition, studies of olfactory perception in persons with dementia-related diseases occurring mainly in the later years are also reviewed (i.e., Alzheimer's disease and parkinsonism). The reader is referred elsewhere for age-related developmental studies of human olfactory structure and function (Doty, 1986*a;* Engen, 1982; Pyatkina, 1982; Self *et al.*, 1972).

Richard L. Doty and James B. Snow, Jr. • Smell and Taste Center and Department of Otorhinolaryngology and Human Communication, School of Medicine, University of Pennsylvania, Philadelphia, Pennsylvania 19104.

2. HUMAN OLFACTORY PERCEPTION IN LATER LIFE

Age-related changes are observed in a number of olfactory tests, including those that measure the ability to identify odors, the absolute sensitivity to odors, the hedonic responses to odors, and the ability to perceive differences in the intensity of odors across suprathreshold stimulus concentrations. Since such tests do not always measure independent aspects of smell function (e.g., a marked decrement in absolute sensitivity would be expected to influence odorant pleasantness and the ability to identify odors), it may be that only differential performance on such tests has the potential for providing specific information about the underlying brain mechanisms involved. Unfortunately, investigators rarely administer more than one type of olfactory test to the same set of subjects. Thus, even circumstantial data are not available to determine whether the observed alterations represent one or more underlying disabilities. What is clear, however, is that changes on all these types of tests are particularly evident after the sixth decade of life, and that large individual differences are present (Murphy, 1985; Schiffman, 1979; Weiffenbach, 1984).

2.1. Odor Identification

Although a number of studies have suggested that older persons, on the average, have difficulty in identifying or recognizing odorants (Murphy, 1985; Schemper, Voss and Cain, 1981; Schiffman, 1977), practical considerations have generally limited such studies to only a few stimuli and to relatively small numbers of "young" and "old" subjects. Thus, unlike the case of vision and audition (where standardized tests have been administered to thousands of persons across the entire age span), the nature and degree of age-related changes in the ability to identify odors was, until recently, unknown. As a result of the development of a rapidly administered microencapsulated test of olfactory function (the University of Pennsylvania Smell Identification Test or UPSIT), the odor identification ability of thousands of subjects has now been assessed and the period in which this ability begins to wane established (Doty *et al.*, 1984*a,b*). The results of such testing show that, on average, (1) peak performance in odor identification occurs in the third through fifth decades of life and markedly declines after the seventh decade, (2) nonsmokers outperform smokers, and (3) women outperform men, particularly in the later years of life (Fig. 1). More than one half the subjects between the ages of 65 and 80 evidence major impairment, whereas more than three-fourths of those over the age of 80 years do. The poor scores in the older age range are unlikely due to losses in memory, *per se*, since (1) the memory load on the UPSIT probably does not exceed the span of immediate attention and (2) UPSIT scores of elderly subjects do not significantly correlate with scores on the Wechsler Memory Scale (Doty *et al.*, 1984*a*). Interestingly, sex differences observed in odor identification ability are present to the same relative degree in several cultural groups, including American Blacks, American Whites, American Koreans, and Native Japanese (Doty *et al.*, 1985*a;* Doty, 1986*a*).

2.2. Odor Detection

As in the case of odor identification, impairment in the ability to detect low concentrations of odorants is generally found in the later years (Chalke *et al.*, 1958; Fordyce,

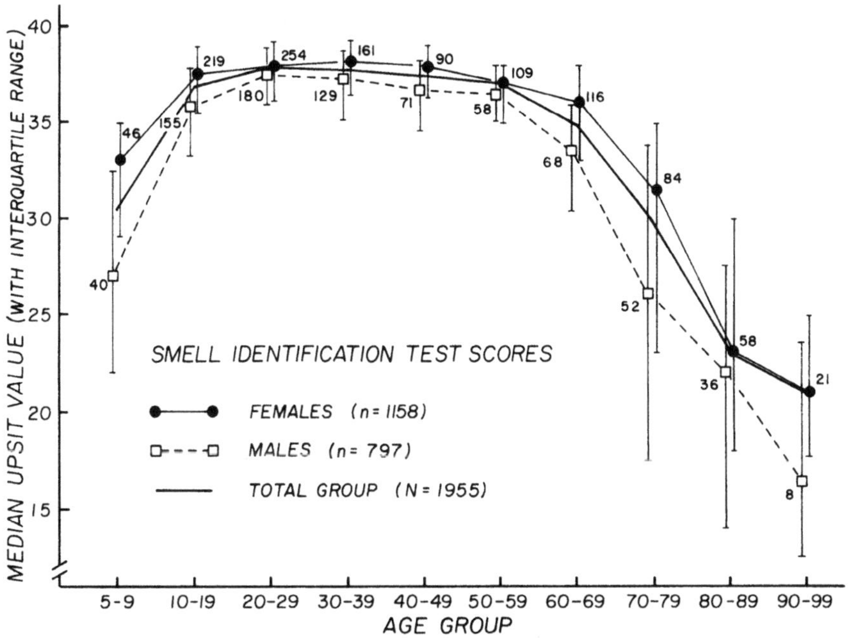

Figure 1. Scores on the University of Pennsylvania Smell Identification Test as a function of age in a large heterogeneous group of subjects. Numbers by data points indicate sample sizes. (From Doty *et al.*, 1984.)

1961; Kimbrell and Furchtgott, 1963; Murphy, 1983; Schiffman *et al.*, 1976; Vestrom and Amoore, 1968). Since phenyl ethyl alcohol odor detection thresholds correlate rather strongly with UPSIT scores (Doty *et al.*, 1984*a*), it is not surprising that such thresholds follow an age-related function similar to that noted in Fig. 1 when plotted inversely on the *y* axis (cf. Deems and Doty, 1987). The function presented in Fig. 1 may be representative of age-related changes in the thresholds to many odorants, since detection thresholds are correlated in most cases with one another; i.e., persons who evidence low sensitivity to one odorant typically also show low sensitivity to others, and those who evidence high sensitivity to an odorant typically show high sensitivity to others. Thus, a "general olfactory acuity" factor may exist, analogous to the general intelligence factor derived from items of intelligence tests (Yoshida, 1984; see also Koelega and Koster, 1974; Punter, 1983; Stevens and Cain, 1985).

2.3. Suprathreshold Odor Intensity Perception

Although Rovee *et al.* (1975) reported stability over a wide age range in the growth rate of the perceived intensity of propanol across a suprathreshold concentration series, more recent studies using other odorants do not find such stability. For example, using a procedure in which numbers were assigned in proportion to the perceived odor intensity (i.e., a magnitude estimation procedure), Murphy (1983) found the average rate of growth of intensity of menthol as a function of concentration twice as great for 10 young adults (18–26 years of age) as for 10 elderly ones (66–93 years of age). As noted by Murphy, however. menthol has both olfactory and trigeminal stimulative properties (cf. Doty *et al.*, 1978), making it unclear whether this phenomenon is due to olfactory stimulation, trigeminal stimulation, or some combination of both.

In an important methodological study, Stevens *et al.* (1982) asked 20 young (aged 18–25 years) and 20 elderly (aged 65–83 years) subjects to provide intensity magnitude estimates to various concentrations of isoamyl butyrate (a relatively nonirritating odorant) and CO_2 (a trigeminal stimulus with minimal or no odor qualities), as well as to a low-pitched noise. The estimates of the auditory stimulus, presented in the same test session as the olfactory stimuli, provided a means of normalizing the subjects' scales of measurement, termed cross-modality matching (see Stevens and Marks, 1980). These investigators found the standardized magnitude estimation functions of young and old persons for each odorant to be nearly parallel. However, the function of the older subjects for both odorants was displaced downward (i.e., evidenced a lower *y* intercept), suggesting that older persons have a proportional loss of smell function across a broad range of stimulus concentrations. A similar finding was reported by Stevens and Cain (1985), in which the tastant NaCl was used as the nonolfactory matching stimulus. This experiment found the aforementioned constant percentage reduction of olfactory strength across the concentration levels of all six odorants examined (isoamyl butyrate, benzaldehyde, *d*-limonene, pyridine, ethanol, isoamyl alcohol), suggesting that the deficit is present for odorants ranging widely in chemical structure, psychological quality, and hedonic tone.

Age-related decrements in suprathreshold smell perception have also been reported for odorants presented retronasally (i.e., to the olfactory receptors from inside the oral cavity, as during chewing and swallowing). Thus, Stevens and Cain (1986) noted that 20 young subjects (18–24 years of age) perceived the overall intensity of orally sampled solutions of ethyl butyrate to be much weaker when the nose was occluded than when it was open. This disparity was not evident in a group of 20 elderly subjects (aged 67–83 years), presumably because of age-related alterations in the olfactory system proper. Since the intensity of retronasally perceived odor is largely dependent upon mouth movements, such as those that occur normally during deglutition (cf. Burdach and Doty, 1987), some age-related alterations in flavor perception could be due to changes in behaviors that dramatically alter pressure–flow relationships within the nasopharynx, including the speed and amount of chewing or swallowing.

2.4. Odor Discrimination

As might be expected from these studies, the ability to distinguish qualitatively among odorants, as measured by techniques such as multidimensional scaling (which represents relative perceptual differences among stimuli as distances among spatial coordinates), is impaired in many elderly persons. For example, Schiffman and Pasternak (1979) had sixteen 19- to 25-year-olds and sixteen 72- to 78-year-olds rate 91 pairs of 14 commercial food flavors on a 5-inch "same–different" rating scale. The multidimensional scaling procedure applied to the 32 similarity matrices for all subjects yielded a two-dimensional solution in which two main clusters emerged (simulated fruit flavors and simulated meat flavors). Analysis of the spaces of individual subjects suggested that some of the elderly subjects could not discriminate among many of the odorants, since a number of stimuli normally found in disparate sectors of the multidimensional space were located near one another.

2.5. Odor Pleasantness

The sense of smell has received considerable attention from the perspective of hedonic measurement, in part because subjects spontaneously use hedonic descriptors in judging odors, e.g., good, bad, pleasant, unpleasant (cf. Harper *et al.*, 1968; Schiffman, 1974) and because this sense is closely associated with the determination of flavor, the monitoring of foodstuffs, and, at least in the case of most nonhuman mammals, sexual behavior (cf. Doty, 1974, 1986*c*). Although alterations in the perceived pleasantness of an odor can be theoretically independent of alterations in its perceived intensity, there is often an association between these two dimensions. For instance, certain odorants judged pleasant at low concentrations are judged even more pleasant at high concentrations, whereas some odorants judged unpleasant at low concentrations are typically judged more unpleasant at high concentrations. However, strict monotonicity is rarely the rule, and the subject's basic odor preference is often idiosyncratic. For example, persons who dislike the odor of licorice (anethole) report it to be even more unpleasant at higher concentrations, whereas those who like this odor find it more pleasant at higher concentrations (Doty, 1975).

If an odorant is perceived as less intense by an elderly person than by a younger one, its perceived pleasantness would also be expected to be correspondingly altered, depending on the form of the intensity–pleasantness relationship for the odorant in question. It is therefore not surprising that Murphy (1983) found increases in menthol concentration to produce much larger increments in estimates of odor pleasantness by young than by elderly subjects, in light of her observation of a smaller increase in perceived intensity in the older individuals. Likewise, the observations by Springer and Dietzmann (Engen, 1977) that diesel fumes are less offensive to older than to younger persons conceivably reflects age-related lessened nasal chemosensitivity.

3. HUMAN OLFACTORY PERCEPTION IN AGE-RELATED DISEASES

Recent studies have demonstrated that perhaps most people suffering from a number of dementia-related diseases, including Alzheimer's disease and parkinsonism, evidence olfactory dysfunction (Corwin *et al.*, 1985; Doty and Reyes, 1986; Doty *et al.*, 1986, 1987; Koss *et al.*, 1986; Liss, 1984; Mair *et al.*, 1986; Ward *et al.*, 1983; Warner *et et al*, 1986; see also the earlier observations of Richard and Bizzini, 1981, and Waldton, 1974). Furthermore, there is evidence that such dysfunction is accompanied by structural (and biochemical) alterations in olfactory-related brain regions, although cause and effect relations have not yet been determined. For example, in the case of Alzheimer's disease, neurofibrillary tangles and neuritic plaques are disproportionately found within limbic structures associated with olfactory function (Fig. 2 and 3); (Esiri and Wilcock, 1984; Esiri *et al.*, 1986; Hooper and Vogel, 1976; Pearson *et al.*, 1985; Reyes *et al.*, 1985, 1987). Pearson *et al.* (1985) summarize the olfactory involvement as follows:

> The invariable finding of severe and even maximal involvement of the olfactory regions in Alzheimer's disease is in striking contrast to the minimal pathology in the visual and sensorimotor areas of the neocortex and cannot be without significance. In the olfactory system, the sites that are affected—the anterior olfactory nucleus, the uncus, and the medial group of amygdaloid nuclei—all receive fibers directly from the olfactory bulb. These observations at least raise the possibility that the olfactory pathway is the site of initial involvement of the disease.

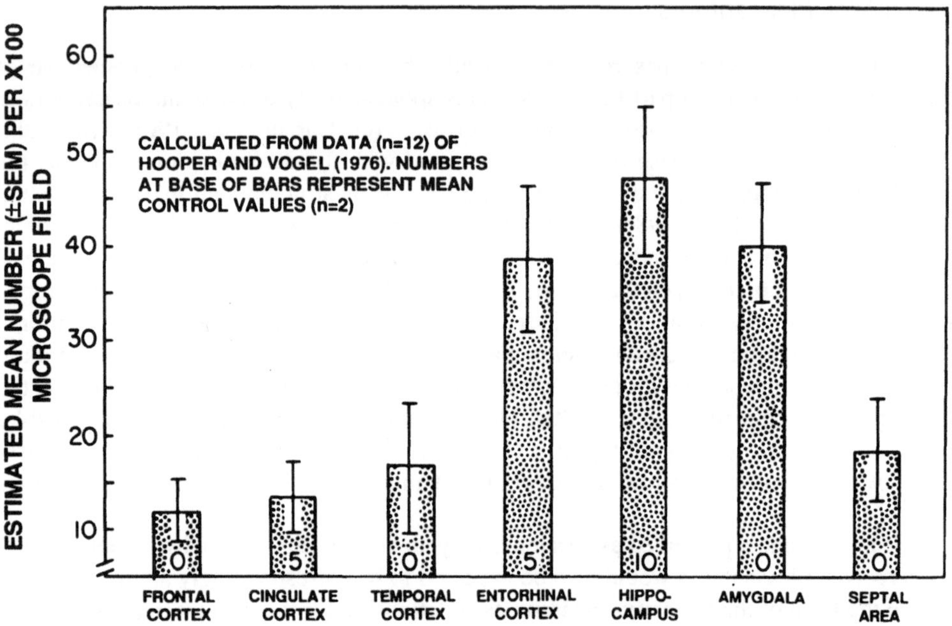

Figure 2. Estimated mean number (±SEM) of neurofibrillary tangles per X100 microscope field in seven brain regions. Calculations based on assignment of their estimated ratings as follows: 4+ = 70; 3+ = 50; 2+ = 25; and 1+ = 10. Original ratings reflected the following: 44+ = ⩾60 tangles; 3+ = 41–60; 2+ = 21–30; 1+ = 1–20.

Such findings gain even more significance in the context of theories that some dementia-related diseases may be related to environmental toxins or viruses (e.g., Esiri, 1982), and evidence that (1) the olfactory system is a major conduit of such materials into the central nervous system (CNS) (Shipley, 1985; Stroop *et al.*, 1984; Tomlinson and Esiri, 1983; Monath *et al.*, 1983) and (2) inoculation of rodents with some viruses results in necrosis of the olfactory neuroepithelium, the olfactory bulbs and tracts, and the prepyriform cortex (cf. Goto *et al.*, 1977; Reinacher *et al.*, 1983).

3.1. Alzheimer's Disease

Recently we and others have administered the University of Pennsylvania Smell Identification Test to patients with Alzheimer's disease and, in all such cases, the same general level of average olfactory deficit was observed (Doty and Reyes, 1987; Doty *et al.*, 1987; Koss *et al.*, 1986; Warner *et al.*, 1986). In one of these studies (Doty *et al.*, 1987), the UPSIT was administered to 34 patients who satisfied stringent criteria for the clinical diagnosis of Alzheimer's disease. According to the criteria of Cummings (1983), 10 patients were at stage 1, 21 at stage 2, and 3 at stage 3 of the disease. Thirty-four control subjects were individually matched to these patients on the basis of age, gender, and ethnic background. Since nine Alzheimer patients were unable to score 35 or more on the Picture Identification Test (PIT), a test identical in content and format to the UPSIT except that pictures, rather than odors, serve as stimulus items (cf. Vollmecke and Doty,

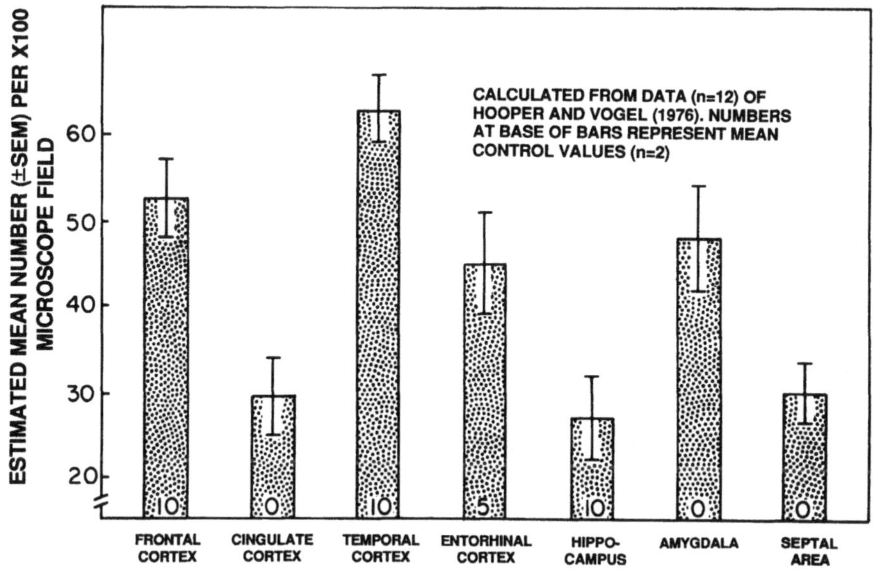

Figure 3. Estimated mean number of (±SEM) of neuritic plaques per X100 microscope field in seven brain regions. Calculations based on assignment of their estimated ratings as follows: 4+ = 70; 3+ = 50; 2+ = 25; and 1+ = 10. Original ratings reflected the following: 4+ = ⩾60 tangles; 3+ = 41–60; 2+ = 21–30; 1+ = 1–20.

1985), they were eliminated from the study as being too demented to comprehend the nonolfactory components of the UPSIT. Thus, 25 stage 1 and 2 Alzheimer patients and 25 matched normal controls comprised the final study group.

The average UPSIT scores are presented in Fig. 4 as a function of subject age at the time of testing. The UPSIT scores of the 25 Alzheimer patients differed markedly and consistently from those of their matched controls (Wilcoxin matched-pairs signed-rank test $T = 3$, $p < 0.001$). Indeed, only two patients evidenced UPSIT scores above those of their respective controls. Furthermore, it is of interest that the average UPSIT scores of stage 1 and 2 patients did not differ significantly and that little or no average decline occurred in these scores across the age range examined. However, the expected decline in

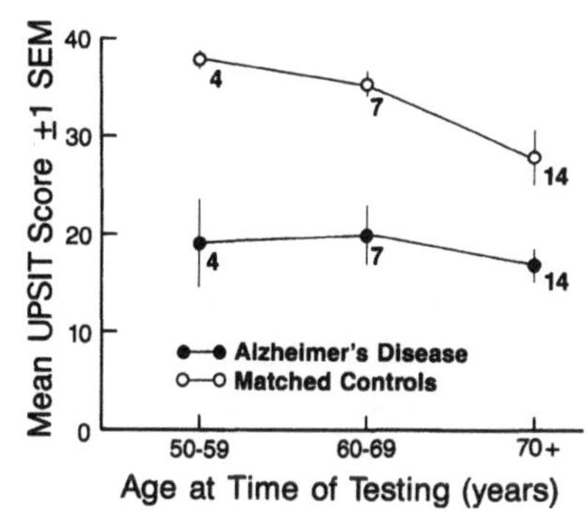

Figure 4. Mean University of Pennsylvania Smell Identification Test (UPSIT) scores as a function of age in Alzheimer's disease patients and in matched normal controls. Numbers by data points indicate sample sizes. (Data adapted from Doty *et al.*, 1987*a*.)

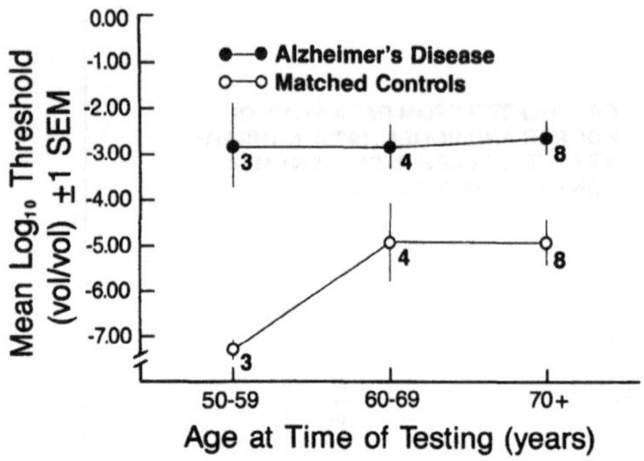

Figure 5. Phenyl ethyl alcohol detection thresholds as a function of age in Alzheimer's disease patients and in matched normal controls. Numbers by data points indicate sample sizes. (Data adapted from Doty *et al.*, 1987*a*.)

UPSIT scores was present in the normal controls (cf. Doty *et al.*, 1984*a*). These data suggest that the olfactory deficit is present to a high degree early in the disease process.

In addition to the UPSIT, a forced-choice single staircase detection threshold test was administered to 15 of patients and to an equivalent number of age-, race-, and gender-matched controls. This test, which incorporates an odorant with relatively low trigeminal stimulative properties (phenyl ethyl alcohol) (cf. Doty *et al.*, 1978), uses the geometric mean of the last four of seven staircase reversals as the threshold estimate (for details, see Doty *et al.*, 1984*b;* Ghorbanian *et al.*, 1983). As can be observed in Fig. 5, the Alzheimer patients evidenced significantly higher detection thresholds than their matched controls (Wilcoxon $T = 1$, $p < 0.001$), with only one patient evidencing a threshold value below that of his control (and in this case the difference was less than 2/10ths of a log concentration unit). As with the UPSIT scores, no change in the test values of the Alzheimer patients was apparent as a function of disease stage or age at the time of testing, even though an age-related increase in the threshold values was present in the control group.

Although additional studies are needed to provide conclusive information on this issue, the cross-sectional data collected in our laboratory support the notion that little change occurs across time in either UPSIT scores or olfactory threshold values of patients with Alzheimer's disease. If borne out by longitudinal testing, this observation, along with our preliminary finding of no significant difference between the test scores of Stage 1 and 2 Alzheimer patients, implies that olfactory dysfunction occurs early in the disease process and is maintained at a constant level from that time on.

Another point of interest from these preliminary studies is that the olfactory deficit observed in most Alzheimer cases is not one of total anosmia (although some exhibit this problem). The deficit may reflect hyposmia, however, since average detection threshold values are markedly elevated (Fig. 5). On the other hand, it is conceivable that Alzheimer's disease patients have difficulty comparing intensities of successively presented odorants in a forced-choice task (e.g., as a result of loss of short-term olfactory memory), and that their sensitivity would prove normal if tested by a procedure not requiring successive discrimination (such as a single-choice signal detection procedure).

Whatever the nature of the deficit reflected in the olfactory threshold measure, it is clear that most Alzheimer patients we have evaluated are initially unaware of an olfactory problem. Thus, only two responded affirmatively before testing to the question, "Do you suffer from smell and/or taste problems?" Whether this lack of awareness relates mainly to sensory or cognitive problems remains to be determined.

3.2. Parkinson's Disease

As in the case of Alzheimer's disease, a number of studies report that olfactory dysfunction is present in patients with Parkinson's disease (PD), although the nature of the problem, including its frequency and severity, is only now being established by the use of standardized quantitative tests and relatively large sample sizes.

Early suggestion that olfactory dysfunction is common among patients with parkinsonism comes from several sources. In a pioneering study, Anasari and Johnson (1975) administered an amyl acetate threshold test to 22 parkinsonians and reported that 10 (45%) evidenced decreased olfactory sensitivity. A half-decade later, Korten and Meulstee (1980) interviewed 80 Parkinson patients and noted that 40 reported "being unable to smell adequately." However, no quantitative testing was performed. More recently, Ward *et al.* (1983), Corwin *et al.* (1985), and Quinn *et al.* (1987) have confirmed the presence of olfactory deficits in parkisonism. In the most extensive of these studies, Ward et al. (1983) found 49% of 72 Parkinson patients unable to identify the odor of fresh ground coffee, and 35% the odor of cinnamon (in Lifesaver candy), even though written response alternatives were provided. This study included 14 pairs of monozygotic twins and 6 pairs of same-sex dizygotic twins. Since the olfactory ability of the nonparkisonian twin was apparently normal in all twin pairs, the olfactory anomaly appears to be acquired rather than inherited.

To further establish the nature of the olfactory problems associated with Parkinson disease, we administered the UPSIT and PIT to 93 parkinsonians who had symptoms ranging in duration from 3 months to 55 years (Doty *et al.*, 1988). Clinical ratings of 11 neurologic symptoms (3 bilateral) were obtained at the time of testing. Since 12 of the PD patients failed to achieve a score of 35 or higher on the PIT, the final study group consisted of 81 patients (46 men and 35 women) and their matched controls. In addition, 38 patients and 38 controls were administered the phenyl ethyl alcohol odor detection threshold test, and odor identification was retested in 24 patients after intervals ranging from 5 months to three years.

As can be seen in Fig. 6, the PD patients evidenced marked decrements in both odor identification and detection ability. No relationship was present between the magnitude of

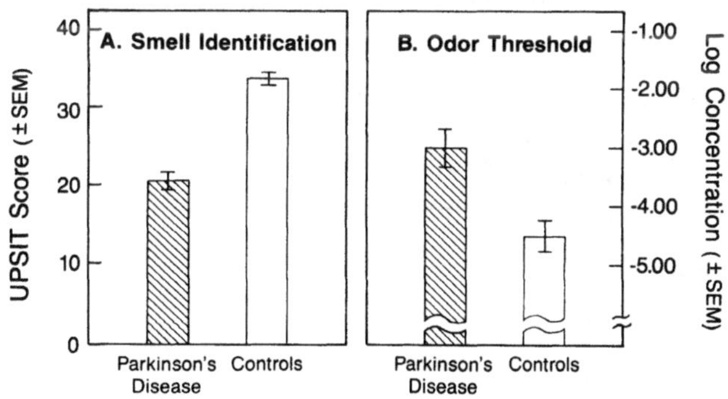

Figure 6. Mean University of Pennsylvania Smell Identification Test (UPSIT) scores (A) and phenyl ethyl alcohol detection threshold values (B) in Parkinson's disease patients and in matched normal controls. (From Doty *et al.*, 1988.)

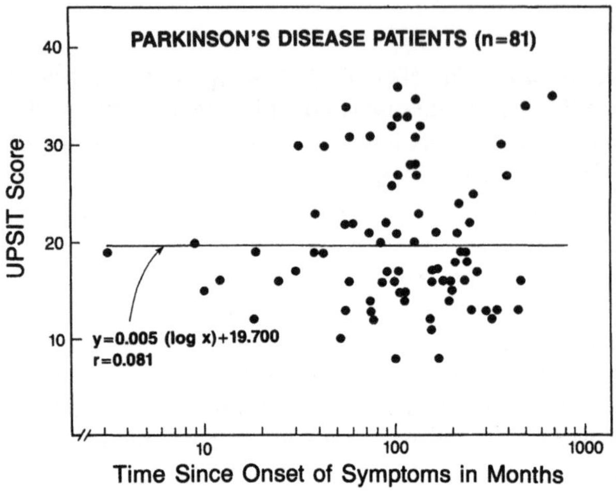

Figure 7. University of Pennsylvania Smell Identification Test (UPSIT) scores of Parkinson's disease patients as a function of disease duration. (From Doty *et al.*, 1988.)

the olfactory test scores and the time since diagnosis (see Fig. 7 and 8). This finding, and the fact that no significant longitudinal changes occurred in the test scores of the 24 patients who were retested, suggests that the olfactory deficit is relatively stable across time.

Since the PD test scores were remarkably similar to those observed in our earlier study of patients with Alzheimer's disease (Doty *et al.*, 1987), we statistically compared the olfactory test scores of the PD patients to those of the Alzheimer's disease patients. To achieve this assessment, we first matched each PD patient to an Alzheimer's disease patient on the basis of ethnic background, age, gender, and smoking habits, and then subjected the UPSIT and threshold scores to separate analyses of covariance with disease type as a factor and PIT score as a covariate (see Doty *et al.*, 1988). The covariate was important because many of the PIT scores of the Alzheimer's disease patients fell below those of the parkinsonians within the 35 to 40 PIT score range, making it impossible to directly match all of the subjects on this cognitive variable.

No significant difference between the UPSIT test scores of the PD and Alzheimer's disease patients was found, although the PIT covariate was highly significant, indicating

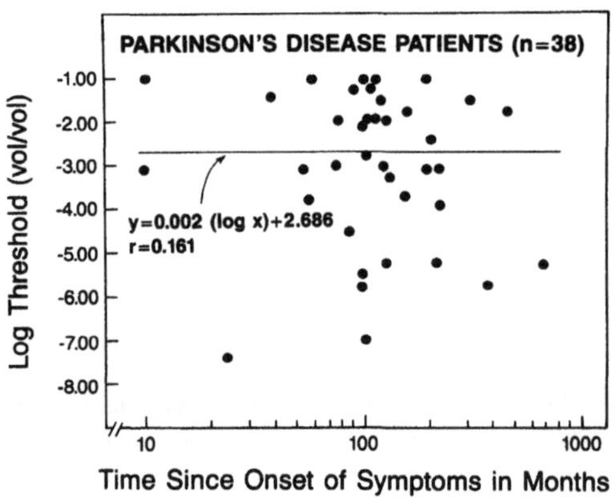

Figure 8. Phenyl ethyl alcohol detection thresholds of Parkinson's disease patients as a function of disease duration (From Doty *et al.*, 1988.)

that—even within the PIT range from 35 to 40—the Alzheimer's disease patients performed more poorly than the PD patients on the PIT. A similar result was found for the detection threshold data, although the covariate did not reach significance at the 0.05 probability level (likely because of the smaller sample size).

To explore the nature of the relationships among the olfactory, cognitive, and neurological measures of the PD patients, we subjected their intercorrelations to a principal components factor analysis with varimax rotation. Six factors were extracted with Eigen values > 1.00 which accounted for 73.4% of the total variance. This analysis demonstrated that the olfactory test scores were independent of the neurologic and cognitive measures, as both the UPSIT and detection threshold values loaded heavily on a single factor which received no strong loadings from any other variable (see Table 1 in Doty *et al.* 1988). Thus, whatever the basis of the olfactory deficit, it appears to be independent of the severity or degree of the motoric or cognitive symptoms of the disease.

We found no pattern of smell loss to the various items of the UPSIT in the PD patients, suggesting that the odor identification problem is a general one. Furthermore, the olfactory deficit appears not to reflect total anosmia in most cases, since only 13% of the 38 patients who received detection threshold testing were unable to detect the highest odorant concentration presented, a figure in close correspondence to a 17% anosmia rate reported earlier by Ward *et al.* (1983). This observation is further supported by the finding that all but one of 41 PD patients who were asked whether or not an odor was present on each UPSIT item answered affirmatively to 35 or more of the items, even though the majority were unable to identify most of the odors or felt that the perceived sensation did not correspond to the response alternatives.

In summary, a large majority of patients with parkinsonism have olfactory dysfunction, as measured by tests of odor identification and detection. This disorder seems to be (1) stable over time, (2) present relatively early in the disease process, (3) general in nature, (4) independent of motoric and cognitive aspects of the disease, and (5) indistinguishable, at present, from the olfactory disorder observed in Alzheimer's disease.

4. AGE-RELATED ALTERATIONS IN THE STRUCTURE AND FUNCTION OF THE NOSE AND THE OLFACTORY SYSTEM

4.1. Airflow and General Nasal Considerations

Before an odorant can be perceived, it must reach the sensory receptor cells of the olfactory neuroepithelium. In humans, such cells are located in the superior region of the nasal cavity and occupy areas of the cribriform plate, dorsal portions of the superior turbinate, and the superior part of the nasal septum. Because the olfactory cleft is quite narrow (often less than 1 mm), alterations in the highly vascularized nasal respiratory epithelium—particularly that located on the turbinates—can alter the airflow patterns into the region and, in some cases, either increase or decrease the amount of odorized air reaching the receptors. Thus, age-related alterations in factors that influence airway patency (e.g., mucosal thickness, turbinal engorgement, inflamation, polyposis) may account for at least some cases of decreased odor perception observed in elderly individuals.

Surprisingly little research has been conducted on age-related alterations in the nasal epithelium and their potential influence on airflow to the olfactory receptors, despite the fact that numerous changes are known to occur in aging human skin and in the respiratory system (see Fowler, 1985; Kligman *et al.*, 1985). The few data available suggest that age-related alterations occur within the nasal cavity. For example, using the ^{133}Xe washout method, Bende (1983) observed a −0.51 correlation between age and blood flow within the nasal mucosa. Hasegawa and Kern (1977) found an increase in nasal patency and a less prominent nasal cycle in persons over the age of 40 (see also Nishihata, 1984). Other workers have reported age-related atrophy and decreases in the vascular elasticity of the nasal epithelium (e.g., Somlyo and Somlyo, 1968). A decline in the nasal mucocilliary clearance rate of some elderly persons was reported by Sakakura *et al.* (1983).

Although a relationship between nasal obstruction and olfactory sensitivity has been shown in children (Ghorbanian *et al.*, 1983), we know of only one empirical study that addresses, in adults, relationships among the variables of age, olfactory sensitivity, and nasal airflow in adults (Murphy *et al.*, 1985). In this research, nasal airway resistance and olfactory detection thresholds to *n*-butanol were measured in 12 men and 12 women between the ages of 18 and 26 years and in an equivalent number of men and women between the ages of 65 and 84 years. Nasal resistance was significantly decreased in the elderly compared to the young, as was olfactory sensitivity. This finding, although correlative, is in accord with observations of Schneider and Wolf (1960), who suggested that olfactory sensitivity is lowest at the two extremes of the nasal engorgement continuum, i.e., when the nose is very congested or very uncongested. Presumably a moderate degree of engorgement results in the shunting of proportionately more air above the superior turbinate into the receptor region, although empirical information on this point is needed.

4.2. Olfactory Neuroepithelium

As with all sensory systems, morphological changes occur in the olfactory system with age. Unlike other sensory systems, however, the primary sensory neurons are particularly susceptible to insult from viruses and environmental toxins, as they are exposed rather directly to the outside environment. This fact may be related evolutionarily to the remarkable plasticity of such neurons, which can be reconstituted under some circumstances if damaged or destroyed (for reviews, see Farbman, 1986; Graziadei-Monti and Graziadei, 1979).

In general, the cellular patterns and zonal distributions of the nuclei of the supporting and sensory cells of the olfactory neuroepithelium seen in the human fetus and neonate are altered in older individuals (although large individual differences are apparent). For example, Nakashima *et al.* (1984) examined, by light microscopy, the olfactory neuroepithelia of five aborted human fetuses and 21 adults ranging in age from 20 to 91 years at autopsy and noted marked degeneration of receptor cells, particularly in the specimens from the elderly persons. The zonal distribution of the basal, supporting, and sensory receptor cells was frequently disturbed. Another characteristic of the adult olfactory region is the intercalation of respiratory epithelium with olfactory neuroepithelium (Fig. 9), suggesting that damaged olfactory neuroepithelium is, at least at times, replaced with respiratory epithelium. Although degeneration of the olfactory neuroepithelium progresses with aging, it can be found in young adults. It seems likely that repeated viral and perhaps bacterial infections and exposure to toxic substances in the environment account for this progressive degeneration of the olfactory neuroepithelium. Furthermore, the olfactory neuroepithelium of older individuals may be particularly susceptible to the influences of viral and perhaps bacterial infections, since upper respiratory infections are the most common cause of anosmia in patients over 50 years of age studied at the University of Pennsylvania Smell and Taste Center.

Aging is also accompanied by a regression of the vessels in the neuroepithelium and by the loss of the cellularity of the lamina propria adjoining the basement membrane. Pigment granules accumulate in the supporting cells of adults (Naessen, 1971). These granules are not seen in the olfactory neuroepithelia of fetuses, infants, and young children. As Naessen suggests, it is possible that these granules represent the cell's inability to deal effectively with its own metabolic wastes and the products of neuronophagic activity.

Structural studies of the olfactory neuroepithelia of aging rodents reveal cellular alterations suggestive of reduced protein synthesis and general cellular metabolism. For example, in one series of observations, Dodson and Bannister (1980) noted, across the 6- to 30-month age range of albino housemice, a gradual reduction in perikaryal size, a reduction in the per cell amounts of granular and agranular endoplasmic reticulum, and an increase in the numbers of secondary lysosomes. In humans, the supporting cells of the older individuals accumulated considerable amounts of cellular debris in their basal processes (Nassen, 1971). Furthermore, while the number of receptors per unit of epithelium volume did not fall appreciably, the number of cells observed in mitosis were fewer in the older animals, and the rate of cell migration away from the basal layer decreased appreciably. These phenomena likely relate to Matulionis's (1982) observation that, unlike younger animals, 24 to 26-month-old adult mice fail to evidence replacement of their olfactory receptor cells following intranasal zinc sulfate lavage.

4.3. Olfactory Bulb

The fact that the destruction of the receptor elements of the olfactory neuroepithelium extends to the glomerular level was used by Smith (1942) to estimate age-related losses of human olfactory receptors. By assessing the number and form of glomeruli present in 205 olfactory bulbs of 121 individuals at autopsy, Smith concluded that loss of olfactory nerves begins soon after birth and continues to occur throughout life at approximately 1% per year. However, a reevaluation of Smith's data (Fig. 1 in Smith, 1942) using medians rather than means leads one to the conclusion of no major loss until the fifth decade of life, implying that this widely cited statistic may be misleading. Although considerable variability among bulbs was seen at all ages examined, marked differences between men and women were not apparent in the data of this pioneering study.

The widespread alterations in bulbar structures with age are perhaps best exemplified in a series of quantitative anatomical studies performed by Hinds and associates in the rat. Using the Sprague-Dawley strain, Hinds and McNelly (1977) measured the volume of main olfactory bulb components (including the glomerular, external plexiform, internal granular, and olfactory nerve layers) at 3, 12, 24, 27, and 30 months of age. In addition, the size and number of mitral cells were measured in both the main and accessory areas of the bulb. Although developmental increases in the layer volumes were noted during the

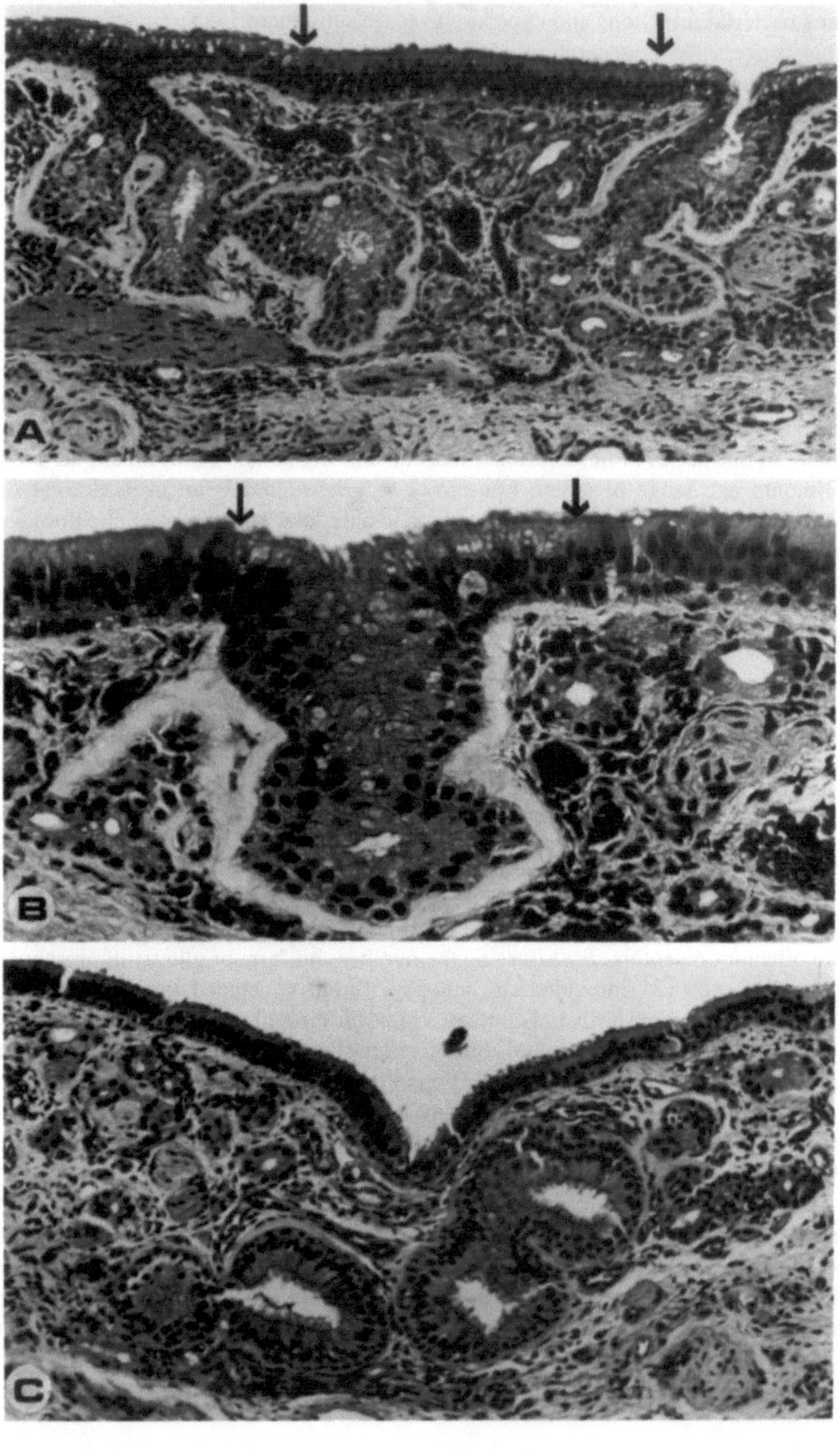
A
B
C

first 24 months, decreases occurred after that time. A sharp decrease in mitral cell numbers was noted, along with an increase in the volume of individual mitral cell dendritic trees and in the perikaryal and nuclear size.

Hinds and McNelly (1981) subsequently replicated most of these findings in the Charles River strain of rat (although no loss in mitral cell number was observed in the older animals) and concurrently evaluated alterations in the olfactory neuroepithelium. A comparison of regression lines for changes in number of septal olfactory receptors with that of the size of mitral cell bodies suggested that the decline in receptor number began several months before the decline in mitral cell size. Thus, the latter alterations may be due to the decline in neuroepithelial receptor cells. Interestingly, a significant increase in the number of synapses per receptor was present in the oldest group evaluated (33 months), possibly reflecting a compensatory increase in the relative numbers of synapses per receptor in the surviving receptors.

4.4. Higher Centers

There is now strong evidence that age-related diseases such as Alzheimer's disease are invariably associated with severe neurochemical and anatomical changes in higher olfactory centers, such as structures within the ventromedial temporal lobe (Pearson *et al.*, 1985; Simpson *et al.*, 1984). Whether aging *per se* results in significant changes in higher olfactory regions is not yet established. However, it is likely, since (1) some plaques and tangles are present in a number of the regions of the olfactory system of many older individuals who evidence no obvious symptoms of dementia, and (2) neuronal loss occurs in selected layers and regions of the aging human cerebral cortex (for reviews, see Duara *et al.*, 1985; A. H. Tomlinson *et al.*, 1968).

Despite such possibilities, recent studies of the rat piriform cortex suggest that it is surprisingly stable with age compared with the olfactory bulb and olfactory neuroepithelium. In a study of the cells and synapses of the piriform cortices of rats aged 3, 12, 18, 24, 30, and 33 months, Curcio *et al.* (1985) found no significant changes in the volumes of cortical laminae Ia and Ib or in the numerical and surface densities of the synaptic apposition zones in layer Ia (which are formed mainly by mitral cell axons). Although age-related changes in nuclear volume, soma volume, or numerical density of layer II neurons were not present, a modest (18%) decline in the proportion of layer Ia occupied by dendrites and spines was observed. This decrease was accompanied by an increase in the proportion of glial processes but not by any alteration in the proportion of axons and terminals.

5. SUMMARY AND CONCLUSIONS

It is clear from the studies reviewed that olfactory function is altered in a large percentage of elderly individuals. This alteration can be detected by a variety of psycho-

Figure 9. Respiratory epithelium in olfactory region of adult human. (A) Ciliated and goblet cell-containing respiratory epithelium has invaded degenerated olfactory neuroepithelium (between arrows). Arrows indicate junction of respiratory and olfactory epithelia (H & E, ×100). (B) Note glandlike invagination (arrows) of respiratory epithelium into lamina propria (H & E, ×200). (C) Glandlike respiratory epithelium with large lumina in lamina propria (H & E, ×100). (From T. Nakashima *et al.*, 1984.)

physical tests, including those involving odor identification, odor detection, and the rating of odor pleasantness. Furthermore, it is now clear that olfactory dysfunction accompanies a number of age-related diseases, including Alzheimer's disease and parkinsonism.

It appears that age-related olfactory dysfunction is associated with altered nasal resistance and degenerative processes within the olfactory neuroepithelium, olfactory bulbs, and higher olfactory centers. It is probable that the structural and functional bases of such changes are multiple, interacting, and complex (Doty *et al.*, 1984*b*). For example, age-related occlusion of the foramina of the cribriform plate may occur in some individuals, thereby eliminating the connections between the olfactory epithelium and olfactory bulb (Krmpotic-Nemanic, 1969). It is hoped that more refined tests of olfactory function will permit us to differentiate, in the near future, among the different types of underlying pathological and pathophysiological states that produce disturbances in olfactory perception.

ACKNOWLEDGMENTS. This work was supported by grant NS 16365 from the National Institute of Neurological and Communicative Disorders and Stroke.

REFERENCES

Anasari, K. A., and Johnson, A., 1975, Olfactory function in patients with Parkinson's disease, *J. Chron. Dis.* **28:** 493–497.

Bende, M., 1983, Bloodflow with ^{133}Xe in human nasal mucosa in relation to age, sex, and body position, *Acta Otolaryngol. (Stockh.)* **96:** 175–179.

Brody, J. A., and Brock, D. B., 1985, Epideminologic and statistical characteristics of the United States elderly population, in: *Handbook of the Biology of Aging* (C. E. Finch and E. L. Schneider, eds.), pp. 3–26, Van Nostrand Reinhold, New York.

Burdach, K. J., and Doty, R. L., 1987, The effects of mouth movements, swallowing and spitting on retronasal odor perception, *Physiol. Behav.* **41:** 353–356.

Chalke, H. D., Dewhurst, J. R., and Ward, C. W., 1958, Loss of sense of smell in old people. *Public Health* **72:** 223–230.

Corwin, J., Serby, M., Conrad, P., and Rotrosen, J., 1985, Olfactory recognition deficit in Alzheimer's and Parkinsonian dementias, *IRCS Med. Sci.* **13:** 260.

Cummings, J. L., 1983, Cortical dementias: Alzheimer's and Pick's diseases, in: *Dementia, a clinical approach* (J. L. Cummings, ed.), pp. 35–72, Butterworth, Boston.

Curcio, C. A., McNelly, N. A., and Hinds, J. W., 1985, Aging in the rat olfactory system: Relative stability of piriform cortex contrasts with changes in olfactory bulb and olfactory epithelium, *J. Comp. Neurol.* **235:** 519–528.

Dodson, H. C., and Bannister, L. H., 1980, Structural aspects of ageing in the olfactory and vomeronasal epithelia in mice, in: *Olfaction and Taste*. Vol. VII (H. van der Starre, ed.) pp. 151–154, IRL Press, London.

Deems, D. A., and Doty, R. L., 1987, Age-related changes in the phenyl ethyl alcohol odor detection thresholds: A normative study, *Trans. Penn. Acad. Opthalmol. Otolaryngol.* **39:**646–650.

Doty, R. L., 1974, A cry for the liberation of the female rodent, *Psychol. Bull.* **81:** 159–172.

Doty, R. L., 1975, An examination of relationships between the pleasantness, intensity and concentration of 10 odorous stimuli, *Percept. Psychophys.* **17:** 492–496.

Doty, R. L., 1983, *The Smell Identification Test* ™ *Administration Manual*, Sensonics, Philadelphia.

Doty, R. L., 1986*a*, Cross-cultural studies of taste and olfaction in humans, in: *Chemical Signals in Vertebrates*. Vol. IV: *Ecology, Evolution, and Comparative Biology* (D. Duvall, D. Muller-Schwartze, and M. R. Silverstein, eds.), pp. 673–684, Plenum, New York.

Doty, R. L., 1986*b*, Gender and endocrine-related influences upon olfactory sensitivity, in: *Clinical Measurement of Taste and Smell* (H. L. Meiselman and R. S., Rivlin, eds.), pp. 377–413, Macmillan, New York.

Doty, R. L., 1986*c*, Odor-guided behavior in mammals, *Experientia* **42:** 257–271.
Doty, R. L., 1986*d*, Ontogeny of human olfactory function, in: *Ontogeny of Olfaction in Vertebrates* (W. Breipohl, ed.), pp. 3–17, Springer-Verlag, Berlin.
Doty, R. L., and Reyes. P. F., 1987, Olfactory dysfunction in Alzheimer's disease: A summary of recent findings, *Ann. NY Acad. Sci.* **510:**260–262.
Doty, R. L., Brugger. W. E., Jurs, P. C., Orndorff, M. A., Snyder, and P. J., Lowry, L. D., 1978, Intranasal trigeminal stimulation from odorous volatiles: Psychometric responses from anosmic and normal humans, *Physiol. Behav.* **20:** 175–185.
Doty, R. L., Shaman, P., and Dann, M., 1984*a*, Development of the University of Pennsylvania Smell Identification Test: A standardized microencapsulated test of olfactory function, *Physiol. Behav.* **32:** 489–502.
Doty, R. L., Shaman, P., Applebaum, S. L., Giberson, R., Sikorski, L., and Rosenberg, L., 1984*b*, Smell identification ability: Changes with age. *Science* **226:** 1441–1443.
Doty, R. L., Applebaum, S. L., Zusho, H., and Settle, R. G., 1985*a*, A cross-cultural study of sex differences in odor identification ability, *Neuropsychologia* **23:** 667–672.
Doty, R. L., Newhouse, M. G., and Azzalina, J. D., 1985*b*, Internal consistency and short-term test-retest reliability of the University of Pennsylvania Smell Identification Test, *Chem. Senses* **10:** 297–300.
Doty, R. L., Reyes, P. F., and Gregor, T., 1986, Olfactory dysfunction in Alzheimer's disease, *Chem. Senses* **11:** 595.
Doty, R. L., Reyes, P., and Gregor, T., 1987, Presence of both odor identification and detection deficits in Alzheimer's disease, *Brain Res. Bull.* **18:** 597–600.
Doty, R. L., Deems, D. A., and Stellar, S., 1988, Olfactory dysfunction in Parkinsonism: A general deficit unrelated to neurologic signs, disease stage of disease duration, *Neurology* (in press).
Duara, R., London, E. D., and Rapoport, S., 1985, Changes in structure and energy metabolism of the aging brain, in: *Handbook of the Biology of Aging* (C. E. Finch and E. L. Schneider, eds.), pp. 595–616, Van Nostrand Reinhold, New York.
Engen, T., 1977, Taste and smell, in: *Handbook of the Psychology of Aging* (J. E. Birren and K. W. Schaie, eds.), pp. 554–561, Van Nostrand Reinhold, New York.
Engen, T., 1982, *The Perception of Odors*, Academic, New York.
Esiri, M. M., 1982, Viruses and Alzheimer's disease, *J. Neurol. Neurosurg. Psychiatry* **45:** 759–760.
Esiri, M. M., and Wilcock, P. K., 1984, The olfactory bulb in Alzheimer's disease, *J. Neurol. Neurosurg. Psychiatry* **47:** 56–60.
Esiri, M. M., Pearson, R. C. A., and Powell, T. P. S., 1986, The cortex of the primary auditory area in Alzheimer's disease, *Brain Res.* **366:** 385–387.
Farbman, A. I., 1986, Prenatal development of mammalian olfactory receptor cells, *Chem. Senses* **11:** 3–18.
Fowler, R. W., 1985, Ageing and lung function, *Age Ageing* **14:** 209–215.
Fordyce, I. D., 1961, Olfactory tests, *Br. J. Indust. Med.* **18:** 213–215.
Gorbanian, S. N., Paradise, J. L., and Doty. R. L., 1983, Odor perception in children in relation to nasal obstruction, *Pediatrics* **72:** 510–516.
Goto, N., Hirano, N., Aiuchi, M., Hayashi, T., and Fujiwara, K., 1977, Encephalopathy of mice infected intranasally with a mouse hepatitis virus, JHM strain, *Jpn. J. Exp. Med.* **47:** 59–70.
Graziadei, G. A., M. and Graziadei, P. P. C., 1979, Studies on neuronal plasticity and regeneration in the olfactory system: Morphologic and functional characteristics of the olfactory sensory neuron, in: *Neural Growth and Differentiation* (E. Meisami and M. A. B. Brazier, eds.), pp. 373–396, Raven, New York.
Harper, R., Bate-Smith, E. C., and Land, D. G., 1968, *Odour Description and Odour Classification*, American Elsevier, New York.
Hasegawa, M., and Kern, E. B., 1977, The human nasal cycle, *Mayo Clin. Proc.* **52:** 28–34.
Hinds, J. W., 1968*a*, Autoradiographic study of histogenesis in the mouse olfactory bulb. 1. Time of origin of neurons and neuroglia, *J. Comp. Neurol.* **134:** 287–304.
Hinds, J. W., 1968*b*, Autoradiographic study of histogenesis in the mouse olfactory bulb. 2. Cell proliferation and migration, *J. Comp. Neurol.* **134:** 305–322.
Hinds, J. W., and McNelly, N. A., 1977, Aging of the rat olfactory bulb: Growth and atrophy of constituent layers and changes in size and number of mitral cells, *J. Comp. Neurol.* **171:** 345–368.
Hinds. J. W., and McNelly, N. A., 1979, Aging in the rat olfactory bulb: Quantitative changes in mitral cell organelles and somato-dendritic synapses, *J. Comp. Neurol.* **184:** 811–820.

Hinds, J. W., and McNelly, N. A., 1981, Aging in the rat olfactory system: Correlation of changes in the olfactory epithelium and olfactory bulb, *J. Comp. Neurol.* **203:** 441–454.

Hinds, J. W., and McNelly, N. A., 1982, Capillaries in aging rat olfactory bulb: A quantitative light and electron microscopic analysis, *Neurobiol. Aging* **3:** 197–207.

Hooper, M. W., and Vogel, F. S., 1976, The limbic system in Alzheimer's disease, *Am. J. Pathol.* **85:** 1–13.

Kimbrell, G. M., and Furchtgott, E., 1963, The effect of aging on olfactory threshold, *J. Gerontol.* **18:** 364–365.

Klingman, A. M., Grove, G. L., and Balin, A. K., 1985, Aging human skin, in: *Handbook of the Biology of Aging* (C. E. Finch and E. L. Schneider, eds.), pp. 820–841, Van Nostrand Reinhold, New York.

Koelega, H. S., and Koster, E. P., 1974, Some experiments on sex differences in odor perception, *Ann NY Acad. Sci.* **237:** 234–246.

Korten, J. J., and Meulstee, J., 1980, Olfactory disturbances in Parkinsonism, *Clin. Neurol. Neurosurg.* **82:** 113–118.

Koss, E., Weiffenbach, J. M., May, C., Haxby, J. V., and Friedland, R. P., 1986. Variability of olfactory dysfunction in Alzheimer's disease, *Soc. Neurosci. Abs.* **12:** 945.

Krmpotic-Nemanic, J., 1969, Presbycusis, presbystasis, and presbyosmia as consequences of the analogous biological process, *Acta Otolaryngol. (Stockh.)* **67:** 217–223.

Liss, L., 1984, Change in dietary habits and olfactory degeneration in Alzheimer's disease, *J. Am. Gerontol. Soc.* **34:** 908.

Mair, R. G., Doty, R. L., Kelly, K. M., Wilson, C. S., Langlais, P. J., McEntee, W. J., and Vollmecke, T. A., 1986, Multimodal sensory discrimination deficits in Korsakoff's psychosis, *Neuropsychologia* **24:** 831–839.

Matulionis, D. H., 1982, Effects of the aging process on olfactory neuron plasticity, in: *Olfaction and Endocrine Regulation* (W. Breipohl, ed.), pp. 299–308, IRL Press, London.

Monath, T. P., Croop, C. B., and Harrision, A. K., 1983, Mode of entry of a neurotropic arbovirus into the central nervous system: Reinvestigation of an old controversy, *Lab. Invest.* **48:** 399–410.

Murphy, C., 1983, Age-related effects on the threshold, psychophysical function and pleasantness of menthol, *J. Gerontol.* **38:** 217–222.

Murphy, C., 1985, Cognitive and chemosensory influences on age-related changes in the ability to identify blended foods, *J. Gerontol.* **40:** 47–52.

Murphy, C., Nunez, K. Withee, J., and Jalowayski, A. A., 1985, The effects of age, nasal airway resistance and nasal cytology on olfactory threshold for butanol, *Chem Senses* **10:** 418 (abst.).

Naessen, R., 1971, An enquiry on the morphological characteristics and possible changes with age in the olfactory region of man, *Acta Otolaryngol. (Stockh.)* **71:** 49–62.

Nakashima, T., Kimmelman, C. P., and Snow, J. B., Jr., 1984, Structure of human fetal and adult olfactory neuroepithelium, *Arch. Otolaryngol. (Stockh.)* **110:** 641–646.

Nishihata, S., 1984, Aging effect in nasal resistance, *Nippon Jibiinkoka Gakkai Kaiho* **87:** 1654–1671.

Pearson, R. C. A., Esiri, M. M., Hiorns, R. W., Wilcock, G. K., and Powell, T. P. S., 1985, Anatomical correlates of the distribution of the pathological changes in the neocortex in Alzheimer disease, *Proc. Natl. Acad. Sci. USA* **82:** 4531–4534.

Punter, P. H., 1983, Measurement of human olfactory thresholds for several groups of structurally related compounds, *Chem. Senses* **7:** 215–235.

Pyatkina, G. A., 1982, Development of the olfactory epithelium in man, *Z. Mikrosk. Anat. Frosch.* **96:** 361–372.

Quinn, N. P., Rossor, M. N., and Marsden, C. D., 1987, Olfactory threshold in Parkinson's disease, *J. Neurol. Neurosurg. Psychiatr.* **50:** 88–89.

Reinacher, M., Bonin, J., Narayan, O., and Scholtissek, C., 1983, Pathogenesis of neurovirulent influenza A virus infection in mice: Route of entry of virus into brain determines infection of different populations of cells, *Lab. Invest.* **49:** 686–692.

Reyes, P. F., Golden, G. T., Variello. R. G., Fagel, L., and Zalewska, M., 1985, Olfactory pathways in Alzheimer's disease (AD): Neuropathological studies, *Soc. Neurosci. Abst.* **11:** 168.

Reyes, P. F., Golden, G. T., Fagel, P. L., Fariello, R. G., Katz, L., and Carner, E., 1987, The prepiriform cortex in dementia of the Alzheimer type, *Arch. Neurol.* **44:** 644–645.

Richard, J., and Bizzini, L., 1981, Olfaction et demences, *Acta Neurol. Belg.* **81:** 333–351.

Rovee, C. K., Cohen R. Y., and Shlapack, W., 1975, Life-span stability in olfactory sensitivity, *Dev. Psychol.* **11:** 311–318.

Sakakura, Y., Ukai, K., Majima, Y., Murai, S., Harada, T., and Miyoshi, Y., 1983, Nasal mucociliary clearance under various conditions, *Acta Otolaryngol. (Stockh.)* **96:** 167–173.

Schemper, T., Voss, S., and Cain, W. S., 1981, Odor identification in young and elderly persons: Sensory and cognitive limitations, *J. Gerontol.* **36:** 446–452.

Schiffman, S., 1974, Physicochemical correlates of olfactory quality, *Science* **185:** 112–117.

Schiffman, S., 1977, Food recognition by the elderly, *J. Gerontol.* **32:** 586–592.

Schiffman, S., 1979, Changes in taste and smell with age: Psychophysical aspects, in: *Sensory Systems and Communication in the Elderly,* (J. M. Ordy, and K. Brizze, eds.), pp. 227–246, Raven, New York.

Schiffman, S., and Pasternak, M., 1979, Decreased discrimination of food odors in the elderly, *J. Gerontol.* **34:** 73–79.

Schiffman, S. S., Moss, J., and Erickson, R. P., 1976, Thresholds of food odors in the elderly, *Exp. Aging Res.* **2:** 389–398.

Schneider, R. A., and Wolf, S., 1960, Relation of olfactory acuity to nasal membrane function, *J. Appl. Physiol.* **15:** 914–920.

Self, P. A., Horowitz, F. D., and Paden, L. Y., 1972, Olfaction in newborn infants, *Dev. Psychol.* **7:** 349–363.

Sheridan, M. N., Langaow, T., and Coleman, P. D., 1983, Quantitative electron microscopy of dendrites in layers I and II of entorhinal cortex of aging rats, *Soc. Neurosci. Abst.* **9:** 931.

Shipley, M. T., 1985, Transport of molecules from nose to brain: Transneuronal anterograde and retrograde labeling in the rat olfactory system by wheat germ agglutinin–horseradish peroxidase applied to the nasal epithelium, *Brain Res. Bull.* **15:** 129–142.

Simpson, J., Yates, C. M., Gordon, A., and St. Clair, D. M., 1984, Olfactory tubercle choline acetyltransferase activity in Alzheimer-type dementia, Down's syndrome and Huntington's disease, *J. Neurol. Neurosurg. Psychiatry* **47:** 1138–1139.

Smith, C. G., 1935, The change in the volume of the olfactory and accessory olfactory bulbs of the albino rat during postnatal life, *J. Comp. Neurol.* **61:** 477–508.

Smith, C. G., 1937, Pathologic change in olfactory nasal mucosa of albino rats with "stunted" olfactory bulbs, *Arch. Otolaryngol. (Stockh.)* **25:** 131–143.

Smith, C. G., 1942, Age incidence of atrophy of olfactory nerves in man, *J. Comp. Neurol.* **77:** 589–595.

Somlyo, A. P., and Somlyo, A. V., 1968, Vascular smooth muscle. I. Normal structure, pathology, biochemistry, and biophysics, *Pharmacol. Rev.* **20:** 197–272.

Stevens, J. C., and Cain, W. S., 1985, Age-related deficiency in the perceived strength of six odorants, *Chem. Senses* **10:** 517–529.

Stevens J. C., and Cain, W. S., 1986, Smelling via the mouth: Effect of aging, *Percept. Psychophys.* **40:** 142–146.

Stevens, J. C., and Marks, L. E., 1980, Cross-modal matching functions generated by magnitude estimation, *Percept. Psychophys.* **27:** 379–389.

Stevens, J. C., Plantinga, A., and Cain, W. S., 1982, Reduction of odor and nasal pungency associated with aging, *Neurobiol. Aging* **3:** 125–132.

Stroop, W. G., Rock, D. L., and Fraser, N. W., 1984, Localization of herpes simplex virus in the trigeminal and olfactory systems of the mouse central nervous system during acute and latent infections by in situ hybridization, *Lab. Invest.* **51:** 27–38.

Tomlinson, B. E., Blessed, G., and Roth, M., 1968, Observations on the brains of demented old people, *J. Neurol. Sci.* **11:** 205–242.

Tomlinson, A. H., and Esiri, M. M., 1983, Herpes simplex encephaltitis: Immunohistological demonstration of spread of virus via olfactory pathways in mice, *J. Neurol. Sci.* **60:** 473–484.

Uyematsu, S., 1921, A study of the cortical olfactory cortex. Based on two cases of unilateral involvement of the olfactory bulb, *Arch. Neurol. Psychiatry* **6:** 146–156.

Venstrom, D., and Amoore. J. E., 1968, Olfactory threshold in relation to age, sex or smoking, *J. Food Sci.* **33:** 264–265.

Vollmecke, T., and Doty. R. L., 1985, Development of the Picture Identification Test (PIT): A research companion to the University of Pennsylvania Smell Identification Test, *Chem. Senses* **10:** 413–414 (abst.).

Waldton, S., 1974, Clinical observations of impaired cranial nerve function in senile dementia, *Acta Psychiatry Scand.* **50:** 539–547.
Ward, C. D., Hess, W. A., and Calne, D. B., 1983, Olfactory impairment in Parkinson's disease, *Neurology (NY)* **33:** 943–946.
Warner, M. D., Peabody, C. A., Flattery, J. J., and Tinklenberg, J. R., 1986, Olfactory deficits and Alzheimer's disease, *Biol. Psychiatry* **21:** 116–118.
Weiffenbach, J. M., 1984, Taste and smell perception in aging. *Gerontology* **3:** 137–146.
Yoshida, M., 1984, Correlation analysis of detection threshold data for "standard test" odors, *Bull. Fac. Sci. Eng. Chuo Univ.* **27:** 343–353.

Index

MIX
Papier aus verantwortungsvollen Quellen
Paper from responsible sources
FSC® C105338

If you have any concerns about our products,
you can contact us on
ProductSafety@springernature.com

In case Publisher is established outside the EU,
the EU authorized representative is:
Springer Nature Customer Service Center GmbH
Europaplatz 3, 69115 Heidelberg, Germany

Printed by Libri Plureos GmbH
in Hamburg, Germany